International Conference and Workshop
on Risk Analysis in Process Safety

This volume is one of a series of publications available from the Center for Chemical Process Safety. A complete list of titles appears at the end of this book.

International Conference and Workshop on Risk Analysis in Process Safety

October 21–24, 1997
Atlanta Airport Marriott Hotel
Atlanta, Georgia

SPONSORED BY

Center for Chemical Process Safety
of the American Institute of Chemical Engineers

U.S. Environmental Protection Agency

Health & Safety Executive, U.K.

European Federation of Chemical Engineering

Copyright © 1997

American Institute of Chemical Engineers
345 East 47th Street
New York, New York 10017

All rights reserved. No part of this publication may be reproduced, stored in a retrieval system, or transmitted in any form or by any means, electronic, mechanical, photocopying, recording, or otherwise without the prior permission of the copyright owner.

ISBN 0-8169-0737-4

It is sincerely hoped that the information presented in this document will lead to an even more impressive safety record for the entire industry; however, neither the American Institute of Chemical Engineers, its consultants, CCPS Subcommittee members, their employers, their employers' officers and directors warrant or present, expressly or by implication, the correctness or accuracy of the content of the information presented in this document. As between (1) American Institute of Chemical Engineers, its consultants, CCPS Subcommittee members, their employers, their employers' officers and directors and (2) the user of this document, the user accepts any legal liability or responsibility whatsoever for the consequence of its use or misuse.

Contents

Risk Acceptance Criteria and Risk Judgment   1
    Arthur Dowell, III
    Wiliam G. Bridges

**Training in Support of Process Hazards Analysis:
A Specialty Chemicals Manufacturer's Approach**   3
    Tom Hoppe and H. Ray Wheeler

**Layer of Protection Analysis: A New PHA Tool
After Hazop, Before Fault Tree Analysis**   13
    Arthur M. (Art) Dowell, III

Risk Criteria for Use in Quantitative Risk Analysis   29
    Brian Greenwood, Louise Seeley, and John Spouge

**Practical Risk and Operability Analysis for New Products
during Research and Development**   41
    Brian R. Cunningham and Mary Ann Pickner

Playing the Killer Slot Machine (a Tutorial on Risk)   53
    Donald K. Lorenzo and William G. Bridges

**Rhône-Poulenc Inc. Process Hazard Analysis
and Risk Assessment Methodology**   61
    Rodger M. Ewbank and Gary S. York

Finding an Appropriate Level of Safeguards   75
    Michael D. Moosemiller and William H. Brown

Contents

**Resource Optimization by Risk Mapping: Extending
Risk Analysis to Allocate Plant and Company Resources**   85
   Robert W. Johnson, Thomas I. McSweeney, and Jeffrey S. Yokum

[handwritten: Computer program alt to IRA see p. 243]

**A Risk Assessment Methodology for Evaluating
the Effectiveness of Safeguards and Determining
Safety Instrumented System Requirements**   111
   Andrew M. Huff and Randal L. Montgomery

[handwritten: see p. 629 / see LPS 8.2]

**A Simple Problem to Explain and Clarify
the Principles of Risk Calculation**   127
   Dennis C. Hendershot

[handwritten: how to do LPS 12.2]

Risk Analysis and Risk Management   147

Dennis Hendershot
Keith Cassidy

**A Context-Specific Approach Toward
Human Reliability Assessment**   149
   Joseph R. Fragola

**Risk Based Evaluation and Human Factors Review
for Petrochemical Unit Startup Options**   171
   Philip M. Myers and Philip G. Brabezon

**New Consequence Modeling Chapter for *Guidelines for
Chemical Process Quantitative Risk Analysis*, Second Edition**   181
   Daniel A. Crowl

[handwritten: ie Chapter 2]

Benefits of Plant Layout Based on Realistic Explosion Modeling   191
   Jan C. A. Windhorst

**Consequence Modeling for the Insurance Industry—
New Philosophies and Methods**   205
   Doug Scott and Sanjeev Mohindra

[handwritten: ADL software]

**PSM Verification of Relief Device Sizing in the Case
of Reactive Chemical Service**   213
   G. Bradley Chadwell

Contents

**Accounting for Common Cause Failures When Assessing
the Effectiveness of Safeguards** 223
Henrique Paula and Emma Daggett

Risk Acceptance Criteria and Risk Judgment 241

Arthur Dowell, III
William G. Bridges

Benefits of Quantifying Process Hazards Analyses 243
Thomas I. McSweeney

**An Integrated Quantitative Decision Approach for Risk Management
Problem Solving** 259
Paul C. Chrostowski and Sarah A. Foster

CCPS Equipment Reliability Database: A Solid Foundation 273
Harold W Thomas

**Using Quantitative Risk Analysis in Decision Making:
An Example from Phenol-Formaldehyde Resin Manufacturing** 285
Don Schaechtel and David Moore

**The Synergy of the Business Planning Process,
Safety Management, and the Acceptance of Risk** 299
John Kimball

Risk Analysis and Risk Management 305

Dennis Hendershot
Keith Cassidy

**How Does the Public Perceive Risks
from Major Industrial Hazard Sites?** 307
Ann Brazier and Gordon Walker

The General Duty Clause under the Clean Air Act: *def'n of hazard assessment*
Issues of Implementation 323
   John Ferris

Development of Integrated Fixed Facility *ADL reconnends 15-20mile segments*
and Transportation Risk Criteria *for calculating Risks* 331
   Lisa M. Bendixen

The RMP as a Tool for Risk-Based Decision Making 341
   Craig Matthiessen and Lyse Helsing

Risk Management and Hazard Reduction Practices *linking to Emerg Response*
in the Ammonia Refrigeration Industry *Practical ideas on Risk management* 349
   D. R. Kuespert and M. K. Anderson

Improving Safety, Environmental Protection, and Reliability
through Integrated Operational Risk Management 377
   Glenn B. DeWolf, Theresa M. Shires, and Keith Leewis *too high level, no details*

Maximizing the Economic Value of
Process Safety Management 397
   Joseph Fiksel and Kenneth Harrington

Integrating Quality Management Principles into
the Risk Management Process 407
   Kevin Mitchell and Jatin N. Shaw

WORKSHOP A
Regulations (EPA and OSHA) 419

   Craig Matthiessen

Importance of Scenario Selection in Preparing
Hazard Assessments for EPA's RMP Rule 421
   J. Ivor John and Henry Ozog

Joint EPA/OSHA Chemical Accident Investigation 435
   Armando Santiago and Breeda Reilly

Contents

**OSHA Enforcement Actions for and
the Defense of Facility Siting Citations** — 443
Mark S. Dreux

WORKSHOP B
Root Causes and Failure Data Bases — 447

Brian D. Berkey

**Guidelines for Developing Equipment System Taxonomies
and Data Field Specifications** — 449
Bernard J. Weber

[handwritten: CCPS committee work, just starting database]

**An Ethylene Decomposition Event at Lyondell Polymers'
Victoria Texas High Density Polyethylene Facility** — 461
Darrel E. Black

[handwritten: Specific C₂H₄ failure]

**Lessons Learned Databases—A Survey of Existing Sources
and Current Efforts** — 471
G. Bradley Chadwell and Susan E. Rose

[handwritten: governmental databases mostly]

**Incident Database and Macroanalysis to Help Set
Safety Direction** — 491
John A. McIntosh, III and Sarah Rogers Taylor

[handwritten: P&G's ILP]

WORKSHOP C
Risk Acceptance Criteria (Including Cost–Benefit)
versus Corporate Liabilities — 511

David Moore

**A Risk-Based Approach to Addressing Recommendations
from Process Hazard Analysis Studies** — 513
David A. Moore and Gregory L. Hamm

[handwritten: Chevron methodology]

Risk-Based Judgments in Process Hazard Analyses — 527
Donald K. Lorenzo

[handwritten: FMEA scoring; Exec Director of Training - JBF; Example of Cost/Benefit Calculation]

Risk Acceptance Criteria and Risk Judgment Tools Applied Worldwide within a Chemical Company ... 545
   William G. Bridges and Tom R. Williams

Startup Challenges for New Risk Management Programs ... 559
   Chuck Fryman

WORKSHOP D
Methodology for Comparing Risk Assessment ... 565
   Patrick J. McNulty

An Internet Thesaurus/Dictionary for Analyzing Risk Assessment Processes, Laws, and Regulations ... 567
   A. J. Ignatowski, I. Rosenthal, and L. D. Helsing

Evaluation of a Proposed Thesaurus/Dictionary for Risk Assessment Using an Industrial Quantitative Risk Analysis ... 581
   Dennis C. Hendershot and Stanley J. Schechter

Use of the OECD Dictionary/Thesaurus to Encode Delaware's Law for Process Safety ... 593
   Patrick J. McNulty, Robert A. Barrish, and Richard C. Antoff

Risk Assessment for Toxic Catastrophe Prevention: New Jersey's Risk Assessment Method Culminates in an Appropriate Risk Reduction Plan ... 603
   Reginald Baldini and Peter Costanza

WORKSHOP E
Risk Perceptions and Communications (Regulatory, Industry, Public) ... 615
   Vin Boyen

Community Participation in Risk Acceptance Criteria ... 617
   Richard G. Runyon

Contents xi

**The East Harris County Manufacturers Association
for EPA RMPlan Communications: Progress 1994–1997**   621
    S. E. Anderson, J. W. Coe, Steve Arendt, and David Hastings

---

WORKSHOP F — also see p. 111
Technical Considerations in Choosing Multiple Levels of Safeguards   629

    Jan Windhorst

**Over-pressure Protection by Means of a Designed System
Rather Than Pressure Relief Devices**   631
    Jan C. A. Windhorst

**Safe Handling of Flammable Liquids
in Process Vessels: A QRA Approach**   651
    Gary F. Darnell, Peter N. Lodal, and Jasbir Singh

Multiple Safeguarding Selection Criteria *getting rid of standards, likes FTA*
or How Much Safety Is Enough?   667
    R. P. Stickles, S. Mohindra, and P. J. Bartholomew /George mehlam

---

POSTER SESSION   675

Methodology for Focusing a Transportation Risk Analysis   677
    Paul E. McCluer

Environmentally Sensitive Flares   691
    John F. Straitz III

**PHAzer: An Intelligent System for Automated
Process Hazards Analysis**   707
    Rajagopalan Srinivasan and Venkat Venkatasubramanian

**Strategic Financial Risk Assessment for Railcar
Business Acquisitions**   717
    Philip M. Myers and Richard S. Morgan

**Risk-Based Decision Making for Fire and Explosion
Loss Control Strategies: A Practical Approach**   729
   Thomas F. Barry

**Pressure Relief System Documentation: Equipment Based
Relational Database is Key to OSHA 1910.119 Compliance**   751
   P.C. (Pat) Berwanger, R.A. (Rob) Kreder, and A. A. (Aman) Ahmad

Integrated Safety Analysis Project   761
   Robert W. Johnson and Mark Elliott

**Air Modeling Issues Associated with the
Risk Management Program**   781
   Geoffrey D. Kaiser

LUNCHEON ADDRESSES   793

Meeting the Needs of Our Stakeholders   795
   Jack Weaver

Luncheon Presentation   799
   Hans Pasman

**Quality Assurance, Uncertainty, and Expert Judgment
in Risk Analysis**   803
   Jim McQuaid

International Conference and Workshop
on Risk Analysis in Process Safety

Risk Acceptance Criteria and Risk Judgment

Chairs **Arthur Dowell, III**
Rohm and Haas
William G. Bridges
JBF Associates

Safety Training @ CIBA

(semi + batch process)

Training in Support of Process Hazards Analysis: A Specialty Chemicals Manufacturer's Approach

Tom Hoppe and H. Ray Wheeler
Ciba Specialty Chemicals

ABSTRACT

Part of the mission of the Additives Division's Central Safety Laboratory and Corporate Safety is to provide training in the setting of critical processing limits and in defining the probability and severity of exceeding those limits. The setting of those limits within the risk assessment process is both a legal (OSHA's Standard on Process Safety Management of Highly Hazardous Chemicals, EPA's Risk Management Program) and an internal Ciba Specialty Chemicals (CSC) requirement.

A key issue affecting the quality of these risk assessments is the establishment of realistic critical limits and realistic probability and severity ranking. These parameters drive the type of measures required and subsequently affect both the residual risk and the cost of obtaining that risk level.

A team's or individual's ability to set realistic limits and ranking can be significantly improved with a better understanding of the basic data and of the methodologies used in the overall assessment of risk. For this reason, a number of workshops in basic data interpretation and process safety engineering standards using case studies and accepted process hazards methodologies have been provided to our engineers and chemists over the last seven years.

An overview of the workshops is presented along with an analysis of the successes and failures of the program in providing CSC with consistent results in establishing uniform acceptable risk criteria.

Introduction

In 1991 it was decided to standardize process safety training throughout Ciba-Geigy. One of the objectives of this effort was to minimize variations in the determination of acceptable risk factors. Initially a basic course was established which was called here in the United States the **"General Course in Process Safety."** The agenda was established at Corporate Headquarters in Basel, Switzerland. The quality of the course was periodically monitored by Central Safety personnel.

This course runs for approximately three days and is structured around a case study. The case study is normally selected from a local chemical process.

Participants are divided into teams and required to perform a risk analysis. Prior to performing the risk analysis, modules on basic data categories such as industrial hygiene, flammability, thermal process safety are given in an attempt to establish a base line of understanding of the basic data. Modules on legal requirements (OSHA's PSM) are also given. This course has been given over 50 times since 1991 to over 1000 Ciba-Geigy employees.

Coinciding with the promulgation of OSHA's PSM standard, Ciba-Geigy, USA, initiated an extensive auditing program to help insure compliance with internal and external environment and safety regulations. Compliance with process safety information and the risk analysis elements of the PSM standard were part of the audits. During these audits it became clear that the understanding of basic data related to the hazardous properties and reactivity of chemicals along with the ability to set appropriate critical limits needed to be improved. To improve the understanding of these basic data a number of courses on **Explosion Technology, Electrostatics, and Inherent Safer Designs for Batch and Semi-Batch Processing (Thermal Process Safety)** were offered. Each of these courses normally run from 4 to 6 hours.

In 1994 Ciba-Geigy experienced a serious near miss when approximately 8000 lb of epichlorohyrdrin was added to a reactor over three minutes instead of the three hours called for by the process. Only the initiation of an emergency quench system saved the almost certain catastrophic failure of a 3000-gal reactor. One of the casual factors cited by the incident investigation team was improper setting of critical processing limits. A recommendation from the team was to establish an **Advanced Course in Process Safety** which would focus on establishing critical processing limits. This new course was made available to Ciba-Geigy personnel in 1995. This course runs approximately 6 hours.

Overview of Courses

The General Course on Process Safety

The objective of this course is to provide an introduction into how to perform a risk analysis using a " what-if check list" method developed by Ciba-Geigy. The focus is on semi-batch operations which is the type of reactor used for the vast majority of our chemical synthesis. The course is divided into four modules; regulatory requirements, basic data (toxicity, flammability, thermal process safety), a risk analysis of a semi-batch reactor and a risk analysis of a continuous process using the HAZOP methodology. There are five break out sessions where teams of five to six people perform the various tasks of a risk analysis (establish critical limits, search for deviations, determine the severity and probability of the deviation and develop measures and assess the residual risks). The teams are required to demonstrate their understanding of the material by presenting their results to all participants at the end of each team session. The agenda is as follows:

Day 1

- Introduction/DMSO Explosion
- Regulatory Aspects of Risk Analysis
- Internal Requirements of Risk Analysis
- Industrial Hygiene
- Flammability
- Presentation of Case Study (Semi-Batch)
- Team Session 1—Critical Limits of Process for Hazard Criteria of Flammability, Toxicity, Interaction of Chemicals, Ecotoxity

Day 2

- Thermal Process Safety
- Team Session 2—Critical Limits of Process for Hazard Category Reactivity/Thermal Stability
- Search for Hazards (Deviations Based on What-If Check List)
- Team Session 3—Search for Hazards Semi-Batch
- Risk Evaluation (Probability, Severity, Measures, Residual Risk)
- Team Session 4—Risk Evaluation Semi-Batch

Day 3

- Continuous Plant
- HAZOP Method
- Team Session 5—HAZOP of Continuous Process
- Automated Process Control Safety

Training in Basic Data

Three courses are offered in this area; electrostatics, explosion technology and inherent safer designs in batch and semi-batch processing. As a result of our general training in risk analysis, it became apparent that one of the problems the teams were having in setting realistic and consistent critical limits stemmed from the fact that the application of certain types of data in the assessment of risk were only encountered infrequently during every day operations. For example, although engineers may have used heat of reaction (ΔH) and heat generation data to determine heat exchanger sizing, they infrequently use it for the determination of hazard potential. To illustrate this point during training the instructor in the General Course in Process Safety uses a batch distillation as a case study. He/she asks the following question: " What potential (ΔH) in units of energy per weight of reaction mass would cause you concern if this energy was released rapidly and in an uncontrolled manner?" This is a fundamental question because it addresses the potential severity of the operation from a reactivity criteria. In the over 50 times we have given this course we seldom got a response to that question. It became apparent that training in basic data was required to improve the quality of our risk analysis.

The agenda for the basic data course on Electrostatics is provided as an example of the type of subjects covered in a workshop.

Electrostatics

- Charge generation/dissipation/accumulation
- Types of discharges (spark, conical, propagating, brush)
- Ignitabilty of dusts and solvent vapors
- Measures against ignition hazards due to electrostatic discharges
- Handling of solvents and gases, suspensions, and combustible solids

The Advanced Course in Process Safety

The objective of this course is to challenge the participants in the setting of realistic critical processing limits. Three case studies are used, the drying and transfer of the powder with an minimum ignition energy ≤1 mJ, the continuous regeneration of t-butyl hydrogen peroxide from t-butyl alcohol and the semi batch production of a polymer. The fundamental requirement of this course is to fill out **Form 1—EVALUATION OF BASIC DATA, PROCESS CONDITIONS, CRITICAL LIMITS** (see the facing page). Normally a risk analysis team would start with a blank form 1. The hazard categories and critical limits columns would be blank. Audit findings indicated large variations in how individual teams approached filling out this form. In some cases, teams would place a chemical in the hazard category column and simply list data in the critical limits column which had no bearing on the protection concept. For example, in the hazards categories column a flammable solvent such as toluene would be listed. In the critical limits column its flash point would be entered. However, for the process under going the risk analysis operation above the flash point was necessary for the chemistry. Therefore using the flash point as a processing limit was not possible.

Obtaining Consistent Critical Processing Limits Through Training

In order to help teams do a better job the Advance Course on Process Safety was offered. During this course Form 1 is filled out for a powder transfer, a semi-batch and a continuous process. Form 1 is given to the participants partially filled out in order to ensure a clearer understanding of what is meant by hazard categories and to give guidance to the types of critical processing limits that normally apply to the hazard category.

BASIC DATA FOR RISK ANALYSIS Form No.: 1

PRODUCT: XYZ, Case Study UNIT: AUTHOR:	DATE: 10/21/97
CRITERIA (Hazard Categories)	**CRITICAL LIMITS/CONCLUSIONS/ PROCESS CONDITIONS**
Reactivity	Process Temperature Limits: Dosing Rate: Agitation: Pressure: Time:
Thermal Stability	Max. Allowable Temperature: *decision required*
Toxicity	TLV, Special Properties e.g., Carcinogenic, sensitizer
Flammability (solvents)	Limiting O_2, Temperature Limit (Flash Point—5°C) Electrical Classification
Fire/Explosion (dusts)	Grounding, tip speeds, transfer rates, volume limits, materials of construction
Ecotoxicity	

The training session is designed to create active dialog among participants. There is a minimum amount of lecture. The participants are given the basic data, PIDs and a process description. The instructor then facilitates a discussion where the data pertaining to each hazard category are discussed and a consensus on the corresponding critical limits established.

The instructor is normally an individual who has extensive experience in setting critical limits and fully understands **"the acceptable risk culture of the organization."**

Obtaining a Consistent Assessment of Probability and Severity

One important factor in obtaining a uniform level of acceptable risk is to provide the team with clear understanding of the probability and severity of the hazard. In the CSC system this is approached in two ways. First, examples are given for low, medium and high categories for both probability and severity. These examples are discussed in detail during training modules offered in The General Course in Process Safety. During the training the teams are required to use these examples on a case study to determine severity and probability rankings. It has been our experience that consensus on severity ranking are easier for the teams to arrive at than consensus on probability. Quantification of probability is not done because accurate equipment/service failure data are not available and the level of improvement of the overall analysis for the additional resources applied makes the return on investment questionable. This is an inherent problem for specialty chemical manufacturers where multiproduct processing lines are the norm.

EXAMPLES OF SEVERITY CATEGORIES

Category	People	Effect on Environment	Material Value
Low	Slight injury	Short-Term noise pollution	Minor machine damage, loss of a batch
Medium	Injury without permanent disability	Water discoloration, smell	Plant damage without prolonged plant shutdown
High	Injury with permanent disability	Dead fish, defoliation, poisoning of waste treatment plant	Loss of plant or building

Typical Ratings of Probability

Probability	Technical Failure	Human Error	External Influence
High	• Failure of on-line analytical probes, (pH, REDOX, O_2)	• Mix-up of products in identical packing • Wrong interpretation of verbal instructions	• Frost, rain
Medium	• Failure of measuring elements: pressure temperature, flow • Solenoid or control valve failure • Leakage at flanged connection with flat gasket	• Mix-up of products delivered in drums/ bags • Wrong interpretation of written instructions	• Prolonged power failure • Transportation accident
Low	• Failure of redundant elements • Leakage at grooved flange	• Mix-up of products delivered from fixed piping. • Wrong interpretation of doubled checked written instructions	• Aircraft crash on chemical plant

In the second approach, basic data forms related to **Reactivity (Form 0f) and Thermal Stability (Form 0h)** are structured to insure certain plausible worst case scenarios are included and " rules of thumb" are used to help define probability and severity.

BASIC DATA FOR RISK ANALYSIS/ Reactivity Form No.: 0f

PRODUCT: XXZ, Case Study UNIT: AUTHOR:		DATE: 10/21/97
DESCRIPTION OF SYNTHESIS REACTION		(USE ONE SHEET PER STEP)
Batch size at the start at the end	1039 gallon 1123 gallon	10134 kg 10454 kg
Potential		
Heat of reaction Q_R = 300 kJ/kg	Heat capacity C_P = 1.84 kJ/kg/K	Adiabatic Temperature Rise (Ratio Q_R/C_P) ΔT_{ad} = 163 °C
Highest theoretical attainable temperature in case adiabatic conditions occur: 193 °C Total gas evolution: l/kg l/batch Source of data: CST Data Ref : 560-97 Consequences of allowing adiabatic reaction (Check appropriate boxes) ☐Harmless temperature rise ☒ Boiling (Bp = 140°C) ☐ Gas Release ☒Critical temperature rise ☒ Decomposition ☐ Pressure build up ☒ Other: The adiabatic time to maximum rate at 190 °C is less than 5 hrs.		
DYNAMIC ASPECTS Maximum gas evolution rate: Not Available l/kg/hr, ft³/lbm/hr l/batch/hr, ft³/batch/hr Maximal heat release rate: 45 W/kg, 470 kW/batch,		
CONTROL OF REACTION ☐Batch reaction (no feed during reaction) catalytically initiated ☐Batch reaction (no feed during reaction) thermally initiated ☒Semi-batch reaction (with feed of at least one reactant) Feed controlled by (valve, diaphragm...): valve and feed line diameter		
Accumulation of reactants: For semi-batch reaction, in case of cooling failure, the accumulated reactant will result in a non controllable temperature rise. Data correspond to feed rate as defined in the process. • The maximal accumulation occurs at feed time 2 hours (e.g. at end of feed) • Accumulation in % of reactant fed: 50% • Accumulation corresponds to an energy of 150 kJ/kg • Adiabatic temperature rise due to accumulation 82 °C • Source of data: CST Data Ref: 560-97		

MAXIMAL TEMPERATURE OF SYNTHESIS REACTION
(Maximum temperature which can be reached in case adiabatic conditions occur if only the main reaction is considered)
$\vert$ M T S R = 112 °C
Starting from: 30°C
BEHAVIOR AT BOILING POINT (Fill in only if the Bp will be reached in case of cooling failure)
Nature of boiling solvent: Heat of vaporization (ΔHv):   kJ/kg solvent, BTU/lbm solvent
Amount of solvent which can be evaporated after reaching the boiling point: kg solvent, lbm
Heat Release Rate at Bp: BTU/lbm   Boiling Rate: lbm solvent/hr
W/kg   kg solvent/hr
CONSEQUENCES   (Tick all appropriate Boxes, taking in account the control by feed if available)
☒ No critical consequence   ☐ Boiling   ☐ Gas release
☐ Decomposition   ☐ Other   ☐ Pressure build up
DESCRIPTION: Adiabatic time to maximum rate at 112°C is >24 hrs. The boiling point of the solvent is 140°C.

Inherent in the structure of form 0f is the requirement that the risk analysis team initially evaluate the energy potential of the process for batch reactor operating in an adiabatic mode. This requirement insures that the worst case, representing a complete loss of heat transfer, is evaluated. Next the form requires that the semi-batch mode is evaluated. Again an adiabatic system is assumed for the consequence analysis. Finally, the behavior of the runaway at the boiling point is accessed. By using this approach, a consistent use of a series of plausible worst case scenarios is assured.

The second form, 0h, provides guidance in the assessment of the probability and severity of initiating any secondary reactions. In the severity and probability sections the risk analysis team is asked to fill in either high, medium, or low. Although it is always difficult to talk in terms of absolutes, an attempt is made here to provide guidance to the teams by providing concepts and/or data that are as independent as possible of measurement systems and plant design, i.e., the adiabatic temperature increase and the adiabatic time to maximum rate.

As can be seen in form 0h, the probability of initiating a secondary reaction is considered low if the processing temperature plus the adiabatic temperature increase from the wanted reaction (MTSR) brings the reaction mass to a temperature region where the TMR for the unwanted reaction is 24 hours. The severity of the reaction is based on the adiabatic temperature increase where a ΔT_{ad} of <50°C is considered low and a ΔT_{ad} of >200°C is considered high.

Training in Support of Process Hazards Analysis

BASIC DATA FOR RISK ANALYSIS/ Thermal Stability

Form No.: 0h

PRODUCT: XYZ, Case Study
UNIT:
AUTHOR: **DATE:** 10/21/97

CHARACTERISTIC DATA OF DECOMPOSITION REACTIONS
(USE SEVERAL SHEETS IF NECESSARY)

Reaction Mass No.: Process step: 9
Label: Rx Mass 1 hour after the addition of Z at 30° C

Severity:

    Energy potential of relevant decomposition reactions: 769 kJ/kg

    Heat capacity of reaction mass 1.84 kJ/(kg.K)

    Adiabatic temperature rise 418 °C

    Boiling Point of Reaction mass (if relevant): 140 °C, °F

    Gas evolution m³/batch, ft³/batch

    Known decomposition products:

    Source of data: CST Data Ref: 560-97

    Assessment of severity: ☐ Low ☐ Medium ☒ High

PROBABILITY:

    Decomposition is not critical below (TMR_{ad} > 24 h): 120°C

    Decomposition becomes critical above (TMR_{ad} < 8 h): 160°C

    Maximum Temperature of Synthesis Reaction (MTSR): 112°C

    This temperature is the highest which can be reached after loss of control of the desired reaction only see form 0f

    TMR_{ad} at MTSR: >24h

    Autocatalytic Character: ☒ No ☐ Yes

    Source of data: CST Data Ref: 560-97

    Assessment of probability: ☒ Low ☐ Medium ☐ High

BEHAVIOR AT BOILING POINT (Fill in only if the Bp will be reached in case of cooling failure)

Nature of boiling solvent: Heat of vaporization (ΔHv): kJ/kg solvent, BTU/lbm solvent

Amount of solvent which can be evaporated after reaching the boiling point: kg solvent lbm

Heat Release Rate at Bp: W/kg, BTU/lbm Boiling Rate: kg solvent/hr, BTU solvent/hr

CONSEQUENCES: (e.g. Thermal explosion, gas generation, pressure build up, vapor pressures)

Thermal runaway will occur if polymerization/decomposition reaction is initiated. Pressure potential of reaction is not significant at temperatures of less than 140°C. Because sufficient solvent is available to absorb the heat release, initiation of the decomposition reaction is very unlikely.

Conclusions

Significant variations in the quality of the risk analysis performed at Ciba-Geigy during early 1990s were observed. These variations were more a problem of the chemists and engineers not understanding basic data and the application of the basic data to establishing critical processing limits than to a misunderstanding of risk analysis methodology. To address this problem a series of training courses were introduced which emphasized basic data and its relationship to critical processing limits. As a result of this training more consistent risk analysis and therefore acceptable risk factors are being experienced within CSC.

ACKNOWLEDGMENTS

The authors would like thank both the management of Novartis and Ciba Specialty Chemicals for their commitment to process safety for giving us permission to use the risk analysis forms published with this paper. We would also like to recognize the former members of Ciba-Geigy's Central Safety Department who over many years of hard work put together the lion's share of the training systems and forms described.

- difficult to provide laboratory demos of disasters
possibly use videos instead

Layer of Protection Analysis: A New PHA Tool After Hazop, Before Fault Tree Analysis

Arthur M. (Art) Dowell, III
Senior Technical Fellow, Hazard Analysis, Rohm and Haas Company,
PO Box 1915, Deer Park, TX 77536-1915
E-Mail: chedowe@rohmhaas.com

ABSTRACT

How do you know how many safeguards are enough to prevent or mitigate a chemical process impact event? What integrity level should be chosen for a Safety Instrumented (interlock) System (SIS)?

Building on the CCPS (Center for Chemical Process Safety) *Guidelines for Safe Automation of Chemical Processes*, this paper describes a new PHA (Process Hazard Analysis) tool called Layer of Protection Analysis (LOPA). Starting with data developed in the HAZOP (HAZard and OPerability analysis), and suggested screening values, the methodology accounts for the risk reduction of each safeguard. The mitigated risk for an impact event can be compared with the corporation's criteria for unacceptable risk. Additional safeguards or independent protection layers can be added. The required integrity level for any SIS safeguards can be determined.

LOPA focuses the risk reduction efforts toward the impact events with the highest risks. It provides a rational basis to allocate risk reduction resources efficiently.

LOPA can be easily applied after the HAZOP, but before fault tree analysis.

Introduction

In the Safety Life Cycle outlined in ISA-S84.01-1996 (ISA, 1996), steps are included to determine if a SIS (Safety Instrumented System) is needed and to determine the target SIL (Safety Integrity Level) for the SIS. The SIL is defined by the PFD (Probability of Failure on Demand) of the SIS (1). S84.01 gives guidance on building an SIS to meet a desired SIL; Green and Dowell (1995) outline how to set standard SIS designs.

How does one determine what SIL is appropriate for a particular process? Companies and individuals have struggled with qualitative ways to make this determination. It was frequently inconsistent and was often very upsetting. For example:

Portions of this paper will be published in ISA Tech/97 and the Journal of Loss Prevention. Used by gracious permission.

ENGINEER:	"Why is this existing interlock SIL 2?"
RISK ANALYST:	"I don't know off the top of my head. What does the documentation say?"
ENGINEER:	"It was set in a safety review. And you were there!"
RISK ANALYST:	"Beats me! It doesn't look like it should be SIL 2 when I look at it now."

Undesired events and their causes are identified in a Process Hazard Analysis, such as HAZOP or What-If. For an undesired event, several methods are in use in the process industries to determine the required SIL.

1. The modified HAZOP (HAZard and OPerability analysis) method in CCPS (1993) and in the informative annex of S84.01 really depends on the team comparing the consequence and frequency of the impact event with similar events in their experience, and then choosing an SIL. If the event being analyzed is worse or more frequent, then they would choose a higher SIL. It is very much in the experience and judgment of the team. Thus, the SIL chosen may depend more on whether a team member knows of an actual impact event like the one being analyzed, and it may depend less on the estimated frequency of the event.

2. The safety layer matrix listed in CCPS (1993) and in the informative annex of S84.01 (p49) uses categories of frequency, severity, and effectiveness of the protection layers. The categories are described in general terms and some calibration would be needed to get consistent results. The matrix was originally developed using quantitative calculations tied to some numeric level of unacceptable risk (Green, 1993).

3. The consequences-only method (mentioned in S84.01) evaluates only the severity of the unmitigated consequence. If the severity is above a specified threshold, a specified SIL would be required. This method does not account for frequency of initiating causes; it assumes all causes are "likely". It is recognized that this method may give a higher required SIL than other methods. The perceived trade-off is reduced analysis time. On other hand, for events whose causes have a high frequency, this method could give a lower SIL.

4. The fault tree analysis (FTA) method (ISA, 1996) quantitatively estimates the frequency of the undesired event for a given process configuration. If the

TABLE I
Safety Integrity Level (SIL) (ISA, 1996)

Safety Integrity Level (SIL)	Probability of Failure on Demand Average Range (PFD avg)
1	10^{-1} to 10^{-2}
2	10^{-2} to 10^{-3}
3	10^{-3} to 10^{-4}

frequency is too high, an SIS of a certain SIL is added to the design and incorporated into the FTA. The SIL can be increased until the frequency is low enough in the judgment of the team. FTA requires significant resources.
5. This paper describes a new method, Layer of Protection Analysis.

What Analysis Is Really Needed?

Each method to determine SIL attempts to deal with the following issues, either explicitly or implicitly:
- the severity of each consequence—fires, injuries, fatalities, environmental damage, property damage, business interruption, etc.
- the likelihood, or frequency, of each initiating cause of the undesired event—challenge occurs x times per year.
- the capability of non-SIS layers of protection—no layer of protection is perfect; for example, a pressure relief valve may fail to open 1 out of 100 times it is challenged.
- the frequency of the mitigated event compared to a target frequency — if the frequency of the mitigated event is low enough, the risk is viewed as tolerable. The more severe the consequences, the lower the target frequency.

> **Non-SIS Layer of Protection**—Any Independent Protection Layer that prevents the impact event. Includes:
> - Relief Valves, Rupture Disks
> - Evacuation Procedures
> - Process Design (e.g., vessel maximum allowable working pressure is greater than the maximum pressure generated by the initiating cause.)
> - Basic Process Control System (when control loop or logic can prevent the impact event)
> - Operator Response to Alarms

Inconsistency in determining SIL often comes from a lack of clarity for the frequency of the initiating cause and the target mitigated event frequency for which the risk is viewed as tolerable. These issues may be handled implicitly with individual team members having a different perception of the frequencies and the risk level that is tolerable. Some methods listed in the introduction do not deal with the causes explicitly, some do not deal with the frequencies of causes explicitly, and some do not deal with the target frequency for a risk level that is tolerable. Yet each team member is doing some sort of intuitive, internal analysis that asks:

- How bad is it?
- How often could it be caused?
- How effective will the layers of protection be?
- Is the mitigated event frequency intolerable or not?

Some companies have published guidelines for the risk the process imposes on the community (Renshaw, 1990), industrial neighbors, and employees. These guidelines can be used to establish criteria for the SIL evaluation as shown later in this paper.

On the other hand, many companies have not published guidelines for the risk the process imposes on the community, industrial neighbors, and employees. However, for various process configurations, decisions are still made to apply further risk reduction via design change or additional IPLs, or not to apply additional risk reduction (i.e., risk is tolerable). This information can be converted to targets for use in determining SIL. The target could take the form of the number of IPLs and the SIL value required for a given consequence severity and challenge frequency.

What is needed is a way to determine the required SIL rationally and consistently among individuals, teams, projects, and companies.

Layer of Protection Analysis (LOPA)

LOPA is built on concepts from chapter 7 of CCPS (1993). This paper is based on more than five years' use of the technique.

LOPA uses a multi-disciplined team, like a HAZOP team. Knowledgeable representatives are needed from:

- Operations—operator, foreman
- Management
- Process Engineering
- Control Engineering
- Instrument/Electrical (craftsman, foreman, or engineer)
- Risk Analysis (hazard evaluation specialist)

At least one person must be skilled in the LOPA methodology. One of the team members should be skilled as a meeting/team facilitator.

A HAZOP (or other hazard identification procedure) is done first. HAZOP tables usually list Deviations, Causes, Consequences, Safeguards, and Recommendations. The HAZOP table may also include estimates of the Frequency for each Cause and Severity for each Consequence. With these estimates a risk matrix can be used to estimate Risk for a Cause-Consequence pair (Fryman, 1996). Figure 1 shows the HAZOP information and the LOPA information in graphical form. The solid lines show the sequence of the HAZOP or LOPA development. The dotted lines show how HAZOP information is transferred to the LOPA. A sample LOPA table is shown in Figure 2.

Layer of Protection Analysis: A New PHA Tool 17

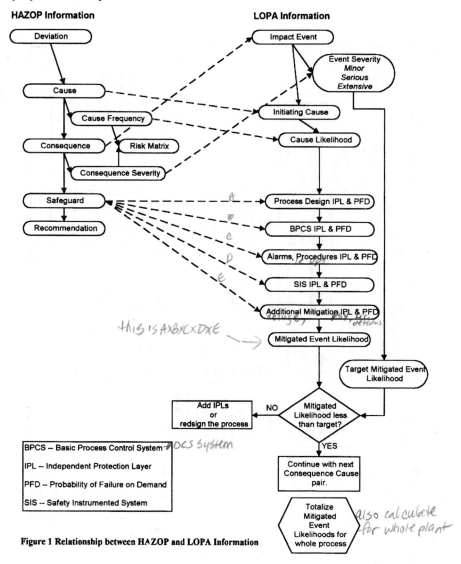

Figure 1 Relationship between HAZOP and LOPA Information

Impact Event Classification

Each Impact Event from the Hazard Identification is classified for Severity Level and Maximum Target Likelihood for the impact event using 2. The Impact Event, Severity Level, and Maximum Target Likelihood are written into column 1 of the Layer of Protection Analysis form, Figure 2.

used for Sy since 1992

Figure 2 Layer of Protection Analysis

Process: _____ Company Plant: _____ Asset Number _____ Drawing Number _____ Page 14 of 14
Interlock Number _____ Meeting Dates: _____
Sample — Work in Progress

#	Impact Event & Severity	Initiating Cause	Challenge Likelihood /yr	Process Design	Independent Protection Layers					IPLs	Mitigated Event Likelihood /yr
					BPCS (DCS)	Alarms, Procedures	SIS (PLC, relays)	Additional Mitigation: Pressure Relief Fire Protection System Restricted Access Etc...		
1.	Catastrophic rupture of distillation column with shrapnel, toxic release E Maximum Target Likelihood = 1E-8 /yr	Loss of cooling tower water to condenser, once every 10 years	1E-1	Column, condenser, reboiler, and piping maximum allowable working pressures are greater than maximum possible pressure from steam reboiler 1E-2	Logic in DCS trips steam flow valve and steam RCV on high pressure or high temperature. No credit since not independent of SIS.	High column pressure and temperature alarms can alert operator to shut off the steam to the reboiler (manual valve). 1E-1	Logic in PLC trips steam flow valve and steam RCV on high pressure or high temperature (dual sensors separate from DCS). (SIL 3) 1E-3	Pressure relief valve opens on high pressure. 1E-2	3	1E-9 calculated
2.	Toxic release from distillation column relief valve S Maximum Target Likelihood = 1E-6 /yr	Loss of cooling tower water to condenser, once every 10 years	1E-1	none	Logic in DCS trips steam flow valve and steam RCV on high pressure or high temperature. No credit since not independent of SIS.	High column pressure and temperature alarms can alert operator to shut off the steam to the reboiler (manual valve). 1E-1	Logic in PLC trips steam flow valve and steam RCV on high pressure or high temperature (dual sensors separate from DCS). (SIL 3) 1E-3		2	1E-5 higher than corporate target (Additional Prevention / Mitigation needed)
3.										
4.										

Notes: Severity Levels: **E** - extensive; **S** - Serious; **M** - Minor
1E-8 equals 1×10^{-8}

Likelihood value(s) are events per year, other numerical values are probabilities of failure on demand. Participants: XXXX
BPCS is Basic Process Control System, PLC is Programmable Logic Controller, IPLs is Independent Protection Layers.
file: C:\MY DOCUMENTS\AMD WORKR&H\CCPS_ATL\CCPS97C.DOC Date Printed: May 26, 1997

Layer of Protection Analysis: A New PHA Tool

TABLE 2
Impact Event Severity Levels and Target Mitigated Event Likelihoods

Impact Event Level	Consequence	Target Mitigated Event Likelihood, events per year	Basis
Minor (M)	Impact initially limited to local area of event with potential for broader consequence if corrective action not taken.	Depends on the economics of life cycle cost of additional layers of protection versus cost of the impact events	
Serious (S)	Impact event could cause any serious injury or fatality onsite or offsite	1.00×10^{-6}	Corporate Risk Criteria
Extensive (E)	Impact event that is five or more times worse than a **Serious** event.	1.00×10^{-8}	Two orders of magnitude less than Serious

[handwritten: For F-N checking of acceptable frequencies, they lump all E, all S, all M and calc frequency in each category + make judgement of frequency against standards]

Initiating Cause

For each Impact Event, the team lists all the Initiating Causes in column 2 of Figure 2. Note that a HAZOP Consequence may be listed in several sections of the HAZOP. It's important to gather all the Causes. The remaining calculations are carried out for **each** Initiating Cause for **each** Impact Event.

Initiating Cause Likelihood

For each Initiating Cause, the team fills in the Challenge (Initiating Cause) Likelihood in column 3, Figure 2, with units of events per year. Typical Initiating Cause Likelihoods are shown in 3. The team uses its experience to estimate the Initiating Cause Likelihood. The Initiating Cause Likelihood is also called the frequency of the challenge.

TABLE 3
Typical Initiating Cause Likelihood

Initiating Cause	Likelihood
Control loop failure	1.0×10^{-2} events per year
Relief valve failure	1.0×10^{-2} events per year
Human Error (trained, no stress)	1.0×10^{-2} events per number of times task was done
Human Error (under stress)	0.5 to 1.0
Other initiating events	Use experience of personnel, e.g., CTW pumps trip twice a year, total power failure once every two years.

Rules for IPLs

1. Each protection layer counted must be truly **independent** of the other protection layers. That is, there must be no failure that can deactivate two or more protection layers.
2. The frequency reduction for an IPL is two orders of magnitude, i.e., 10^{-2} PFD (that is, the **availability** is 99%).
   - **Exception**: Risk reduction for **Operator Response to Alarms** is one order of magnitude, i.e., 10^{-1}.
   - If an IPL is believed to be more reliable (lower value for PFD), a Quantitative method should be used to confirm the PFD. (For example, if the team desires to improve the unavailability of risk reduction logic in the BPCS (Basic Process Control System) by adding additional sensors or final elements, the impact event should be reviewed by a quantitative method such as fault tree.)
3. The IPL is **specifically** designed to prevent or mitigate the consequences of a potentially hazardous event.
4. The IPL must be **dependable**; it can be counted on to do what it was intended to do.
5. The IPL will be designed so it can be **audited** and a system to audit and maintain it will be provided.
6. If the initiating event is caused by a failure in the Basic Process Control System (BPCS), the BPCS cannot be counted as an IPL.
7. Alarms that are annunciated on the BPCS are not independent of the BPCS; if the BPCS is counted as an IPL, then such alarms cannot be counted as an IPL.
8. A control loop (PID loop) in the BPCS whose normal action would compensate for the initiating event can be considered as an IPL. For example, an initiating cause for high reactor pressure could be failure of a local upstream pressure regulator; the normal action of the reactor pressure controller would be to close the inlet PV, thus providing protection against the impact event.

Independent Protection Layers and Probability of Failure on Demand

The team lists all the Independent Protection Layers that could prevent the Initiating Cause from reaching the Impact Event. The IPLs may be different for different Initiating Causes (columns 4–7, Figure 2). The team determines which protection layers are independent.

The team assigns a PFD (Probability of Failure on Demand) to each Independent Protection Layer, typical values are shown in 4.

The IPLs and their PFDs are written in columns 4-7 of Figure 2.

Layer of Protection Analysis: A New PHA Tool 21

TABLE 4
Typical Independent Protection and Mitigation Layer PFDs

Independent Protection Layer	PFD
Control loop failure	1.0×10^{-2}
Relief valve failure	1.0×10^{-2}
Human Error (trained, no stress)	1.0×10^{-2}
Operator Response to Alarms	1.0×10^{-1}
Vessel pressure rating above maximum challenge from internal and external pressure sources	10^{-2} or better, if vessel integrity is maintained (i.e., corrosion understood, inspections and repairs in place)
Other events	Use experience of personnel, e.g., CTW pumps trip twice a year, total power failure once every two years.

Additional Mitigation

The team lists Additional Mitigation layers and assigns a PFD to each layer. A mitigation layer reduces the severity of the impact, but may not prevent all aspects of the event. Examples of mitigation layers include: relief valves, rupture disks, overflows to safe location, sensors to detect a release and an evacuation procedure, sensors and automatic deluge system. Again, each layer must be independent. The Additional Mitigation layers and their PFDs are written in column 8, Figure 2.

The team should be sure to understand the severity of the consequence of the mitigated event. An unmitigated event might be vessel rupture with toxic release. It could be mitigated to toxic release from a relief valve. If the severity of release from the relief valve is serious or extensive, it should be entered into the LOPA as another impact event.

Mitigated Event Likelihood

The team calculates the Mitigated Event Likelihood by multiplying the Initiating Cause Likelihood (column 3, Figure 2) by the PFDs of the IPLs (columns 5–8) and enters the number in column 10. The Intermediate Event Likelihood has units of events per year. The Intermediate Event Likelihood is compared with the Target Mitigated Event Likelihoods shown in 2.

If the Mitigated Event Likelihood is less than the Target Mitigated Event Likelihood, there are probably enough IPLs to meet the Corporate Risk Criteria and additional IPLs may not be required. (However, further risk reduction may be desirable.)

If the Mitigated Event Likelihood is more than the Target Mitigated Event Likelihood, then additional risk reduction is probably needed. The team should seek to reduce the risk, first by applying inherently safer concepts, and then by applying additional layers of protection. The LOPA table would be updated for the design changes.

Number of IPLs

The number of Independent Protection Layers is entered in column 9, Figure 2. **Serious** and **Extensive** Impact events normally require at least two IPLs.

SIS Needed

If the team finds that an SIS is needed to meet the Target Mitigated Event Likelihood, the team enters the SIS description in column 7 and assigns it a PFD. The SIL is entered in column 7, Figure 2.

The team should use an SIS only if other design changes (using inherently safer concepts) cannot reduce the Mitigated Event Likelihood to less than the target (CCPS, 1996). Avoid using safety interlocks (added-on features). If possible, use built-in features (inherent) to reduce risk.

The team continues the iterative process of increasing the number of protection layers and recalculating the Mitigated Event Likelihood until the Mitigated Event Likelihood is less than the Target Impact Event Likelihood.

Add Up All The Risk

After all the impact events are analyzed and tabulated in the LOPA Table in Figure 2, the team adds up **all** the Mitigated Event Likelihoods for Serious and Extensive Impact Events for each affected population group.

The Risk of Fatality for each affected population is calculated by the following formulas or their equivalents:

Fire:

    Risk of Fatality = (Mitigated Event Likelihood of Release)
                                        × (Probability of Ignition)
                                        × (Probability of person in Area)
                                        × (Probability of Fatal Injury in the Fire [usually 0.5])

Toxic Release:

    Risk of Fatality = (Mitigated Event Likelihood of Release)
                                        × (Probability of person in Area)
                                        × (Probability of Fatal Injury in the Release)

The team uses the Risk Analyst expertise and the knowledge of the team to adjust these equations for the conditions of the release and the work practices of the affected populations.

> **Example:** The team found the likelihood of a release that could lead to a large fire was $2*10^{-5}$ per year. The probability of ignition is taken as 0.5. The operator is in the area where the fire could occur for about 20 minutes each hour, so the probability the operator is in the area at the time of the fire is $20/60 = 0.33$, round to 0.3. The probability of fatal injury if a person is in a large fire is taken as 0.5.
>
> Substituting in the equation above,
>
> Risk of fatality = (Mitigated Event Likelihood of Release)
> × (Probability of Ignition)
> × (Probability of person in Area)
> × (Probability of Fatal Injury in the Fire)
>
> = ($2*10^{-5}$ per year) × (0.5) × (0.3) × (0.5)
>
> = $1.5*10^{-6}$

Corporate Risk Criteria Test

The total risk from all impact events for the affected population should be compared to the Corporate Risk Criteria.

- If the total risk does not meet the criteria for the affected population, then the team should seek to reduce the risk, first by applying inherently safer concepts, and then by applying additional layers of protection. Such design changes will require an update to the LOPA table.
- If the total risk is less than the criteria for the affected population and additional risk reduction can be achieved by some additional cost, the Team should recommend those additional risk reduction features to the business (Renshaw, 1990).
- If the total risk is **substantially** less than the criteria for the affected population, then no further risk reduction is needed.

The objective is to be sure the **total risk from the facility** meets the Corporate Risk Criteria. The team should remember that employees and the community may have risk from other parts of the unit, from other projects, and from other units. That additional risk must be considered against the Corporate Risk Criteria.

Sample Problem

Part of a sample problem for Layer of Protection Analysis is shown in Figure 2. The system under study is an atmospheric distillation column with a steam reboiler and a cooling tower water condenser.

Impact Event I

The HAZOP identified high pressure as a deviation. One consequence of high pressure in the column was catastrophic rupture of the column, if it exceeded its design pressure. In the LOPA, this impact event is listed as Extensive for Severity Class, since there is potential for five or more fatalities. The Maximum Target Likelihood for Extensive impact events is 1×10^{-8}/yr. The impact event, its class, and Maximum Target Likelihood are written in column 1 of Figure 2.

Note that Figure 2 uses an alternate notation for scientific numbers for better legibility at smaller font sizes (1×10^{-8} = 1E-8).

The HAZOP listed several Initiating Causes for this impact event. One initiating cause was loss of cooling tower water to the main condenser. The operators said this happened about once every ten years. The Initiating Cause is written in column 2 of Figure 2, and the Challenge Likelihood is written in column 3 (1/10 yr = 1×10^{-1}).

The LOPA team identified one Process Design IPL for this impact event and this cause. The maximum allowable working pressure of the distillation column and connected equipment is greater than the maximum pressure that can be generated by the steam reboiler during a cooling tower water failure. Its PFD is 1×10^{-2}. This design feature is listed in column 4 of Figure 2.

The Basic Process Control System for this plant is a Distributed Control System (DCS). The DCS contains logic that trips the steam flow valve and a steam RCV on high pressure or high temperature of the distillation column. This logic's primary purpose is to place the control system in the shut-down condition after a trip so that the system can be restarted in a controlled manner. It is listed in column 5, Figure 2, since it can prevent the impact event. However, no PFD credit is given for this logic since the valves it uses are the same valves used by the SIS—the DCS logic does not meet the test of independence for an IPL.

High pressure and temperature alarms displayed on the DCS can alert the operator to shut off the steam to the distillation column, using a manual valve if necessary. This protection layer meets the criteria for an IPL—the sensors for these alarms are separate from the sensors used by the SIS. The operators should be trained and drilled in the response to these alarms. This information is recorded in Figure 2, column 6, with the PFD of 10^{-1}.

SIS logic implemented in a PLC will trip the steam flow valve and a steam RCV on high distillation column pressure or high temperature using dual sensors separate from the DCS. The PLC has sufficient redundancy and diagnostics

Layer of Protection Analysis: A New PHA Tool 25

such that the SIS has a PFD of 10^{-3} or SIL 3. This information is written in column 7 of Figure 2.

The distillation column has Additional Mitigation of a pressure relief valve designed to maintain the distillation column pressure below the maximum allowable working pressure when cooling tower water is lost to the condenser. Its PFD is 10^{-2}. This information is recorded in column 8, Figure 2.

The number of independent protection layers is 3. This value is entered in column 9 of Figure 2.

The Mitigated Event Likelihood for this cause-consequence pair is calculated by multiplying the Challenge Likelihood in column 3 by the IPL PFDs in columns 4, 6, 7, and 8:

Challenge Likelihood		Process Design		Alarms, Procedures		SIS		Relief Valve		Mitigated Event Likelihood
1×10^{-1}/yr	×	(1×10^{-2})	×	(1×10^{-1})	×	(1×10^{-3})	×	(1×10^{-2})	=	1×10^{-9}/yr

The Mitigated Event Likelihood is entered in column 10 of Figure 2. The value of 1×10^{-9} is less than the maximum target likelihood of 1×10^{-8} for extensive impact events.

Note that the relief valve protects against catastrophic rupture of the distillation column, but it introduces another impact event—a toxic release. The toxic release is entered on the Layer of Protection Analysis form as Impact Event #2.

Impact Event 2

The toxic release from the distillation column is classed as a Serious event. The impact event description, severity, and maximum target likelihood are entered in column 1 of Figure 2.

The Initiating Cause and Challenge Likelihood are the same for Impact Events 1 and 2. The information in columns 2 and 3 in Figure 2 is copied into the row for Impact Event 2.

The process design IPL of Impact Event 1 can protect against the relief valve release only if the relief valve set pressure is greater than the maximum pressure from the steam reboiler. For this example, the relief valve set pressure is less than the maximum pressure produced by the steam reboiler. Thus, there is no process design IPL for this impact event.

The Impact Event 1 information in the IPL columns of BPCS, Alarms, Procedures, and SIS also applies to Impact Event 2. Columns 5, 6, and 7 are thus duplicated.

The pressure relief valve does not prevent the release. There is no additional mitigation for this event.

The number of IPLs for this event is 2. This is written in column 9 of Figure 2.

The Mitigated Event Likelihood for this cause-consequence pair is calculated by multiplying the Challenge Likelihood in column 3 by the IPL PFDs in columns 6 and 7:

Challenge Likelihood		Alarms, Procedures		SIS		Mitigated Event Likelihood
$(1 \times 10^{-1}/\text{yr})$	$\times$	(1×10^{-1})	$\times$	(1×10^{-3})	$=$	$1 \times 10^{-5}/\text{yr}$

The Mitigated Event Likelihood is entered in column 10 of Figure 2. The value of 1×10^{-5} is more than the maximum target likelihood of 1×10^{-6} for extensive impact events. The team should consider if the design could be changed to be inherently safer to avoid the toxic release. Additional independent protection layers may be needed. A scrubber or flare could be added to treat the release from the relief valve. Alternately, the relief valve set pressure could be increased to the maximum allowable working pressure of the equipment.

Add Up All The Risk

After all the impact events and all the cause have been analyzed and recorded in the layer of protection analysis form, the team will add up all the Mitigated Event Likelihoods for all the Serious and Extensive Impact Events. The Risk of Fatality will be calculated as described above in this paper and compared with the Corporate Risk Criteria to be sure the distillation column and the other processing units do not impose intolerable risk on affected populations.

LOPA Advantages

LOPA focuses greater risk reduction efforts on Impact Events with high severity and high likelihood. It ensures that all the identified Initiating Causes are considered, and it confirms which Independent Layers of Protection are effective for each Initiating Cause. LOPA can be used to allocate risk reduction resources efficiently, so that one Impact Event is not left with too little protection, while another is overly protected.

LOPA encourages thinking from a system perspective. Formerly, interlocks were labeled by the sensor, as in "High Reactor Pressure." LOPA shows the Layers of Protection for different Impact Events stemming from the same Initiating Cause: for example, "catastrophic rupture of the reactor" and "release of reactor contents through the relief valve."

LOPA gives clarity in the reasoning process and it documents everything that was considered. While this method uses numbers, judgment and experience are not excluded. In some cases, the team's "gut feel" was uncomfortable with the number calculated, so it went back and reviewed the assumptions for the frequency of the initiating event. The method makes the input from "gut feel" explicit, rather than implicit.

In addition, LOPA offers a rational basis for managing Layers of Protection that may be taken out of service — e.g., interlock bypass.

LOPA is more quantitative than the qualitative hazard consequence and likelihood categories often used to estimate risk rankings in a HAZOP, but it is less work than Fault Tree Analysis or Quantitative Risk Analysis.

ACKNOWLEDGEMENTS

To the CCPS and ISA committees who wrote the *Guidelines for Safe Automation of Chemical Processes* and the ISA-S84-01, respectively.

To Dallas Green, David Patlovany, Rich Sypek, and Mieng Tran, who sharpened my thinking as we wrote internal interlock guidelines.

To W. H. Johnson, Jr., who gives excellent training in LOPA.

To Paul Gruhn, who asks excellent questions.

DISCLAIMER

Although we believe the information contained in this paper is factual, no warranty or representation, expressed or implied, is made with respect to any or all of the content thereof, and no legal responsibility is assumed therefore. The examples shown are simply for illustration, and as such do not necessarily represent Rohm and Haas Company guidelines. The readers should use data, methodology, and guidelines that are appropriate for their situations.

REFERENCES

Center for Chemical Process Safety (CCPS) (1993). *Guidelines for Safe Automation of Chemical Processes*. New York: American Institute of Chemical Engineers.

Center for Chemical Process Safety (CCPS) (1996). *Inherently Safer Chemical Processes: A Life Cycle Approach*. New York: American Institute of Chemical Engineers.

Fryman, C. (1996). "Managing HazOp Recommendations Using an Action Classification Scheme," AIChE Spring National Meeting, New Orleans, LA, February 25–29, 1996.

Green, D. L. (1993). Personal communication to A. M. Dowell, III, June, 1993.

Green, D. L., and A. M. Dowell, III (1995). "How to Design, Verify, and Validate Emergency Shutdown Systems." *ISA Transactions* 34, 261–272

Instrument Society of America (ISA) (1996). ISA-S84.01-1996. *Application of Safety Instrumented Systems to the Process Industries*. Research Triangle Park, NC: Instrument Society of America.

Renshaw, F. M. (1990). "A Major Accident Prevention Program" *Plant/Operations Progress* 9,3 (July), 194-7

LIST OF ACRONYMS

BPCS	Basic Process Control System
CCPS	Center for Chemical Process Safety
DCS	Distributed Control System
FTA	Fault Tree Analysis
HAZOP	HAZard and OPerability Analysis
IPL	Independent Protection Layer
ISA	International Society for Measurement and Control
LOPA	Layer of Protection Analysis
PFD	Probability of Failure on Demand
PHA	Process Hazard Analysis
PLC	Programmable Logic Controller
QRA	Quantitative Risk Analysis
RCV	Remote control valve
SIL	Safety Integrity Level
SIS	Safety Instrumented System (also sometimes called Safety Interlock System)

Risk Criteria for Use in Quantitative Risk Analysis

Brian Greenwood and Louise Seeley
Det Norske Veritas
16340 Park Ten Place, Suite 100
Houston, TX 77084

John Spouge
Det Norske Veritas
Palace House, 3 Cathedral Street
London SE1 9DE, United Kingdom

ABSTRACT

Many international regulatory agencies as well as international companies have established risk criteria for use in land planning and/or managing industrial risks. Since many major corporations have made a commitment to use Chemical Process Quantitative Risk Analysis (CPQRA) and to abide by the results of the analyses, many processing sites across the US and abroad are already taking action to meet their risk criteria.

This paper presents individual and societal risk criteria used around the world to make judgments on the acceptability of new and existing processing facilities. Concluding remarks are made to summarize the various criteria and to generalize the findings into a single set of risk criterion.

1.0. Introduction

Purpose

This paper reviews the subject of risk criteria for onshore industrial installations. Risk criteria are used to evaluate the significance of risk estimates, and to help decide whether particular risk reduction measures are necessary. Such decisions may be influenced by operational, social, political, environmental and economic factors as well as risks, so risk criteria can only be used as guidelines, not as inflexible rules. The criteria suggested here are intended to be used when there are no applicable company or government risk criteria.

Three different but complementary approaches to risk criteria are presented:

- Individual risk criteria
- Societal risk criteria
- Cost-benefit analysis

Definitions

Risk criteria are the standards which are used to translate numerical risk estimates (e.g., 10^{-7} per year) as produced by a QRA into value judgments (e.g., "negligible risk") which can be set against other value judgments (e.g., "high economic benefits") in a decision-making process.

There have been several interpretations of the terminology of risk criteria, in which the terms "acceptable," "tolerable" and "justifiable" sometimes refer to different levels of risk and sometimes are used interchangeably. The definition used in this paper is that an *activity* as a whole, comprising a package of risks and benefits, may be regarded as "acceptable." Its *risks* alone, which are always borne with some reluctance, would then be regarded as "tolerable."

Existing Official Risk Criteria

Government authorities world-wide specifying numerical risk criteria include:

- Ministry of Housing, Physical Planning and Environment (VROM) in the Netherlands.
- Health and Safety Executive (HSE) in the United Kingdom.
- Coordinating Committee for Potentially Hazardous Installations (CCPHI) in Hong Kong.
- Department of Planning (DP) in the Australian State of New South Wales.
- Environmental Protection Authority (EPA) in Western Australia.
- Major Industrial Accidents Council of Canada (MIACC).
- County of Santa Barbara, California, USA.

In addition, several major companies have established risk criteria. Some of these are presented in the following sections.

2.0. Individual Risk Criteria

Individual risk criteria are used to ensure that individuals living or working near to a hazardous activity do not bear an excessive risk. They may also be used for land-use planning, and to help protect hospitals, underground stations, office blocks etc, which are difficult to evacuate in an emergency.

Definition of Individual Risk

Individual risk is the risk experienced by a single individual in a given time period. It reflects the severity of the hazards and the amount of time the individual is in proximity to them. It is not significantly affected by the number of people present.

Individual risk is defined formally by the I.Chem.E (1992) as the frequency at which an individual may be expected to sustain a given level of harm from the realization of specified hazards. It is usually taken to be the risk of death, and usually expressed as a risk per year.

Individual risk may be calculated in various ways, and although each is consistent with the above definition, the results may differ substantially.

Definition of FAR

Individual risks for workers are commonly expressed as a fatal accident rate (FAR), which is the number of fatalities per 10^8 exposed hours. FARs are typically in the range 1–30, and are more convenient and readily understandable than individual risks per year, which are typically in the range 10^{-4}–10^{-3}. The number of 10^8 exposed hours is roughly equivalent to the number of hours at work in 1000 working lifetimes. The FAR is defined as:

$$\text{FAR} = \frac{\text{Fatalities} \times 10^8}{\text{Person hours exposed}}$$

FARs have the advantage that they allow comparison of activities with completely different durations (e.g., driving a car, flying in a helicopter, working on a plant or in an office). However, they may be misleading because they represent a rate of risk *per unit time in the activity*, and hence cannot necessarily be added together like individual risks per year.

Individual Risk Criteria for Workers

Individual risk criteria for workers are normally expressed in the form of annual risks of death, and are typically based on the experienced risk of death in other industries. For example, the UK HSE criterion for maximum tolerable risks to workers of 10^{-3} per year is based on the risk experienced in deep-sea fishing (HSE, 1992). These have been applied to workers in high risk groups in the chemical and offshore industries, and are in general relatively easy to meet because they are influenced by previous poor accident experience in the industry.

Some companies have expressed criteria in the form of FARs, typically based on previous company experience. For example, Statoil use a FAR target of 5 onshore, equivalent to an annual risk of approximately 10^{-4} per year. These are applied to average workers in the company, and are in general relatively strict, because they are based on more limited accident experience which usually does not include any major accidents.

A summary of individual risk criteria for workers used by various companies and authorities around the world is given in Table 1. They all refer to risk of death.

Individual Risk Criteria for Members of the Public

Individual risk criteria for members of the public are normally expressed in the form of annual risks of death. They may be based on observed risks of death, e.g., the Dutch individual risk criteria for new plants of 10^{-6} per year were originally based on 1% of the risk of death from all causes in the lowest risk age group. Alternatively they may be derived from criteria for workers in the industry, e.g., the UK HSE criteria for the public are set a factor of 10 below those for workers.

A summary of individual risk criteria for members of the public used by various authorities around the world is given in Table 2. They all refer to risk of death, except the HSE criteria for new housing developments near existing installations, which refer to a "dangerous dose" rather than risk of death. They are roughly equivalent to risks of death a factor of 3 lower.

The Dutch criteria are recognized to be quite strict, and have caused their industry considerable difficulty in meeting them. Many plants probably fall only just inside the maximum tolerable criterion. The UK HSE criteria are probably the most lenient, particularly since their risk calculations take account of people being indoors and of escape action which may reduce the risks by an order of magnitude. However, the HSE also apply the ALARP (As Low As Reasonably Practicable) principle, which ensures that very few plants approach the maximum tolerable criterion.

TABLE 1
Individual Risk Criteria for Workers

Authority and Application	Maximum Tolerable Risk (per year)	Negligible Risk (per year)
Health & Safety Executive, UK (Existing hazardous industry)	10^{-3}	10^{-6}
Shell (Onshore and offshore) (approx.)	10^{-3}	10^{-6}
BP (onshore and offshore)	10^{-3}	10^{-5}
Norsk Hydro (Onshore plants)	10^{-3}	
ICI (Onshore plants)	3.3×10^{-5} (FAR = 2)	
Statoil (Onshore plants)	8.8×10^{-5} (FAR = 5)	

TABLE 2
Official Individual Risk Criteria for the Public

Authority and Application	Maximum Tolerable Risk (per year)	Negligible Risk (per year)
VROM, The Netherlands (New plants)	10^{-6}	Not used
VROM, The Netherlands (Existing plants or combined new plants)	10^{-5}	Not used
VROM, The Netherlands (Transport)	10^{-6}	Not used
Health & Safety Executive, UK (Existing hazardous industry)	10^{-4}	10^{-6}
Health & Safety Executive, UK (New nuclear power stations)	10^{-5}	10^{-6}
Advisory Committee on Dangerous Substances, UK (Existing dangerous substances transport)	10^{-4}	10^{-6}
Health & Safety Executive, UK (New housing near existing plants)	10^{-5} (dangerous dose)	10^{-6} (dangerous dose)
Hong Kong Government (New plants)	10^{-5}	Not used
Department of Planning, New South Wales (New plants and housing)	10^{-6}	Not used
Environmental Protection Authority, Western Australia (New plants)	10^{-6}	Not used
Santa Barbara County, California, USA (New plants)	10^{-5}	10^{-7}

(handwritten annotations: "assumes no movement; vulnerability = 1" next to VROM New plants row; "allows credit for movement" next to HSE New housing row)

Suggested Individual Risk Criteria

The most comprehensive and widely used criteria for individual risks are the ones proposed by the UK HSE as follows:

Maximum tolerable risk for workers	10^{-3} per year
Maximum tolerable risk for members of the public	10^{-4} per year
Negligible risk	10^{-6} per year

These criteria were developed for application to *existing activities*, and are rarely exceeded. Their normal effect is to place the most hazardous activities in the ALARP region, and they are therefore regarded as a back-stop to a cost-benefit assessment.

For *new activities* or for more safety-conscious companies, criteria are proposed which are an order of magnitude more strict, as follows:

Maximum tolerable risk for workers	10^{-4} per year (FAR = 5)
Maximum tolerable risk for members of the public	10^{-5} per year
Negligible risk	10^{-7} per year

These criteria lie within the range of criteria used around the world.

These should be applied to the sum of industrial risks imposed on the most exposed individual, assumed to have average vulnerability to the hazards, to be present for a realistic proportion of time, to be outside, and to attempt to escape when an accident occurs. For risks in the middle band, ALARP considerations would determine the need for risk reduction.

The maximum tolerable criterion for members of the public is roughly 10 times less than the risk of death on the road. The negligible criterion is roughly equal to the risk of being killed by lightning.

3.0. Societal Risk Criteria

Societal risk criteria are used to limit the risks to local communities or to the society as a whole from the hazardous activity. In particular, they are used to limit the risks of catastrophes affecting many people at once. Societal risks include the risk to every exposed person, even if they are only exposed on one brief occasion.

Definition of Societal Risk

Societal risk is defined by the I.Chem.E (1992) as the relationship between the frequency and the number of people suffering a given level of harm from the realization of specified hazards. It is usually taken to refer to the risk of death, and usually expressed as a risk per year.

Societal risks may be expressed in the form of:

- FN curves, showing explicitly the relationship between the cumulative frequency (F) and number of fatalities (N).
- Annual fatality rates, in which the frequency and fatality data is combined into a convenient single measure of societal risk. This is also known as potential loss of life (PLL) per year.

FN Criteria Slopes

FN curves derived from historical accidents in hazardous industrial activities tend to show slopes of about −1 (Fernandes Russell, 1987). World-wide FN curves for natural disasters are significantly less steep.

An FN curve criterion with a slope of −1 therefore reflects the current frequency of high-fatality accidents relative to low-fatality ones in hazardous industries. The ACDS criteria are of this type. A steeper criterion, such as the Dutch Government's, represents a desire to achieve a reduction in the ratio of high-fatality to low-fatality accident frequencies. A slope of −2 is likely to produce criteria which are very difficult to meet in some cases which would be accepted on subjective grounds.

When FN curve criteria with slopes exceeding −1 lead to the adoption of risk reduction measures, the greatest expenditure per fatality averted tends to be devoted to high fatality accident causes. If the same expenditure was devoted to lower-fatality accident causes, it would in principle be able to save more lives. Thus, reducing the relative risk of high-fatality accidents leads to some additional loss of life for a given expenditure. This is one argument against FN criteria with slopes steeper than −1.

Upper Limit on Fatalities

A reduction in the ratio of high-fatality to low-fatality accidents might also be achieved by using an absolute limit on the maximum number of fatalities an activity may cause. This approach is used in Hong Kong, on the basis that a finite risk of a 1000-fatality accident would be politically unacceptable. However, this approach seems to disregard the essential probabilistic nature of risks. It also tends to focus debate about the risk analysis onto the largest accidents modelled. These are invariably the most uncertain parts of the analysis, and are too rare for the debate to be resolved satisfactorily. This approach also involves some additional loss of life, compared to what could be achieved if the same resources were devoted to risk reduction for all sizes of events.

Existing Societal Risk Criteria

Societal risk criteria have not been as widely used as individual risk criteria because the concepts and calculations involved are much more difficult. However, their value is becoming recognized, especially for transport activities, but also as complementary to individual risk criteria in general. A summary of the few existing official societal risk criteria is given in Table 3.

Suggested Societal Risk Criteria

Tentative suggestions for general societal risk criteria for typical industrial plants and complexes are shown in Figure 1. These could be applied to risks to

TABLE 3
Official Societal Risk Criteria

Authority	FN Curve Slope	Maximum Tolerable Intercept with $N=1$	Negligible Intercept with $N=1$	Limit on N
VROM, The Netherlands*	-2	10^{-3}	—	—
Hong Kong Government (New plants)	-1	10^{-3}	10^{-5}	1000
Advisory Committee on Dangerous Substances, UK (Existing ports)	-1	10^{-1}	10^{-4}	—

*The societal risk criteria line actually starts at $N = 10$ and a frequency of 10^{-5}.

members of the public, including workers on all plants in the area. If they are found to allow risks substantially greater than occur at present, consideration could be given to using stricter criteria (e.g., the Dutch Government criteria as risk targets).

4.0. Cost–Benefit Analysis

Cost–benefit analysis (CBA) is a technique for comparing the costs and benefits of a project, developed to help appraise public sector projects. In risk assessment, it is usually used to assess additional safety measures on a project by comparing:

- The cost of implementing the measure.
- The benefit of the measure, in terms of the risk-factored cost of the accidents it would avert.

In order to make this comparison, the costs and benefits must be expressed in common units. Traditionally this has been in monetary units, which involves the contentious valuation of risks to human life in monetary terms.

The purpose of CBA is to show whether the benefits of a measure outweigh its costs, and thus indicate whether it is appropriate to implement the measure. CBA cannot provide a definitive decision, because factors other than risks and costs may be relevant, but it provides an important guide.

CBA is not the only technique for making decisions about safety measures. Subjective judgements may be more appropriate for detailed operational measures; compliance with detailed international or national regulations may be appropriate for many issues during design. However, when considering major measures affecting remote but potentially catastrophic risks, cost–benefit analysis provides a uniquely valuable input.

Risk Criteria for Using Quantitative Risk Analysis

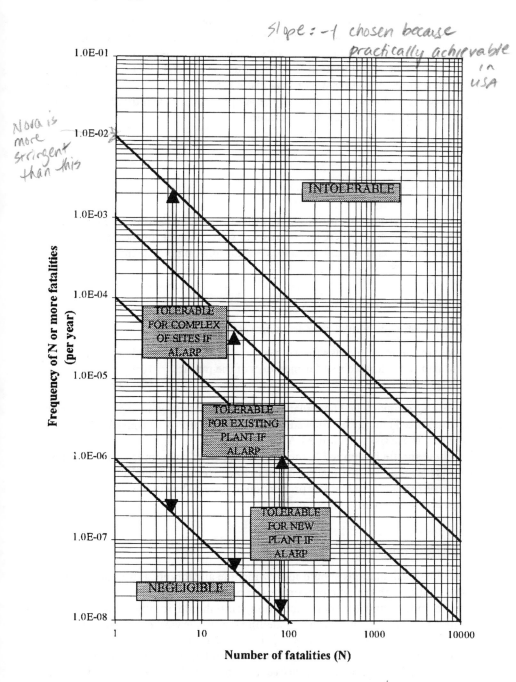

FIGURE 1: Suggested Societal Risk Criteria for Industrial Plants by DNV

Measures whose cost is less than $5 million per fatality averted are considered to be cost-effective and should in general be adopted. Measures whose cost is in the range $5–$100 million per fatality averted should be considered for adoption and may be appropriate if the individual or societal risks are high in the ALARP region. Measures whose cost exceeds $100 million are not considered cost-effective and would not normally be adopted unless the individual or societal risks were considered to be intolerable.

5.0. Conclusions

The following criteria are proposed for evaluating the risks from onshore industrial activities in cases where no company or government criteria are applicable. They should be used as guidelines for decision-making, not as inflexible rules.

Individual Risk Criteria

The criteria for individual risks from *new activities* are:

Maximum tolerable risk for workers	10^{-4} per year
Maximum tolerable risk for members of the public	10^{-5} per year
Negligible risk	10^{-7} per year

Existing activities (i.e. those planned before the intention to apply the criteria was announced) might be allowed to impose risks up to 10 times higher, provided that these risks are also ALARP.

These should be applied to the sum of industrial risks imposed on the most exposed ("at-risk") individual, assumed to have average vulnerability to the hazards, to be present for a realistic proportion of time, to be outside, and to attempt escape action when an accident occurs.

Societal Risk Criteria

Societal risk calculations should ideally include all workers as well as members of the public, with realistic proportions of people being present, realistic splits between indoor and outdoor population, and should ideally model realistic escape action. In reality, accurate modelling for on-site workers is often difficult, so the calculations may cover only off-site population.

Tentative criteria for public off-site societal risk are given in Figure 1. Transport activities affecting the area should be included when comparing with these criteria. If on-site risks are included, it may be appropriate only to apply the criteria for 10 or more fatalities.

Cost–Benefit Analysis

Unless the individual and societal risks are clearly negligible, cost–benefit analysis should be used to ensure that the risks are as low as reasonably practicable (ALARP). Risk reduction measures should be adopted unless their cost is grossly disproportionate to the risk-factored cost of accidents avoided.

Measures whose cost is less than $5 million per fatality averted are considered to be cost-effective and should in general be adopted. Measures whose cost is in the range $5–$100 million per fatality averted should be considered for adoption and may be appropriate if the individual or societal risks are high in the ALARP region. Measures whose cost exceeds $100 million are not considered cost-effective and would not normally be adopted unless the individual or societal risks were considered to be intolerable.

6.0. References

ACDS (1991): "*Major Hazard Aspects of the Transport of Dangerous Substances,*" Health and Safety Commission, Advisory Committee on Dangerous Substances, HMSO.

Ale, Dr. Ben (1997), "*Risk Criteria and Europe,*" National Institute for Health and Environment, The Netherlands, Risk 2000 Conference, 24 April 1997, London, England.

Beaumont, J. (1995): "*Clyde and Seillean,*" Presentation to "Safety Case Preparation, The Industry Responds," Fire and Blast Information Group Technical Review Meeting, The Steel Construction Institute, Ascot, UK, July 1995.

Bottelberghs, P.H., (1995): "*QRA in the Netherlands,*" Conference on Safety Cases, IBC/DNV, London, February 1995.

Dalvi, M.Q. (1988): "*The Value of Life and Safety: A Search for a Consensus Estimate,*" Department of Transport, London.

DoT (1995): "*Road Accidents Great Britain 1994—The Casualty Report,*" Department of Transport, HMSO.

Dutch National Environment Policy Plan (1989): "*Premises for Risk Management,*" Second Chamber of the States General, session 1988 to 1989, 21137, no5.

Fernandes Russell, D.P. (1987): "*Societal Risk Estimates from Historical Data for UK and Worldwide Events,*" School of Environmental Science, University of East Anglia.

Fleishman, A.B. and Hogh, M.S. (1989): "*The Use of Cost Benefit Analysis in Evaluating the Acceptability of Industrial Risks,*" International Symposium on Loss Prevention and Safety Promotion in the Process Industries, Oslo, June 1989.

HSE (1981): "*Canvey - A Second Report. A View of Potential Hazards from Operations in the Canvey Island/Thurrock Area 3 Years after Publication of the Canvey Report,*" Health and Safety Executive, HMSO.

HSE (1989a): "*Quantified Risk Assessment: Its Input to Decision-Making,*" Health and Safety Executive, HMSO.

HSE (1989b): "*Risk Criteria for Land-Use Planning in the Vicinity of Major Industrial Hazards,*" Health and Safety Executive, HMSO.

HSE (1992): *"The Tolerability of Risk from Nuclear Power Stations,"* Health and Safety Executive, HMSO.
HSE (1996): *"Use of Risk Assessment within Government Departments,"* Health and Safety Executive, HMSO.
I.Chem.E. (1992): *"Nomenclature for Hazard and Risk Assessment in the Process Industries,"* Institution of Chemical Engineers, Rugby, UK.
Jones-Lee, M.W. (1989): *"The Economics of Safety and Physical Risk,"* Basil Blackwell, Oxford.
Kennedy, B. (1993) *"ALARP in Practice— An Industry View,"* Offshore Safety Cases Conference, HSE, Aberdeen, April 1993.
Lees, F.P. (1996): *"Loss Prevention in the Process Industries,"* 2nd edition, Butterworth-Heinemann, Oxford.
NRPB (1986): *"Cost-Benefit Analysis in the Optimisation of Radiological Protection,"* National Radiological Protection Board, ASP9, Chilton, HMSO.
Royal Society (1992): *"Risk: Analysis, Perception and Management,"* Report of a Royal Society Study Group, The Royal Society, London.
Tveit, O.J. (1995): *"Risk Acceptance Criteria—10 Years' Experience from an Oil Company,"* SRD Members' Conference, "How Safe Is Safe?," Warrington, September 1995.

Practical Risk and Operability Analysis for New Products during Research and Development

Brian R. Cunningham and Mary Ann Pickner
The Lubrizol Corporation, 29400 Lakeland Blvd., Wickliffe, OH 44092

ABSTRACT
Addressing process safety, pollution prevention, and product stewardship issues early in the research and development stage of new chemicals increases total life cycle savings opportunities. A two-step process has been developed to assess and communicate opportunities for improvements starting at the invention stage and continuing through commercialization. Developing a comprehensive list of issues and establishing levels of concern for them based on the desired operating culture of the company are keys to its success. Significant benefits of this approach include its simplicity and its ability to feed manufacturing plant concerns into preliminary R&D and commercialization decisions. evelopment work to improve levels of inherent safety and pollution prevention can also e supported by providing a basis for comparing life cycle costs with and without the proposed improvements. This approach is also suitable for assessment and prioritization of existing plant-scale processes that may be candidates for process improvement efforts.

Introduction

Risk analysis is widely recognized as an important aspect of ensuring that chemical process and material hazards are properly controlled. Many approaches for conducting risk analysis have been developed over the years, and much progress in advancing quantitative risk analysis is ongoing.

Risk analysis also takes time. This time is a sound investment when analyzing large scale, commercial processes, but can be difficult to justify early in research and development projects, when there is a high level of uncertainty whether or not the potential new product or process will ever be commercialized. This creates somewhat of a paradox, since identification of material and process risks in early stages of R&D provides the best opportunity to incorporate inherently safer options and apply pollution prevention principles while both the product and process are still being defined. In addition to providing cost effective risk reductions, pursuing these options during research and development can also result in processes that are simpler and less expensive to develop than the originally conceived process. The challenge for implementing a risk analysis method in research and development is therefore one of balancing effectiveness with resource constraints.

Competing Forces Regarding Risk Analysis Early in R&D Projects	
Early analysis and action provide the most cost effective benefits. **VS.**	Scarce resources cannot be wasted on "false start" projects.
	Timely new product introduction requirements cannot accept delays from exhaustive review procedures.

Key Considerations for an Effective Risk Analysis Methodology for Research and Development

An effective risk analysis methodology must ensure that the analysis includes a balanced view across many dimensions, including but not limited to employee health and safety, process safety, pollution prevention, product stewardship, and economic considerations. Without a balanced perspective, decisions can be made that may incrementally reduce risks from one dimension, while greatly increasing risks in a dimension that was not considered.

It is also essential that any process and material risk assessment conducted at the beginning of new product introduction is guided by corporate risk tolerance philosophies and the thresholds for levels of concern at the company's manufacturing facilities. This ensures that the priorities established for risk reduction during process development are those that will have the greatest impact on the safety, environmental, and operability performance of the commercial process. dditional benefits of incorporating manufacturing facility perspectives early in d evelopment are the prevention of last-minute project delays caused by a poor fit f the developed process in the plant and the avoidance of follow-up process mprovement rework to correct process weaknesses that were not adequately ddressed initially.

Pressures to reduce both the time to commercialize new products and the consumption of resources for research and development are increasingly prevalent throughout the chemical industry. As a result, an effective risk analysis method must account for the lack of detailed, specific knowledge at the early stages of new product introduction, must quickly focus on the highest priorities for improvement and be simple enough to be conducted in a short timeframe. ince commercial development support also progresses in parallel to the develo pment of the process, an effective risk analysis method must also be able to ighlight the pertinent issues that may have a significant economic impact in erms of capital, product registration, and operating costs as early as possible so that informed decisions can be made regarding the product's viability.

Not all risks associated with producing a new material can be reduced through process development or improvement. If a product requires a hazardous raw material or certain process conditions such as very high temperatures and pressures, then the only options are to accept the costs of managing the risks or terminating the project

---

Key Considerations for an Effective Risk Analysis Methodology for R&D Projects

- Complete in a short timeframe with minimal resource commitment
- Balanced view across many dimensions
- Guided by corporate risk tolerance and manufacturing thresholds for levels of concern
- Focused on highest priorities
- Incorporate economic impact of risks in commercial decision-making
- Account for lack of detailed knowledge early in the R&D cycle
- Updated and communicated as better information is generated ing development

---

Overview of a Risk Analysis and Improvement Process for Research and Development

The analysis methodology we have developed addresses the considerations discussed above and has three formalized links to our existing process for introducing new materials. These links ensure that the analysis is an integral part of our normal operating environment instead of an added-on task that can sometimes be overlooked or forgotten. The analysis is initially conducted jointly by a synthesis chemist and a process development engineer as part of the transition from the invention stage in synthesis to process development. An updated analysis is conducted as development progresses, with direct involvement of a manufacturing plant engineer and is formally reviewed again at the "commitment to commercialize" stage. The final analysis is summarized in the technology transfer report from process development to the manufacturing facilities to highlight the improvements that have been made, the improvements that were attempted but were not able to be achieved within the overall project constraints, and most importantly, the aspects of the process that contain the greatest residual risks.

> **Integration into the Normal Operating Environment**
>
> - Initial assessment as part of the transition from invention to development
> - Updated assessment reviewed as part of the commitment to commercialize decision
> - Overall summary in technology transfer report to manufacturing facilities

There are two basic aspects of the methodology: process assessment and process improvement. The process assessment section identifies the levels of concern, based on corporate and manufacturing facility criteria, for several material and process issues. The process assessment is kept simple by assigning a level of concern as 1, 2, or 3 as outlined in the box below:

> **Process Assessment Levels of Concern**
>
> 1 = Improvement is not Warranted
> 2 = Significant Improvement Opportunity Exists
> 3 = Improvement Must Be Pursued in Development

Levels of concern were chosen as the basis for this assessment in order to quickly determine which material and process issues need to be addressed. More detailed and more quantitative methods could be applied (at the expense of time and complexity) but the main message is still whether or not action needs to be taken. This basis also allows the assessment to seamlessly highlight production operability concerns along with other risk criteria.

Guidelines for determining the appropriate levels of concern are provided for each process or material related issue incorporating the corporate risk tolerances and manufacturing facility perspectives. These help the synthesis chemist and process development engineer assign appropriate levels of concern within the context of the guidelines without requiring detailed analysis. These guidelines will be discussed in more detail later.

The process improvement section provides a structure for brainstorming ideas for reducing the highest levels of concern based on established principles of pollution prevention and inherent safety (CCPS, 1993). This helps provide a starting point for targeting specific improvements on the most critical risk issues very early in the process development plan.

The joint effort between the synthesis chemist and process development engineer (and later between the process development engineer and the manufacturing plant engineer) helps improve the depth of understanding of work done in the previous stage of the new material's life cycle and provides better continuity for each transition toward commercialization. It also helps to increase the awareness within the R&D organization of manufacturing plant concerns which can help further improve future projects. The structure provided by the established list of issues and the guidelines for assessment provide an added benefit of ensuring that engineers and chemists with little experience will address all of the pertinent issues.

Overview of Key Characteristics of the Methodology

- Identify material and process issues that are important to *your* operations
- Establish guidelines for rating levels of concern that reflect *your* operating philosophy and include broad representation of input from within your organization to ensure the validity of them from many perspectives
- Include operability as well as safety and environmental criteria in the guidelines to increase the value and the buy-in of the operating culture to the assessment process
- Focus the attention on the message of the assessment, not the mechanics
- Use established pollution prevention and inherent safety principles to guide improvement ideas
- Minimize the potential for oversights by inexperienced personnel

---

Description of the Methodology in Practice

Examples of the completed worksheets used in this methodology are shown in Tables 1 and 2. The Process Assessment worksheet requires entry of the raw materials, major process steps, byproducts, and the intended product across the top of the columns. The selected process and material related issues are presented down the left side. Each cell in the table is filled in with an entry of 1, 2, or 3 based on the fit of the process being assessed with the guidelines provided for each particular issue.

TABLE I
Process Assessment Worksheet

process flow to product ⟶
A + B − C   C   T↑ ⊕   D   −E E

MATERIAL AND/OR PROCESS ISSUES	MATERIALS USED OR GENERATED, OR MAJOR PROCESS STEP							
	Organic Acid A	Aqueous Base B (50%)	A + B − C	Reaction Product C	Heat to 300°F to Strip Water	Aqueous Distillate	Coupling Agent D	2C + D − E	Product (E)
Volume of Waste Streams Generated (Include Vent Streams)	1	3	1	1	1	1	1	1	1
Regulatory Issues with Materials or Wastes (Include Vent Streams)	1	2	1	1	2	2	2	1	1
Conversion of Raw Materials to Desired Product	1	1	1	1	1	1	1	1	1
Conversion of Energy to Desired Product	1	1	1	1	2	1	1	1	1
Acute or Chronic Human Toxicity	2	3	2	2	1	1	2	1	1
Acute or Chronic Environmental Toxicity	2	3	2	2	1	1	1	1	1
Material / Process Stability	2	2	3	2	1	1	1	1	1
Material Storage Requirements	2	2	1	1	1	1	1	1	1
Inadvertent Mixing	3	3	1	1	1	1	1	1	1
(Vapor) Pressure	1	1	2	2	1	1	1	1	1
Process Temperatures	1	1	1	1	1	1	1	1	1
Chemical Release / Spill Impact	3	3	3	2	1	1	1	1	1
Fire	2	1	1	2	1	1	1	1	1
Explosion / Overpressurization	2	2	2	2	1	1	1	1	1
Fit with Existing Plant Capabilities/ Material Mix	2	2	2	1	1	1	1	1	1
Transportation Incident Off-Site Impact	3	3	1	1	1	1	1	1	1

Level 1 = Improvement not Warranted
Level 2 = Significant Improvement Opportunity Exists
Level 3 = Improvement Must Be Pursued during Development

TABLE 2. Process Assessment Worksheet

This is how to address level 3 concerns

MATERIALS USED OR GENERATED OR PROCESS STEP	RECOMMENDED IMPROVEMENT APPROACHES					
	ELIMINATE or REDUCE the Amount of Material Used on the Site or Generated by the Process	REPLACE the Raw Material or Waste with a Less Hazardous One	REUSE the Waste Stream to Produce Products (or Consume as Fuel)	Use LESS HAZARDOUS Process Conditions or Form of a Material (or Produce a Less Hazardous Waste)	CHANGE the Basic Process to Eliminate or Reduce Wastes or Process Hazards or Increase the Conversion to the Desired Product	SIMPLIFY the Process to make Operating Errors Less Likely, and More Forgiving of Errors that are Made	Design Facilities and Processes to MINIMIZE the IMPACT of a Release of Hazardous Material or Energy
Organic Acid A (Hydrolysis side reaction)				Are other acids of the same family suitable for this product but less sensitive to water?	Can the reaction be run in a non-aq. environment?		
Aqueous Ease B (Polymerization sensitivity)	Use a non-aq. form of this material to eliminate most of the aq. waste?	Can we purchase a derivative, pre-coupled material (2B + D)?	Purchase solid base and reuse water from the previous batch to make the aq. solution?		Can the reaction be run in a non-aq. environment?		
A + B -> C (Sensitivity to Feed Ratio)		<------	<------	<------	Can the reaction be run in a non-aq. environment? Change the rxn sequence: 2A+(2B+D)->E?	Use an in-line mixer to convert most of the material before it gets into the large batch reactor?	

In this example, the shaded idea, purchasing a pre-coupled version of the base, greatly reduced the material and process hazards while also providing a simpler and more robust process.

Development of the guidelines we used for the material and process issues involved engineers from process development and manufacturing plants, as well as representatives from corporate environmental, process safety, employee health and safety, toxicology, and transportation to ensure that a broad perspective was obtained and to achieve buy-in to the assessment ratings that will be assigned. The guidelines were intentionally written to provide a context for the assessment of new processes and materials since we recognized that we would not be able to anticipate every possible new product characteristic or process condition that would be encountered for every new material. Examples of the guidelines for determining the levels of concern for the issues of Process Temperature and Fit With Existing Plant Capabilities/Material Mix follow.

Examples of Guidelines for Establishing Levels of Concern

Process Temperature

Processes that require temperatures above or below the capabilities of steam and process water can create implementation and/or potential safety problems.

Level 1 = process water and/or steam are sufficient for heating and cooling for process temperatures between 90–325°F

Level 2 = hot oil is required for process temperatures of 325–400°F

Level 3 = refrigeration is required (Z°F) or hot oil is required to raise the process temperature above 400°F

Fit with Existing Plant Capabilities/Material Mix

The uniqueness of hazards associated with new materials, process equipment, or process steps to a facility can impact both the safety and the efficiency of operations within a plant. Introducing new products or processes to plants that already have the capability of handling their special needs will generally be better than introducing them to plants where the capability would have to be established. If several level 2's or any levels 3's were identified in the previous sections, assess this section to temper the difficulty of implementation.

Level 1 = materials are already handled on site, equipment with available capacity exists, and people are already properly trained; new process equipment and/or methods are well known to the chemical process industry.

Level 2 = expansion/modification of process equipment or increased inventories are needed for hazardous materials or additional "type materials" of a current hazardous chemical family are added to the product mix.
Level 3 = extraordinarily new process equipment and/or methods are required, or hazardous materials must be introduced at the site for the first time

Materials or major process steps with any level 3 concerns are translated to the rows in the Process Improvement Worksheet (Table 2). This table contains space to write in ideas for possible approaches for eliminating sources of risk or reducing the potential consequences based on established pollution prevention and inherent safety principles. These ideas provide a reference for structuring the process development plans toward the issues that will have the greatest opportunity for simplifying the process and reducing risk (and the associated risk management requirements) at the manufacturing plant scale. The completed worksheets from this initial assessment are included in the documentation for the transition from invention to development as the first of the three formalized links to our existing process for new product introduction.

An updated assessment used at the commitment to commercialize decision stage is the second link to our existing new product introduction process. As the project progresses, communication between the process development engineer and the manufacturing plant engineer leads to a reassessment of the improved process before the final commitment to commercialize the new product is made. t this point, some process modifications have already been implemented and mproved process knowledge has been developed, which allow specific issues to e explored at a more rigorous level since plant personnel are involved in the assessment. The commitment to commercialize the product can therefore be made with a very clear understanding of the unresolved risks associated with the product. In some cases the business incentives may be such that they can justify implementing the project and supporting the costs to manage these risks. In other cases, the decision may be made to delay the commercialization until further risk reduction occurs or to cancel the project.

Inclusion of a summary in the technology transfer report to manufacturing facilities represents the third and final phase of this methodology. The summary serves two primary functions. The first is documentation of the Level 3 concerns that have been designed out of the original process. The second is communication of the residual Level 3 concerns that could not be eliminated, accompanied by a description of the work that was done and the reasons additional work was not conducted. Secondary benefits of the summary include the opportunity to audit the effectiveness of the development process at reducing the inherent ocess and material issues as well as to provide a basis for discussion with plant personnel to review the appropriateness of the guidelines in representing their levels of concern for specific materials and processes.

Example of the Methodology for a Semi-Batch Process

The following reaction sequence, loosely based on one of our products, serves as the basis for how the process assessment and process improvement worksheets (Tables 1 and 2) are used:

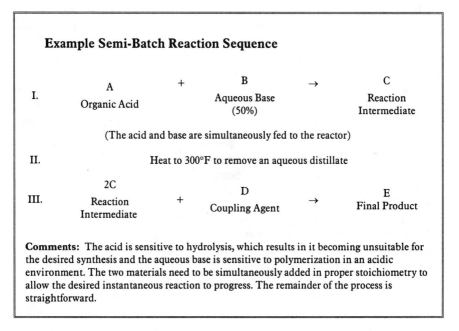

Summary

This approach to risk analysis early in the research and development cycle resolves the competition between the earlier mentioned forces of needing to obtain the greatest and most cost effective benefits vs. the needs to consume minimal resources and to avoid delays in new product introduction. Two people working one to two hours can easily complete the initial assessment and improvement worksheets. The time spent on this activity yields immediate savings in terms of a better transition between invention and development and a more focused initial development plan.

We also found that including plant operability concerns along with the more direct risk-related concerns into the guidelines was a key step in obtaining support from R&D and manufacturing facilities for this effort. This support was critical in helping to institutionalize the methodology into our existing R&D operating culture. This approach should also readily transfer to other organizations by incorporating their operating philosophy and concerns into the issues and guidelines for assessment rankings.

Lastly, this methodology can also be used effectively at manufacturing facilities as a screening tool for determining candidates for process improvement, or screening processes for potential worst case scenario reporting under the EPA's RMP Rule.

Reference

Center for Chemical Process Safety, "Guidelines for Engineering Design for Process Safety," American Institute of Chemical Engineers, New York, 1993.

	per task	per year
upset to person	€	€
protection (must relate effective to person)	.9ε	.96
Risk	2E-7	.01
Reward	$10	$70K

Playing the Killer Slot Machine (a Tutorial on Risk)

Donald K. Lorenzo and William G. Bridges
JBF Associates, Inc., 1000 Technology Drive, Knoxville, TN 37932-3353

ABSTRACT

A new casino has opened in Las Vegas, and its slot machines pay well on every pull of the handle—except for a chance that on any pull of the handle, the machine may be charged with 50,000 volts and the player will be fried. Would you play the game? This paper explores how the acceptability of risk changes under a variety of circumstances. It also explores how these same principles apply to hazard analysis teams that are judging the acceptability of engineered and administrative controls, and whether or not to generate recommendations.

Tutorial

As you step into the casino lobby, your eye is immediately drawn to the mob of people milling around an enormous, gold-plated slot machine. Then, you notice a neon sign above the machine flashing "FREE SPIN" and television cameras filming the event. "Must be some new game show," you think as you join the throng. You work your way closer to the stage, and you see why everyone is clamoring to pull the handle—this machine pays off with every spin. In the excitement, you start jumping up and down, hoping to be noticed and selected by the master of ceremonies. Finally, your effort is rewarded, and a liability waiver is thrust into your hand as you are being pulled toward the stage. Your signature is required, they say, before you can play the game. Reading hurriedly, your eyes focus on:

> **WARNING: ANY SPIN RESULTING IN ALL LEMONS ON THE PAY LINE CAUSES DEATH**

Will you sign the waiver?

This is a classic risk-based decision. There is no right or wrong answer, only a decision based on the individual's (or organization's) risk acceptance criterion. In this example, the negative consequences are severe. For some people, that fact alone is enough to dissuade them. But, most of us willingly engage in activities,

such as driving or flying, with potentially fatal consequences. What other factors influence our willingness to play the slot machine?

- **How many wheels are on the machine?** Each wheel is essentially a layer of protection. More people would be willing to play a machine with ten wheels than a machine with only one wheel.
- **What percentage of each wheel's positions are lemons?** Each lemon represents a failure of that protective layer. The lower the percentage of lemons, the more dependable the protection, and the more people would be willing to play.
- **Are all the wheels spinning?** Periodic maintenance is required to prevent machine breakdowns and to repair problems that cause failures to occur. Management may decide to save money by locking some wheels in the "lemon" position. Obviously, more people would be willing to play a machine in good working order than one in disrepair, and more people would be willing to play a machine whose status could be determined immediately before play than a machine whose condition is unknown.
- **Do all the wheels spin independently?** A multiwheel machine appears to offer more layers of protection than a single-wheel machine. However, there may be common-cause failure mechanisms (e.g., same manufacturer, power supply, location, or maintenance) that effectively lock two (or more) of the wheels together. Thus, if one wheel stops on "lemon," one (or more) of the other wheels will also stop there. People will prefer a machine with independent wheels over one whose wheels are interlinked.

Answers to these four questions allow you to calculate the odds of death if you play the slot machine. Some people initially willing to play the game now decide they are not willing to play. Others, however, need more information before deciding.

- **Are the consequences immediate?** The prospect of immediate death is more daunting than death at some later time. The longer death is deferred, the more likely something else (e.g., old age) may cause death anyway. Advances in medical technology may render a currently fatal injury/disease survivable in the future. So, more people are willing to play a machine with delayed consequences than one with immediate consequences.
- **Whose death results from the bad spin?** It seems only fair that the person taking the risk should suffer the consequences of losing the gamble. If others are at risk, the acceptability of the gamble changes. People would generally be less willing to play a machine if it killed their children rather than themselves on a losing spin. On the other hand, people are more willing to play if the potential fatality is a faceless stranger with whom they have no personal connection.
- **What is the benefit of playing?** Even with a complete understanding of risk, the prospect of being rewarded influences people's acceptance of the risk. If there were no payout for nonlemon spins, few would play the machine. If all nonlemon spins paid $1 million, more people would play. If all nonlemon spins paid $1 billion, even more people would play. If all nonlemon spins paid a random amount ranging from zero to $1 billion, some people would play, based on their personal expectation of an "average" payout.
- **Have there been previous incidents?** The charred remains of a previous player would undoubtedly give pause to the most eager volunteer. Conversely, watching previous players haul buckets of money away would embolden others to play in the false belief that bad things won't (or can't) happen to them. The success experienced by previous players does not "prove" there is no danger. In fact, a careful observer should note how often players were paid for near-miss spins of lemons on all but one or two wheels. Such spins may indicate that the machine is malfunctioning or that the odds are worse than you thought. Even a machine working perfectly will randomly kill a player from time to time. Nevertheless, more people will play a machine that paid its previous player than one that fried its previous player.

Considering the answers to these questions, you must now decide whether to play the slots. You understand the likelihood of various outcomes and the severity of both positive and negative consequences. Accept the risk or get off the stage.

"Take Me to the Real World! But Is It Safer?"

Consider the following real-world situation. A chemical process contains a reactor that produces a proprietary material by reacting chlorine with a caustic solution. The reactor normally operates at atmospheric conditions, and the reactor is vented to a scrubber to remove any excess chlorine. During the chlorination, operators are required to periodically take a sample from the surface of the liquid slurry in the reactor. A hatch is typically left open on top of the reactor to (1) allow taking of the sample and (2) provide plenty of air flow during continuous venting of the reactor. This is a small operation, and there is only one operator in the chlorination control room during each shift.

The key accident scenarios of interest are (1) overchlorination, which will drive the pH low enough to start a decomposition of the product in the reactor, and (2) low level in the reactor, which will allow chlorine to escape (unreacted) to the atmosphere. Even worse, if scenario number one occurs, the decomposition will start at the bottom of the reactor (where the chlorine enters), and the decomposition gases formed (essentially a chlorine gas bubble) will quickly lift almost all of the slurry out of the reactor (into the scrubber and out the open hatch into the process building), resulting in the low-level scenario (number two). Since operators may be near by (or on the floor below) when an over-chlorination occurs, the resulting consequences could be quite severe, possibly even fatal.

The company has many safeguards against these scenarios. A pH sensor and an oxidation-reduction sensor continuously monitor the degree of chlorination, and the operator periodically (twice each hour) performs a direct titration on a sample drawn from the reactor (to verify the calibration of sensors, which tend to drift). The operator performs other tasks, besides sampling, near the open hatch. The chlorine flow to the reactor can be stopped quickly by closing a manual quarter-turn valve at the reactor or by closing a remotely actuated valve from the control room. The chlorine spargers inside the reactor are visually checked every 3 months to help ensure even distribution of chlorine in the bottom of the reactor (to reduce the chance of localized over-chlorination). There are chlorine gas detectors within the building near the reactors; and these are checked and calibrated every month. The control room is kept under positive pressure using air drawn from the roof of the facility (three stories above the room). All employees working in the area (the shift chlorination operator, support personnel who may be visiting, maintenance personnel, etc.) are required to have escape respirators with them at all times, and operators are required to wear full-face respirators when taking samples through the hatch and any other time they are working near the hatch of the reactor. Finally, there is rigorous training, including drills twice each year, on emergency shutdown procedures, evacuation, and rescue.

The average pay for a chlorination operator (the employee who is primarily at risk) is between $25 and $40 per hour. The plant is nonunion and is located in a very rural area where this is one of the highest paid jobs in the area. The company has been one of the most stable employers for the past 20 years.

Playing the Killer Slot Machine (a Tutorial on Risk)   57

You find out that the company is hiring a chlorination operator. Now, assuming you were 20 years old, lived nearby, needed a job, and wanted to work in a production setting, would you PLAY THE GAME (take the job) if offered to you? Having trouble deciding? Well, then let's break the problem down into more bite-size pieces to help make the decision easier. And for illustration purposes, think about the problem in terms of the Killer Slot Machine.

First, consider the possible negative consequences (ZAP!!) if you get a PAY LINE with all lemons: Chemical pneumonia, poisoning, or asphyxiation caused by inhaling a large breath of chlorine. Note that chlorine can be quickly lethal at concentrations above 50 ppm in air, unless medical attention is prompt and effective.

Next, consider the possible positive consequences ($$$!!) if you get a PAY LINE with less than all lemons: A day's wage of at least $200! Cool! But, is there one "pull" of the handle each day? Two each day? That depends on the number of possible initiating events each day, which is tough to estimate without considerably more data and evaluation. Let's assume we perform the analysis. We determine that the number of pulls each day is about 20, based primarily on the number of times we are potentially exposed to a lethal breath of chlorine, which we estimate to be the 20 times we are near the open hatch of the reactor to take a sample and make other adjustments to the chlorinator during each 8-hour shift. (There are many other exposure scenarios, including over-chlorination while you're not near the hatch, a leaking or broken pipe, etc. We are going to assume those additional risks are negligibly small.) Based on 20 pulls per shift and wages of $200 per shift, the average payout per pull of the slot machine handle is at least $10 (pay received for vacations, holidays, process downtime, etc., increases the actual payout to about $11 per pull). There are approximately 47 workweeks for new operators, with an average 5-day workweek, so YOU will get to pull the handle on this slot machine about 4,700 times each year!

Next, consider the odds of each consequence: Before you decide to play (i.e., before you take the job), you probably want to know the likelihood of the negative consequences per pull, right? To estimate this, you need to determine how may "wheels" are on the slot machine and the likelihood of getting a lemon on each "wheel." Based on the information given, Table 1 is a summary of the "wheels" and "lemons" analysis of one specific accident scenario, which (as you will recall from event tree basics) is one pathway from initiating event through safeguard success or failures to a specific consequence.

Finally, determine the overall payout and odds: Even though the preceding arguments are oversimplifications for the sake of this example, you must nevertheless make the decision about accepting the job offer before the offer is withdrawn. The bottom-line odds for this particular slot machine are summarized in Table 2.

TABLE I
Percentage of Lemons on the Initiating Event and Safeguard Wheels

Wheel	Description	Percent of Lemons on This Wheel
1 (Initiating Event)	Automatic sensors prevent overchlorination, assuming good sparging and considering the titration by operators to check the calibration. Failure of this control system is the most likely initiating event	0.0005% (Probability = 5E–6) This is based on a failure rate of 1 overchlorination every 2 years, 350 total operating days a year, and 5 minutes exposure per sample or batch adjustment
1 (Initiating Event)	Uneven distribution from the sparger can cause a localized pocket of material to over-chlorinate, triggering the remainder of the reactor to decompose as well. Failure of the sparger is the second most likely initiating event	
2 (Safeguard)	Annunciation by chlorine gas detectors will alert you to a large release of chlorine (this will allow you to take action to limit the chlorine release and evacuate)	100% (Probability = 1.0) This safeguard applies only when you are on a different floor or when you are on the reactor floor but are not taking a sample at the time of the release
3 (Safeguard)	Quick shutdown of the chlorine feed to the reactor will limit the release	100% (Probability = 1.0) This safeguard protects others, not you. With only an escape respirator within arm's reach (or already on), most operators would run for the door and try to find a way to shut off the chlorine flow from outside the building
4 (Safeguard)	A personal respirator will protect you for a few minutes. You should wear it each time you take a sample or do other work near the hatch. You must wear it correctly to get a good seal each time	5% (Probability = 5E–2) There are many reasons or excuses why you might not wear your respirator as prescribed
5 (Safeguard)	Effective rescue may save you if you are not wearing your respirator and the release occurs while you are near the hatch	80% (Probability = 8E–1) If your respirator fails, you will probably die before rescuers can help you

Playing the Killer Slot Machine (a Tutorial on Risk)

TABLE 2
Summary of Risk Considerations in Accepting the Job Offer

Description of Risk Factor	Risk Factor Per Pull of Handle	Risk Factor Per Person-year (4,700 pulls)
1. Chances of an event that could result in a catastrophic release of chlorine from the reactor (**unmitigated negative risk**)	Probability = 0.0005% or 5E–6 (Not too bad) (Odds of a lemon on Wheel 1)	Probability = 2.35% or 2.35E–2 (Pretty serious risk factor, but this is mitigated by safeguards as described below)
2. Chances of not surviving the large release, given the release does occur (**mitigation factors**)	Probability = 4% or 4E–2 (Combined odds of lemons on both Wheel 4 and Wheel 5)	Probability = 4% or 4E–2 (Combined odds of lemons on both Wheel 4 and Wheel 5)
3. **Mitigated risk of negative consequence**	Probability = 0.00002% or 2E–7	Probability ≈ 0.1% or 1E–3
4. **Chances of getting paid**	Probability ≈ 100% or 1.0	Probability ≈ 100% or 1.0
5. **Pay**	At least $10	$52,000 to $83,200 (With fringe benefits, but without overtime pay or other incentives)

So, with the prospect of working 40 years until retirement, you will have a 4% chance of dying by this accident scenario. If you play and win, you will earn about $3 million during that time. Otherwise, your family will probably receive a hefty settlement (at least in U.S. courts), your life insurance benefits, and a percentage of your earnings, although they will lose you!

Now decide: Will you play? I'd take the job! What if there were five chlorination reactors you had to simultaneously operate? Would you still play? Maybe. What if you evaluate the many other killer slot machine outcomes during a work shift and the cumulative risk from all the machines is five times higher. Would you still play? Hmm, maybe you'd need more safeguards.

A Couple More Questions

If you were on the hazard review team for this process, would you recommend additional safeguards for the overchlorination scenario? As an operator, I probably would if I thought it wouldn't cost too much (such as getting the most reliable style of escape respirator available). As a manager representing the interest

The Process Safety Team utilized process mapping techniques to develop the appropriate inputs, process activities and outputs for each type of process hazard analysis. Proposed process hazard analysis processes were developed for the following three categories: existing units, new capital projects and revalidations. The processes were presented at RPI's Health Safety and Environmental Conference in April, 1996 and later issued corporate-wide in the Process Hazard Management Program (PHMP). PHMP contains requirements and suggested recommendations for process safety management activities for all RPI North American facilities. It also contains guidance on how to meet process safety objectives.

This paper describes the PHMP recommended risk assessment methodology for existing unit process hazard analysis. It does not include details on how to do HAZOP's or What-ifs since those methodologies are described quite well by others. Within Rhône-Poulenc, the objectives of an existing unit process hazard analysis are to identify potential hazards, objectively rate the severity, develop the scenarios leading to the potentially hazardous event and qualitatively rate the potential risk considering the protective and preventive measures in place. After this is completed efforts are begun to determine whether, and to what extent, such risks may be mitigated. Development of this methodology and its subsequent introduction and usage has aided in the consistent determination of risk within the company. As more experience is gained with the methodology even more improvements are expected.

FIVE BASIC STEPS IN CONDUCTING A PHA

Scope and Objectives

Establish the unit (process) boundaries to be addressed by the PHA. Establish the severity vs. consequence classifications to be used during the PHA. Specific additional site or Enterprise consequence considerations that are to be addressed will be established. Team membership, qualifications and schedule considerations are to be determined at this time. Establish hazard identification methodology.

Preliminary Safety Analysis

1. DESCRIPTION OF THE TOOL

The Preliminary Safety Analysis (PSA) is a general hazard identification technique. It focuses on the hazardous materials, reactions, processing steps, and any overall site hazards. The PSA identifies potentially hazardous uncontrolled releases of mass or energy. Available process data are recorded and missing information is noted for further investigation. This type of analysis relies heavily on the process knowledge and experience of the analyst and all team members.

2. PARTICIPANTS

A PSA should be conducted with an experienced study leader in a meeting of a multidisciplinary team knowledgeable in the proposed process. Typical participants include engineers, chemists, operations, maintenance, and process safety personnel. A PSA can be accomplished with as few as two technical people with a safety background or with less experienced personnel but at the risk of sacrificing completeness.

3. INPUTS REQUIRED

- Material Safety Data Sheets and available Material Technical Sheets
- Reaction Sheets or Process Chemistry description
- Process flow diagrams
- Mass and energy balances including waste and emission streams
- Site plan and equipment layout.
- Historical data, especially involving upsets or accidents.

4. SCHEDULING ISSUES

Most projects and processes for which required inputs have been assembled can be reviewed by an experienced process safety team within one day.

5. METHODOLOGY

A PSA formulates a list of hazards and generic hazardous situations following the completion of the forms for *Materials, Reactions, Processing Steps and Unit Overall*. **Appendix A** contains an overview of the methodology including the abbreviations for each hazard as described on the PSA Hazard Codes page.

6. OUTPUTS

The outputs are qualitative in nature but with some indication of severity of potential consequences. They include:

- Completed materials table
- Completed tables for reactions
- A completed processing steps table
- A completed unit overall table
- A description of the major elements of the proposed design with respective impacts on general issues of health, safety, and environmental protection. (More applicable to new projects.)
- A list of potential hazards of the project or process based upon the available design details
- Recommendations to reduce, prevent, or protect against potential hazards in the subsequent design phases. (More applicable to new projects.)
- Identification of gaps in process safety and environmental information and recommendations for further study.

Hazard Identification

1. DESCRIPTION OF THE TOOL

Hazard identification can utilize various techniques to identify causes and consequences of deviations in a process. Techniques available are HAZOP, What-If and Brainstorming. All of these techniques can be successfully applied to various processes. Selection will depend upon the scope and objectives determined for the PHA. Hazard identification also includes evaluating facility siting and human factor issues using available checklists. Techniques are described below.

2. PARTICIPANTS

A hazard identification leader must be experienced in the application of the technique selected. The team is to be multi-disciplinary with members experienced in the operations, maintenance and technology of the process. Team size will vary depending upon the plant and process. Optimum team size is five to seven persons.

3. INPUTS REQUIRED
- Chemical and reaction hazard information
- Equipment design
- P&ID's
- Standard operating procedures
- Material and energy balances
- Site plan and layout drawings
- Preliminary Safety Analysis documentation
- Historical data including accidents and incidents

4. SCHEDULING ISSUES

Hazard identification is generally very time intensive. Typically reviews will average 4–8 hours per P&ID. Schedules should be arranged to maintain team integrity for the entire analysis.

5. METHODOLOGY

A. Hazard and Operability Analysis (HAZOP) The Hazard and Operability Analysis or HAZOP is a technique used to identify and evaluate safety hazards and operability problems in a process. The Rhône-Poulenc focus is on identification of process safety hazards. A leader experienced in the HAZOP process systematically guides a multidisciplined team through the plant design using a set of guide words. These guide words are applied to specific points in the plant design to identify potential deviations from the intended operation.

The HAZOP starts by dividing the process into sections or nodes for review. The following words and their meanings are listed below:

Deviations:	Differences from the design intention that are used to stimulate the brainstorming activity for identifying process hazards. (Refer to the attached list.)
Causes:	Credible reasons for the deviations to occur.
Consequences:	Undesirable result(s) of the deviation (fire, explosion, toxic release, etc.)
Severity:	Rating of the hazard consequences (see **Appendix B**)
Comments:	A column for comments or actions that may be required for follow-up
Risk Sheet:	A reference to risk sheets that are opened for severity 1 and 2 hazards.

B. *What-If/Brainstorming Analysis* The What-If and Brainstorming analyses are loosely structured, creative techniques that identify potential hazards by posing questions derived from a review team's creative imagery. The methods are typically performed by a team knowledgeable in the process being reviewed. The team uses either technique to creatively identify the various deviations that can occur in the process. In the What-If technique questions are posed as "What-If this event happened? (i.e., What-If the pressure regulator failed?, What-If the operator charged the materials in the wrong order?) The Brainstorming technique is similar except that questions need not be limited to What-If. This method may also be used to focus on specific events and their potential causes.

The following words and their meanings are listed below:

Question: The What-If or Brainstorming deviation posed by the team

Consequences: Undesirable results of the question (fire, explosion, toxic release)

Severity, Comments and Risk Sheets: Same as HAZOP meanings above

C. *Facility Siting* A Facility Siting review of the studied process is included as an integral part of the hazard identification. The HAZOP, What-If and Brainstorming techniques generally focus on process chemistry, hardware and operator errors during the review. The purpose of the Facility Siting review is to discover possible hazards associated with global design issues such as plant layout, location of control rooms, effects of the surrounding area or emergency response issues. The hazard identification team will analyze these issues during their meetings. The process hazard analysis final report will contain results of any findings from the Facility Siting review. A Facility Siting review for use on existing processes is available. A more detailed Facility Siting review for use with new processes and facilities is also available.

D. *Human Factors* A Human Factors review of the studied process is also included as an integral step of the hazard identification process. The traditional hazard identification techniques discussed above typically identify potential errors as a result of human error. However, they typically do not look at underlying issues and external influences such as deficient procedures, inadequate supervision, lack of training, poor physical work environment and deficient human-machine interfaces. The hazard identification team will analyze these issues during their review to identify potential hazards. Any findings or recommendations are to be included in the process hazard analysis final report. A Human Factors review checklist is available for use.

Risk Assessment

1. DESCRIPTION

A risk assessment evaluates the severity and probability of a sequence of events and assigns a qualitative risk rating. Key elements of this methodology consist of identifying and/or rating:

- Final event with potential hazardous consequences
- Consequences of a final event
- An initial failure that could lead to the final event
- Probability of the initial failure
- A sequence of events that logically leads from the initial failure to the final event
- Safeguards that prevent the scenario from being completed, or reduce the severity of the final event
- Probability that the scenario would be completed
- The Risk of the final event

2. PARTICIPANTS

Risk assessments should be conducted with an experienced study leader in a meeting of multidisciplinary team knowledgeable in the process under review. Typical participants include:

- Study leader experienced in the methodology, typically process safety
- Production manager/engineer
- Technical representative typically a chemist or process engineer
- Operator(s)

3. INPUTS REQUIRED

- Hazard Identification Documentation
- Material Safety Data Sheets and Material Technical Sheets
- Maintenance records
- Reaction Sheets or Process Chemistry
- Piping & Instrumentation Diagrams
- Mass and energy balances including waste and emission streams

- Site plan and equipment layout
- Incident and Accident Reports regarding upsets or incidents
- Standard Operating Procedures

4. SCHEDULING ISSUES

Risk assessment is generally time intensive. Available data and complexity of the technology are important factors in determining the actual time required. A subgroup can assemble the scenarios prior to team review to improve productivity.

— Sometimes 1 unit takes 2 days, some 2 months

5. METHODOLOGY

A Risk Evaluation Sheet is used to document the risk assessment for a given scenario. **Appendix B** contains an overview of the risk assessment methodology. The appendix details the information used to assess the risk. The discussion follows the usual sequence of activities and where the information is recorded on the form.

Report

The report of the process hazard analysis will contain the following:

- Scope and objectives of the PHA
- Team members and qualifications
- Date(s) conducted
- Summary of findings and recommendations
- Attachment with PSA sheets, hazard identification sheets and risk evaluation sheets.

Appendix A
OVERVIEW FOR PSA FORMS

MATERIALS

THE materials review focuses on chemicals used in the process including raw materials, intermediates, additives, solvents, catalysts, effluents, inerts, and products. Physical properties are reviewed for inherent hazards and compatibility with materials of construction. Mixtures are investigated to determine if any potentially explosive, flammable, reactive or toxic effects exist.

REACTIONS

A reaction table should be filled out for each significant chemical or physical reaction involved in the process. The table requires discussion and documentation of reaction material and energy balance, critical parameter control, potential upsets and incompatibilities. Potential hazards and their prevention or protection is included.

PROCESSING STEPS

Each step or unit operation is discussed and documented. Criteria for startup, normal, upset, and shutdown conditions are reviewed against maximum system parameters such as temperature, pressure, flow etc.

UNIT OVERALL

The Unit Overall Table provides a summary of the key issues identified during the materials, reactions, and processing steps analyses. This is where the interface and reliability of major process components upon support and utility systems and their control is to be considered. This stage of the PSA should review potential hazards due to operating philosophies, administrative or automatic control schemes, human factors, ergonomics and citing. Environmental issues such as waste streams, vent emissions, and permitting also need consideration.

Preliminary Safety Analysis—Hazard Codes

F   Fire, i.e. combustion of condensed materials (leaking of flammable liquids, layers of dust, etc.)

E   Explosion of a purely physical origin (high pressure circuit going into a tank, sudden evaporation of a liquid, etc.)

EG   Gas or vapor explosion, deflagration of oxidant/fuel mixtures

EP   Dust explosion, deflagration of powders suspended in air

ET   Thermal explosion, main cause is a chemical reaction out of control (temperature runaway, gas formation, exothermic decomposition)

D   Detonation explosion occurring in special conditions, either in gaseous phase with certain pure compounds (acetylene) and mixtures (hydrogen–air) or in condensed phase with certain compounds (peroxides) and mixtures (nitro derivatives) with a high energy content.

T   Toxicity or corrosive to human tissue covers hazards due to immediate or acute health affects or future effects (chronic toxicity, carcinogenicity, etc.)

P   Physical hazards from a non-process-related event such as moving machinery, rotating equipment or elevated work.

Appendix B
OVERVIEW/INSTRUCTIONS FOR RISK EVALUATION SHEET

General Risk Assessment Form Information

Much of the information at the top of the form is used to identify the general area of the analysis:

- Plant, Unit, Area—Used to identify the process being reviewed
- Sheet #—An identification aid for the risk areas
- Equipment—Reactor, tank, etc.
- Mtls.—A list of the materials involved

- Serial No.—A second method to index the sheet (optional)
- Other References—At times the risk is similar to that of other scenarios where reference to other Risk Sheets might be useful.

Final Event

The typical final events being considered include fires, explosions, and releases of materials having toxic or corrosive properties. Included are runaway reactions that could result in the rupture of a vessel with chemical releases and possible shrapnel. The final events were determined during the hazard identification step of the process hazard analysis,

Consequences and Severity of the Final Event

The consequences of the final event will be determined by analyzing the impact of the event on personnel, the community, and the environment. Calculations of the magnitude of the consequences of the event are performed to assist in determining the severity rating. Specifically, radiation levels from fires, pressure generated from explosions, and concentration of toxic materials from releases need to be calculated. These consequence analyses should be conducted prior to the team meeting. Consequence calculations are based upon the following guidance.

Severity	Personnel	Environmental
High (1)	Irreversible Health Effects	Pollution Off-site
Medium (2)	Reversible Health Effects	Pollution On-site
Low (3)	Minor effects	Minor effects

Severity ratings are made without regard to safeguards. Safeguards are addressed as part of the risk rating. The following table lists criteria for High (1), Medium (2), and Low (3) Severity Ratings:

Severity	Fires	Explosions	Toxic Release[a]
Measurement[b]	Radiation (R) (BTU/hr/ft^2)	Pressure (P) (psig.)	Concentration (C)
Evaluated at *(From source)*	100 feet[c]	100 feet[c]	Fence line
High (1)	$R > 3{,}000$	$P > 2$	$C >$ ERPG 3 or $C >$ IDLH
Medium (2)	$1{,}500 < R < 3{,}000$	$0.3 < P < 2$	ERPG 2 $< C <$ ERPG 3 or $0.1 \times$ IDLH $< C <$ IDLH
Low (3)	$R < 1{,}500$	$P < 0.3$	$C <$ ERPG 2 or $C < 0.1 \times$ IDLH

from consequence modelling PHAST

[a] Use IDLH values only when ERPG values are not available. When neither ERPG nor IDLH values are available seek expert assistance to determine technically sound alternatives.

[b] In addition other considerations and classifications at the request of the Enterprise can be addressed. Examples of these are:
- Odors
- Nuisance Effects
- Other Water, Air, Land Damage
- Wildlife Damage
- Permit Violations
- Regulation Violations
- Community Considerations

did not show us their severity rankings

[c] Evaluate both radiation and pressure generated at the lesser of 100 feet or distance to targets of concern.

Initial Failure

Initial failures are identified by reviewing the hazard identification documentation. See Hazard Identification for methodologies. The following table lists typical initial failures:

Type of Failure	Typical Examples
An instrument failure	Sensor, control valve, instrument air, electrical failure
An operator failure	Incorrect valving, Incorrect timing of cooling or heating steps
Improper sequences	Charging materials in reverse order
Omitting key material	Catalysts, reactive chemicals
Incorrect amounts	Overcharging reactive materials, undercharging materials

Probability of Initial Failure

The probability of the initial failure is assigned without regard to subsequent safeguards according to the following table:

only qualitative

Rating	Description	Frequency
High (1)	Frequent—has happened or is predicted to occur often.	More often than once in five years.
Medium (2)	Possible—could occur within the life of the plant.	Less often than once in five years, more often than once in 50 years.
Low (3)	Rare—not anticipated to occur within the life of the plant.	Less often than once in 50 years.

The probability of the initial failure is usually a "2" for either instrumentation failure or operator failures. If there is a history of a frequent instrument failure or if the operator is in a high stress situation, the rating might be a "1."

Event Sequence

A detailed, logical step by step sequence of events leading from the initial failure to the final event needs to be constructed. The event sequence focuses on the process dynamics that must occur. As the scenario is entered into the Risk Evaluation Sheet, number the steps remembering that step 1 was the initial failure.

Hardware and procedural safeguards will be recorded under the columns for **Existing Safety Devices to Minimize the Failures/Consequences** or **Existing Procedures to Minimize the Failures/Consequences**.

Safeguards

Once the sequence is completed, consider the hardware and/or procedures in place to prevent the chain of events or to limit the impact of the final event. Enter the hardware actions under the **Existing Safety Devices to Minimize the Failures/Consequences** and enter appropriate procedures under **Existing Procedures to Minimize Failure/Consequences**. Number each safeguard with the corresponding step in the scenario. For example, if the third step is "temperature rises to 300 degrees," and a temperature controller or interlock will stop a feed flow that is causing the temperature increase, enter "3. Temperature controller 100–101 will close valve FV 100–101B stopping the reactant flow."

Numbering the safeguards in this manner helps in the risk rating step.

Probability of the Final Event

When assessing probability of the final event, consider the probability of the initial failure and the effect of all safeguards, both hardware and procedures. A sequence that requires many failures would be rated as rare or a "3". Refer to the previous probability assessment chart for ratings to be used.

Risk

The risk rating is based on an evaluation of the severity rating and probability of the final event. Refer to the following definitions:

Rating	Definition
High (1)	Undesirable—requires high priority action
Medium (2)	Improvement required
Low (3)	Acceptable

In general, there are only four cases that result in risk ratings of "1" or "2." See the following table for guidelines to Risk Ratings as a function of Severity and Probability of Final Event:

Probability	RISK RATING		
1	1	2 *	3
2	1 **	2	3
3	3	3	3
	1	2	3
	Severity		

* Possibly a 1 based on Enterprise considerations
**Possibly a 2 based upon Enterprise considerations

[handwritten: 3 × 3 risk matrix]

[handwritten: R.P. had discussions about whether this was 1 or 2 but decided that sooner or later it would happen thus risk=1]

[handwritten: Risk=1 must do something within 6m–1y]

Rhône-Poulenc Inc. Process Hazard Analysis and Risk Assessment Methodology 73

RPNA BUSINESS CONFIDENTIAL
RISK EVALUATION SHEET

PLANT:		SHEET #	
UNIT:	AREA:		
EQUIP.:	MTLS.:		
SERIAL NO.:		DATE:	
CONSEQUENCES OF THE FINAL EVENT		SEVERITY:	☐
1. INITIAL FAILURE:		PROBABILITY OF INITIAL EVENT	☐
EVENT SEQUENCE RESULTING IN THE FINAL EVENT	EXISTING SAFETY DEVICES TO MINIMIZE THE FAILURES/CONSEQUENCES	EXISTING PROCEDURES TO MINIMIZE THE FAILURE/CONSEQUENCES	
FINAL EVENT TO BE PREVENTED:	PROBABILITY OF FINAL EVENT ☐	RISK ☐	
POSSIBLE IMPROVEMENTS OR RECOMMENDATIONS:			
OTHER REFERENCES:			

Finding an Appropriate Level of Safeguards

Michael D. Moosemiller and William H. Brown
Det Norske Veritas, Houston, Texas

ABSTRACT

There is much discussion of the appropriate level of safeguards for given situations during process design reviews, process hazards analyses, and in the wake of incidents. Some companies have adopted rather dogmatic rules to simplify decision making. For example, you often hear the statement "We don't consider double jeopardy events." This statement assumes all hazards and safeguards are equivalent, lumping together nuisances with catastrophes and low reliability safeguards with high reliability protections. It also flies in the face of experience, in that major losses are primarily a result of multiple failures.

On the other hand, it is possible to have too many or too good safeguards, thereby making a process so idiot-proof as to ultimately result in operators who are unable to handle any non-routine situations. In principle, mitigation decisions should be made on a risk basis, which would include consideration of the reliability/availability of the safeguards, and the consequences of their failure. This paper describes appropriate criteria for decision making, and attempts to topple some of the more arbitrary rules that operators commonly use.

1.0 Putting Too Much Credence in Safeguards

In conducting hazard reviews, many people will make comments such as "We don't consider double jeopardy events," "We've never had that happen here, so we don't think it's credible," and so on. Some of this thinking is valid; in most cases it probably isn't worth guarding against something that a group of people having 100+ years of operating experience have not run across. However, of the catastrophic events that come to mind, how many have *not* been double or more jeopardy?

It is therefore not necessarily appropriate to dismiss multiple-cause events out of hand. Other assumptions that people often make:

- *Relief valves always work*—Really? What is the actual historical reliability of relief valves:

    (a) 99.9%   (b) 99.5%   (c) 99%   (d) 90%   (e) 86%

The "best average' answer may be (b) 99.5%, and can be inferred from an average of nuclear industry[1] and offshore oil[2] data, assuming a typical number of demands/tests. However, another source[3] gives probabilities for failure to open within 10% overpressure of only 99%, 90% and 86% in "clean", "average" and "dirty" services, respectively. These may seem low to many people (or perhaps not).

The fact that catastrophic vessel failures don't occur regularly is perhaps more a tribute to the overdesign factors that are used, the fact that most overpressure events are not gross overpressures, and that many relief situations manage to vent themselves through blown gaskets or other lesser events. Nonetheless, to assume blindly that a relief valve will work is not a healthy attitude, particularly if the consequences of an equipment failure are truly catastrophic. For cases where gross overpressure can occur (2–5 times design pressure), redundant overpressure protection may be justified.

- *Check valves always work*—Really? What is the actual historical reliability of check valves (per demand):

    (a) 99%   (b) 97.5%   (c) 90%

    There is not much data available on this subject. A "best estimate," if there is such a thing, is probably (b), based on old nuclear industry data and assuming the check valve has been in service for some years. People's expectations on the performance of check valves are probably more realistic than for relief valves, but vary widely.

    Some facilities require their process hazards analysis teams to assume that a check valve will not work on demand, while other groups are instructed to assume that they will always work. There certainly is room for different standards in this case—most people would probably agree that check valves will be much less reliable in dirty service or in vapor service than in clean (unreactive), liquid service.

- *A catastrophic failure of an atmospheric storage tank is not credible*—What is the actual historical catastrophic failure rate for atmospheric storage tanks:

    (a) One in 500,000 years   (b) One in 50,000 years   (c) One in 5,000 years

    In this case, the best answer is probably (b) One in 50,000 tank-years.[4] So there is good reason to discount the possibility of a failure ... or is there? Perhaps, if we have only one tank, and its failure is contained in a dike. Consider, though, a large petrochemical company which has

Finding an Appropriate Level of Safeguards

perhaps 1,000 tanks at its facilities. Then over the life of the company (greater than 30 years, one hopes), the odds are that the company *will* experience a catastrophic tank failure.

No problem, you say—your tanks are surrounded by dikes which are designed to handle full tank volume. The fact is, most dikes are *not* designed to contain a catastrophic failure—much of the wave of liquid simply washes over the dike wall. So consider your vulnerabilities; if the tank contents are particularly nasty, or if you are located in a sensitive location (e.g. upstream of a city's water intake, or near an office building), additional precautions such as improved dike design or lower inventories may be appropriate.

- *Our operators will always respond promptly to an alarm*—What is the actual expected reliability in response to a dedicated alarm, but which signals a large number of false alarms and can easily be deactivated:

(a) 99.9%   (b) 99.5%   (c) 99%   (d) 90%

The "best answer" is probably (b) 99.5% for simple systems, although this requires a good deal of judgment. There are several approaches to quantifying human factors reliability, the major ones of which are discussed in Lees[5]. The value above was arrived at by using the HEART technique. Human error rates increase rapidly with job complexity or stress.

- *"It's never happened here, so we don't consider it credible"*—Have you heard this refrain, or something like it, before? Then using the same logic, before the first time any of the accidents that ever occurred at the site took place, they too were "not credible"?

2.0 Safeguard Overkill

The discussion above was *not* about saying that industry does not take its hazards seriously. Which of the following industries has the best, and worst, injury and illness records:

(a) Chemical   (b) Wholesale and Retail Trade   (c) Food

The best record is enjoyed by (a), the chemical industry, and the worst is (b), the wholesale and retail trade industry.[6] On the whole, the chemical and oil industries have a record to be proud of. Unfortunately, the nature of the activities undertaken by the industries means that when something really bad happens, it can be spectacular, involve the public, and be very photogenic to boot.

So, do we provide safeguards to control these risks?

Of course!

Do we do everything possible to eliminate the risk?

Absolutely not!

Besides being an exercise in futility, attempting to remove all risks through added safeguards can be counterproductive. Consider the following real-life examples:

CASE 1
Two fluid catalytic cracking units at the same site, one old and virtually non-instrumented by current standards, one new with tremendous instrumentation. Result? The operators working on the old unit are able to respond to virtually any situation. The operators on the new unit have to shut it down if there is any situation which is out of the ordinary, since they don't know how to run the unit if the computer doesn't do it for them.

CASE 2
A failure of a vessel used for an exothermic reaction due to overpressure takes place, with distortion of the vessel and nearby piping. The site response - replace the vessel with one which is rated for much higher pressure. The net effect? Replace relatively non-energetic failures with catastrophic (albeit less frequent) failures.

CASE 3
The operator of an HF alkylation unit installs a deluge system, the purpose of which is to knock down any HF which is released. The problem? The deluge system costs millions of dollars, mitigates a risk estimated to be an average of one fatality in 10,000 years, and is of a design which laboratory experiments indicate may be expected to remove 20% of the HF released, even assuming that it is activated almost immediately.

In summary, we don't want uncontrolled risks, nor do we desire to achieve the impossible dream of zero risk. What we seek is a happy medium.

3.0 Using Risk as a Basis for Selecting the Appropriate Level of Safeguards

Risk—The frequency of an event multiplied by its consequences

Risk may be expressed in a variety of terms, all of which must contain the elements of frequency and consequence. Examples include:

- Lost dollars per operating day
- Fatalities per year
- Spill of 1,000 barrels at least once in the life of the facility

Finding an Appropriate Level of Safeguards 79

We have stated above that while everyone wants to control risks, it is practically impossible to eliminate them altogether, and that the effort to do so can be counterproductive. So do we at least want to *minimize* risk? Again we say NO! What we want to do is to *optimize* risk. This leads us away from clearly nonproductive pursuits such as building concrete bunkers to minimize the impact from meteors, and away from something as comfortable or uncomfortable as gut feeling, and towards something which at first may seem difficult to get a handle on.

The effort is worth it however, and is not as onerous, complicated or expensive as some people would have you believe. The following section describes a logical process to use to evaluate the merits of risk mitigation or prevention options.

4.0 Cost/Benefit Analysis

For the purposes of this approach we will conveniently assume that political pressures do not exist. In this utopia, a quantitative cost/benefit analysis is warranted whenever a significant outlay of money is expected. It is also preferable that the cost of doing the analysis does not exceed the cost of the proposed expenditure! Depending on circumstances, the following are suggested:

4.1. Proposal to Spend 25,000

Use an easy approach, such as a risk/cost matrix. Remembering that risk is a combination of frequency and consequence, this results in a three-dimensional function, but it is simple enough to cut it down to size. First consider a risk matrix such as many people use to prioritize issues in process hazards analyses:

C O N S E Q U E N C E	5 Catastrophic	5	6	7	8	9
	4 Serious	4	5	6	7	8
	3 Major	3	4	5	6	7
	2 Low	2	3	4	5	6
	1 Minimal	1	2	3	4	5
		1 Rare	2 Very Low	3 Low	4 Moderate	5 High

F R E Q U E N C Y

where the consequences and frequency are defined as follows:

CONSEQUENCE

1	(Minimal)	$10,000 loss, minor injury, or nonreportable environmental release
2	(Low)	$10,000–$100,000 loss, serious injury, or reportable release
3	(Major)	$100,000–$1MM loss, permanently disabling injury, or major release
4	(Serious)	$1–10MM loss, fatality, or permanently disabling injury to public, or catastrophic release
5	(Catastrophic)	$10MM loss, multiple fatalities, fatality to public, or irreparable environmental damage

FREQUENCY

1	(Rare)	Very unlikely to happen in life of the plant, but may have happened elsewhere (once in 1,000 years)
2	(Very Low)	Unlikely to happen in life of the plant, but credible (1/1,000 to 1/100 years)
3	(Low)	Could happen in the life of the plant (1/100 years to 1/10 years)
4	(Moderate)	Can be expected to happen more than once in the life of the plant (1/10 years to once per year)
5	(High)	Will happen frequently (once per year)

This risk value can be overlaid onto a normalized value for the cost, as follows:

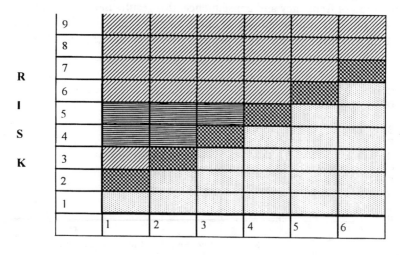

The costs in this table are defined below. Note that these values should include any capital costs, plus operating/maintenance, training, etc. expenses:

COST
- 1—Less than $1,000
- 2—$1,000 to $10,000
- 3—$10,000 to $100,000
- 4—$100,000 to $1MM
- 5—$1 to 10MM
- 6—>$10MM

The results of this matrix indicate the worthiness of the proposal:

  Do it

  Maybe (see below)   ⎱ may use QRA to decide
                                           ⎰ if worth doing

  Maybe (management discretion)

Don't

Some points about this matrix:

1. The "Maybe (see below)" area is described with some level of uncertainty because the issue may be related to a catastrophic consequence/rare likelihood event back at the risk matrix. In such a case, a more rigorous frequency or consequence analysis may be appropriate to better define the risk. For example, an event having a frequency of 1 and a consequence of 5 may or may not be significant, depending on how rare "rare" is—both a Bhopal-type event and a meteor hit would fall into this category; the management response to these two possibilities is likely to be different.
2. For low-cost/low-consequence events that are in the "maybe" range, it may be tempting to construct a more precise cost matrix with frequencies ranging from 1/100 years to >1/month or so and consequences ranging only up to $25,000 or so. In practice, it is just as simple to perform the actual risk/cost analyses since the values are generally much more precisely known. First evaluate the risk (consequence in $, or $ equivalent multiplied by frequency in events per year). Then do a simple cost–benefit analysis—do I get more out of this than the cost of implementing and operating the fix?
3. Usually a "6" cost would suggest that we should just get out of the business. It could be usefully applied to a corporate-wide program implementation, however.
4. All these matrices make some "value of life" and similar judgments, which are likely to be different from one company to the next.

4.2. Proposal to Spend More Than $25,000

As the stakes become higher, so too are the tools which should be brought to bear to analyze the situation. The first step is still to perform the cost matrix analysis above. Often with "high ticket" items, the results will land in the "maybe" category, in which case a more sophisticated analysis is warranted. It is usually the case in this situation that either the frequency and/or the consequence cannot be judged accurately based on personal knowledge.

A frequency analysis may be as simple as looking up a "generic" failure rate in a standard text, or as complicated as doing a highly detailed fault tree. In our experience, the costs of doing this analysis range from free (from us, if it's a simple generic answer) to upwards of $25,000. Normally a single-event frequency estimate is something under $1,000.

A consequence analysis will usually involve dispersion modeling of some sort. Crude (and generally, very conservative) results can be obtained from public sources such as the EPA RMP lookup tables with minimal effort. However, for the purposes of judging the worth of a major expenditure, the use of more sophisticated models is appropriate. For a single event, the cost of an analysis will generally range from $200 to $2,000, depending on the nature of the chemical being modeled and the detail in results necessary for the cost analysis.

Where both frequency and consequence are unclear, a quantitative risk assessment (QRA) may be appropriate. Historically, QRAs have been applied to entire units or facilities rather than individual equipment items, as a means of identifying the major risk contributors at a location. If the risks are significant, identification of the primary contributors to that risk usually point the direction toward the most effective risk reduction measures. QRAs may cost anywhere from $5,000 to $100,000 or more, depending on the number of equipment items, etc. included in the analysis, and so are generally reserved for situations in which a company expects to be undertaking expensive risk mitigation measures.

In any event, the costs of the analysis should not be significant compared to the costs of implementing any suggested prevention or mitigation efforts. Contrary to popular belief, such analyses are very cost-effective, generally ranging from 0.1% to 10% of implementation cost—and as often as not, risk reduction costs will be drastically lowered or eliminated by identifying more cost-effective solutions than were originally envisioned.

5.0. The Need for Good Reliability and Consequence Data

The best source for the likelihood of an event and its consequences usually resides in the experience of the people at the plant. In unusual cases, either the probability of the event is so low, or its consequences so severe, that it is not within the realm of experience of the plant personnel. The issue of good data is problematic for several reasons, including:

Reliability Data

- Plant personnel often don't have access to reliability data
- Existing equipment reliability data sources are based on limited data, and can vary greatly from one source to the next

On this latter point, please note that a consortium of companies operating under CCPS oversight is in the process of developing an improved equipment reliability data base for the process industry. For more information, see the paper by Hal Thomas of Air Products and the workshop presentation by Bernie Weber of DNV, both at this conference.

Consequence Data

- Publicly available models are often overly conservative (e.g. EPA RMP)
- Other models (e.g., Dow Fire & Explosion Index) can be misinterpreted, resulting in gross understatement of potential effects

There are very good proprietary models and consulting services available, and accuracy is worth the price when considering major expenditures.

6.0. Examples of Equipment Reliability, Put in Context

6.1. Catastrophic Pressure Vessel Failure

Expected failure rate—Once in 150,000 years[7]
Expected consequences—estimate 30 fatalities plus $200,000,000 monetary loss)
Average risk = 0.0002 fatalities per year ($3,300 per year at $10,000,000 per fatality)

6.2. Check Valve Failure

Assume failure rate of 2.5% per demand, with 2 demands per year

CASE 1
Back flow results in $1,000 pump damage

  Risk = $50 per year

CASE 2
Back flow results in uncontrolled reaction in supply tank, with loss of $100,000

  Risk = $5,000 per year

Retrofitting double or even triple check valves, or installing more mechanically sophisticated backflow preventers, would probably not be justified in the first case, but would be in the second case.

- assemble nuclear weapons for the nation's stockpile
- disassemble nuclear weapons being retired from the stockpile
- evaluate, repair, and retrofit nuclear weapons
- demilitarize and sanitize components from dismantled nuclear weapons
- provide interim storage for plutonium from the dismantled weapons
- develop, fabricate, and test chemical explosives and explosive components.

Pantex is a U.S. Department of Energy (DOE)-owned, contractor-operated facility with approximately 3,000 employees. The Plant has more than 425 buildings and comprises two contiguous tracts of land: 10,230 acres of land owned by DOE and 5,850 acres leased from Texas Tech University, which are primarily used as a safety and security buffer.

The Need

Historically, regulatory drivers have been the critical foundation for prioritizing DOE work and cost validation. Recent budget pressures, combined with a loss of Congressional confidence in the DOE, have resulted in a loss of DOE funds. Consequently, resources are no longer available to allow full and immediate implementation of all prior DOE activities.

This prospect led Pantex Environmental Management (EM) staff to develop a plan to identify, evaluate, and fund those activities that provide the greatest reduction in risk to public health, worker health and safety, and the environment. The previous EM decision making process lacked a mechanism to integrate risk information to establish priorities among competing EM Program tasks, which include both Waste Management (WM) activities and Environmental Restoration (ER) projects.

The result was an application of Battelle's risk mapping process. This tool was found to support both risk management and budgetary decisions. Specifically, risk mapping provides objective, detailed information that is necessary to

- provide stakeholders with a baseline risk assessment for Pantex work
- allocate resources to provide the greatest reduction in risk
- defend against loss of budget for activities critical to the facility's mission, including environmental, health and safety functions
- identify detailed impact of loss in funding
- gain customer support for associated work scope reductions.

The risk mapping of Pantex EM activities has subsequently been extended to prioritization of plant projects on a site-wide basis. This paper will only present the results of the EM risk mapping (waste management and environmental restoration activities).

Risk Mapping

Risk mapping is a tool to manage risk, optimize resource allocations, and adjust project schedules based on cost and risk information. It combines an order-of-magnitude integrated risk analysis approach with cost data and importance measures. Thus, risk mapping extends safety and environmental risk analysis efforts to have true business value to an organization. The risk mapping information is stored in a computerized database that interfaces with project management software.

In this paper, the concepts and application of risk mapping will be presented in the following order. The framework of risk mapping will be presented first, followed by the order-of-magnitude approach used to obtain the cost and risk parameters. The results obtained by this methodology will be presented next, illustrated by the risk mapping of waste management and environmental restoration activities at the Pantex Plant. Using the results of risk mapping to optimize resources will then be discussed. The order-of-magnitude integrated risk analysis approach used in risk mapping is presented in an Appendix.

Risk Mapping Framework

Risk mapping can be understood by viewing the importance of a plant activity from a reverse perspective: if the plant activity is discontinued or delayed, what effect will that have on the facility in terms of both costs and risks? The magnitude of the effect is weighed against the costs and risks involved in performing the activity, and when this comparison is made for all plant activities, a prioritization of the activities can be drawn. This prioritization can take into account not only safety and environmental costs and risks, but also other impacts such as business interruption and product liabilities. The methodology is flexible enough to view costs and risks at many levels of detail, from individual projects and activities to entire plant sites.

Costs and Risks

How should a true cost–benefit analysis be performed for an option that has the potential for accidental losses or other liabilities that may or may not be incurred? Most analyses in the past either have looked at expected (budgeted) costs versus expected benefits, or have looked at expected risk reduction costs versus risk-reduction benefits. A true cost–benefit analysis must look at both costs *and* risks to capture all the information relevant to an investment decision.

The difference between costs and risks can be summarized as follows:

- **Costs** are *expected* expenditures that can be included in a budget or financial forecast for an economic time frame of interest. This time frame for most organizations is a fiscal year.

- **Risks** represent expenditures or liabilities that are potential but not *expected* within the same economic time frame; hence, they are not generally included in a budget or financial forecast. A probability exists that the expenditure or liability will actually be incurred within each time frame of interest. Thus, the expense will be zero if the loss incident does not occur. However, the expense or liability can be very high if it does occur, and can have a significant impact on a business. Examples of risks include fires, explosions, accidental releases of hazardous materials, business interruptions, project delays, noncompliance fines, and product liabilities.

Note that it is possible to move liabilities from the risk category to the cost category by buying *insurance* for the risk. For example, without fire insurance, the risk of a fire represents a potential loss that must come out of company profits if realized. The purchase of adequate fire insurance transfers the risk to the insurance company, and the cost of the insurance premiums can be included as a budgeted expense. Likewise, technologies such as predictive maintenance can turn unexpected equipment-failure expenditures into more manageable, ongoing testing and maintenance costs.

Combining Costs and Risks

To combine costs and risks, they must both be in the same units of measure. Since costs are generally in monetary units, and decisions are generally made on an economic basis, it follows that risks must also be converted to monetary values. This is standard practice in the insurance industry, and is becoming more prevalent with other risk analysis and risk management activities in industry as well.

Risk is defined as a combination of the likelihood of occurrence and the severity of consequences of unexpected loss incidents. To combine risks with costs, the risks are put into units of *dollars per year*. (U.S. dollars are assumed throughout this presentation; the same methodology of course applies to any other monetary unit.) Focusing on a single risk, such as a business interruption due to an extended plant-wide power failure, this unit of measure is obtained from the definition of risk by observing that

(Incident likelihood)(Consequence severity) = Risk

(Incidents per year)(Dollars per incident) = Dollars per year

The "Dollars per year" risk measure is thus an annualized liability or loss rate.

Note that "Incidents per year" is generally a fractional number, since the loss incidents are not expected to be realized within the one-year time frame considered. Using the plant-wide power failure example: if such a power failure occurred several times a year, it would soon become an expected event that could be factored into budgeted costs (assuming nothing is done to reduce the power failure frequency). On the other hand, if the likelihood of occurrence is one chance in 10 per year of operation, then the likelihood term becomes 0.1 incidents per year. Assuming an average severity of consequences of $100,000 for an extended plant-wide power failure, the risk measure becomes (0.1)($100,000) = $10,000/year.

The risk can now be combined with budgeted costs by looking at the costs on the same one-year time scale. For example, if the total budgeted cost to operate the above facility is $300,000 per year, then the cost plus risk is $300,000/year + $10,000/year = $310,000/year. If other risks are present, they should also be added.

The significance of cost plus risk becomes apparent by comparing two facilities with similar operating budgets (including raw material costs, transportation costs, etc.) to make the same product. One of the facilities may have a significantly higher risk, such as due to unreliable utilities or closer off-site neighbors. If so, then the cost plus risk, or total long-term expenditures, for the higher-risk facility should be taken into account when making business or risk management decisions with respect to the two facilities. Likewise, if two projects having similar benefits are contemplated, then the project with the lowest cost plus risk should be given higher priority.

Net Elimination Liability

The framework of cost plus risk can be extended to both the "cost" and the "benefit" side of a cost–benefit analysis. Although it may not be immediately evident, this can be accomplished by asking how to prioritize activities in the face of declining resources. If something must be eliminated because budgets are being cut, then what should be eliminated that will have the least negative impact? Or, if resources must come from somewhere in a fixed budget to accomplish objectives such as adding new risk-reduction safeguards, from where should these resources be taken, such that the least negative impact will be realized?

The **"benefit"** side of the equation includes both cost reduction and risk reduction. If an activity is eliminated, then its budgeted *costs* will not be incurred, and those funds then become available for other purposes. In addition, any *risks* involved in the performance of the activity will cease to exist. The net "benefit," then, is equal to the annual budgeted cost plus the annualized performance risk, both of which are discontinued if the activity is eliminated.

The **"cost"** side of the equation includes both expected negative *impacts* and risk increases that result from eliminating the activity. If an activity is eliminated, then there may be various negative impacts that can be expected. For example, if a compliance-driven activity is eliminated, then a noncompliance penalty might be expected. If an activity in the mainstream of the business is eliminated, then a major business interruption may be expected. Elimination of a safeguarding activity (prevention, protection or mitigation function) may not have an *expected* negative impact within the one-year time scale of interest; however, its elimination is likely to pose a long-term *risk increase*. For example, plant operations can continue without any expected negative impact upon elimination of a wet-pipe sprinkler system. However, its elimination will clearly pose an increase in the facility's fire risk, owing to the greater severity of consequences in the unexpected event of a fire starting in the sprinkler-protected area. The net "cost" of eliminating an activity is hence equal to the expected negative impacts plus the total risk increase.

Rather than using a cost–benefit ratio as the means of combining and comparing the benefits and costs, the *difference* between the "cost" and the "benefit" is used. This difference, termed the Net Elimination Liability, gives in absolute terms the net cost plus risk incurred if an activity is eliminated. The Net Elimination Liability (NEL) can be expressed as

$$\text{NEL} = \text{elimination impact} + \text{elimination risk increase} - \text{budgeted cost} - \text{performance risk}$$

For the purposes of displaying results, the following abbreviations will be used for the respective terms in the NEL equation:

$$\text{NEL} = \text{ECost} + \text{ERisk} - \text{PCost} - \text{PRisk}$$

Thus, the larger the NEL for an activity, the less desirable it is to eliminate. It is possible for an activity to have a negative NEL. If an activity has a negative NEL, then it may actually be desirable to eliminate the activity, considering both costs and risks. Such an activity should obviously have highest priority for elimination if budget cuts are necessary. The resources made available by eliminating the activity can be put to better use elsewhere.

Net Delay Liability

In most cases involving budget cuts, many activities can be delayed until the next budgeting cycle (e.g., fiscal year) rather than being altogether eliminated. This is especially true for plant projects, which may have slack time or otherwise not be on a critical path. For this reason, an additional prioritization measure is used, which is designated the Net Delay Liability (NDL). The parameters are the same as for the NEL, except that the impacts and risk increases associated with delaying an activity can often be much less than their NEL counterparts. Hence, the NDL is determined by the equation

$$\text{NDL} = \text{delay impact} + \text{delay risk increase} - \text{budgeted cost} - \text{performance risk}$$

As for the NEL, for the purposes of displaying results, the following abbreviations will be used for the respective terms in the NDL equation:

$$\text{NDL} = \text{DCost} + \text{DRisk} - \text{PCost} - \text{PRisk}$$

Thus, the larger the NDL for an activity, the less desirable it is to delay. Activities often have negative NDLs, particularly activities with zero or very low delay impacts.

Cost and Risk Magnitudes

To facilitate the risk analysis part of risk mapping, cost and risk parameters are obtained and documented on an order-of-magnitude basis. Further, to simplify the display and combination of cost and risk parameters, only the exponents of the magnitudes are used. This is illustrated for annual cost and risk values in Table 1. For example, an annualized risk of $100,000 per year is recorded as a 5.0, since $100,000 per year is equal to 10^5 per year, and only the exponent "5" is used. The purpose for displaying cost and risk parameters by magnitudes rather than by their dollar-value equivalents is discussed and demonstrated in the Appendix, which gives details of the order-of-magnitude risk analysis approach.

To illustrate, a parallel can be drawn between the cost/risk magnitudes and the severity of earthquakes as measured by the Richter scale. A tremendous range is covered over several magnitudes. An earthquake measuring 3.0 on the Richter scale may not even be recognized as an earthquake, whereas an 8.0 earthquake can cause widespread damage and much loss of life. Likewise, on the scale of most industrial facilities, a magnitude 3.0 loss ($1,000) may go unnoticed to plant management, whereas a magnitude 8.0 loss ($100,000,000) would be catastrophic.

TABLE I
Cost and Risk Magnitudes

Magnitude	Cost or Risk
9.0	$1,000,000,000 per year
8.0	$100,000,000 per year
7.0	$10,000,000 per year
6.0	$1,000,000 per year
5.0	$100,000 per year
4.0	$10,000 per year
3.0	$1,000 per year
2.0	$100 per year
1.0	$10 per year
0.0	$1 per year
−1.0	$0.10 per year

Final cost and risk values, and some intermediate values, are recorded and displayed to the nearest tenth of a magnitude. This is essentially equivalent to one significant figure; finer resolution is clearly not warranted. A table illustrating tenths of a magnitude between 4.0 and 5.0 and their corresponding cost or risk values is given in Table 2.

Risk Mapping Results

Presented in this section are the results of the recent risk mapping efforts at the Pantex Plant. All waste management and environmental restoration activities at the site were analyzed. The results will be presented separately for the sake of clarity, although one additional purpose of the risk mapping was to combine the WM and ER results into a single prioritization, since both are administered by the same department.

Waste Management Results

Waste management activities at the Pantex Plant are similar to those at many chemical facilities, with the addition of waste streams that contain low levels of radioactivity. Thus, the major waste types at the Pantex Plant are hazardous waste, low-level waste, mixed low-level waste (containing both hazardous and radioactive components), and nonhazardous waste. Hazardous waste families

TABLE 2
Tenths of a Magnitude between 5.0 and 4.0

Magnitude	Cost or Risk
5.0	$100,000 per year
4.9	$79,000 per year
4.8	$63,000 per year
4.7	$50,000 per year
4.6	$40,000 per year
4.5	$32,000 per year
4.4	$25,000 per year
4.3	$20,000 per year
4.2	$16,000 per year
4.1	$13,000 per year
4.0	$10,000 per year

include solvents, acids, caustics, solvent-contaminated soils, explosives and explosive-contaminated solids, recyclable metals, paint sludge and residues, and batteries. Only minimal on-site treatment is performed, before storage and eventual shipment to various off-site contractors.

Waste handling operations are performed by the primary site contractor, Mason & Hanger Corporation. Waste management activities such as storage area inspections and pollution prevention are the responsibility of Battelle Memorial Institute. Personnel from both organizations were involved in the data collection phases of the risk mapping.

It was found beneficial to develop flow charts for each of the major waste types, in order to define the baseline of what WM activities are performed in what sequence. For example, the flow chart for hazardous waste (after waste characterization, pick-up, and data input into the waste tracking system) is shown in 1. Each of the activities in the flowchart with a solid upper-left corner had deviations that could increase risks, and consequently was analyzed for performance risk.

The NDL results for all Waste Management activities are shown numerically in Table 3 and graphically in Figure 2. These results indicate that, although many activities have highly positive NDL values, a few activities have negative NDLs. These activities with negative NDLs should be considered the most likely candidates to be delayed until next fiscal year if budget cuts become necessary. Review of the corresponding NEL values would indicate whether it might be desirable to eliminate one or more activities altogether. The activities with the most negative NDLs (Downgrading mixed "legacy" waste, and Reclassification/recharacterization of waste) indicate that the cost and risk involved in reclassifying certain wastes to a less hazardous status exceed the reduced disposal costs if reclassified.

Note that the results in the "PRisk" performance risk column of Table 3 represent the total risk, in dollar-per-year magnitudes, such as would be obtained from an order-of-magnitude risk analysis without the other risk mapping factors. The methodology for order-of-magnitude risk analysis is discussed in the Appendix, as well as in the paper by Johnson and Elliott, "Integrated Safety Analysis Project," included in the proceedings of this conference.

Examination of the values in the "PRisk" column indicates that Waste Characterization, with a risk magnitude of 6.2, poses the highest performance risk. This is from the underlying risk scenarios pertaining to mischaracterization of wastes, with resulting potentials for worker and public health effects, regulatory noncompliance, and stakeholder concerns (adverse publicity). Several other activities have performance risks within the same magnitude.

Environmental Restoration Results

Risk mapping of environmental restoration activities is different from risk mapping of waste management activities in several important respects. Waste man-

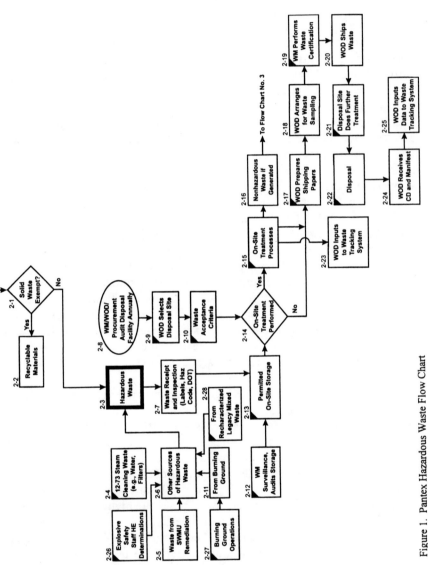

Figure 1. Pantex Hazardous Waste Flow Chart

TABLE 3
Waste Management Net Delay Liability Results

Activity	PCost	PRisk	DCost	DRisk	Net Delay Liability
WM/ER & WOD Pre-audit Checks	5.3	5.0	6.0	7.0	$11,000,000
Shipping manifest, procedures, and certification	5.4	4.4	7.0	4.3	$11,000,000
Off-site treatment	5.2	5.6	7.0	0.0	$11,000,000
Waste characterization	4.4	7.0	4.3	7.0	$ 9,000,000
Disposal, receipt of Certificate of Destruction	6.5	4.8	7.0	0.0	$ 8,000,000
Division Waste Coordination	5.1	4.5	6.1	6.8	$ 7,000,000
Waste Certification	5.1	3.5	5.3	6.8	$ 4,000,000
Generator accumulation activities at <90-day and <55-gal sites	4.7	6.1	6.3	6.5	$ 4,000,000
On-site permitted storage	5.2	4.4	6.2	6.3	$ 3,000,000
Audits/Surveillances of storage areas	4.6	2.8	6.1	6.3	$ 3,000,000
12-73 Building (steam cleaning)	5.0	5.2	4.3	6.5	$ 3,000,000
Waste receipt and inspection	5.2	5.5	6.3	5.4	$ 1,900,000
Off-site waste acceptance criteria	5.2	6.0	6.5	5.0	$ 1,800,000
Explosive safety examination of HE level	4.0	5.1	6.0	0.0	$ 900,000
Waste Operations provides container	4.8	5.1	5.1	5.9	$ 800,000
Burning ground operations	4.9	5.3	6.0	0.0	$ 700,000
RCRA-B Permit Training	5.1	5.5	6.0	3.9	$ 700,000
Spill response	5.3	5.7	6.1	3.9	$ 500,000
Waste Tracking System	5.6	4.4	4.1	5.8	$ 300,000
On-site treatment	4.8	3.9	5.0	5.0	$ 140,000
Existing MERF Determination	4.2	3.0	4.1	5.1	$ 110,000
Nevada Test Site Waste Acceptance Criteria requirements	4.6	5.2	5.5	0.0	$ 110,000
Preliminary Waste Characterization	4.5	2.0	5.1	0.0	$ 90,000
12-42 waste staging facilities	5.0	4.6	5.3	4.4	$ 80,000
Disposal site selection	5.0	4.2	4.3	5.1	$ 30,000
DWC approves MERF	4.2	2.1	3.5	0.0	($ 14,000)
Generator requests container (doesn't cover biohazard waste)	4.6	5.1	5.1	0.0	($ 14,000)
On-site transportation	5.0	5.1	5.3	0.0	($ 17,000)
Waste Minimization/Pollution Prevention	5.4	5.3	4.1	5.5	($ 130,000)
Recycling opportunities	4.9	5.2	3.5	0.0	($ 200,000)
Reclassification/recharacterization of waste	5.3	6.1	5.5	0.0	($ 1,100,000)
Downgrading "legacy" waste (mixed)	5.4	6.1	4.5	0.0	($ 1,400,000)

agement activities are similar to a continuous chemical process, with relatively constant waste streams. By contrast, ER activities are project-based, with time-varying costs and risks as restoration proceeds from site characterization through cleanup to closeout. For ER risk mapping, cost data and performance risk scenarios were developed by year where necessary, to take into account the time-varying costs and risks. The following is a discussion of the ER activities at the Pantex Plant and the risk mapping results.

Past activities at the Pantex Plant have resulted in contamination of the near-surface soils and perched groundwater. The types of contamination include nitrate residues from explosives manufacture, chromium from cooling water treatment, PCBs from transformer leaks, pesticides from weed control, accidental spills of hazardous materials, leaking underground storage tanks, low-level radioactive waste from explosives testing, and soils contaminated at other locations and disposed of at Pantex. The major concern is the potential for contamination of the Ogallala aquifer. This aquifer is a major source of water for farming and potable water. The well fields for the city of Amarillo are located a few miles north of the site. Under the Pantex site, a perched aquifer is above the Ogallala aquifer. A major ER effort is to pump and treat the perched aquifer to minimize the risk of the contamination spreading off site to the Ogallala aquifer.

The sites of contamination have been identified and consist of approximately 150 "solid waste management units" (SMWUs). Many of these units have similar contamination and consequently, for planning and budgeting purposes, all the SMWUs have been placed into one of 15 SMWU groupings. DOE, which funds the restoration activity, requires that an Activity Data Sheet be prepared for each of the 15 budgetary categories. The Activity Data Sheets contain information on the risk levels before, during and after the planned remediation in each of the seven impact categories described in the Appendix. One of the purposes of this prioritization activity was to obtain better estimates of these risks for each of the Activity Data Sheets.

Plans to remediate the site must be approved by the Texas Natural Resources Conservation Commission, or TNRCC. A site cannot be considered fully remediated until TNRCC has approved the final closure report. This commission also has established three closure levels:

Risk Reduction 1	Closure requires the remediation to natural background levels
Risk Reduction 2	Closure requires the remediation to established federal/state compliance standards or calculated health-based medium-specific concentrations.
Risk Reduction 3	Closure requires the remediation of affected media to the maximum extent practicable.

For each SMWU or SMWU grouping, three-party negotiations between DOE, TRNCC, and the site operating companies (Mason & Hanger Corporation and Battelle Memorial Institute) determined the risk reduction level for closure.

In most cases, these agreements, while not formally signed by all the participants, have been established and the SMWUs are being cleaned up to the closure level negotiated.

In addition to the TNRCC closure levels, the remediation activities must meet the requirements specified in the U.S. National Environmental Policy Act, (NEPA), administered by the U.S. Environmental Protection Agency (EPA). Under this act, the site contractor may be required to perform a Resource Conservation and Recovery Act (RCRA) Facility Assessment (RFA), RCRA facility investigations (RFIs), interim corrective measures (ICMs), Corrective Measures Studies (CMSs), Corrective Measures Implementations (CMIs), and finally a RCRA Final Investigation Report (RFIR) for closure. Since the site is also listed on the National Priorities List (NPL), it is also required to meet all the requirements of the Comprehensive Environmental Response, Compensation and Liability Act (CERCLA), as well as the corrective actions requirements of RCRA, in an integrated report. Failure to meet the requirements of these acts can result in fines and penalties. Under RCRA, also administered by the U.S. EPA, a waste-generating permit is also negotiated with each site, primarily to prevent additional contamination of the site. This permit frequently includes the cleanup milestones agreed to under NEPA, enabling the governmental agencies such as TNRCC to revoke the waste-generating permit if the schedule is not met. If the site cannot generate waste, it is unlikely that it can perform any of its programmatic activities. Thus, revoking the waste-generating permit effectively shuts down the site. This ability to revoke the permit and levy fines if a cleanup milestone is not met becomes important when prioritizing risks, particularly for the Compliance and Mission Impact severity categories.

The basic strategy for risk mapping the ER activities was the same as for the WM activities. The severity of consequences of eliminating ER activities tended to be higher than for WM activities, particularly where an ER activity remediated a SMWU grouping or delayed a particular cleanup activity. Since contamination exists at the site, "elimination" of an ER activity means leaving the site in its current state. If the three-party agreement commits the contractor to clean up to Risk Reduction 1 or 2, then "elimination" means violating the agreement and, if the milestone is part of the RCRA waste generating permit, running the risk of being shut down. The response might be less severe for sites where no cleanup technology exists and the three-party negotiations have agreed that the SMWU should be closed to Risk Reduction 3. However, even here, the negotiations typically require some stabilization or cleanup. Thus, for ER activities, the impact of elimination tended to have very high "Mission Impact" risks and significant "Compliance" risks. Permanent elimination is not considered an option for any of the ER activities.

While the risk impact of elimination was very high, if there are funding cuts, "delay" can be an effective option to the decision-maker. Table 4 shows the ER activities, prioritized by Net Delay Liability.

Resource Optimization by Risk Mapping

One of the reasons why delay of some ER activities is feasible is that many of the SMWUs are lightly contaminated and, in many cases, the site contractor has performed Voluntary Corrective Actions (VCAs) prior to formally entering the NEPA process. By using this strategy, the initial NEPA report is the RFIR report recommending closure of the site. Sending in one NEPA report recommending closure results in savings of several hundred thousand dollars. One risk associated with VCAs is that TRNCC may require additional cleanup. This may force the project into the formal reporting requirements specified by NEPA.

TABLE 4
Environmental Restoration Net Delay Liability Results

Activity	PCost	PRisk	DCost	DRisk	Net Delay Liability
Groundwater Monitoring	5.9	6.0	8.1	0.0	$120,000,000
General Management	5.7	5.4	8.0	0.0	$110,000,000
Stakeholder	4.8	3.0	8.0	6.2	$100,000,000
Quality Assurance Support	4.8	6.3	8.0	5.3	$100,000,000
Program Mgt Support Remediation Tech/Reg Compliance Support	5.3	5.1	6.3	7.2	$17,000,000
Groundwater IED/GIS	5.8	4.0	7.0	0.0	$10,000,000
Security Support	4.9	1.3	6.5	0.0	$3,000,000
General Records/Building Management	4.8	4.4	5.9	6.4	$3,000,000
Misc HE/Rad Sites Remediation Tech/Reg Compliance Support	4.0	4.3	6.3	0.0	$2,000,000
Finalize Misc HE/Rad Sites RFIR	4.2	3.2	6.0	0.0	$1,000,000
Groundwater Corrective Measures Study	5.1	5.0	6.0	0.0	$800,000
Finalize Burning Grounds RFIR	4.3	4.0	5.8	0.0	$600,000
Firing Sites Assessment Tech/Reg Compliance Support	4.3	3.0	5.3	0.0	$200,000
Finalize Cooling Tower Zone 12 RFIR	4.2	3.2	5.3	0.0	$180,000
Old Sewage Treat Plant Remedn Tech/Reg Compliance Support	3.7	3.3	5.2	0.0	$170,000
Burning Grounds Remediation Tech/Reg Compliance Support	3.6	2.0	5.1	0.0	$120,000
Finalize Ditches and Playas RFIR	5.0	4.1	5.3	0.0	$90,000
Finalize Old Sewage Treatment Plant RFIR	4.5	3.2	5.0	0.0	$70,000
Firing Sites Remediation Tech/Reg Compliance Support	4.2	4.3	5.0	0.0	$60,000
Landfills Remediation Tech/Reg Compliance Support	4.2	4.6	5.0	0.0	$50,000
Hi Pri Potential Rel Sites Remedn Tech/Reg Compliance Support	4.0	1.0	4.6	0.0	$30,000
USTs at Other Locations Remedn Tech/Reg Compliance Support	3.7	0.0	4.0	0.0	$6,000
FTA Burn Pits Assessment Tech/Reg Compliance Support	3.4	3.0	3.3	0.0	($1,800)
Leaking USTs Remediation Tech/Reg Compliance Support	3.7	1.0	3.0	0.0	($4,000)
Misc Chemical Spills Assessment Tech/Reg Compliance Support	3.6	3.0	3.0	0.0	($4,000)
Cooling Tower Zone 12 Assessment Tech/Reg Compl Support	3.7	3.0	3.0	0.0	($5,000)
Finalize Hi Priority Potential Release Sites ICM Closure Reports	4.2	3.5	4.0	0.0	($7,000)
FTA Burn Pits ICM/VCA (ICM Closure Report)	3.6	3.7	3.0	0.0	($8,000)
Misc Chemical Spills Remediation Tech/Reg Compliance Support	3.9	3.5	3.3	0.0	($9,000)
Landfills Assessment Tech/Reg Compliance Support	4.3	2.0	4.0	0.0	($9,000)
Finalize FTA Burn Pits RFIR	4.0	2.3	3.0	0.0	($10,000)
Cooling Tower Zone 12 Remedn Tech/Reg Compliance Support	3.7	4.3	4.0	0.0	($18,000)
Finalize Firing Sites RFIR	4.4	2.5	4.0	0.0	($18,000)
Finalize Miscellaneous Chemical Spills RFIR	4.5	-2.0	4.0	0.0	($20,000)
Ditches and Playas Remediation Tech/Reg Compliance Support	3.9	5.1	5.0	0.0	($20,000)
Leaking USTs CAP Report	4.4	3.4	3.0	0.0	($30,000)
Finalize High-Priority Potential Release Sites RFIR	4.6	3.0	4.0	0.0	($30,000)
USTs at Other Locations CAP Report	4.5	3.4	3.0	0.0	($40,000)
Supplemental Verification Sites Remedn Tech/Reg Compl Support	4.3	4.3	3.0	0.0	($40,000)
FTA Burn Pits Remediation Tech/Reg Compliance Support	3.9	4.6	3.3	0.0	($50,000)
Finalize Landfills RFIR	4.8	2.5	4.0	0.0	($50,000)
Finalize Supplemental Verification Sites RFIR	4.9	3.2	3.0	0.0	($70,000)
Landfills Corrective Measures Study	5.3	4.5	4.0	0.0	($200,000)
Cooling Tower Zone 12 ICM/VCA	5.5	5.2	5.1	0.0	($400,000)
Landfills Treatability Study	0.0	5.7	5.0	0.0	($400,000)
Groundwater Sitewide Closure Support	0.0	5.8	5.0	0.0	($500,000)
Groundwater Treatability Study Phase III	6.0	5.9	6.0	0.0	($700,000)
Firing Sites ICM/VCA	5.8	4.9	3.0	0.0	($700,000)
Misc HE/Rad Sites ICM/VCA	5.9	5.4	3.0	0.0	($1,100,000)
Ditches and Playas ICM Phase II	6.1	6.2	6.0	0.0	($1,700,000)

Delay in funding these VCA efforts would be expected to force the project down the more formal and expensive RCRA process. Thus, there is a significant cost penalty for delay. On the other hand, since there is no RCRA fine for not performing VCAs, these activities, although they save hundreds of thousands of dollars, tend to be at the bottom of the Net Delay Liability priority list.

Another entry at the bottom of Table 4 is the Phase II Interim Corrective Measure (ICM) remediation of the ditches and playas. This has lowest priority because the ditches and playas are being closed to Risk Reduction 3, and treatment will not change that to a lower Risk Reduction level. The delay cost is lower than the performance costs and risks.

Groundwater site-wide closure support was also low on the list. This task had already been delayed as indicated by the zero for the Performance Cost (i.e., not funded for the current fiscal year). The delay cost was smaller than the estimated risk associated with performing the task, placing it toward the bottom of the list. The determining factor for all low-priority activities is the low severity of delay consequences relative to the performance costs and risks.

Three reasons are evident for activities having high Net Delay Liabilities (i.e., at the top of Table 4):

- Some ER projects are following the formal NEPA process and, if the funding were delayed for a year, the risk of the regulating agency shutting down the site for violation of its RCRA Waste Generation Permit was judged to be high.
- Another reason was more costly remediation in the future. This was the case for Groundwater Remediation. If this activity was delayed, the contaminated plume would spread further and additional costs would be incurred in the future.
- Some activities provide significant protection against incidents occurring while conducting other activities. For example, program management support provides oversight and the quality assurance function that qualifies laboratories analyzing the soil samples used to determine the level and extent of contamination.

Recently, a funding shortfall occurred for site remediation at Pantex. The decision maker did not have the results of the ER risk mapping available to aid selection of activities to eliminate or delay. Nevertheless, 5 of the 6 activities that were actually delayed were at the bottom of Table 4, which would be the expected candidates for delay based on the risk mapping. The sixth activity was midway up the table. When asked why this sixth activity was delayed, the manager stated that it was his assessment that, while the activity was important to perform, there was some slack in the schedule and as long as it was funded next year, it would not pose any problem to the program. Thus, although the results in Table 4 were not used in these delay decisions, this situation supported the validity of the risk mapping approach for identifying activities that should receive first consideration if cuts or delays are necessary.

Use of Risk Mapping Results for Resource Optimization

The value of risk mapping goes well beyond prioritizing candidate activities for delay or elimination.

- The ranking of performance risks indicates where resources, such as for additional safeguards or inspections, should be focused for continuous risk reduction. Rather than hoping for a place in next year's capital budget or trying to find increasingly scarce available funds, risk mapping shows what activities can be delayed or eliminated and those activity resources redirected toward risk reduction.
- New project ideas can be evaluated relative to existing plant projects, to judge the priority that should be placed on each new project. In many cases, it will be evident that the new project should proceed, even if an already-funded project with a lower priority will need to be delayed or eliminated.
- Project management is engaged by linking the risk mapping database (Figure 3) to project scheduling software such as Microsoft Project® using an ODBC driver. The database contains all the fields for each activity that are necessary for project management. Project schedules can in this way be automatically updated whenever an activity is delayed or eliminated. Activity dependencies, where delay or elimination of one activity will affect other activities, are linked within the task structure of the project.

Efforts are currently underway to analyze Pantex waste management activities at a lower level of detail. This will allow ranking of risks and prioritization of resources and scheduling at the individual waste stream level rather than at the activity level. The risk mapping methodology is flexible enough to view and combine work elements at different levels of resolution, such as by major programs, projects, or subprojects. Further development of the methodology is expected to progress towards automatic optimization of processes or projects based on combined costs and risks.

APPENDIX
Risk Mapping Terms and Order-of-Magnitude Risk Analysis

Detailed in this appendix is each of the risk-based terms that are used in risk mapping. They are presented in order of increasing complexity, since each builds on features of the preceding term:

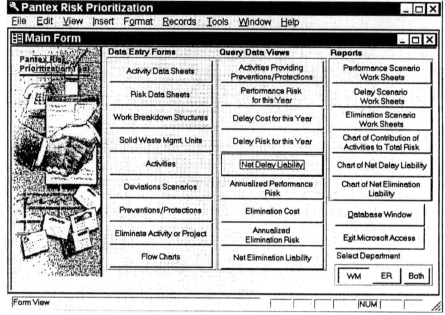

FIGURE 3. Risk Mapping Database

- PCost (Performance Cost)
- ECost (Elimination Cost) and DCost (Delay Cost)
- PRisk (Performance Risk)
- ERisk (Elimination Risk Increase) and DRisk (Delay Risk Increase).

Database input is described in parallel with each parameter, to give a picture of the procedure for actual entry of the cost and risk parameters into the risk mapping database and the calculations automatically performed by the database.

PCost (Performance Cost)

The Performance Cost for each activity is the currently budgeted amount expected to be spent to accomplish the activity. This expected cost could change throughout the year, so it needs to be updated when significant cost-factor changes occur.

The time frame for performance costs must be the same time frame as for the risk measures. This time frame is generally one year, with the budgeted costs matching the company's fiscal-year calendar. Figure A-1 shows an input screen for entering budgeted costs into the risk mapping database by year. The input screen is for the "Waste Characterization" waste management activity, with a budgeted cost of US $609,000 for fiscal year 1997. Budget data have not yet been entered for later years.

Resource Optimization by Risk Mapping 103

Note that the *magnitude* of the Performance Cost is determined most easily by taking the base ten logarithm of the $609,000 value, or *5.8*. This annual cost magnitude is in the "PCost" column of Table 3 for the Waste Characterization activity.

Other information is also entered on the Figure A-1 activity data form:

- Flow chart identifying number (1-28)
- Responsible department (WM)
- Dates that risk scenarios were developed and risk parameters were determined
- Risk Data Sheets that are associated with this activity
- Work breakdown structure (WBS) numbers associated with this activity.

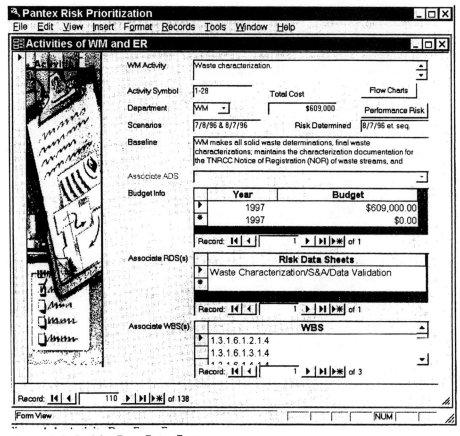

Figure A-1. Activity Data Entry Form

Risk mapping of different facilities will warrant different means of identifying each work activity and showing its relationship to the facility's organizational structure.

ECost (Elimination Cost) and DCost (Delay Cost)

Whereas the Performance Cost is the expected cost to perform an activity, the Elimination Cost is the expected cost if the activity is not performed. It is an indirect measure of the activity's importance. When combined with the Elimination Risk Increase (as described later), it indicates the cost + risk impact of not performing the activity. The Delay Cost differs only from the elimination cost in the permanence of not performing the activity. The Elimination Cost is the expected cost if the activity is eliminated altogether. The Delay Cost is the expected cost if the activity is delayed until the next fiscal year.

Values for Elimination Cost and Delay Cost are determined on an order-of-magnitude basis. This is done by consulting a table that gives the severity of consequences in the seven "performance categories" currently used in the U.S. Department of Energy's Risk Data Sheet (RDS) process. Table A-1 gives these seven categories (Public Health & Safety, etc.) and descriptions of typical consequences in each category as they relate to Impact Magnitude.

This approach of combining diverse impact types (public health and safety, business loss, etc.) is not new. It can be traced back at least as far as the Rapid Ranking method employed by ICI (Gillett, 1985). The approach here compares the impact severities using order-of-magnitude exponents for simplicity of risk analysis and presentation, as will be illustrated in the next section of this Appendix.

The establishment of the Impact Magnitudes table is a critical part of the risk mapping process. It must be agreed upon by all parties immediately responsible for the management of the activities under study, since this table indicates how parallels are drawn between the diverse impact types. As such, it can be viewed as a risk management function.

The seven categories in Table A-1 were selected to be consistent with the DOE RDS performance categories. However, it is possible to work with fewer categories, since many of the impacts can be equated to dollar-denominated liabilities, such as compliance fines and penalties. Figure A-2 illustrates how a simpler Impact Magnitudes table can be constructed.

Some of the impacts in Figure A-2 cover more than one order of magnitude. This may be desirable in order to give the persons conducting the risk mapping a broader range within which to choose where a given impact would best fit. Note also that severities are not limited to the impact magnitude integers alone. The illustration was given above how a performance cost of $609,000 equates to a 5.8 magnitude. Likewise, using the example from Figure A-2 of Site S&H impacts, a single fatality may be assigned a 6.5 impact magnitude. If the fatality is not cer-

Resource Optimization by Risk Mapping

TABLE A-1
Impact Magnitudes for Pantex WM/ER Risk Mapping

Impact Magnitude	Public Health & Safety	Site Personnel Health & Safety	Environmental	Compliance	Mission Impact	Mortgage Reduction	Social, Cultural, Economic
7	Immediate or eventual loss of life or permanent disability		Catastrophic damage to environment (widespread, long-term or irreversible effects)		Serious negative impact on ability to accomplish a major program mission	~$10,000,000 avoidable cost due to degraded infrastructure, loss of capital investment or opportunity for cost savings	
6		Injuries or illnesses, permanent total disability, chronic or irreversible ill-nesses, extreme overexposure or death	Significant damage to environment (widespread, short-term or localized long-term or irreversible effects)	Major noncompliance with laws, enforcement agreements or compliance agreements		~$1,000,000 avoidable cost as above	Significant adverse impact to social, economic, or cultural value with no possible mitigation
5	Excessive exposure and/or injury	Injuries or illnesses resulting in permanent partial disability, temporary total disability or serious overexposure		Major noncompliance with executive orders, directives, codes, standardsModerate negative impact on ability to accomplish a major program mission		~$100,000 avoidable cost as above	Moderate adverse impacts that would cause notable damage to social, cultural, economic value
4		Injuries or Illnesses resulting in hospitalization, temporary illnesses, disability, slight overexposure	Minor to moderate damage to environment (localized or short-term)			~$10,000 avoidable cost as above	
3	Moderate to low-level exposure					~$1,000 avoidable cost as above	

NOTE: Impact magnitude is increased by 1 in Public H&S and Site Personnel H&S categories if several persons are affected.

Impact Magnitudes

	Public S&H	Site S&H	Environ-mental	Cost, Loss, or Liability
8	Fatality or permanent health effect	Many fatalities	Widespread and long-term or permanent	$100,000,000
7		Fatality or permanent health effect		$10,000,000
6	Severe injury or multiple injuries		Widespread and short-term or localized and long-term	$1,000,000
5	Injury or hospitalization	Severe injury or multiple injuries		$100,000
4	Exposure above limits	Recordable injury	Localized and short-term	$10,000

FIGURE A-2. Simplified Impact Magnitudes Table

tain to occur, but perhaps a person is in the area of a hazard only 30% of the time, then this might be reduced to a 6.0 magnitude. On the other hand, if two or three fatalities might be expected, then a 7.0 magnitude could be selected.

PRisk (Performance Risk)

The Performance Risk must be considered to accurately prioritize activities, since risks are true liabilities even though not certain to occur. The Performance Risk can be analyzed by any valid quantitative risk analysis methodology; however, the order-of-magnitude risk analysis approach described here was used for several reasons:

- Order-of-magnitude precision has been found to be sufficient for risk mapping purposes, due to the wide range of costs and risks involved as illustrated by the difference of over $12,000,000 in Net Delay Liabilities in Table 3
- Using impact and risk-factor magnitudes greatly simplifies the recording, calculation and presentation of risk mapping results
- The Impact Magnitude table provides an absolute-measure scale against which diverse impacts can be compared and combined.

The second point becomes apparent by viewing Figure A-3, which is a form used to input Performance Risk parameters directly into the risk mapping database. The RISK for one scenario, shown at the right-center of the form, is determined by combining the two adjacent factors of Scenario Frequency and Total

Impact. Adding the base-ten logarithms of two numbers is the same as multiplying the two numbers. Since RISK is determined by multiplying the likelihood of occurrence (Scenario Frequency) and the severity of consequences (Total Impact) of an undesired event, adding the Scenario Frequency magnitude of –2.0 with the Total Impact magnitude of +6.0 gives the scenario RISK magnitude of +4.0. This is far simpler, both in calculation and in presentation, than multiplying (0.01 incidents per year) by ($1,000,000 liability per incident) to get an annualized liability of $10,000 per year. However, it is evident that the risk mapping facilitator must have a thorough understanding of the risk-calculation factors to avoid confusing frequencies and probabilities, for example, when only using magnitude numbers without their units of measure.

Figure A-3 shows the basics of how a Performance Risk scenario is analyzed. The risk analysis is similar in form to a quantified Hazard and Operability (HAZOP) study, with several important distinctions:

- Scenarios are analyzed on a year-to-year basis, to capture time-varying risks for a given activity
- Each scenario starts with a deviation from intended activity, as does a HAZOP study; however, the analysis does not trace each deviation back to individual causes
- All risk parameters are quantified on an order-of-magnitude basis
- The severity of consequences is assessed across all impact categories for each deviation scenario
- The risk mapping analyzes "prevention" and "protection" safeguards separately and explicitly.

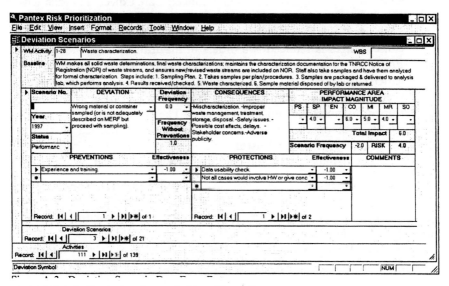

FIGURE A-3. Deviation Scenario Data Entry Form

To elaborate on the final point, preventions reduce the likelihood of the deviation occurring, whereas protections reduce the likelihood of the consequences occurring, given that a deviation occurs. It was found necessary to document preventions as well as protections in the risk mapping, in order to assess the ERisk (Elimination Risk Increase) and DRisk (Delay Risk Increase) for activities providing a prevention function. To illustrate, there is likely to be no short-term, certain impact on a facility if a preventive maintenance activity is eliminated. However, it is clear that this would increase the risk of various losses resulting from equipment failures. Hence, to accurately map the importance of the preventive maintenance activity, the deviation scenarios must capture what loss scenarios will be less likely to occur by continuing the preventive maintenance activity.

The deviation scenario in Figure A-3 can be described as follows, to illustrate the order-of-magnitude risk analysis:

- **Deviation** and **Deviation Frequency:** The deviation, "Wrong material or container sampled," is expected to occur at a frequency magnitude of **0.0**, which is $1 \times 10^{0.0}$ deviations per year, or once a year
- **Preventions:** The prevention activity of "Experience and training" was assigned an effectiveness of **–1.0**, or one order of magnitude risk reduction
- **Frequency without Preventions:** If the prevention activity of "Experience and training" were discontinued, the deviation frequency would therefore be expected to occur one order of magnitude more frequently, or at a frequency magnitude of **1.0**
- **Protections:** Two protection activities, "Data usability check" and "Not all cases would involve HW or give concern" are expected to each have a one order of magnitude effectiveness
- **Scenario Frequency:** With both protections in place, the frequency of the consequences is reduced two orders of magnitude, from the Deviation Frequency magnitude of 0.0 to the Scenario Frequency magnitude of **–2.0**, which corresponds to $1 \times 10^{-2.0}$ incidents per year or one chance in 100 per year of operation of realizing the consequences
- **Impact Magnitudes and Total Impact:** The greatest impact magnitude, **6.0**, is in the Compliance (CO) performance area; impacts in all other performance areas are at least an order of magnitude less severe, and hence do not contribute significantly to the Total Impact of **6.0**.

These performance risk scenarios have been effectively developed and analyzed using a facilitated team review approach, similar to a HAZOP study. The persons most knowledgeable with each activity are interviewed to review the baseline description, develop the deviation scenarios, and decide on the risk parameters and the elimination and delay impacts. Note that this risk mapping approach extends Process Hazard Analyses (PHAs) such as HAZOP studies, with relatively minimal additional effort, so that they can give much better support to cost and risk decisions.

ERisk (Elimination Risk Increase) and DRisk (Delay Risk Increase)

The final factors *ERisk* and *DRisk* pertain to the risk increase if a prevention or protection activity is eliminated or delayed. Thus, for activities that do *not* provide a prevention or protection function to any other activity, the ERisk and DRisk will be zero.

ERisk and DRisk are calculated automatically by the risk mapping database. This is done by setting the Effectiveness for one Prevention or Protection activity at a time to zero throughout the entire risk mapping database, and calculating the total risk increase.

An example to illustrate DRisk would be a decision not to test and inspect relief valves for one fiscal year. The Delay Cost (DCost), which is the certain, expected impact, of not testing the facility's relief valve for a year, may be very low, since the facility can continue operation without testing the relief valves. However, the Delay Risk Increase (DRisk) may be quite high, since a large number of risk scenarios may have credited emergency relief as protection, and the delaying of relief valve testing will make the emergency relief less effective.

References

Gillett, J.E. (1985). "Rapid Ranking of Process Hazards." Process Engineering, February, page 19.

Johnson, R.W. and Elliott, M. (1997) "Integrated Safety Analysis Project." International Conference and Workshop on Risk Analysis in Process Safety, Atlanta, Georgia, October.

safeguards required to prevent or mitigate hazardous events. Risk acceptance criteria that are too stringent may result in costly and unneeded safeguards; risk acceptance criteria that are too lenient expose a company to substantial risks that can lead to serious, and possibly devastating, losses.

Unfortunately, in the U.S., neither mandated methods for defining acceptable risk nor voluntary standards that provide acceptable risk guidance for the process industry are specified. Even if a practical definition of acceptable risk (or method for determining acceptable risk) were available, the definition would most likely not be tolerable or practical for all companies and hazards. For example, a large corporation can typically survive (and thus accept) a higher loss of revenue (e.g., lost production, equipment damage) as a result of an event. A smaller, privately held company most likely cannot withstand the same loss.

Various methods are used for defining risk acceptance criteria. Companies may use criteria established by government agencies (which are often inconsistent), or they may compare process risks to some common risks encountered in daily living, such as flying on an airplane or driving a car. Some companies adopt criteria used by others when such criteria are available; however, most develop their own risk acceptance criteria. Areas to consider when defining risk acceptance criteria include:

- Plant personnel hazards,
- Impact to the community and public relations,
- Environmental damage, and
- Property damage and revenue loss.

A basic understanding of risk is needed to understand how to establish risk acceptance criteria and to analyze risk. Risk can be defined as the product of frequency of occurrence of an event (F) and the severity of the resulting consequences (C), as shown in the equation below.

$$\text{Risk} = F \times C$$

Figure 1 is a log–log plot depicting this relationship. The solid line is the "rational" risk level (constant slope) representing a company's willingness to accept the same amount of loss (e.g., dollars) per year, regardless of the severity of the loss. In other words, an average loss of $10,000 per year resulting from pump seal leaks is no more or less acceptable than an average loss of $1,000,000 once every 100 years resulting from a catastrophic accident. In reality, because of a company's aversion to catastrophic events, a company's acceptable risk profile tends to diverge from the "rational" risk profile when the severity of consequences are high. This aversion can be modeled as a continuously decreasing acceptability of risk with increasing consequences (as depicted with the dashed curve), or as a discontinuous jump(s) to a lower risk level(s) (as depicted with the dashed line).

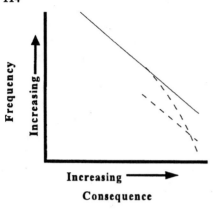

FIGURE 1. Acceptable Risk Profile

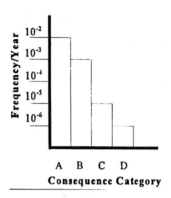

FIGURE 2. Risk Histogram

Using the linear risk profile shown in Figure 1 to determine the acceptable frequency of an accident with a specific consequence can become a laborious task when analyzing the wide range of hazards associated with a process plant. A nonlinear risk profile makes the task even more difficult. To simplify the risk assessment task, consequence categories are defined and risk acceptance criteria for each consequence category are defined in terms of an acceptable frequency level for each category. The frequency level of each category should correspond to the maximum acceptable frequency of occurrence of the most severe consequence within that category. The result is a simple risk histogram, such as depicted in Figure 2. The intersection of each bar on the histogram with the frequency axis defines the acceptable frequency level for each consequence category.

The methodology in this paper involves translating the company's risk acceptance criteria into a frequency threshold score (F_t, which is congruent with the scoring system explained later in this paper) for each consequence category. The worked example at the end of this paper illustrates how a company can translate its risk acceptance criteria into frequency thresholds.

2. Identifying and Characterizing Hazardous Events of Interest

Companies that have completed PHAs of their processes have identified a range of accident scenarios and consequences of these scenarios. From this information, each company can determine which accident scenarios result in a hazardous event with undesirable effects. Generally, the effects to people (both on site and off site), property, and the environment are considered when determining which hazardous events have undesirable effects. These events will be referred to as "hazardous events of interest." In a HAZOP or what-if/checklist analysis, the hazardous events of interest are usually tabulated in the "Consequences" category of the analysis documentation tables.

The next step in the methodology is to determine the appropriate consequence category for each hazardous event of interest. Once the consequence category is determined, the corresponding frequency threshold score (F_t) is assigned using the risk acceptance criteria defined in the first step. The results can be documented in a table such as Table 1. (Table 1 provides a means for assembling information on hazardous events of interest and is the documentation tool used in this analysis. The data presented in Table 1 are extracted from the example at the end of the paper.) Hazardous events of interest and their corresponding frequency thresholds (F_t) are recorded in columns 1 and 2 of Table 1, respectively.

TABLE I
Table for Documenting the Risk Assessment Analysis

1	2	3	4	5	6	7	8	9
Hazardous Event of Interest	Consequence Category/ Frequency Threshold (F_t)	Initiating Event (IE)	Initiating Event Score (F_i)	Safeguards PFD Score (S_{pfd})	Effectiveness Score (E_s)	Reduced Frequency Score (F_r)	Additional Protection Required (S_{add})	SIS and SIL Recommended
Potential Furnace Explosion	D/1	Gas FCV fails open	6	Product Discharge TAHH-1.5 and Furnace Tube Skin TAH -1.5	3	3	2	Yes— independent TAHH shutdown with SIL 1
		Natural gas supply pressure high	6	Furnace Temperature Controller TIC -1.5	1.5	4.5	3.5	Yes— independent PAH shutdown with SIL 3
		TIC fails— low signal	7	Furnace Tube Skin TAH- 1.5	1.5	5.5	4.5	Yes— independent TAHH shutdown with SIL 2 and flame scanner/ interlock with SIL 2

3. Identifying and Determining the Frequency of Initiating Events

Once the hazardous events of interest are identified, identifying and determining the frequency of the initiating event(s) that lead to each hazardous event of interest is necessary. An initiating event is an event that, if left unmitigated, leads to a hazardous event. In a typical HAZOP or what-if/checklist analysis, the initiating events can be found in the "Causes" or "Questions" column. For instance, a valve failure that could potentially result in a rupture of the line, is an initiating event. The initiating event(s) for each hazardous event of interest is listed in column 3 of Table 1.

Next, the initiating event's frequency of occurrence must be estimated. Estimating and accepting a single value for a frequency of occurrence such as "occurs only once every 100 years" can be difficult, especially without analytical data to support the estimate. Estimating the frequency as a range (i.e., between 10 and 100 years) is often easier. However, performing analyses with ranges can be cumbersome. To simplify the analysis, the methodology presented in this paper uses a scoring system that correlates frequency ranges with discrete frequency scores as shown in Table 2.

Using this scoring system simplifies the calculations and makes data selection for events easier. (It also simplifies selection of the frequency thresholds.)

TABLE 2
Frequency Scores for Various Ranges of Initiating Event Frequencies

Expected Time between Occurrences	Frequency Range	Probability of at Least One Occurrence in 100 Years of Experience (10 systems for 10 years)	Frequency Score (F_i)
<0.01 y (~4 days)	>100/y	~100%*	10
0.01 y (~4 days) to 0.1 y (~37 days)	100/y to 10/y	~100%*	9
0.1 y (~37 days) to 1 y	10/y to 1/y	~100%*	8
1 y to 10 y	1/y to 0.1/y	~100%*	7
10 y to 100 y	0.1/y to 0.01/y	~100% to ~60%	6
100 y to 1,000 y	1×10^{-2}/y to 1×10^{-3}/y	~60% to 10%	5
1,000 y to 10,000 y	1×10^{-3}/y to 1×10^{-4}/y	10% to 1%	4
10,000 y to 100,000 y	1×10^{-4}/y to 1×10^{-5}/y	1% to 0.1%	3
100,000 y to 1,000,000 y	1×10^{-5}/y to 1×10^{-6}/y	0.1% to 0.01%	2
≥1,000,000 y	$\leq 1 \times 10^{-6}$/y	≤0.01%	1

* Multiple occurrences expected.

To use the scoring system, an approximate frequency of occurrence of each initiating event is estimated. (Be conservative and move to the higher score when faced with a frequency that is near the boundary between two scores.) Published failure data, such as *Guidelines for Process Equipment Reliability Data*, can aid in estimating frequencies when historical or analytical information for an initiating event is incomplete. Column 4 in Table 1 lists the frequency score for each initiating event (F_i). The next step in the process is to identify the safeguards relevant to each initiating event and to evaluate their effectiveness in reducing the event's likelihood.

4. Identifying Safeguards and Evaluating Safeguard Effectiveness

Safeguards include special process designs, process equipment, administrative procedures, the basic process control system (BPCS), and/or planned responses (either automated or initiated by human actions) to imminent adverse process conditions. (Note: Opinions differ regarding the BPCS as a safeguard when determining SIS requirements. *Guidelines for Safe Automation of Chemical Processes* discourages the inclusion of the BPCS because the BPCS itself may be the cause of an upset. Each company will need to establish guidelines for deciding if the BPCS should be included as a safeguard.) Several issues need to be considered when identifying which safeguards are effective in preventing an initiating event:

- Independence of safeguards. Do potential common cause failures of safeguards exist?
- Common cause of safeguard failure and initiating event. Are the safeguards independent of initiating events?
- Adverse effects caused by safeguards. Does the activation of a safeguard cause a hazardous event (e.g., a PSV opening may prevent a vessel rupture but may vent a toxic gas to the atmosphere)?
- Auditability of the safeguard. Can the safeguard's performance be periodically verified?
- Automation of the safeguard. Is an operator required to perform multiple, simultaneous actions when responding to multiple safeguards?
- Specificity of the safeguard. Is the safeguard intended to prevent and/or mitigate the hazardous event?

The methodology in this paper evaluates situations in which multiple safeguards exist by only including those safeguards that are **independent** of each other. For instance, a process alarm and an interlock whose input signals are supplied from the same field device should not be considered as two independent safeguards because of the potential for a common cause failure. The effect of common cause failures on multiple "non-independent" safeguards can be assessed by performing a common cause failure analysis to determine the unavailability of the safeguards. (However, common cause failure analysis is beyond the scope of this paper.)

To evaluate the effectiveness of a safeguard, its PFD must be determined. The PFD of a safeguard is a measure of the probability that the safeguard will not perform its intended function when challenged by an initiating event. Therefore, the PFD can be combined with the frequency of the initiating event to determine the frequency of the hazardous event. The following equation represents this relationship:

$$F_{\text{hazardous event}} = F_{\text{initiating event}} \times \prod PFD_{\text{safeguard}}$$

To use this relationship to determine the effectiveness of safeguards, a PFD range must be estimated for each safeguard. PFD ranges can be derived from actual plant data (if available) or published data such as *Guidelines for Process Equipment Reliability Data*. Once a PFD range is determined, a score is assigned to each safeguard. (Again, be conservative and move to the lower score when a probability of failure falls near the boundary between two scores.) Table 3 illustrates probability ranges and number of failures based on 1,000 challenges and correlates these ranges to a PFD score (S_{pfd}). (Note: The scores in the PFD and frequency range tables must be consistent [i.e., a tenfold reduction in probability or frequency equates to a one unit change in score].)

A total effectiveness score (E_s) for the safeguards responding to an initiating event is determined by the sum of the PFD scores:

$$E_S \sum_{i=1}^{N} S_{pfd}$$

The effectiveness score (E_s) is then used to calculate a reduced frequency score (F_r) for the hazardous event of interest resulting from a particular initiating event. The reduced frequency score is calculated by subtracting the safeguard effectiveness score from the frequency score for a given initiating event:

$$F_r = F_i - E_s$$

Note: If a passive safeguard (e.g., blast wall or containment dike) exists (or is to be implemented) that reduces the severity of the consequence of an event rather than reducing the frequency of occurrence, then the methodology (beginning at Step 2) should be repeated using a higher frequency threshold score, F_t, corresponding to a lower severity of consequence category. An active safeguard (e.g., pressure relief valve [PSV], TAHH, or administrative control) should be considered only as reducing the frequency of occurrence of an event.

In the next step, the reduced frequency score (F_r) is used to determine if the risk acceptance criterion (now defined as a frequency threshold score, F_t) is exceeded. For each initiating event, columns 5, 6, and 7 of Table 1 are used to document S_{pfd}, E_s, and F_r, respectively.

A Risk Assessment Methodology for Evaluating the Effectiveness of Safeguards

TABLE 3
Probability of Failure Scores for Various Ranges of Demanded Event Probabilities

Probability Range (PFD)	Number of Failures (or Chance of Failure if Less than One) Based on 1,000 Challenges	Probability of Failure on Demand Score (S_{pfd})
~1	~1,000	0
~1 to ~0.3	~1,000 to 300	0.5
~0.3 to 0.1	300 to 100	1
0.1 to ~0.03	100 to 30	1.5
~3×10^{-2} to 1×10^{-2}	30 to 10	2
1×10^{-2} to ~3×10^{-3}	10 to 3	2.5
~3×10^{-3} to 1×10^{-3}	3 to 1	3
1×10^{-3} to ~3×10^{-4}	100% to 30% chance of one occurrence	3.5
~3×10^{-4} to 1×10^{-4}	30% to 10% chance of one occurrence	4
1×10^{-4} to ~3×10^{-5}	10% to 3% chance of one occurrence	4.5
~3×10^{-5} to 1×10^{-5}	3% to 1% chance of one occurrence	5
1×10^{-5} to ~3×10^{-6}	1% to 0.3% chance of one occurrence	5.5
~3×10^{-6} to 1×10^{-6}	0.3% to 0.1% chance of one occurrence	6
$\leq 1 \times 10^{-6}$	$\leq 0.1\%$ chance of one occurrence	6.5

5. Determining the Additional Safeguards Required to Meet the Frequency Threshold Criteria

To determine the need for additional safeguards, the frequency threshold (F_t) for the hazardous event of interest is subtracted from the reduced frequency score (F_r):

$$S_{add} = F_r - F_t$$

A zero or negative value means the existing safeguards reduce the likelihood of the hazardous event of interest below the acceptable frequency threshold (i.e., the company risk acceptance criteria are met). A large negative value suggests that the safeguards may be overdesigned. A positive score means additional safeguards are needed. Using Table 3, S_{add} can be correlated to the PFD required for an additional safeguard(s).

Achieving the appropriate amount of additional protection required to meet the company risk acceptance criteria can be accomplished by various methods. Several alternatives should be considered, such as:

- Can the process be redesigned (e.g., inherently safer design, passive safeguards) to reduce the severity of the consequences (thus reducing the risk acceptance criteria)?
- Can maintenance procedures be developed or revised to improve safeguard availability?
- Can process equipment, such as control valves, be replaced with a more reliable design?
- Can operator response be reasonably enhanced?
- Can non-SIS safeguards be implemented (e.g., relief valves)?
- Can SIS safeguards be implemented (e.g., high pressure shutdown)?

If an SIS is implemented, ISA-S84.01 stipulates that the required safety integrity level (SIL) be determined. The required SIL is easily determined from the methodology. S_{add} (which represents the PFD score [S_{pfd}] for the SIS) can be correlated to SIL levels in ISA-S84.01. Table 4 provides a correlation between S_{add} and SILs. The SIL is the performance criteria by which the SIS is designed, installed, and operated. The SIL and the safety function that the SIS must perform to reduce the likelihood of the hazardous event of interest (e.g., isolate the source of pressure upon detection of high pressure) are the bases for the conceptual and detailed designs of the SIS.

The following is a worked example to illustrate the use of the methodology.

Worked Example Using the Methodology

The risk acceptance criteria for this example are expressed as the maximum acceptable frequency of occurrence for specific consequence severity levels. The first step in establishing these criteria is to identify consequence categories and define severity levels. Table 5 illustrates typical categories based on the severity of the consequences to the personnel, community, environment, and/or facility. Next, the maximum acceptable frequency of occurrence for events in each category must be established, and the thresholds selected must be consistent with boundaries in the frequency categories for initiating events (i.e., Table 2). A typical risk histogram depicting a company's acceptable risk profile is shown in

TABLE 4
SIS Score Correlation

S_{add}/S_{pfd} Score*	SIL	Probability of Failure on Demand
0.5–2	1	10^{-1}–10^{-2}
2.5–3	2	10^{-2}–10^{-3}
3.5–4	3	10^{-3}–10^{-4}

* When the S_{add}/S_{pfd} is at the upper boundary of a range, strongly consider moving to the next higher SIL.

Figure 2. *The frequency values in Figure 2 are chosen for discussion purposes only and are not meant to be recommended acceptance criteria for these types of consequences. Each company must establish its own acceptance criteria commensurate with its risk tolerance.* The frequency threshold (F_t) is determined by converting the maximum acceptable frequency into a score using Table 2. For example, category A events have a maximum acceptable frequency of 10^{-2}/y, which falls within the frequency range of 1×10^{-2}/y to 1×10^{-3}/y. This range correlates to a score of 5. Table 6 shows the frequency thresholds consistent with the risk acceptance criteria used in this example.

TABLE 5
Example Severity of Consequences Categories

Category	Personnel	Community	Environment	Facility
A	Minor or no injury, no lost time	No injury, hazard, or annoyance to public	Recordable event with no agency notification or permit violation	Minimal equipment damage at an estimated cost of less than $100,000, and with no loss of production
B	Single injury, not severe, possible lost time	Odor or noise annoyance complaint from the public	Release that results in agency notification or permit violation	Some equipment damage at an estimated cost greater than $100,000, or minimal loss of production
C	One (or more) severe injury	One (or more) minor injury	Significant release with serious offsite impact	Major damage to process area(s) at an estimated cost greater than $1,000,000, or some loss of production
D	Fatality or permanently disabling injury	One (or more) severe injury	Significant release with serious offsite impact, and more likely than not to cause immediate or long-term health effects	Major or total destruction to process area(s) estimated at a cost greater than $10,000,000, or a significant loss of production

The next step is to identify the hazardous events of interest. Two deviations from a sample HAZOP analysis of an ethylene dichloride (EDC) and vinyl chloride monomer (VCM) pilot operation will be used in this example[1]. Tables 7 and 8 provide the results of the sample HAZOP. Table 7 lists selected deviations examined for the VCM furnace, along with the causes, consequences, and safeguards. The HAZOP recommendations are summarized in Table 8.

---

1 HAZOP analysis was extracted from Chapter 14 of *Guidelines for Hazard Evaluation Procedures, Second Edition with Worked Examples*.

TABLE 6
Example Risk Acceptance Criteria

Severity Category	Maximum Acceptable Frequency	Frequency Threshold (F_t)*
A	10^{-2}/yr	5
B	10^{-3}/yr	4
C	10^{-5}/yr	2
D	10^{-6}/yr	1

* In this methodology, the frequency threshold for a given consequence severity is directly related to the company's risk acceptance criteria. The frequency is typically lower for higher consequence severities.

Table 7
EDC and VCM Pilot Plant HAZOP Table (with selected deviations)

| \multicolumn{6}{l}{5.0 Furnace— VCM Furnace (NORMAL OPERATION—RAISE EDC TO 900°F, 160 PSIG TO MAKE VCM; FLOW—1,200 LB/HR)} |

Item No.	Deviation	Causes	Consequences	Safeguards	Actions
5.2	High flow—natural gas	Gas FCV fails open	Possible loss of flame and potential explosion	Natural gas supplier has been very reliable over 15 years	1 2 3
		Natural gas supply pressure high	High process temperature in the VCM furnace (Item 5.7)		
		TIC fails—low signal	Potential tube damage and fire if tube ruptures. Excess by-products in product stream (Item 5.12)	TIC controls gas supply	
5.7	High temperature	High natural gas flow (Item 5.2)	High pressure in the VCM furnace tubes (Item 5.9)	Furnace tube skin TAH	2
		Low EDC flow (Item 5.4)	High production of by-products during EDC cracking. Potential furnace tube damage and possible furnace fire if tube ruptures	Product discharge TAH and TAHH with hard-wired shut-down of plant Furnace tubes designed to withstand very high temperatures	

TABLE 8
Sample Action Items from the EDC and VCM Pilot Plant HAZOP Analysis

List Number	Action to be Considered
1	Consider a PAH and high pressure shutdown interlock for natural gas using a positive isolation valve (Item 5.2)
2	Consider installing an independent TAHH and interlock that shuts down the furnace on high furnace discharge temperature (Items 5.2, 5.7)
3	Consider installing a flame scanner and interlock that shuts down the furnace upon loss of flame (Item 5.2)

The first step in using the HAZOP analysis is to identify the hazardous events of interest, which in this case are those potentially resulting in a fire and/or explosion. The hazardous events of interest potentially resulting in a fire and/or explosion for two deviations (5.2 and 5.7) are listed in column 1 of Table 9.

A frequency threshold, F_t, is established for each hazardous event of interest based on the company's risk acceptance criteria. To accomplish this, Table 5 is used to determine which consequence severity category characterizes the hazardous event of interest. Then, using Table 6, the frequency threshold is determined. The proper category for the potential explosion is category D (because fatality or disabling injury can possibly occur). For category D events, the acceptable frequency criterion is 10^{-6} events/year, which translates into a frequency threshold of 1. For the event potentially resulting in a furnace fire, the consequence is a category C event. This information is documented in column 2 of Table 9.

Next, the initiating events for each hazardous event of interest are identified and documented in column 3. A frequency range is estimated for each initiating event and then translated into an initiating event frequency score (F_i) using Table 2. Column 4 in Table 9 is used to document the frequency scores. Based on published failure rate data, the frequencies of the gas FCV failing open and the TIC failing are estimated to be 0.1 to 0.01 and 1 to 0.1 events/year, respectively. These frequency ranges convert into frequency scores of 6 and 7, respectively. Actual plant history data are used to determine the frequency of occurrence for high natural gas supply pressure and low EDC flow, and then the ranges are converted into the frequency scores in Table 9.

To continue the analysis, the safeguards are evaluated to determine which initiating events, listed in the HAZOP table, they effectively prevent or mitigate. In Table 7, for the initiating event "Gas FCV fails open," there are two safeguards: the *product discharge* TAHH and the *furnace tube skin* TAH. (Note: The *product discharge* TAH is not independent of the TAHH because a single transmitter provides the signal for both alarms; therefore, the *product discharge* TAH has not been included as a safeguard.) For the initiating event "TIC fails — low signal," the only independent safeguard is the *furnace skin* TAH. (Note: There is a potential common cause failure [i.e., failure of the input signal] for the TIC, the *product discharge* TAH, and the *product discharge* TAHH, so no credit is given for these alarms as independent safeguards.)

Next, a safeguard failure score is estimated for each safeguard from PFD data. For safeguards that require human intervention (e.g., alarms requiring operator response to correct the situation), the PFD for the human response is used versus the PFD for the equipment. The furnace skin tube TAH will require operator diagnosis and corrective action. Thus, the PFD for an operator not detecting equipment deviations (PFD = 10^{-1} to 10^{-2}) is used versus the PFD for an alarm (PFD = 10^{-3} to 10^{-5}). Using Table 3, a PFD score (S_{pfd}) is determined for each safeguard based on its estimated PFD, and then the scores are summed to yield a total effectiveness score (E_s) for each initiating event. The S_{pfd} and E_s are documented in columns 5 and 6, respectively, of Table 9.

The reduced frequency score (F_r) of each initiating event is determined by subtracting the total effectiveness score (E_s) from the initiating event frequency score (F_i). E_s and F_i for the gas FCV failing are 6 and 3, respectively. The result is an F_r of 3 as documented in column 7. This step is performed for each of the initiating events and the resulting F_r is compared to the F_t for the hazardous event of interest to determine if the risk criterion is met. For the gas FCV failing, the result of the comparison is 2 (F_r of 3 minus F_t of 1) and is documented in column 8 (S_{add}). Because S_{add} is not less than or equal to zero, the safeguards do not provide sufficient prevention for this initiating event. S_{add} represents the additional prevention needed to meet the frequency threshold for this hazardous event of interest.

The decision must be made whether to implement additional protection as a non-SIS safeguard or an SIS. If a non-SIS safeguard is selected, the additional protection score is converted back into a PFD range using Table 3. This PFD range is the design requirement for the non-SIS safeguard. For this example, SISs will be implemented. The additional protection score is used to determine the SIL needed for the SIS. To sufficiently defend against the hazardous event of interest (potential furnace explosion) and the initiating event (TIC fails), additional protection yielding a score of 4.5 is needed. Table 4 illustrates that a single SIS with SIL 3 will not provide sufficient protection by itself. Therefore, at least two SISs will be required (i.e., TAHH and flame scanner/interlock recommended during the HAZOP, Table 8). The sum of the SIL scores for the two SISs must equal or exceed 4.5. This is achieved by implementing an SIL 2 TAHH shutdown (with a PFD score of 2.5 to 3) and an SIL 2 flame scanner/interlock (with a PFD score of 2.5 to 3), yielding a combined PFD score of 5 to 6.

This example provides another issue to consider. The TAHH will defend against multiple initiating events (i.e., gas FCV fails open, TIC fails, high natural

TABLE 9
Example Risk Assessment Analysis

1	2	3	4	5	6	7	8	9
Hazardous Event of Interest	Consequence Category/ Frequency Threshold (F_t)	Initiating Event (IE)	Initiating Event Score (F_i)	PFD Score (S_{pfd})	Effectiveness Score (E_s)	Reduced Frequency Score (F_r)	Additional Protection Required (S_{add})	SIS and SIL Recommended
Potential Furnace Explosion	D/1	Gas FCV fails open	6	Product Discharge TAHH-1.5 and Furnace Tube Skin TAH-1.5	3	3	2	Yes—independent TAHH shutdown with SIL 1
		Natural gas supply pressure high	6	Furnace Temperature Controller TIC-1.5	1.5	4.5	3.5	Yes—independent PAH shutdown with SIL 3
		TIC fails—low signal	7	Furnace Tube Skin TAH-1.5	1.5	5.5	4.5	Yes—independent TAHH shutdown with SIL 2 and flame scanner/interlock with SIL 2
Potential Furnace Fire	C/2	High natural gas flow	7 (based on worst-case initiating event, TIC Fails—low signal)	Furnace Tube Skin TAH-1.5	1.5	5.5	3.5	Yes—independent TAHH shutdown with SIL 2 and flame scanner/interlock with SIL 2
		Low EDC flow	6	Product Discharge TAHH-1.5 and Furnace Tube Skin TAH-1.5	3	3	1	Yes—independent TAHH shutdown with SIL 1

gas flow) with differing SILs. The highest SIL required will be the design criterion selected when implementing the TAHH. This will ensure adequate protection for all initiating events.

At this point, the analysis is complete for all initiating events and all shutdown (SIS) recommendations made in the HAZOP. In addition to confirming the recommendations, the analysis table becomes an important document in the process design activities.

Conclusion

The methodology presented in this paper provides a systematic approach for using existing qualitative hazard analyses to help companies better understand the risks involved in their processes and the adequacy of their safeguards. The methodology presented begins with establishing risk acceptance criteria by which the effectiveness of existing safeguards can be judged, and then it determines the need for additional safeguards. The methodology also provides a technique for determining the performance criteria (either the PFD or SIL) for the safeguards. Establishing the need for an SIS and its SIL is necessary before engineers can begin to implement the steps mandated in ISA-S84.01.

References

AIChE/CCPS, *Guidelines for Hazard Evaluation Procedures, Second Edition with Worked Examples*, New York, (1992).

AIChE/CCPS, *Guidelines for Process Equipment Reliability Data, with Data Tables*, New York, (1989).

AIChE/CCPS, *Guidelines for Safe Automation of Chemical Processes*, New York, (1993).

ISA-S84.01-1996, *Application of Safety Instrumented Systems for the Process Industries*, (February 15, 1996).

A Simple Problem to Explain and Clarify the Principles of Risk Calculation

Dennis C. Hendershot
Rohm and Haas Company, Engineering Division, PO Box 584, Bristol, PA 19007

When a thought is too weak to be expressed simply, it should be rejected.
—Marquis de Luc Vauvenargues
Refléxions et Maximes (1746)

ABSTRACT
Many texts and case studies on quantitative risk analysis have been published. They usually explain the frequency and consequence calculations in detail, but do not explain how these results are combined to produce specific measures of risk. The risk measures seem to appear from the underlying data as if by magic. In this example problem, the background, frequency, and consequence data for a risk analysis are highly simplified, so that the actual risk calculations can be understood easily. The example problem has been extremely useful in explaining the principles of Chemical Process Quantitative Risk Analysis (CPQRA) calculations to engineers, plant management, and other customers of CPQRA studies. The example also illustrates the complexity of risk. Even though the problem is extremely simple and uses trivial models, a large number of valid, but numerically different, risk estimates can be generated. Even for this very simple example, there is no single answer to the question "What is the risk?"

Introduction

In *Flatland* (Abbott, 1884), Edwin Abbott uses a highly simplified, two-dimensional universe to explain the concepts of multidimensional geometry. Similarly, Dionys Burger's *Sphereland* (Burger, 1965) uses the same approach to explain multidimensional, nonlinear geometry. Both books explain complex ideas by reducing them to simple geometry, illustrating the ideas and facilitating an understanding of how they apply to our world. Abbott and Burger also use their books as a vehicle for social commentary on their societies (Victorian England and modern Europe, respectively). This approach (without the social commentary) will be used in this paper to explain the concepts of Chemical Process

Quantitative Risk Analysis (CPQRA), particularly the methods used to combine incident frequency and consequence estimates to produce various measures of risk. A CPQRA study in a universe much simpler than ours will be described, allowing us to concentrate on the risk calculations rather than on complex incident frequency and consequence calculations.

The example in this paper has evolved over several years, and has proven useful in explaining the concepts of CPQRA to engineers and managers at Rohm and Haas plants which are the subject of CPQRA studies. The simple example allows an understanding of how the various risk measures are calculated and what they mean. It also clearly illustrates the complexity of the concept of risk—this example has many numerically different risk measures which can be calculated. An understanding of the different risk measures is important when using CPQRA and quantitative risk estimates as a risk management tool. All users of the risk estimates must understand the risk measure, and ensure that the numbers used in any risk comparisons are calculated on the same basis.

Background and General Information

Riskland is a very simple universe, where most phenomena occur as simple step functions. The Riskland Chemical Company (RCC) operates in this universe, and has determined that a CPQRA study is appropriate as a part of the process risk management program for one of its hazardous installations. RCC follows the general CPQRA procedure as outlined by CCPS (1989), shown schematically in Figure 1.

In the Riskland universe, the following apply:

- All hazards originate at a single point.
- Only two weather conditions occur. The atmospheric stability class and wind speed are always the same. Half of the time the wind blows from the northeast, and half of the time it blows from the southwest.
- There are people located around the site. The specific population distribution will be described later in the example, when the information is needed.
- Incident consequences are simple step functions. The probability of fatality from a hazardous incident at a particular location is either 0 or 1.

These simple conditions, and the description of the impact zones of incidents as simple geometric areas, allows easy hand calculation of various risk measures. The techniques used to derive the risk measures from the underlying incident frequency and consequence information are the same as for a complex CPQRA study using sophisticated models intended to represent our world as accurately as possible. The concepts are the same; the difference is in the complexity of the models used, the number of incidents evaluated, and the complexity of the calculations.

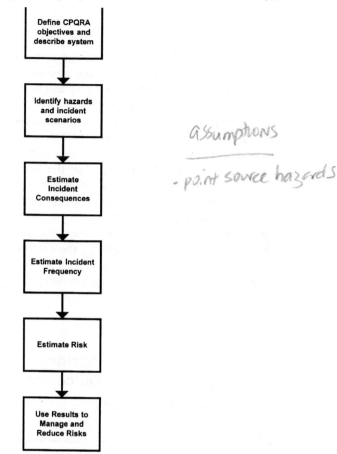

FIGURE 1. CPQRA Procedure

Incident Identification

RCC applies appropriate incident identification techniques, including historical information (plant and process specific, as well as generic industrial experience), checklists, and one or more of the hazard identification methodologies described in the *Guidelines for Hazard Evaluation Procedures* (CCPS, 1992). This is perhaps the most critical step in a CPQRA, because any hazards not identified will not be evaluated, resulting in an underestimate of risk. RCC's hazard identification and process safety reviews identify only two hazardous incidents which can occur in the facility:

    I. An explosion resulting from detonation of an unstable chemical.
   II. A release of a flammable, toxic gas resulting from failure of a vessel.

Incident Outcomes

The identified incidents may have one or more outcomes, depending on the sequence of events which follows the original incident. For example, a leak of volatile, flammable liquid from a pipe might catch fire immediately (jet fire), might form a flammable cloud which could ignite and burn (flash fire) or explode (vapor cloud explosion). The material also might not ignite at all, resulting in a toxic vapor cloud. CCPS (1989) refers to these potential accident scenarios as *incident outcomes*. Some incident outcomes are further subdivided into *incident outcome cases*, differentiated by the weather conditions and wind direction, if these conditions affect the potential damage resulting from the incident.

RCC reviewed the identified incidents for their facility to determine all possible outcomes, using an event tree logic model. Incident I, the explosion, is determined to have only one possible outcome (the explosion), and the consequences and effects are unaffected by the weather. Therefore, for IncidentI there is only one incident outcome and one incident outcome case. This can be represented as a very simple (in fact, trivial) event tree with no branches, as shown in Figure 2.

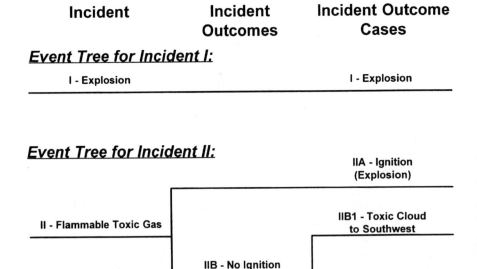

FIGURE 2. Event Trees for the Two Incidents

Incident II, the release of flammable, toxic gas, has several possible outcomes (jet fire, vapor cloud fire, vapor cloud explosion, toxic cloud). RCC determines that, in their facility, only two outcomes can occur. If the gas release ignites there is a vapor cloud explosion. If the vapor cloud does not ignite, the result is a toxic cloud extending downwind from the release point. Because there are only two possible weather conditions in Riskland, three incident outcome cases are derived from Incident II as shown in the event tree in Figure 2.

Consequence and Impact Analysis

Determining the impact of each incident requires two steps. First, a model estimates a physical concentration of material or energy at each location surrounding the facility—for example, radiant heat from a fire, overpressure from an explosion, concentration of a toxic material in the atmosphere. A second set of models estimates the impact that this physical concentration of material or energy has on people, the environment, or property—for example, toxic material dose-response relationships. These models are described in Chapter 2 of the *Guidelines for Chemical Process Quantitative Risk Analysis* (CCPS, 1989).

The application of consequence and impact models to the Riskland facility results in very simple impact zone estimates for the identified incident outcome cases:

- Incident Outcome Case I (explosion)—the explosion is centered at the center point of the facility; all persons within 200 meters of the explosion center are killed (probability of fatality = 1.0); all persons beyond this distance are unaffected (probability of fatality = 0).
- Incident Outcome Case IIA (explosion)—the explosion is centered at the center point of the facility; all persons within 100 meters of the explosion center are killed (probability of fatality = 1.0); all persons beyond this distance are unaffected (probability of fatality = 0).
- Incident Outcome Cases IIB1, IIB2 (toxic gas clouds)—all persons in a pie shaped segment of radius 400 meters downwind and 22.5 degrees width are killed (probability of fatality = 1.0); all persons outside this area are unaffected (probability of fatality = 0).

Figure 3 illustrates these impact zones.

Frequency Analysis

Many techniques are available for estimating the frequency of incidents (CCPS, 1989, Chapter 3), including fault tree analysis, event tree analysis, and the use of historical incident data. RCC applies an appropriate set of models and historical incident and failure rate data and estimates the following frequencies:

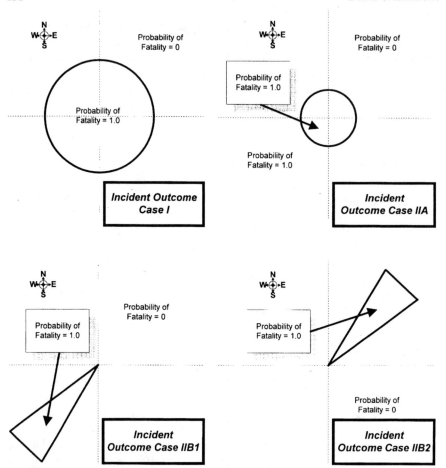

FIGURE 3. Impact Zones for Incident Outcome Cases

- Incident I —Frequency = 1×10^{-6} events per year
- Incident II —Frequency = 3×10^{-5} events per year
- Incident II —Ignition Probability = 33%

These estimates along with the specified weather conditions (wind blowing from the Northeast 50% of the time, and from the Southwest 50% of the time) give the frequency estimates for the four incident outcome cases, as shown in the event trees of Figure 4.

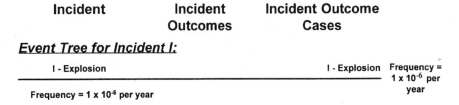

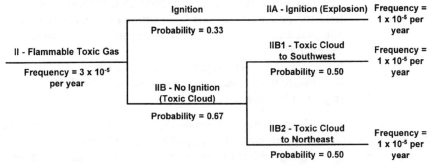

FIGURE 4. Frequency Estimates for Incidents, Incident Outcomes, and Incident Outcome Cases

Individual Risk Estimation

Individual risk is defined by CCPS (1989) as "The risk to a person in the vicinity of a hazard. This includes the nature of the injury to the individual, the likelihood of the injury occurring, and the time period over which the injury might occur."

Individual risk is useful in understanding and managing risk at a location where people might be present. It is also useful in understanding the risk to a particular person, or a group of people, based on knowledge of the geographical location of that person or those people.

In this example, the nature of the injury for both individual and societal risk calculations will be immediate fatality resulting from fire, explosion, or exposure to toxic vapors.

Individual Risk Contours / Geographical Risk

Individual risk at any point is given by the following equations (CCPS, 1989):

$$IR_{x,y} = \sum_{i=1}^{n} IR_{x,y,i} \tag{1}$$

$$IR_{x,y,I} = f_i p_{f,I} \tag{2}$$

where:

- $IR_{x,y}$ = the total individual risk of fatality at geographical location x, y (probability of fatality per year)
- $IR_{x,y,I}$ = the individual risk of fatality at geographical location x, y from incident outcome case I (probability of fatality per year)
- n = the total number of incident outcome cases considered in the analysis
- f_i = frequency of incident outcome case I, (per year)
- $p_{f,I}$ = probability that incident outcome case I will result in a fatality at location x,y

This example problem has been set up so that this calculation is simple, because each incident outcome case has an equal impact (probability of fatality $p_{f,I} = 1$) throughout its geographical impact zone. Therefore, within the impact zone for each incident outcome case, the individual risk from that incident outcome case $IR_{x,y,I}$ is equal to the frequency of that incident outcome case (Equation 2). Outside the impact zone, $IR_{x,y,I}$ is zero.

The simple impact models make it easy to do the calculations graphically. The four impact zones from the four incidents are superimposed on a map of the region of the plant and its surroundings as shown in Figure 5. The total individual risk of fatality at each geographical location is then determined by adding the individual risk from all incident outcome case impact zones that impact that location (Equation 1). For example, in the area labeled "C" in Figure5, application of Equation1 gives the results listed in Table 1.

A similar calculation for the other areas in Figure 5 gives the results summarized in Table 2. Figure 5 is an **individual risk contour** plot for this example problem, with the individual risk values for each area listed in Table 2. Note that for most CPQRAs, the individual risk contours are plotted for orders of magnitude of risk (for example, 10^{-4}, 10^{-5}, etc.). In this example, the specific values of risk calculated are plotted.

Individual Risk Profile, or Risk Transect (taking a route)

The individual risk profile (risk transect) is a graph showing the individual risk as a function of distance from the source of the risk in a particular direction. For the example problem, Figure6 is the individual risk profile in the northeast direction.

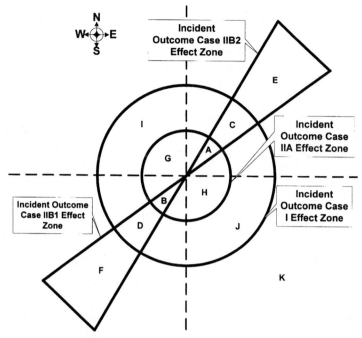

FIGURE 5. Individual Risk Contour Map

TABLE I
Individual Risk Calculation for Area "C" in Figure 5

Incident Outcome Case	f_i (per year)	$P_{f,i}$	IR_i (per year)
I	10^{-6}	1	10^{-6}
IIB2	10^{-5}	1	10^{-5}
$IR = \sum IR_i =$			1.1×10^{-5}

Other Individual Risk Measures

In developing the individual risk contour map and the individual risk transect (Figures 5 and 6), no information about the surrounding population was needed. Figure 5 represents the risk to a person if he were to be at a particular location 100% of the time (8760 hours per year). For the example problem, several other individual risk measures can be calculated with additional data on the population surrounding the plant. Figure 7 shows the location of people in the area surrounding the RCC facility.

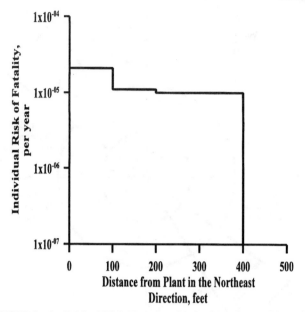

FIGURE 6. Individual Risk Transect in the Northeast Direction.

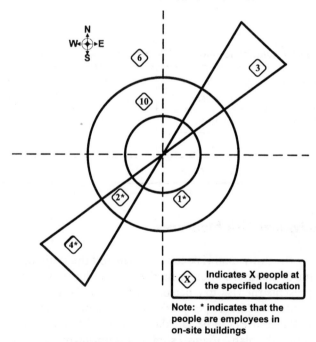

FIGURE 7. Population Distribution

The **maximum individual risk** is the highest value of individual risk at any geographical location. For the example, the absolute maximum individual risk (regardless of whether or not there is any person at that location) is 2.1×10^{-5} per year, at all locations in Regions A and B in Figure 6. The maximum individual risk for any actual person is 1.1×10^{-5} per year, for the two people approximately 200 meters southwest of the facility. Note that these are **not** the people closest to the plant—the person in the southeast quadrant is actually closer to the plant. The prevailing wind directions in this example result in a higher risk to people somewhat farther away from the facility, but located in a direction toward which the wind blows more frequently.

The **average individual risk** is the average of all individual risk estimates over a defined population. It is important to define a population which does not include a large number of people at little or no risk, as this will give a low bias to the result. Average individual risk is given by CCPS (1989) as:

$$IR_{AV} = \frac{\sum_{x,y} IR_{x,y} P_{x,y}}{\sum_{x,y} P_{x,y}} \quad (3)$$

where:

IR_{AV} = average individual risk in the exposed population (probability of fatality per year)

$P_{x,y}$ = number of people at location x, y

TABLE 2
Individual Risk Results

Region (See Figure 5)	Incidents Impacting Region	Total Individual Risk of Fatality (per year)
A	I, IIA, IIB2	2.1×10^{-5}
B	I, IIA, IIB1	2.1×10^{-5}
C	I, IIB2	1.1×10^{-5}
D	I, IIB1	1.1×10^{-5}
E	IIB2	1.0×10^{-5}
F	IIB1	1.0×10^{-5}
G	I, IIA	1.1×10^{-5}
H	I, IIA	1.1×10^{-5}
I	I	1.0×10^{-6}
J	I	1.0×10^{-6}
K	None	0

Applying Equation 3 to the population in the example (Figure 7), averaging only over the population which is subject to risk from the facility (individual risk 0) gives:

$$IR_{AV} = [(3)(10^{-5}) + (1)(10^{-6}) + (2)(1.1 \times 10^{-5}) + (4)(10^{-5})$$
$$+ (10)(10^{-6})] / (3 + 1 + 2 + 4 + 10)$$

$$IR_{AV} = (1.03 \times 10^{-4}) / 20$$

$$IR_{AV} = 5.2 \times 10^{-6} \text{ per year (for the exposed population)}$$

If all people in the area, even those who incur no risk from the facility, are included in the individual risk calculation, the denominator in the above calculation is 30, and the average individual risk is:

$$IR_{AV} = (1.03 \times 10^{-4}) / 30$$

$$IR_{AV} = 3.4 \times 10^{-6} \text{ per year (for the total population)}$$

Another average individual risk which might be of interest is the average individual risk to on-site employees (the people marked * in Figure 7). The average individual risk for the RCC employee population (those people in Regions D, F, and J of Figure 5/Table 2) is:

$$IR_{AV} = [\underbrace{(2)(1.1 \times 10^{-5})}_{\text{Region D}} + \underbrace{(4)(10^{-5})}_{\text{Region F}} + \underbrace{(1)(10^{-6})}_{\text{Region J}}] / (1 + 2 + 4)$$

$$IR_{AV} = (6.3 \times 10^{-5}) / 7$$

$$IR_{AV} = 9 \times 10^{-6} \text{ per year (for the RCC employee population)}$$

The **Fatal Accident Rate (FAR)** is calculated from the average individual risk, and is normally used as a measure of employee risk in an exposed population. Using the average individual risk for the RCC employee population, FAR is calculated from the following equation:

$$FAR = (1.14 \times 10^{4}) IR_{AV} \quad \text{(for the employee population)} \quad (4)$$

where IR_{AV} has units of probability of fatality per year, and FAR has units of fatalities per 10^8 man-hours of exposure. Applying Equation 4 to the example gives:

$$FAR = IR_{AV} (1.14 \times 10^{4})$$
$$= (9 \times 10^{-6})(1.14 \times 10^{4})$$
$$= 0.1 \text{ fatalities} / 10^8 \text{ man-hours of exposure}$$

Societal Risk Calculation

Societal risk measures the risk to a group of people (CCPS, 1989). Societal risk measures estimate both the potential size and likelihood of incidents with multiple adverse outcomes. In this example, the adverse outcome considered is immediate fatality resulting from fire, explosion, or exposure to toxic vapors. Societal risk measures are important for managing risk in a situation where there is a potential for accidents impacting more than one person.

F-N Curve

A common measure of societal risk is the Frequency-Number (F-N) Curve. The first step in generating an F-N Curve for the example problem is to calculate the number of fatalities resulting from each incident outcome case, as determined by:

$$N_i = \sum_{x,y} P_{x,y} P_{f,i} \tag{5}$$

where N_i is the number of fatalities resulting from Incident Outcome Case I

For the example, $p_{f,i}$ in Equation 5 equals 1. Because the impact zones for the example are simple, this calculation can be done graphically by superimposing the impact zones from Figure 3 onto the population distribution in Figure 7, and counting the number of people inside the impact zone. Table 3 summarizes the estimated number of fatalities for the four incident outcome cases.

The data in Table 3 must then be put into cumulative frequency form to plot the F-N Curve:

$$F_N = \sum_i F_i \quad \text{for all outcome cases } i \text{ for which } N_i \geq N \tag{6}$$

where:

F_N = frequency of all incident outcome cases affecting N or more people, per year
F_i = frequency of incident outcome case I, per year

Table 4 summarizes the cumulative frequency results. The data in Table 4 can be plotted to give the societal risk F-N Curve in Figure 8.

TABLE 3
Estimated Number of Fatalities from Each Incident Outcome Case

Incident Outcome Case	Frequency F_i (per year)	Estimated Number of Fatalities
I	1.0×10^{-6}	13
IIA	1.0×10^{-5}	0
IIB1	1.0×10^{-5}	6
IIB2	1.0×10^{-5}	3

TABLE 4
Cumulative Frequency Data for F-N Curve

Number of Fatalities N	Incident Outcome Cases Included	Total Frequency F_N (per year)
3 +	I, IIB1, IIB2	2.1×10^{-5}
6 +	I, IIB1	1.1×10^{-5}
13 +	I	1.0×10^{-6}
>13 +	None	0

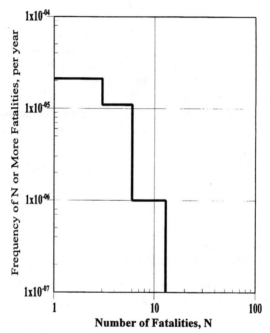

FIGURE 8. Societal Risk F-N Curve for the Example Problem

Other Societal Risk Measures

Other societal risk measures can also be calculated for this example. The **average rate of death (ROD)** is the estimated average number of fatalities in the population from all potential incidents. The ROD is calculated using Equation 7:

$$ROD = \sum_{i=1}^{n} f_i N_i \qquad (7)$$

where ROD is the average rate of death, fatalities per year

Applying the data in Table 3 for the estimated number of fatalities resulting from each incident outcome case to Equation 7:

$$ROD = (1.0 \times 10^{-6}/yr)(13) + (1.0 \times 10^{-5}/yr)(0) + 1.0 \times 10^{-5}/yr)(6)$$
$$+ (1.0 \times 10^{-5}/yr)(3)$$
$$= 1 \times 10^{-4} \text{ fatalities per year}$$

API752 (API, 1995) uses **aggregate risk** as a tool for managing the risk associated with occupied buildings in a process plant. Aggregate risk is defined as "societal risk applied to a specific group of people within a facility" (CCPS, 1996). In this example, the people indicated by an asterisk in Figure 7 are RCC employees working in on-site buildings. The aggregate risk calculation considers only this population, and it will be assumed that the people are present all of the time. Table 5 summarizes the number of fatalities for each incident outcome case for the employee population. This data can be put into cumulative frequency form as shown in Table 6, and the resulting aggregate risk curve is shown in Figure 9.

TABLE 5
Estimated Number of Fatalities for the Employee Population in On-Site Buildings from Each Incident Outcome Case

Incident Outcome Case	Frequency F_i (per year)	Estimated Number of Fatalities in the Employee Population in On-Site Buildings, N
I	1.0×10^{-6}	3
IIA	1.0×10^{-5}	0
IIB1	1.0×10^{-5}	6
IIB2	1.0×10^{-5}	0

TABLE 6
Cumulative Frequency Data for Aggregate Risk Curve for Employee Population in On-Site Buildings

Number of Fatalities in Employee Population in On-Site Buildings, N	Incident Outcome Cases Included	Total Frequency F_N (per year)
3 +	I, IIB1	1.1×10^{-5}
6 +	IIB1	1.0×10^{-5}
>6 +	None	0

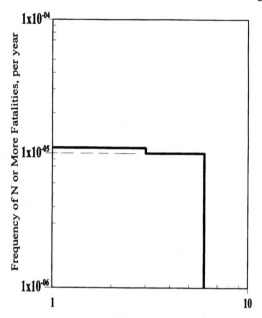

FIGURE 9. Aggregate Risk Curve for Employee Population in On-Site Buildings for the Sample Problem

The **aggregate risk index** (CCPS, 1996) is the average rate of death, as calculated for the people in on-site buildings in a plant. For the example problem, applying Equation 7 using the estimated number of fatalities from each incident outcome case considering the employee population only (the data in Table 5), the aggregate risk index is:

Aggregate Risk Index $= (1.0 \times 10^{-6}/\text{yr})(3) + (1.0 \times 10^{-5}/\text{yr})(6)$
$= 6.3 \times 10^{-5}$ fatalities per year

The **Equivalent Social Cost Index (ESC)** is a societal risk measure which attempts to account for society's aversion to large incidents. The calculation is the same as for the Rate of Death, except that the number of fatalities is raised to a power to increase the contribution of large incidents to the ESC Index:

$$ESC = \sum_{i=1}^{n} F_i N_i^p \qquad (8)$$

where p is the risk aversion power factor ($p > 1$).

Risk aversion power factors of 1.2 and 2 have been suggested (CCPS, 1989). Using these factors, Equivalent Social Cost (ESC) indices for this example, using the total population, are:

$$p = 1.2 \quad ESC = 1.4 \times 10^{-4}$$
$$p = 2.0 \quad ESC = 6.2 \times 10^{-4}$$

The units of Equivalent Social Cost are not meaningful.

Summary of Risk Results

This simple example illustrates the complexity of risk. Although the example considers only the acute risk of fatality, fourteen different measures of risk were calculated, as summarized in Table 7. These measures consider different aspects of risk, and all are valid risk estimates which might be valuable in the appropriate decision making context. For example, a risk study undertaken to determine if the risk to employees in on-site occupied buildings is tolerable, aggregate risk may be the appropriate risk measure. Maximum individual risk to nearby residents and societal risk to the surrounding community might be the best measures to use in order to understand and manage the risk to neighbors.

TABLE 7
Summary of Risk Results for the Riskland Example Problem

Risk Measure	Result
Individual Risk	
Risk Contours	See Figure 5 and Table 2
Risk Transect	See Figure 6
Maximum	2.1×10^{-5} per year
Maximum for Actual Person	1.1×10^{-5} per year
Average, Exposed Population	5.2×10^{-6} per year
Average, Total Population	3.4×10^{-6} per year
Average, Employee Population	9×10^{-6} per year
Fatal Accident Rate (FAR)	0.1 fatalities per 10^8 man-hours of exposure
Societal Risk	
F-N Curve	See Figure 8
Aggregate Risk Curve	See Figure 9
Average Rate of Death	1.0×10^{-4} fatalities per year
Aggregate Risk Index	6.3×10^{-5} fatalities per year
Equivalent Social Cost Index, Total Population ($p = 1.2$)	1.4×10^{-4}
Equivalent Social Cost Index, Total Population ($p = 2$)	6.2×10^{-4}

One can easily envision a number of other risk measures which could be calculated, considering, for example, environmental risk, risk of injury, long term health risk, economic risk, and others. This simple example problem clearly shows that there is no single, simple answer to the question, "What is the risk of this facility?" That question is much too broad.

Application to Risk Management and Decision Making

While this example problem is intended to demonstrate the calculation procedures used to combine frequency and consequence data to produce various specific risk measures, it also illustrates the importance of clearly defining the risk measures to be used in any risk management program which includes quantitative evaluation of risk. When comparing the risk of facilities or design options, it is essential that the risks are calculated on the same basis for the comparison to be meaningful. Similarly, if quantitative guidelines are to be used as a part of an organization's risk management program, the guidelines must be clearly defined in terms of the risk measure to be used, and the calculational procedures used to obtain the measures which will be compared to the guidelines. Failure to clearly define the risk measures used in a risk management program will result in confusion and may lead to inconsistent decisions.

Summary

This example Chemical Process Quantitative Risk Analysis problem is intended to teach the methodology of CPQRA calculations. For a real risk analysis, these calculations are extensive and use much more sophisticated models. The calculations are usually done using computer programs, and the methodology of the calculations may not be clear, even to an analyst who does risk analyses routinely. Should all risk analysts be required to complete at least one simple risk analysis manually, giving them an opportunity to understand what the computer program is doing in a more complex study?

This sample problem is also useful for explaining risk analysis methods and measures to users of risk analysis studies. It explains the meaning of the various risk measures to plant and company management, and to technical staff who must support the work of the risk analyst by providing much of the required data. Before starting a QRA on a facility or plant design, the sample problem can be used to quickly illustrate the kind of input data which will be required, and the format of the results of the completed study.

The example illustrates that risk is complex, and that it can be measured in many different ways. The different measures provide information about different aspects of risk—for example, risk to an individual, risk to a particular group of people, average risk, maximum risk. A complete risk management program may have to consider several of these risk measures, and it is essential that all participants in the risk management process understand the meaning and use of the risk measures considered.

References

Abbott, E. A. (1884). *Flatland*. New York: Barnes & Noble, Inc. (re-published in 1963).

American Petroleum Institute (API) (1995). *Management of Hazards Associated With Location of Process Plant Buildings*. RP 752. Washington, DC: American Petroleum Institute.

Burger, D. (1965). *Sphereland*. New York: Harper & Row, Publishers.

Center for Chemical Process Safety (CCPS) (1989). *Guidelines for Chemical Process Quantitative Risk Analysis*. New York: American Institute of Chemical Engineers.

Center for Chemical Process Safety (CCPS) (1992). *Guidelines for Hazard Evaluation Procedures*. Second Edition, with Worked Examples. New York: American Institute of Chemical Engineers

Center for Chemical Process Safety (CCPS) (1996). *Guidelines for Evaluating Process Plant Buildings for External Explosions and Fires*. New York: American Institute of Chemical Engineers

Hendershot, D. C. (1988). "A Simple Example Problem Illustrating the Methodology of Chemical Process Quantitative Risk Assessment." *AIChE Mid-Atlantic Region Day in Industry for Chemical Engineering Faculty*, April 15, 1988, Bristol, PA.

Theodore, L., J. P. Reynolds, and F. B. Taylor (1989). *Accident & Emergency Management*. New York: John Wiley & Sons.

ACKNOWLEDGMENTS

This example problem has evolved through a number of years of use in explaining the concepts of risk analysis to Rohm and Haas personnel throughout the world, to Rohm and Haas customers, and to outside organizations. It was first developed for presentation at an American Institute of Chemical Engineers "Day in Industry" for college faculty (Hendershot, 1988), and a condensed version was later published by Theodore et al. (1989). I thank all of the engineers and managers from Rohm and Haas Company and other companies who have contributed many suggestions to improve the example over the years. I also thank Art Dowell of Rohm and Haas Texas, Inc., and Bill Bridges of JBF Associates, Inc., for reviewing the manuscript of this paper and providing many constructive comments. The Center for Chemical Process Safety (CCPS) Risk Analysis Subcommittee has reviewed this problem and made a number of suggestions to further enhance it. A version will be included in the Second Edition of the CCPS *Guidelines for Chemical Process Quantitative Risk Analysis*, to be published later this year.

Risk Analysis and Risk Management

Chairs **Dennis Hendershot**
Rohm & Haas Company
Keith Cassidy
Health & Safety Executive, UK

A Context-Specific Approach Toward Human Reliability Assessment

Joseph R. Fragola, IEEE
SAIC New York

ABSTRACT

As individual pieces of equipment become more and more reliable and as maintenance improvements extend their life, the human contribution to the risk of system operation which has always been significant is becoming evermore important. Approaches which address the assessment of human reliability have been around since the late sixties[1], and they have made significant and useful contributions to the identification of the significance of human errors in the risk of operation of nuclear facilities in particular. However there have been significant problems in using these approaches to determine where and to what degree changes in human impacting elements would be cost effective. The reason for this problem is that, just as in the case of early equipment failure assessments, conventional Human Reliability Analysis (HRA) approaches are, to a great degree, context independent. That is the evaluations serve, much as equipment generic data did, to indicate the general magnitude of the risk contribution, but did not attempt to highlight the specific areas and specific value of potential improvements. For this reason plant managers were forced to rely completely on the expert judgment of the HRA analysts in the determination of which changes would be effective. Further, even with this judgment, it was often difficult to justify the risk reduction credit that should be given for changes made to reduce the human error contribution.

Some early attempts[2], were made to address these problems by applying judgmentally based performance shaping factors, but these factors tended to mix the contextual influences of the plant state with those of the plant specific human interface. Also the approach taken toward applying these factors tended to be unsystematic, and often non-reproducible because it was difficult for a single analyst to apply these factors consistently even across a single analysis and therefore almost impossible to maintain any consistency across analyses and across analysts.

The approach presented in this paper was developed from previous work[3] by the author in cooperation with several national and international collaborators in an attempt to address these deficiencies and to allow for the contextual richness of the analysis to be highlighted. The approach attacks the problem of context by separating it into two, to some degree independent, perspectives. That of the accident condition, and that of the plant specific interface. From these points of view the conditions imposed upon the operator in terms the accident process conditions become process specific and therefore somewhat independent of the specific human interface. And, correspondingly, the human performance relevant features of the specific human interface can be evaluated as to their quality in terms of their capability of fulfilling the requirements imposed upon the operator by the accident process. Thus, from these two perspectives human interface capabilities can be discriminated according to the risk importance of their human relevant features in addressing a specific accident sequence of interest.

While experience and expert judgment are likely to be necessary elements in any successful HRA for some time to come it is believed that this approach shows promise in bringing consistency and reproducibility to HRA.

Introduction

Human Reliability remains a young and dynamic field. For this reason the claims of any approach to the performance of an HRA should always be viewed with a healthy dose of skepticism. Authors of these approaches should therefore be very careful to limit the claims made by any approach suggested. So it is with the approach described in this paper. Specifically, this paper makes no claim that any HRA can be routinized and certainly at this point it is not intended to address all HRA related issues. What is claimed is that approach described here, when applied by skilled and experienced HR analysts, will produced estimates of human error risk contribution, for the types of human errors considered within the scope of the approach, that are consistent with the ranges of uncertainty applicable in traditional Probabilistic Safety Assessments[1].

Further because this paper is specifically focused upon the need to consistently consider both the behavioral content of a required human action and the context within which this action is required to take place, it is believed that the application of the approach described herein will produce results which are more consistent throughout a particular analysis, and non repeatable from analyst to analyst and from analysis to analysis.

Purpose of an HRA in a PSA Setting

The type of human reliability analysis addressed here is that performed as a part of an overall PSA. Its purpose therefore is to ensure that the principal risk significant human actions of the operations and maintenance crew are incorporated into the analysis in an appropriate, systematic and reproducible way and in a way which documents the hypothesis suggested, the expert judgments made, the analytical methods and information sources considered all in an appropriate manner. In accordance with this purpose the approach described here is directed at establishing the analytical course to be implemented throughout the human reliability

---

1 Probabilistic assessments of this type are alternately referred to as Probabilistic Risk Assessments (PRAs) or Quantitative Risk Assessments; the term PSA is the preferred term used here to refer to all such assessments.

task for the definition and identification of human actions, the selection of actions for detailed analysis, and the assignment of human error probability estimates for the quantification of systems models and the analysis accident sequences.

HRA Scope and Analysis Development

The approach discussed in this paper is applicable to the identification of and the analysis of human actions included in the fault tree models developed as part of the systems analysis PSA task, as well as to all human actions identified and modeled in the event trees developed as part of the accident sequence analysis task[1]. It also applies to all human recovery actions identified as required to be considered in the quantification of the accident sequences. The approach does not apply to human actions which could initiate accidents because they are addressed elsewhere in the PSA (as explained in the sections that follow). Also the approach described here would require modification and supplementary analysis in order to be applied to true errors of commission (EOCs) or to human actions affected by initiating events which are external to the system undergoing analysis.

Because, as has been stated above, the human contribution to risk is almost always significant if not overwhelming the treatment of human actions in a PSA is a key determinant to the development of a realistic understanding on the part of the analyst and decision maker, of the development and progression of the accident sequences and their involvement in an relative importance to the overall operational risk. For this reason it may be somewhat surprising that the reliability or risk of human actions was often, if not always, ignored in early reliability and risk studies, and continues to be ignored, at least to some degree, in studies performed currently in some industries (notably the space industry). The first PSA performed WASH-1400[5] was originally published in draft version without any consideration of the risk of human action. An American Physical Society[6] review of this draft as well as commentary received from other sources[7] severely criticized this exclusion and an initial attempt at an HRA using an early developed model[8] was hastily prepared and added to the final published version of the report.

The exclusion of human actions although perhaps not defensible even in those early days was more understandable because:

a. Hardware reliability was in such a poor state at the time that it often dominated system problems.
b. Many of the driving industry problems at the time were either unmanned (space launch vehicles and artificial satellites) or when manned were largely automated (manned spacecraft).

---

1 For a readers unfamiliar with the task content of a PSA and their interfaces should refer to the description included in reference[4].

c. Quantitative human reliability analysis approaches were in their infancy, were far from accepted generally, and were very immature as compared to quantitative hardware approaches and even these latter approaches were found to be unacceptable to some industries (particularly the manned space industry).
d. Whatever quantitative reliability approaches existed at this time were largely developed from qualitative bottoms-up analyses (such as Failure Modes and Effects Analyses) and these bottoms-up analyses found it difficult to include human actions.

The development of the PSA scenario based approach; that is a sequence oriented, top down, event tree, fault-tree approach to risk analysis; was inherently more flexible than previous analytical approaches. By virtue of this analytical flexibility human reliability estimates could more easily be combined with estimates of the hardware and even software portions of systems and equipment in operation. In this way, the PSA approach provided a framework wherein the interrelationships between operations and maintenance crews, their associated operated or maintained equipment could be studied in terms of their impact on the progression of accident sequences and the resulting risk.

Even though the development of the PSA scenario based approach allowed for human error estimates to be combined with estimates of other elements of the system to produce overall systems estimates it did nothing to address the problem of how to generate these human error estimates in the first place. However, since whatever model was used to address the problem of human actions in systems must now produce outputs which met the requirements of the overall PSA model the top level requirements of any such model became much more clearly recognized. In particular, the type of human action of concern was human error and the type of output any human error model must produce, to be of use in a PSA setting, was human error probability estimates whose levels of uncertainty were consistent with the overall requirements of a PSA. Moreover the PSA model did not require predictions of the failure probability of a particular hardware component as installed in a particular at a particular time, but rather predictions of the estimate probability of failure of a representative class of components overall mission time and within a range of uncertainty. In the same way human error probability estimates which for a class of human actions overall class of individuals within an acceptable range of uncertainty were also acceptable.

Human Reliability Analysis Top Level Process Flow

As was mentioned the type of HRA approach discussed here is performed within the context of the overall PSA framework. As an integral part of the PSA the HRA Shares in the PSA information set and interfaces with some of the other PSA tasks. The more general interfaces with PSA are depicted in Figure 1 how-

A Context-Specific Approach Toward Human Reliability Assessment

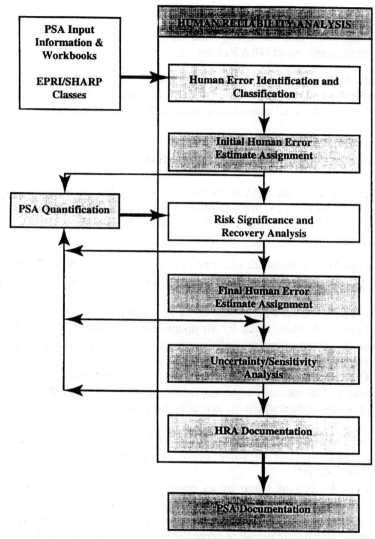

FIGURE 1. Top Level Flow Diagram Indicating Overall PSA Interfaces

ever the figure's primary purpose is to identify the specific subtasks within the HRA task along with the general task flow and task interfaces. The principal subtasks identified in the figure are:

- Definition and Identification of Human Action
- Initial Human Error Estimate Assignment
- Final Human Error Estimate Assignment
- Dependency Analysis

- Recovery Action Identification and Analysis
- Uncertainty and Sensitivity Analysis
- Documentation of HRA Results

These general subtasks are consistent with those defined in the Systematic Human Action Reliability Procedure (SHARP) given as reference[9], and refined in [10], [11]. The individual subtasks are discussed in the sections which follow.

Definition and Identification of Human Actions

The starting point of an HRA is the definition of the human action types to be considered in the analysis. Since the HRA approach discussed here is consistent with the overall SHARP framework the SHARP type definitions are used. These are as follows: *industrial standard*

- **Type 1:** Actions which occur prior to an initiating event and which have the potential for adversely affecting system reliability (also called "latent" errors)
- **Type 2:** Actions which could bring about an initiating event
- **Type 3:** Actions taken by an operator or operations crew during the course of an accident while following procedures in an attempt to mitigate the situation
- **Type 4:** Actions taken by an operator or operations crew during the course of an accident under the mistaken belief that the taken actions are appropriate but which in actuality have the potential for making the situation worse thereby complicating the mitigation process. (So called "well intentioned actions with unintended or undesired consequences," or what at least to some degree inappropriately called "Errors of Commission")
- **Type 5:** Actions taken by an operator or operations crew during the course of an accident which are not unequivocally included in the procedures and have as their objective the recovery of failed equipment or use of alternative means to serve the function of this equipment.

For reasons that are explained more fully elsewhere[3] Type 2 failures are treated using plant specific and generic data and are therefore not typically modelled. However in those cases where such data are not available Type 2 failures must be treated on an individual basis. Type 4 actions as yet have no generally accepted method for treatment and are therefore treated on an individual case basis, or in the case where symptom-based emergency procedures are employed, they can be treated as Type 3 actions.

Therefore the quantification approach discussed here is related to developing human error probability (HEP) estimates for Type 1,3, and 5 and for those Type 4 actions which can be treated as Type 3. Further since the focus of this paper is on the manner in which the context should be taken into account in developing HEP estimates and since the contextual development is the same for Type 3 and for those Type 5 actions which are not uniquely treated, and those Type 4 actions not treated individually this paper will discuss only the development of HEP estimates for Type 1 and Type 3 actions.

Incorporation Identified Type 1 and Type 3 Errors in the Fault Tree Models

Following the process stated above helps to ensure that the complete set of Type 1 errors are identified. However, not all of the identified errors are appropriate to be incorporated into the system models because plant features or procedures minimize the probability that the error will have an accident sequence impact and some of the indentified actions may not be of Type 1. For this reason, the set of identified errors are reviewed against the following set of guidelines to screen out errors of unlikely impact.

Generic Impact Assessment

1. Erroneous positioning is *not* to be included in the modeling of unavailability prior to an accident of the equipment whose state (position, power supply or point of actuation) would be alarmed in a unique way or would undergo surveillance at least every 24 hours.
2. Those human errors which could possibly affect the overall function of a component via its incorrect test or maintenance are *not* to be included when they are already taken into consideration in an implicit way in the failure probability of the affected equipment. This case would include erroneous positioning following test, maintenance, operational realignment or calibrations of other equipment and/or their erroneous calibration.
3. Erroneous positioning prior to an accident of all equipment which, in an accident condition, receive an automatic signal to return to their operation position after actuations and which do not require disablement of this automatic function for positioning are not to be included in the modeling of unavailability.
4. Incorrect positioning prior to an accident of equipment which undergo functional testing of their operation to verify that they are in the correct state after having been altered for test, maintenance, operational realignment or calibration are *not* to be included in the modeling of unavailability.
5. Human recovery actions which the operations crew could implement during an accident are *not* to be included *initially*.

Dependency Analysis

Only these actions which have not been excluded from consideration by one of the preceding guidelines are included in the fault tree models in the form of human errors. They are assigned an event designator indicating a human error prior to an accident. Wherever possible, actions taken on all equipment in a group should be represented by a single basic event associated with its corresponding train, leg or system.

Once the human errors have been included in the model, those which are combined through AND gates are reviewed to determine whether actuations exist which could be considered as a common cause or whether another dependency exists between them. If this common cause possibility cannot be discarded through the application of the previously stated rules, a new basic event is generated whose event designator indicates a common cause human error event prior to an accident.

Development of Contextual Factors in Type I Actions (Latent Human Errors)

Any approach which attempts to account for the influence of the specific in-situ environment on the probability of human error must address two key aspects of that environment. That is it must address the process aspects of the required task and the operational setting provided to the operations and maintenance crew to allow them to successfully complete these tasks. In the case of Type 1 actions the tasks involved are usually related to test, operational re-alignment, maintenance and calibration of equipment and the associated operational actions necessary to "clear" or isolate the equipment to be maintained or calibrated. The boundary equipment are just as important as the affected equipment because the types of latent errors associated with these tasks are usually related to leaving some component in an incorrect or inoperable state. Therefore for maintenance and calibration actions the process environment is simply the set of clearance and isolation requirements for both the installed device being addressed and its supporting and boundary components as they relate to the functional requirements and capabilities of the device. The operational setting is related to the comprehensiveness of the activity as performed in the plant and the verification activities related to providing the assurance that the calibrated device has been properly calibrated and that the maintained component is functioning properly and has been properly returned to service.

The approach recommended here addresses the process environment by the identification in each system independent train and subtrain groupings from the maintenance perspective. Each group of components identified range, from node to node, where the nodes are determined by the furthermost component used to isolate the maintained component from the system. For example if plant procedure requires that a pump be isolated from a fluid perspective both at the source and also at the pump inlet, and at the system inlet as well as at the outlet then the inlet node is the source boundary valve and the outlet node is the system interface boundary valve, and the independent sub-train is all the equipment in between including all supporting equipment. Where cross-ties or interconnections are involved this approach will often lead to double counting of some individual components. This is acceptable because the approach defines the general problem of incorrect positioning or disablement as considered for all components in the subtrain or functional grouping as will be explained later.

As far as the operational setting is concerned the possibility of dependencies, both positive or negative, are assessed by reviewing the maintenance and calibration procedures and by reviewing the actual process as it is conducted in the plant. Additionally the verification procedures and process is reviewed and the frequency of each verification procedure capable of identifying and correcting an initial error is obtained from plant surveillance schedules and from estimates of corrective maintenance.

These two influences are addressed as follows. First a maintenance matrix is constructed with the equipment assumed to be in maintenance listed along the rows of the matrix and the equipment whose state is affected by the maintenance is listed across the columns. This latter group includes the maintained equipment, all boundary equipment, and all sources of supporting resources (such as power, cooling, lubrication). The matrix is completed by addressing each equipment in maintenance in turn and indicating which other equipment are affected to allow for the maintenance action to take place as well as what the effect on the affected equipment is.

Since the latent Human Error concern is not with the maintained equipment but with its boundary equipment the Maintenance Matrix, once constructed, is applied by addressing the affected equipment. Each of these is considered in turn and a list of all the maintained equipment which would require its disablement for maintenance is taken from the row entries in the matrix associated with the column addressing the particular affected equipment. This information is combined for all components within a basic event grouping as well as with other factors to determine what HEP value should be assigned to the group. These other factors include whether the affected equipment is in the same group or not, an estimate of how often it is maintained, and what verification or test might be available to uncover the latent error. An example of a portion of the results of such an analysis is provided in Table 1. The calculation procedure used to generate the unavailabilities with (Ic) and without verification or check (Is) is given in reference [12].

In a similar fashion, and following a similar process, human errors in calibration, test, and due to operation realignments are also identified and once identified all the incorrect positioning and calibration human errors are included in the HRA task notebook or other equivalent document. Each error is

TABLE I
Example Calculation of Type I Human Error Basic Event Probabilities

BASIC EVENT	COMPONENTS IN GROUPING	MAINTENANCE SAME GROUP (Y/N)	ANNUAL FREQUENCY	BASIC HEP	VERIFICATION OR TEST	ANNUAL FREQUENCY	I_s I_c	I_c $I_{c,v}$		EF	VALUE
ETRAMA2EHA	V-1501-6B	MOV-1501-7B (Y) MOV-1501-7D (Y)	1.50E-01 1.50E-01	1.00E-03 1.00E-03	PV-O-231B PV-O-314B	1.20E+01 4.00E+00	1.00E-03 6.98E-02	6.98E-05 2.64E-01	2.35E-05	3	2.94E-05
ETRAJ2B6HA	V-1501-3B	FLT-SSS-9B (N)	0.00E+00	1.00E-02	PV-O-231B PV-O-314B	1.20E+01 4.00E+00	========= =========	========= =========	=========	3	0.00E+00
ETRAMC1EHA	V-1501-15A	CMB-1503A (Y) MOV-1501-16A (N)	2.40E-01 1.50E-01	1.00E-03 1.00E-02	PV-O-231B PV-O-314B	1.20E+01 4.00E+00	4.46E-03 8.88E-02	3.96E-04 2.68E-01	1.35E-04	3	9.12E-04
	V-1501-17A	CMB-1503A (Y) MOV-1501-19A (N) MOV-1501-23A (N) MOV-1501-28A (N) MOV-1501-34A (N) MOV-1501-35A (N) MOV-1501-16A (N) MOV-1501-19A (N)	2.40E-01 1.50E-01 1.50E-01 1.50E-01 1.50E-01 1.50E-01 1.50E-01 1.50E-01	1.00E-03 1.00E-02 1.00E-02 1.00E-02 1.00E-02 1.00E-02 1.00E-02 1.00E-02	PV-O-231B PV-O-314B	1.20E+01 4.00E+00	8.11E-03 2.22E-01	1.80E-03 2.57E-01	5.95E-04		
ETRAMC2EHA	V-1501-15B	CMB-1503B (Y) MOV-1501-16B (N)	2.40E-01 1.50E-01	1.00E-03 1.00E-02	PV-O-231B PV-O-314B	1.20E+01 4.00E+00	4.46E-03 8.88E-02	3.96E-04 2.68E-01	1.35E-04	3	9.12E-04
	V-1501-17B	CMB-1503B (Y) MOV-1501-19B (N) MOV-1501-23B (N) MOV-1501-28B (N) MOV-1501-34B (N) MOV-1501-35B (N) MOV-1501-16B (N) MOV-1501-19B (N)	2.40E-01 1.50E-01 1.50E-01 1.50E-01 1.50E-01 1.50E-01 1.50E-01 1.50E-01	1.00E-03 1.00E-02 1.00E-02 1.00E-02 1.00E-02 1.00E-02 1.00E-02 1.00E-02	PV-O-231B PV-O-314B	1.20E+01 4.00E+00	8.11E-03 2.22E-01	1.80E-03 2.57E-01	5.95E-04		
ETRAMK2EHA	V-1501-4B	MOV-1501-5B (Y) FLT-SSS-9B (Y)	1.50E-01 0.00E+00	1.00E-03 1.00E-03	PV-O-231B PV-O-314B	1.20E+01 4.00E+00	1.00E-03 3.61E-02	3.61E-05 2.57E-01	1.20E-05	3	1.50E-05
KTRAM0BEHA	V-1301-16	MOV-1301-2 (N) MOV-1301-3 (N) CMB-1302 (Y)	1.50E-01 1.50E-01 2.40E-01	1.00E-02 1.00E-02 1.00E-03	PV-O-415	5.00E-01	6.00E-03 5.19E-01	3.12E-03		3	3.89E-03

organized by plant system affected so that the errors can be readily incorporated into the appropriate system model.

Identification of Type 3 Actions in the Event Sequences

In general, Type 3 actions are identified based upon a review of the operations and emergency procedures. This review is directed at identifying those actuations that the operations crew must implement according to these procedures and particularly those which, if not performed properly, could lead to an accident sequence. Since the review is directed at identifying crew actions it requires interface with representative operators and a walkthrough of the procedures with them either on the plant analyzer or simulator, within the control room, or by use of a control room mockup.

Human actions identified in the procedures which require the manipulation of a component or a group of components within a system and whose erroneous performance is accident related are included in models as Type 3 and are identified using an event designator whose failure mode corresponds to a human error during an accident, unless they are considered to be recovery actions. As has been mentioned previously recovery actions are not initially included in any of the models. Recovery actions are those actions which if implemented properly would improve the conditions of an undesired plant state and are performed to recover the functionality of a failed equipment or require the implementation of means which are not clearly included in the procedures. When numerous equipment must be activated simultaneously according to a procedure or when a required action deals with different trains or sub-trains of the same system their complete dependency is assumed and the same event designator is assigned to the entire equipment set. However, the assigned event designator should include an indication of the more global nature of this event and the entirety of the equipment set affected by the designated human action.

When the event trees are constructed often the top event/functional headings will identify a human action which affects the development of the sequences. If the identified human action applies to a single system then the event should be modeled at the system level and included in the system fault tree. However, when the human action identified and modeled represents the actuation of equipment corresponding to different systems then it must be modeled explicitly in the top event/functional headings. These identified human errors should be designated as human error events occurring during an accident. These events represent the cognitive portion of the human action sequence (i.e., the deciding what is to be done portion). The associated manual portion (i.e., implementation of the decided upon action) is modeled at the level of the different systems affected.

Initial Human Error Assignment and Screening

For all Type 1 and Type 3 actions identified initial conservative HEP values are applied according to a procedure given elsewhere[3]. This procedure is not discussed here for the sake of conciseness, because it has been well documented, and because the focus of this paper is on the representation of context. However, once initial human error values are assigned the set of identified human errors are included in the initial quantification of the overall PSA model. A review of the quantification results against the criterion chosen to identify risk significant events (e.g., that and event contributes more than 1% to a sequence which contributes more than 1% of the overall risk) allows risk significant human error events to be identified for final detailed evaluation and human error assignment for input into the later overall PSA quantifications.

Quantification of Type 3 Human Actions

The unavailability estimates to be assigned for the quantification of human errors derived from Type 3 actions correspond to the addition of branches of the Boolean operator action tree used for human action representation. Each of these branches represents a possible error mechanism in the implementation of the actuation and has a distinct treatment, depending upon the model used. Generally, two classes of error mechanism are distinguished; cognitive and manual. Each of these are constituents of the overall human action and the sum of their probabilities constitute the overall HEP. It is worth recalling that, as opposed to Type 1 actions whose unavailabilities are not directly those of the basic event from which they originate (it is also necessary to consider the relative occurrence frequencies), the overall HEP assigned to Type 3 actions is directly equivalent to the sum of the basic event probabilities from which they are derived.

Cognitive Part of Type 3 Actions

The model used for the cognitive process, including the error mechanisms of detection, diagnosis, and response time is given in reference[3].

The result of the application of this model is the development of an estimate for the HEP associated with the cognitive part of an identified action, within the time available to implement the action, and for a given sequence of events. This estimation becomes more specific when a greater understanding is obtained of the various contextual factors (human factors/performance shaping factors; training, procedural quality, man-machine interface, etc.) which influence the actions. This understanding requires comprehensive operations team interviews and includes observations during simulator training exercises.

Interactive graphic simulators, plant emulators, and plant analyzers are of particular assistance in the evaluation of the time-related aspects of such human actions. However, when they are applied, care must be taken to account for the fact that the simulator environment is not the contextual equivalent of the accident environment. Specifically the incredulity effects and vigilance affects must be addressed and accounted for in the process of applying these devices to evaluate accident environment human actions.

Manual Part of Type 3 Actions

The detailed analysis of the manual part of the Type 3 actions is performed using the NUREG/CR-1278 (THERP)[2] methodology, similar to the process indicated for Type 1 actions, with the exception that the HEP for the manual actuations is applied directly, and through the Boolean sum with the cognitive part, directly produces the identified basic event probability.

Development of Contextual Related Performance Shaping Factors

The basis for the evaluation of the contribution of Human Errors to risk is a generically derived small set of Human Error Probabilities (HEPs) which have been distilled from the observations over many years of experience of actual and simulated responses to required action types. These HEP results have been presented according to the broadest of human action types, segregating them into those whose probabilities are essentially independent of their time available for response and those for which the time available is particularly relevant. In this procedure, the time available is particularly relevant. The time independent actions have been referred to particularly in Type 1 Human Errors and in the manual part of Type 3 and Type 5 actions. The time dependent actions have been referred to as the cognitive parts of Type 3 and Type 5 actions. While these designations speak to the character of the human action type required and therefore, at least to some degree distinguish the HEP values appropriate, they only refer to the nature of the required action and not at all to the contextual requirements implied by the action, nor to the degree to which the local context satisfies their requirements. In this sense, this generic evaluation ignores the important characteristics of the action in context and therefore allows individual in-situ conditions to be neither credited nor debited. Furthermore, it also provides no way to credit changes to the in-situ conditions made to improve the probability of correct action being taken. Therefore while it might be possible to estimate the global contribution of Human Error to the overall risk of a system using a generic approach, such an approach makes it impossible to distinguish between the individual performance of specific systems and changes to this contextual environment.

Early HRAs attempted to account for the context specific environment by applying a set of factors which were intended to account for the context and to modify the generic estimate of the HEP accordingly. Unfortunately, as was almost immediately recognized, such an ad hoc application of factors produced estimates which were not repeatable even within reasonable uncertainty bounds. This lack of repeatability occurred with experienced analysts not only from analyst to analyst, but it also occurred for estimates separated in time by the same analyst analyzing the same events.

The variability in these early estimates should not be surprising since there were often variations in the set of factors applied to the same event, the basis for the factors selected, and the level of applicability (i.e., the modification value applied for each factor). In particular the basis for the selection of a factor and its applied value was sometimes the perceived importance of the factor in assuring a correct action while other times the factor might be selected based upon its importance, but its value would be selected based upon the quality of the factor observed in the particular in-situ environment.

For these reasons, the procedure presented below was developed to address the following issues:

1. The development of a standard set of performance factors to be considered in the detailed analysis of a human error event.
2. The development of a scale of values for these factors which could be applied both to the measurement of the importance of each factor in assuring a correct action in a specific sequence and separately, for measuring the quality of the factor as observed in the in-situ environment as compared to an industry norm.

Performance Influencing Factors

Many factors influence human performance. However from a review of nuclear power generating station operational experience and from numerous simulator and in-situ observations, the following set of factors have been selected as the primary general factors that influence operator performance in a nuclear operational setting. Not all these factors are applicable to evaluation of all events and in some cases particular factors not included might be added to the list. However, this has not been found necessary in the studies performed to date so the following developed set of factors represent a checklist for each event undergoing detailed analysis:

1. Procedures
2. Training/Experience
3. Man-machine interface
4. Interactions/Size of Group
5. Communication
6. Workload
7. Stress

Establishment of Ranking Ranges

For each of these factors a range of values was selected within a full scale value for the above set of influencing factors from both the perspectives of importance and quality. In both cases the full scale span for each factor was selected as 100%, in both cases the sum of the rankings across all categories of factors need not add to 100% for simplicity in application. However, in the case of the importance evaluation the factor value chosen more closely represented a rank and therefore the values chosen in this case are normalized to 100% during the evaluation process.

While the number of categories chosen within the full span was certainly not fixed, it was not completely arbitrary either. Previous experience with evaluations had indicated that useful scale had to include two extreme values and a mid-point value. This implied that the number of categories had to be added and that the minimum was three. From this minimum number additional categories could be added at 5, 7, 9 etc. Whatever the number selected the number had to be large enough to allow for a reasonable discrimination between alternatives analyzed and small enough so as not to be overly specific. On the basis of experience 3, 5, and 7 categories were considered to be viable with 5 preferred. This choice of 5 categories to span the scale automatically indicates for a linear scale that each category should have a range of 20 points. That is 0–20, for the first, 20–40 for the second and so on. For both the importance and quality evaluations a higher category and higher value would indicate a more significant influence. That is a higher importance value assigned to Category 5-Communications, would indicate that in the context of the event being evaluated Communications was seen as more critical to the assurance of a correct action, and a lower value less critical. Similarly a higher quality value for Communications indicated that within the specific plant context the communications loop was better (i.e., tighter) than the industry norm.

Establishment of Rank Setting Descriptors

As has been mentioned, one of the previous difficulties with detailed HR evaluations had been their lack of repeatability and consistency. In order to maintain consistency across events even a single analyst must continually strive to establish the same normative contextual structure from one event to another. To do this across a large set of events, especially if their evaluation is well separated in time, requires continual mental recalibration otherwise the natural tendency is to allow evaluations to drift by calibrating not to a standard norm, but to the secondary calibration provided by pairwise comparison to recent previous evaluations. Even if a single analyst were able to maintain his or her own normative scale some method must be applied to ensure that different analysts would use similar norms for calibration as is required for repeatability.

These issues were addressed by attempting to capture, in a few keywords, the character of the values that would be assigned for both importance and quality evaluations according to the procedure developed norm in each category. These keywords were developed by the analyst asking himself in each category what key words would characterize the extreme of the category and then attaching those words to the category. So for example, consider the case of the importance value of "Procedures" in assuring a particular action. The developer was asked what were the extreme cases. That is, when would procedures be least important and when would they be most important. The former case was judged to be when the action required was very simple, direct, and one that was easily memorized, while in the latter case the action would be indirect, complex, without the possibility of memorization. In the former case Procedures were judged to be of minimal importance, given a 0–20 score, and the key words selected were. "Direct and memorized action." In the latter case Procedures were judged to be of maximum importance, given an 80–100 score, and the key words selected were, "Action without possibility of memorization." The intervening three categories and their associated key words were similarly deduced for Procedure importance as: 20–40, simple deduced (reasoned) action; 40–60, Complex deduced (reasoned) action; and 60–80, Complex sequential action.

For the quality ranking a similar approach was taken so that, "Confusing and/or Ambiguous Procedures" were considered the lowest category and scored 0–20, and "Procedures that were assisted by computer" or some other operator aid that provided step by step feedback were considered the highest and scored 80–100. In a similar fashion all the values given in Tables 2 and 3 were developed. When this approach was applied in accordance with the calculational procedure given in [3] the consistency and repeatability of results improved greatly.

Completion of the Human Reliability Analysis

Once the detailed quantification for Type 1 and Type 3 actions have been obtained the HRA tasks are not complete. The events identified must be reviewed for relevant dependencies and these dependencies must be modeled according to the procedures given in [3]. Quantification runs completed with the detailed human error values and the identified dependencies indicate the significant risk contributors without the consideration of recovery. These risk significant events are then investigated further to identify potential recovery actions which would mitigate the impact of the event occurrence. Recovery actions are Type 5 human actions. These actions are modeled in a fashion similar to Type 3 actions when a unique recovery action is able to be identified as described above using the appropriate Time Reliability Correlation as given in [3]. However when operators must improvise a recovery action, the action must be addressed on particular basis, or no credit for recovery is taken. Finally significant human error events are included in the overall Sensitivity and Uncertainty analysis.

TABLE 2
Influencing Factor Scale Based upon the Importance to the Sequence Being Analyzed

[Note: Each sequence should be understood in terms of the requirements it imposes related to each category of influence]

1—Procedures	
0–20	Direct and memorized action
20–30	Simple deduced (reasoned) action
40–60	Complex deduced (reasoned) action
60–80	Compled sequential action
80–100	Sequence of actions without possibility of memorization
2—Training/Experience	
0–20	Customary action with large time frame
20–40	Customary action simple
40–60	Customary action with short time frame
60–80	Unusual action
80–100	Unusual action with very short time frame
3—Man–Machine Interface	
0–20	Required actions clearly differentiated (distinguished without need for feedback)
20–40	Required actions clearly differentiated (distinguished)
40–60	Required actions with need of simple feedback
60–80	Required actions with need of continuous feedback
80–100	Required actions not differentiated (distinguished) and in need of continuous feedback
4—Interactions/Size of Operations Group	
0–20	Requires isolated simple action
20–40	Requires isolated complex action
40–60	Requires multiple simple action
60–80	Requires multiple sequential action independent of time
80–100	Requires multiple sequential actin with short time frame
5—Communications	
0–20	Communications not required
20–40	Direct indication for action in control room, feedback to control room on a deferred basis

40–60	Communications required from control room to plant and vice versa but immediate feedback	
60–80	Continuous communications required from control room to plant with deferred feedback	
80–100	Special detailed non-routine communication required (requires significant discussion to determine proper action)	
6—Workload		
0–20	Simple routine with large time frame	
20–40	Simple actions with short time frame	
40–60	Various actions with large time frame	
60–80	Various actions with short time frame	
80–100	Sequential actions with short time frame	
7—Stress		
0–20	Requires action similar to normal operation	
20–40	Requires action after the recovery of important function	
40–60	Requires action after parameters have reached potential Emergency conditions, but with large time frame	
60–80	Requires actions after parameters have reached potential Emergency condition, but with short time frame	
80–100	Requires action during serious Emergency Conditions	

TABLE 3
Influencing Factor Scale Based upon Quality

1—Procedures		
0–20	No Procedure or confusing and/or ambiguous type procedure	
20–30	Unidrectional Procedure, without check points	
40–60	Multidirectional Procedure, with check points	
60–80	Symptom based procedures	
80–100	Procedure implementation assisted and monitored via a computer aid	
2—Training/Experience		
0–20	Without training or experience	
20–40	Informally instructed during training	
40–60	Within the instruction program	
60–80	Trained or in simulator/analyzer	

| 80–100 | Experienced in genuine conditions |

3—Man–Machine Interface

0–20	Controls and/or annunciators not easily accessible
20–40	Ergonomics not quite adequate
40–60	Adequate ergonomics without integration of information
60–80	Adequate ergonomics with integration of information
80–100	Advanced aid systems to group operations

4—Interactions/Size of Operations Group

0–20	One Isolated operator
20–40	Operator with shared supervision
40–60	Operator with supervision
60–80	Many operators with supervision
80–100	Existence of technical support group

5—Communications

0–20	Ambiguous communications
20–40	Continuous communications required from control room to plant
40–60	One time only communication from control room to plant
60–80	Communications continued to control room
80–100	Communications not necessary

6—Work Load

0–20	Local Actions in adverse conditions or with special skills/tools
20–40	Primarily local actions routine conditions
40–60	Actions primarily within the control room
60–80	Actions confined to the control console
80–100	Single action on the control console

7—Stress

0–20	Serious emergency condition
20–40	Emergency conditions
40–60	Potential emergency conditions
60–80	Low activity conditions
80–100	Normal activity conditions

Conclusions

Human error is, and will continue to be an important contributor to plant operational risk. It is therefore important for its risk contribution to be properly characterized so that decision makers can be presented with an accurate picture of not only the magnitude of global contribution but of the individual elements. Only in this way can a reasonable approach towards human error risk management be developed. Properly capturing the context of the human error contributions has been shown to be critical to properly characterizing the risk and identifying significant risk contributing elements. This paper has attempted to address the problem, and hopefully has contributed thereby towards more effective human error risk management.

ACKNOWLEDGMENT

The approach presented here is an enhancement of that given in [3] as developed by the author with contributions by other over the intervening years since the reference was published. Significant contributions were made by E.M. Dougherty in identifying context as a key element of an effective analysis[13], and by Juan Muñoz of Empresarios Agrupados in Spain who is the co-developer of the context factor ranking approach. Also recently Sergei Kuzin of the Kola units in Russia added comments which suggested modification to the initial factor lists.

References

[1] Swain, A.D., Altman, J.W., and Rook, L.W., "Human Error Quantification: A Symposium," SCR-610, Sandia National Laboratory, Albuquerque, NM, April 1963.
[2] Swain, A.D., and Guttmann, H.E., *Handbook of Human Reliability Analysis with Emphasis on Nuclear Power Plant Applications*, NUREG/CR-1278, Sandia National Laboratory, August 1983
[3] Dougherty, E.M., Jr. And Fragola, J.R., *Human Reliability Analysis: A Systems Approach with Nuclear Power Plant Applications*, Wiley, New York, 1988.
[4] U.S. Nuclear Regulatory Commission, "PRA Procedures Guide—A Guide to the Performance of Probabilistic Risk Assessment for Nuclear Power Plants," NUREG/CR-2300, Washington DC, December 1982.
[5] U.S. Nuclear Regulatory Commission, "Reactor Safety Study—An Assessment of Accident Risks in U.S. Commercial Nuclear Power Plants," WASH-1400 (NUREG-75/014), Washington DC, October 1975.
[6] American Physical Society, "Special Report of Review Group of the Reactor Safety Study," *Reviews of Modern Physics*, 1974, APS, New York.
[7] Union of Concerned Scientists, "Review of the Reactor Safety Study—Draft," 1974, Washington D.C.

[8] Swain, A.D., "Human Reliability in Nuclear Power Plants," Monograph SCR-69-1236, Sandia National Laboratory, Albuquerque, NM, April 1969.
[9] Hannaman, G.W., Spurgin, A.J., and Fragola, J.R., "Systematic Human Action Reliability Procedure (SHARP)," Interim Report, NP-3583, Electric Power Research Institute (EPRI), June 1984.
[10] Muñoz, J., The Human Reliability Results for the Ascó NPP PSA, Proccedings of PSAM II Converfence, Beverly Hills, CA, 1992.
[11] Muñoz, J. and Fragola, J.R., The Human Reliability Analysis for Ascó NPP-PSA, Proceedings of the 17th Inter-RAM Conference for the Electric Power Industry, Hershey, PA, 1990.
[12] Muñoz, J. and Evans, M.G.K, Evaluation of the Contribution of Unrevealed Human Errors to Core Damage Following a Transient, Proceedings of the PSA 1989 Conference, Pittsburg, 1989, pp 136-141.
[13] Dougherty, E.M., Jr., "Context and Human Reliability Analysis," *Reliability Engineering and System Safety, 41* (1993) 25-47, 095-8320/93, 1993 Elsevier Science Publishers Ltd., England.

Risk-Based Evaluation and Human Factors Review for Petrochemical Unit Startup Options

Philip M. Myers and Philip G. Brabezon

Four Elements Inc., 355 East Campus View Blvd., Columbus, Ohio, USA
Four Elements Ltd., 8 Cavendish Square, London, W1M 0ER, UK

ABSTRACT

Nearly twenty-five percent of the largest process industry financial losses occur during startup. Human factors clearly play a significant role in the startup process where many human actions and interactions take place. Risk based evaluations and human factors reviews have been conducted for startup of upgraded petrochemical process units in large complexes. These units previously had been started up and operated from local control rooms with analog displays and pneumatic controls. As part of facility wide automation projects, the controls were to be upgraded and the units controlled with distributed control systems. Ultimately, these petrochemical units may be started up and controlled from remote central control rooms to address facility siting concerns. An additional question remains if a remote central control room is utilized—are risks lower with startup directly from the remote central control room or utilizing a familiar, "local", team-based startup with future "hot cut-over" to the central control room? A methodology was developed to specifically identify startup option differences which are then quantified in terms of the potential for and magnitude of financial loss.

Introduction and Background

Throughout the petroleum and chemical process industries many changes are taking place as a result of facility siting concerns. Often, existing control rooms are located within or at the perimeters of the units they serve. This historic approach promoted team building and problem solving, and resulted in the operators being close to the equipment operated. This also facilitated easy, first hand, visual inspection of the operating equipment, as well as audible cues. For a number of older facilities the most economic solution to address facility siting concerns may be construction of "remote" central control rooms to place personnel and key controls a greater distance from hazards posed by the process. Yet, this poses a "cultural" change for the operators and engineers who must start up, operate, and shutdown the unit. All training, operational experience, staffing arrangements and teams, and methods of communication are often based on past using a "local" or "field" control room based startup. Therefore, use of remote central control rooms are not a panacea and may not be the best solution for all facilities. There are tremendous differences which must be considered in evaluating change to a "remote" central control room—changes that must be carefully planned for and addressed for successful startup and operation of the process if a central control room option is chosen.

Startup Options and Analysis

Consider the following startup options for a petrochemical unit:

- **"Local" Field Information/Instrument Center Startup**: Unit is started up using a newly installed Honeywell TDC 3000 Distributed Control System (DCS) from a single "six-pack" control panel in the Field Information/ Instrument Center. The unit is then switched over to the Remote Central Control Room (RCCR) one control loop at a time after the unit has been stabilized.
- **"Remote" Central Control Room (RCCR) Startup**: Unit is started up using the newly installed Honeywell TDC 3000 Distributed Control System (DCS) and two control panel six-packs directly from the Remote Central Control Room.

Given the safety concerns often expressed by operations personnel and the increased potential for an accident during startup, planning and analysis of the startup options are important. The focus of this particular example is to identify *differences* between the startup options that can result in hazards and risks to people, the environment, assets, and the business, and to determine the safest startup option.

For this particular unit, the consequences of accidental release are well known and potentially severe. Therefore, the study focused on identification of scenarios can have adverse impacts, paying particular attention to human factors issues expected to be significant during startup. Through identification of the differences between the startup options, and determination of the relative likelihood or impacts of certain hazardous events, the key information for decision making is provided. Therefore, the total, absolute risks are not evaluated—only the differences between the startup options.

Approach

The study is carried out in the following phases:

- A review of available information pertaining to startup of the unit is carried out, and a plan and procedure developed for study of the startup options.

- A multidisciplinary team identifies and details the differences between the two startup options.
- Quantitative estimates are made of the likelihood and/or impact of certain key events identified to aid decision making.

A description of the approach for team identification of startup option differences, using the protocol developed, follows.

Identification of Startup Differences

A multidisciplinary study team, constituting members with process, operational, safety, process control, technical, risk management, and human factors skills, is assembled. The team follows a structured approach, developed by Four Elements, for investigating the significance of differences between the two startup options. The procedure involves three steps:

- Identification of the differences between the two options, including Human Factors differences specifically
- Definition and discussion of the startup activities
- Analysis of the impact of the differences on the startup activities

Human Factors and other differences between the two options which could influence the risks associated with the startup process are identified and catalogued. To structure this task, a basic three stage "system control" model is used. The stages are - Monitor, Interpret/Decide and Control and are connected by "communication" links as shown in *Figure 1*.

Through discussion, and with the assistance of a prompt list of important factors developed, the study team reviews the "real world" components involved in each of the three stages and communication links. The prompt list includes human factors and other issues associated with:

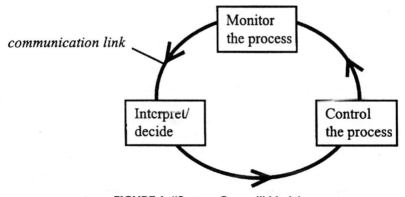

FIGURE 1. "System Control" Model

- displays
- controls
- internal communications
- external communications
- people
- team group interactions
- reference documents
- other interactions
- environment
- hardware
- software

To enable the study team to consider how the differences between the two options could impact risk, the startup activities are divided into primary categories and prioritized based upon the perceived differences in startup risk and complexity. The ranked order of startup activities follows:

- Cold Systems
- Feed and Furnaces
- Utilities
- Crack Gas Compressors
- Fractionation Train

Each of the primary activities is described by operations—including the startup activities included in each category or startup phase, as well as the start and end conditions. Prior to analyzing the potential impact of the differences, the procedure for each startup activity is reviewed in outline form and discussed.

The purpose of doing so is twofold:

- to ensure a common understanding among the study team of the activity, and
- to provoke consideration of the human factors of the activity.

The review and discussion of each startup activity covers:

- the objective of the activity, in terms of the process conditions to be achieved
- the main milestones reached in fulfilling the objective
- the main tasks to be performed. This includes identifying the people involved in the tasks, and the type and frequency of communications.
- any particular problems which could arise during the activity, such as the tendency for key process parameters to deviate from target, critical equipment failures, or leaks from pipework or process equipment.

Through use of the model and prompt list, general differences between the two startup options are revealed. A selection of differences identified are listed in *Table 1*.

TABLE I
Examples of Identified Differences between Startup Options

Issue	No.	"Local" Startup	"Remote" (RCCR) Startup
Displays	1	One TDC six-pack (of control panels).	Two TDC six-packs (of control panels).
	2	Boardmen can "hear" and sense behavior of certain equipment.	Boardmen are too remote from the unit to have any direct sensory feedback.
Controls	3	EIV activation panel not provided at the control panel.	EIV activation at the TDC console.
Internal/External Communication	4	Face-to-face communications between outside operators and boardmen is possible.	Reliance on radio communications between boardmen and operators.
People	5	Additional personnel may be present due to center of activity near the unit, and therefore exposed to explosion risks.	Fewer personnel in the unit and exposed to explosion risks. Boardmen are removed from the unit.
Team Group Interactions	6	Plant operators can easily confer with the boardmen about tasks and responsibilities which were formerly held by the boardmen.	Reliance on experience of the outside operators.
Reference Documents	7	A "master" copy of drawings/procedures can be examined by all.	Notes on RCCR copies of drawings and procedures will not be the same as those made on copies in the unit.
Other Interactions	8	Many people can be in the control room creating distractions.	On-going activities for other units in the RCCR could create distraction.
Environment	9	Full range of alarm sounds can be used, without possibility of confusion.	Similar alarm sounds from other units in the RCCR may create a problem.
Hardware	10	No UPS planned.	UPS for both six-packs.

also - software, etc

For each of the identified key differences between the startup options, possible undesirable outcomes for each startup phase are identified and evaluated in terms of the probability of occurrence and the financial impacts on the business. Due to the confidentiality of the project, the probabilities of occurrence and financial impacts calculated have been genericized.

Table 2 presents a summary of an evaluation carried out for one difference in "displays" planned for the startup options. In the "local" startup option, only one "six-pack" display was planned while in the remote (RCCR) startup two "six-packs" would be used. It was determined that the chance of startup failure is greater than 75% for "local" startup using one "six-pack." That is, the chance of a significant (unacceptable) loss is greater than 75%. The impact of events caused due to an inability to carefully monitor and control the startup activities in the local startup option are as follows:

- Personnel—in the event of a significant release and explosion, there could be severe impacts on personnel, with a number of personnel injured or killed. The liabilities associated with these third party liabilities could be as high as several hundred million to over one billion dollars.
- Equipment—the property damage arising from a significant explosion can run into the hundreds of millions of dollars.
- Environment—impacts on the environment are minor as most of the materials handled are volatile and are not highly toxic or carcinogenic.
- Business—the startup may be delayed, resulting in losses from several hundred thousand to several million dollars.

The cumulative potential loss is clearly unacceptable. However, a simple alternative was identified—purchase of a second "six-pack" display panel at a cost of approximately $250,000 to completely mitigate the problem.

Another significant difference identified is summarized in *Table 3*. Replication of the Emergency Isolation Valve (EIV) panel to be available in the RCCR was not planned in the "local" startup configuration. Therefore, EIV activation could be accomplished only through field mounted panels. Through the use of consequence modeling, human factors data and analysis, and event trees the difference in probability of safe shutdown for each option is evaluated. The resultant chance of non-isolation in an emergency for the "local" startup option is about 70% whereas the chance of non-isolation in the RCCR startup configuration is on the order of 15%. It can be seen from Table 3 that the financial impacts of non-isolation in an emergency are potentially severe, clearly favoring the RCCR startup option.

Other startup differences were identified that clearly favored the "local" startup option. However, in balance, the risks of RCCR startup were determined to be lower. In addition, solutions were identified for the problems revealed in the "remote" startup option which would need to be resolved for ongoing operation and future startups anyway.

The unit has now been successfully started up and operated from the remote central control room.

TABLE 2
Analysis Summary for One "Display" Difference

Specific difference	One six-pack for both "hot"&"cold" sides, vs. Dedicated six-packs for each	
Description	For the local start-up, only one six-pack is installed. Two six-packs would be used in remote start-up from the central control room.	
Scenario	One six pack reduces the ability of the boardmen to monitor the unit. One of the three lower "control" screens must be used to display alarms, leaving five screens to monitor (and control) both hot and cold sides. Inability to properly monitor the unit will make it extremely difficult or impossible to start up. Lack of ability to monitor could also lead to inability to control the unit and lead to hazardous conditions, and a breach of equipment integrity.	
Mitigation factors	None	
Likelihood (per phase)	Utilities	Not significant, due to low level of monitoring needed from the boardmen.
	Cold systems	Extremely high likelihood of monitoring difficulties with only one six-pack. Likelihood of startup failure estimated to be 75% greater than with two six-packs.
	Feed & furnaces	as for cold systems
	Compression	as for cold systems
	Fractionation	as for cold systems
Impacts	Personnel	If a significant release an explosion occur, there could be severe impacts on personnel, with a number of personnel injured or killed. The liabilities associated with these third party liabilities could be as high as several hundred million to over a billion dollars.
	Equipment/ Assets	The property damage arising from a significant explosion can run into the hundreds of millions of dollars.
	Environment	Impacts on the environment are expected to be minor, as most of the materials handled are volatile—those that are not volatile, are not highly toxic or carcinogenic, and are in contained process areas. In addition, the site is not surrounded by any particularly sensitive environments.
	Business	The startup may be delayed costing hundreds of thousands to tens of millions of dollars in lost profits alone.
Conclusion		The risks to personnel, assets, and business are high. Therefore, the unit should not be started up on one six-pack. The local startup could still be a viable alternative with purchase of an additional six-pack at a cost of $250M.

TABLE 3
Analysis Summary for One "Controls" Difference

Specific difference	Equipment EIVs not available in the local startup versus they are available in the RCCR.	
Description	It is not planned to reproduce the EIV panel in the local startup option. Therefore, for the local startup option, the only method of EIV activation is through field mounted panels close to the equipment.	
Scenario	A major release occurs, requiring equipment activation of the EIVs.	
Mitigation factors *(human factor)*	In the local startup option, it is possible that a person could be stationed in the RCCR on standby, with the sole task of responding to a request to activate one or more EIVs—or a radio message could be conveyed to an RCCR boardman from another unit to use the panel.	
Likelihood (per phase)	Utilities	Little effect as there are for the most part no hydrocarbons in the unit during this startup phase.
	Cold systems	• Local startup option: the probability estimate for non-isolation, in sufficient time to prevent escalation, is 0.70 • Remote startup option: the probability estimate for non-isolation, in sufficient time to prevent escalation, is 0.15
	Feed & furnaces	as for cold systems
	Compression	as for cold systems
	Fractionation	as for cold systems
Impacts	Personnel	Multiple persons could be injured or killed in the event a hazardous situation is escalating and the appropriate EIV(s) are not activated. The third party liabilities could be as high as several hundred million to over a billion dollars.
	Equipment/ Assets	Significant property damage could occur, on the order of tens or hundreds of millions of dollars.
	Environment	No significant impact expected.
	Business	The business interruption costs could be significant, easily running to the hundreds of thousands and millions of dollars—and a maximum potential of tens of millions of dollars.
Conclusion		The personnel, asset, and business interruption risks are significantly higher for the "local" startup option, with the chance of being able to successfully isolate the process before a major accident occurs one-third of the predicted success rate for activation from the RCCR. This clearly favors the RCCR startup option.

Summary

A novel, systematic approach was developed to identify and evaluate human factors differences in process unit startup. Through the course of such studies, the project teams have developed a much more cohesive and complete understanding of the configuration and details of each startup option. Whereas the startup options sometimes initially appeared very similar, use of the analysis technique enabled the teams to identify and evaluate considerable differences in human factors and other aspects of the planned startups.

A significant finding was revealed early in the analysis presented here—namely that the unit could not be safely started up in the "local" option as planned. However, an immediate solution was identified at a reasonable cost so as to not eliminate this startup option. In fact, a number of solutions were identified for each of the problems identified for each startup option. Therefore, either startup option could be used safely, though each required a different level of investment. Ultimately, the remote central control room option was determined to be the best alternative, posing the lowest startup risk. In addition, numerous improvements were identified and outlined for incorporation in the "remote" startup option.

The analysis proved to be a powerful tool in identifying and quantifying human factors and other issues associated with process unit startup and control room siting. As the risks are presented in terms of the financial impacts on the business, it also provides a tremendous decision making tool for site and company business managers.

References

[1] Mudan, Dr. Krishna S., Shah, Jatin N., and Myers, Philip M., "Financial Risk Assessment A Uniform Approach to Manage Liabilities," AIChE Summer National Meeting, Boston, MA, August, 1995.

[2] Myers, Philip M. and Mudan, Krishna S., "A Financial Risk Assessment Case Study—Tank Truck Transportation of Chemicals and Petroleum Products," AIChE Spring National Meeting, New Orleans, LA, 1996.

Summary

A novel systematic approach was developed to identify and evaluate human factors differences in process interruptions. Through the course of such studies, the project teams have developed a much more cohesive and complete understanding of the configuration and details of each startup option. Whereas the startup options scenarios initially appeared very similar, use of the analysis technique enabled the teams to identify and evaluate considerable differences in human factors and other aspects of the planned startups.

A significant finding was made. Early in the analysis presented here, panels that the unit could not be safely started up in the "local" option as planned. However, no immediate solution was identified at a reasonable cost as to not eliminate this startup option. In fact, a number of options were identified for each of the problems identified for each startup option. These new safer startup options could be used safely, though each required a different level of investment. Ultimately, the remote central control room option was determined to be the best alternative, posing the lowest startup risk. In addition, numerous improvements were identified and utilized for incorporation in the "remote" startup option.

The analysis proved to be a powerful tool in identifying and quantifying human factors and other issues associated with process unit startup and control room sitting. As the risks are presented in terms of the financial impact on the business, it also provides a tremendous decision resource for staff and company business managers.

References

Arendt, Vincent Dr., Kodner, F., Saah, Jean N., and Myers, Philip M., "Financial Risk Assessment: A Uniform Approach to Manage Liability," AIChE Summer National Meeting, Boston, MA, August 1995.

Myers, Philip M. and Kodner, Krishna S., "A Financial Risk Assessment Case Study of Tank Truck Transportation of Chemicals and Petroleum Products," AIChE Spring National Meeting, New Orleans, LA, 1996.

New Consequence Modeling Chapter for *Guidelines for Chemical Process Quantitative Risk Analysis*, Second Edition

Daniel A. Crowl
Department of Chemical Engineering, Michigan Technological University, Houghton, MI 49931

ABSTRACT

The upcoming second edition of *Guidelines for Chemical Process Quantitative Risk Analysis* will contain a new Chapter 2 on Consequence Analysis. The reasons for a major update to this chapter are (1) to provide much more detail than available in the original edition, (2) to update the models based on improvements in modeling technology, (3) to provide many more worked examples, and (4) to provide spreadsheet implementation of the examples, available on a disk. The chapter discusses the theoretical basis for the models, and provides a "how to" framework for application of the models to risk analysis.

The outline for the new chapter is unchanged from the original.

This paper will discuss the new changes, and show some of the example spreadsheets.

The expanded chapter is a valuable resource for meeting EPA Risk Management Plan (RMP) consequence modeling requirements. The chapter is also an excellent supplement to more detailed CCPS books on vapor cloud dispersion, explosions, etc.

Introduction

The book *Guidelines for Chemical Process Quantitative Risk Analysis* (AICHE, 1989), frequently denoted as the "CPQRA book," was originally published in 1989 by the Center for Chemical Process Safety of AICHE. The book was written by Technica and the Risk Assessment Subcommittee of the AICHE Center for Chemical Process Safety. As stated in the preface of the original edition, this book "builds on the *Guidelines for Hazard Evaluation Procedures* (AICHE, 1985, 1992) to show the engineer how to make quantitative risk estimates for the hazards identified by the techniques given in that volume." The preface also states that "the primary goal of CPQRA is to provide tools for reducing risks in chemical plants handling hazardous materials." The book was directed toward the analysis of acute hazards, not chronic health effects. The book achieved these goals and has become one of the dominant references for CPQRA.

An outline of the original CPQRA book is shown in Table 1.

TABLE I
Outline of Original CPQRA Book

Chapter	Title
1	Chemical Process Quantitative Risk Analysis
2	Consequence Analysis
3	Event Probability and Failure Frequency Analysis
4	Measurement, Calculation, and Presentation of Risk Estimates
5	Creation of CPQRA Data Base
6	Special Topics and Other Techniques
7	CPQRA Application Examples
8	Case Studies
9	Future Developments
Appendix	**Title**
A	Loss of Containment Causes in the Chemical Industry
B	Training Programs
C	Sample Outline for CPQRA Results
D	Minimal Cut Set Analysis
E	Approximation Method for Quantifying Fault Trees
F	Probability Distributions, Parameters, and Terminology
G	Statistical Distributions Available for Use as Failure Rate Models
H	Errors from Assuming That Time-Related Equipment Failure Rates Are Constant
I	Data Reduction Techniques: Distribution Identification and Testing Methods
J	Procedure for Combining Available Generic and Plant-Specific Data

An important part of the original text was Chapter 2 on *Consequence Analysis*. This chapter is the largest chapter in the text, representing 125 pages of the 585 pages in the book. The purpose of this chapter was to "provide an overview of consequence and effect models commonly used in CPQRA." An outline of chapter 2 is shown in Table 2.

Since the publication of the original CPQRA book in 1989, much has occurred in the area of consequence models. A number of symposia (AICHE, 1991; AICHE, 1995; LPS, 1989-1997), books (Crowl and Louvar, 1991; AICHE, 1994; AICHE, 1996; Lees, 1996), and journals (*Process Safety Progress, Journal of Loss Prevention in the Process Industries*) have advanced the technology of consequence analysis. Furthermore, risk analysis has advanced considerably during these years and considerable industrial experience in the application of these techniques has been accrued.

TABLE 2
Outline of Chapter 2 on Consequence Analysis

\multicolumn{2}{l}{The revised chapter has a separate section on jet fire models.}

Section	Title
2.1	Source and Dispersion Models
2.1.1	Discharge Rate Models
2.1.2	Flash and Evaporation
2.1.3	Dispersion Models
2.1.3.1	Neutral and Positively Buoyant Plume and Puff Models
2.1.3.2	Dense Gas Dispersion Models
2.2	Explosions and Fires
2.2.1	Vapor Cloud Explosions and Flash Fires
2.2.2	Physical Explosion
2.2.3	BLEVE and Fireball
2.2.4	Confined Explosions
2.2.5	Pool Fires and Jet Fires
2.3	Effect Models
2.3.1	Toxic Gas Effects
2.3.2	Thermal Effects
2.3.3	Explosion Effects
2.4	Evasive Actions
2.5	Modeling Systems
2.6	References

In 1994, the Risk Assessment Subcommittee decided that a major rewrite of the CPQRA book was in order. The committee further concluded that most of the effort should be directed toward a complete revision of Chapter 2. This revision should 1) provide much more detail on consequence models, including more models and a more complete presentation on the fundamental basis, 2) update the models based on improvements and experience in modeling technology, and 3) provide many more worked examples. It was subsequently decided to implement all of the worked examples by spreadsheet and to provide a disk containing the spreadsheets with the book.

TABLE 3
Comparison of the Original and Revised Chapter 2

Item	Original Chapter	Revised Chapter (Estimated)
Total Pages:	125	200+
Figures:	34	86
Tables:	15	33
Equations:	47	153
Examples:	17	39
Spreadsheets:	0	43

The outline of the original Chapter 2 was maintained, with the exception that a separate section on jet fire models was included. The sections in the revised chapter retain the structure of the original. Each modeling section contains a presentation of the purpose, philosophy, applications, description of the technique, a logic diagram, theoretical foundation, input requirements and availability, output, simplified approaches, and sample problems. A discussion section for each modeling sections contains a presentation on strengths and weaknesses, identification and treatment of possible errors, utility, resources needed, and available computer codes.

Table 3 provides a comparison of the previous and revised versions of Chapter 2. The new work represents more than a doubling of the content. The page count, number of figures and tables, number of equations, and number of worked examples have significantly increased.

A significant feature of the new Chapter 2 is the worked examples and the spreadsheet implementation of the examples. The purpose here was to provide a set of common modeling tools that could be easily used in CPQRA. Table 4 is a summary of the 35 worked examples and spreadsheets in the new chapter. These spreadsheets could be readily used to meet the new requirements of the EPA Risk Management Plan (RMP). A detailed, manual solution to each example, with explanation and discussion, is provided in the text.

Appendix A contains a complete example problem, including the spreadsheet output. This example draws the isopleths for a plume, given a continuous release of material. A statement of the example and a discussion of the solution method is included.

Summary

The new version of Chapter 2 represents a considerable upgrade in content over the original material. A major new feature is the inclusion of a disk containing the spreadsheet solutions for all of the worked examples.

New Consequence Modeling Chapter for Guidelines for CPQRA, Second Edition

Table 4 *Disk or CDRom in Excel or Quattro pro*
Spreadsheets Provided with Revised Chapter 2

Spreadsheet	Problem
2.1	Liquid Discharge through a Hole in a Tank
2.2a	Liquid Trajectory from a Hole
2.2b	Maximum Discharge Distance from a Hole in a Tank
2.3	Liquid Discharge through a Piping System
2.4	Gas Discharge through a Hole
2.5	Gas Discharge through a Piping System
2.6	Two-Phase Flashing flow through a Pipe
2.7	Gas Discharge Due to External Fire
2.8	Isenthalpic Flas Fraction
2.9	Boiling Pool Vaporization
2.10	Evaporating Pool
2.11	Pool Evaporation using Kawamura and MacKay Direct Evaporation Model
2.12	Pool Spread via Wu and Schroy (1979) Model
2.13	Plume Release #1
2.14	Plume Release #2
2.15a,b	Puff Release
2.15c	Puff Release
2.15d	Puff Release
2.16	Plume with Isopleths
2.17	Puff with Isopleths
2.18	Britter MacQuaid Method *Dense Gas*
2.19	Blast Parameters
2.20	TNT Equivalency of a Vapor Cloud
2.21a	TNO Multi Energy Model *for VCE* — *Figures are auto inputted*
2.21b	Baker Strehlow Vapor Cloud Explosion Model
2.22	Energy of Explosion for a Compressed Gas
2.23	Prugh's Method for Overpressure from a Ruptured Sphere
2.24	Baker's method for Overpressure from a Ruptured Vessel
2.25	Velocity of Fragments from a Vessel Rupture
2.26	Range of a Fragment in Air
2.27	BLEVE Thermal Flux
2.28	Blast Fragments from a BLEVE
2.29	Overpressure from a Combustion in a Vessel
2.30	Radiation from a Burning Pool
2.31	Radiant Flux from a Jet Fire
2.32	Probit Correlation to Dose Curve
2.33	Determining the Percentage of Fatalities Given a Fixed Concentration-Time Relationship
2.34	Determining the Percentage of Fatalities for a Moving Puff
2.35	Thermal Flux Estimate Based on 50% Fatalities
2.36	Fatalities due to Thermal Flux from a BLEVE Fireball
2.37	Effect of Blast Overpressure
2.38	Effect due to Projectiles
2.39	Estimation of Evacuation Failure

The revised chapter represents the considerable collective wisdom of practicing professionals as embodied in the CCPS Risk Assessment Subcommittee. The new CPQRA book is expected to be published in early 1998.

References

1. AICHE (1985), *Guidelines for Hazard Evaluation Procedures*, American Institute of Chemical Engineers, New York.
2. AICHE (1989), *Guidelines for Chemical Process Quantitative Risk Analysis*, American Institute of Chemical Engineers, New York.
3. AICHE (1991), *International Conference and Workshop on Modeling and Mitigating the Consequences of Accidental Releases of Hazardous Materials*, American Institute of Chemical Engineers, new York.
4. AICHE (1992), *Guidelines for Hazard Evaluation Procedures, 2nd Edition with Worked Examples*, American Institute of Chemical Engineers, New York.
5. AICHE (1994), *Guidelines for Evaluating the Characteristics of Vapor Cloud Explosions, Flash Fires, and BLEVEs*, American Institute of Chemical Engineers, New York.
6. AICHE (1995), *International Conference and Workshop on Modeling and Mitigating the Consequences of Accidental Releases of Hazardous Materials*, American Institute of Chemical Engineers, New York.
7. AICHE (1996), *Guidelines for Use of Vapor Cloud Dispersion Models*, 2nd Ed., American Institute of Chemical Engineers, New York.
8. Crowl, D. A. and Louvar, J. F. (1990), *Chemical Process Safety, Fundamentals with Applications*, Prentice Hall, New Jersey.
9. Lees, F. P. (1996), *Loss Prevention in the Process Industries*, 2nd Ed., Butterworth-Heinemann, London.
10. LPS (1989-1997), *Loss Prevention Symposium Proceedings*, American Institute of Chemical Engineers, New York.

APPENDIX
Sample Problem with Spreadsheet Output

Example 2.16: Plume with Isopleths

Develop a spreadsheet program to determine the location of an isopleth for a plume. The spreadsheet should have specific cells for the following inputs:

- release rate (gm/sec)
- release height (m)
- spatial increment (m)
- wind speed (m/s)
- molecular weight
- temperature (K)

- pressure (atm)
- isopleth concentration (ppm)

The spreadsheet output should include, at each point downwind:

- dispersion coefficients, both , (m)
- downwind centerline concentrations (ppm)
- isopleth locations (m)

The spreadsheet should also have cells providing the downwind distance, the total area of the plume, and the maximum width of the plume, all based on the isopleth value.

Use the following case for computations, and assume worst case stability conditions:

Release rate:	50 gm/sec
Release height:	0 m
Molecular weight:	30
Temperature:	298 K
Pressure:	1 atm
Isopleth conc:	10 ppm

Solution

The spreadsheet output is shown in Figure A1. Only the first page of the spreadsheet output is shown. The following notes describe the procedure:

1. The downwind distance from the release is broken up into a number of spatial increments, in this case 10-m increments. The plume result is not dependent on this selection, but the precision of the area calculation is.
2. The equations for the dispersion coefficients are fixed based on stability class, in this case F-stability. These columns in the spreadsheet would need to be redefined if a different stability class is required.
3. The dispersion coefficients are not valid at less than 100 m downwind from the release. However, they are assumed valid to produce a complete picture back to the release source.
4. The isopleth location is found by dividing the equation for the centerline concentration, i.e., $\langle C \rangle(x,0,0,t)$, by the general ground level concentration. The resulting equation is solved for y to give

$$y = \sigma_y \sqrt{2\ln\left(\frac{\langle C \rangle(x,0,0,t)}{\langle C \rangle(x,y,0,t)}\right)}$$

where y is the off-center distance to the isopleth (length), $\langle C \rangle(x,0,0,t)$ is the downwind centerline concentration (mass/volume), and $\langle C \rangle(x,y,0,t)$ is the concentration at the isopleth.

Example 2.16: Plume with Isopleths

USER SPECIFIED VALUES:
Release Rate:	50 gm/sec
Release Height:	0 m
Increment:	10 m
Wind Speed:	2 m/sec
Molecular Weight:	30
Temperature:	298 K
Pressure:	1 atm.
Isopleth Conc:	10 ppm

Assumed Stability Class: F   Rural

CALCULATED VALUES:
Max. plume width:	37.34 m
Total Area:	66461 m**2

Distance Downwind (m)	Dispersion Coeff. Sigma z (m)	Dispersion Coeff. Sigma y (m)	Downwind Centerline Concentration (gm/m**3)	Downwind Centerline Concentration (mg/m**3)	Downwind Centerline Concentration (ppm)	Isopleth Location (m)	Negative	Area (m**2)
0	0	0				0	0	0.0
10	0.16	0.40	124.775	124775.2	101695.5	1.7	-1.7	17.2
20	0.32	0.80	31.303	31302.7	25512.7	3.2	-3.2	31.7
30	0.48	1.20	13.961	13960.8	11378.4	4.5	-4.5	45.0
40	0.63	1.60	7.880	7880.2	6422.6	5.7	-5.7	57.4
50	0.79	2.00	5.061	5060.8	4124.7	6.9	-6.9	69.2
60	0.94	2.39	3.527	3526.6	2874.3	8.1	-8.1	80.5
70	1.10	2.79	2.600	2599.9	2119.0	9.1	-9.1	91.3
80	1.25	3.19	1.997	1997.4	1627.9	10.2	-10.2	101.7

FIGURE A1. Spreadsheet Output for Example Problem 2.16: Plume with Isopleth

Equation (1) applies to ground level and elevated releases.
The procedure to determine an isopleth at any specified time is:
a. Specify a concentration, $\langle C \rangle^*$ for the isopleth.
b. Determine the concentrations, $\langle C \rangle(x,0,0,t)$, along the x-axis directly downwind from the release. Use the following equation for the downwind centerline concentration:

$$\langle C \rangle(x, 0, 0) = \frac{G}{\pi \sigma_y \sigma_z u}$$

where $\langle C \rangle(x,0,0)$ is the average concentration (mass/volume), G is the continuous release rate (mass/time), σ_y, σ_z are the dispersion coefficients in the y and z directions (length), u is the wind speed, (length/time)
Define the boundary of the cloud along this axis.
c. Set $\langle C \rangle(x,y,0,t) = \langle C \rangle^*$ in Equation (1) and determine the value of y at each centerline point determined in step b. Plot the y values to define the isopleth using symmetry around the centerline.
5. The plume is symmetric. Thus, the plume is located at .

6. The plume area is determined by summing the product of the plume width times the size of each increment.
7. The maximum plume width is determined using the @MAX function (for Quattro Pro, or its equivalent function in other spreadsheets).
8. For the maximum plume width and the total area, specific cell numbers must be summed for each run.

Benefits of Plant Layout Based on Realistic Explosion Modeling

Jan C. A. Windhorst
Strategic Initiatives, NOVA Chemicals, Red Deer, Alberta T4N 6A1

ABSTRACT

NOVA Chemicals has participated on the CCPS guideline for building siting and was also one of the first companies in North America to publish its risk criteria for fatalities. Design decisions as they come up during plant design need to be safety and economic risk based. Factual hazard awareness and realistic explosion modeling are essential for an informed risk based design decision making process.

This knowledge-based approach requires a multi-disciplined input into the hazard identification process as well as a willingness to reject consequence analysis methods, which are deemed inadequate. The prime purpose of consequence analyses is its use as a realistic design tool, thereby enhancing inherent safety. For this reason vapour cloud explosions cannot be modeled with TNT equivalency methods. The secondary purpose is as a process safety tool to provide input into the Quantitative Risk Analysis.

This paper presents a design study that shows some of the benefits to buildings and certain utility structures that can be accrued with this approach.

Introduction

NOVA Chemicals operates at present two world-scale ethylene plants (E1 and E2) in Joffre, Alberta, with a third plant, E3 (see Figure 1), slated to come on-line around the turn of the millennium. The E3 project is a joint venture with Union Carbide Corporation and once completed, the Joffre site is expected to be the "ethylene capital" of the world. Feedstock is ethane, which is removed from natural gas produced in the Province of Alberta. The vertical integration of the Alberta ethylene economy makes Joffre the focal point for feedstock suppliers and ethylene consumers. This concentration has infra-structural advantages but also carries certain risks; not only to NOVA Chemicals but also to many other stakeholders, such as suppliers, customers, risk financiers, revenue-gathering governments, etc. Risk control and management, especially the avoidance of potential domino effects is of prime interest.

Ethane based ethylene production is a "clean" gas-phase process; byproducts are minimal and mainly LPG type products. Pool fire hazards are therefore

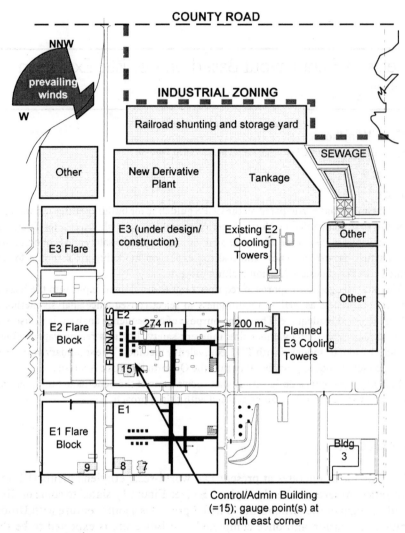

FIGURE 1. Joffre Plant Site

of a lesser concern; on the other hand, jet fires, flash fires and vapour cloud explosions (VCEs) are considered to be main hazards. This paper will focus exclusively on the hazards created by VCEs and does not cover jet and flash fires.

Inherent safety emphasizes the need of early decision making in order to arrive at cost effective solutions. We wanted, therefore, to expedite our quantitative risk analysis (QRA) and explosion consequence modeling for E3 as much as possible. A number of important process safety issues were identified so they could be addressed early on during the conceptual design/process development stage and generate specifications for the detailed design and the development of the plot plan.

Scenarios Development

Initial Scenarios

From the beginning the E3 project aimed at adopting the basic ideas behind the E2's design and layout, since E2 has had a great safety and operational record. E2's basic layout is a "T-shape" which has been rotated counter clockwise over 90 with most of the equipment located along the west-east piperack. Where E3 data was not available we substituted it with E2 data, for postulating and modeling failure scenarios which reasonably speaking would be applicable to E3 and could therefore be incorporated into E3's design.

The following "high level" issues were identified:

- can the "T"-shape layout cause excessive over-pressures in the piperack area, as identified by Baker et al. (see [1]);
- can the E2 cooling towers just east of E3 and the E3 cooling towers just east of E2 be the cause of domino effects;
- criteria for building location;
- can the storage and shunting railroad (RR) yard facilities be a source of high over-pressures? A problem in the RR yard can have an impact on the derivative plant north of E3 with subsequent ramifications for E3 itself, since the RR yard cannot be moved further north without violating zoning restrictions.

Initial Modeling

Modeling started with ethylene dispersion calculations, using PHAST, to find out how *far* clouds could travel when fed from available inventories. Dispersion in an actual process plant is hampered by confinement, while turbulence generating obstacles aid it. This makes it difficult to determine even the approximate isochoric footprint of a dispersed cloud in a process plant. We therefore postulated stoichiometric pancake clouds for consequence modeling purposes. The first two scenarios were investigated by defining a stoichiometric pancake cloud that engulfed the (E?) east-west piperack and adjacent equipment.

Dimensions of the piperack: (1st level) height * width * length = 7.5 * 12.7 * 274 m. Identical dimensions were selected for the equipment to the north and to the south of the piperack. An arbitrary void fraction of 80 percent was given to the three volumes to correct for equipment, resulting in a stoichiometric cloud with a volume of $3 \times 21{,}000$ m^3. Application of the Baker method (see [1]) yielded the following results:

TABLE 1
Baker Method: Piperack and Adjacent Equipment—Ethylene with Some Hydrogen (High Reactivity); No Confinement (3D); and a High Value for Obstacles

Cloud volume = 63,000 m³; Ground reflection; Flammable mass = 4,900 kg 50 percent of flammable mass contributes to percussion		
Flame speed = 0.588 mach		
Over-pressure (bar-g)	Distance (m)	Impulse (kPa ms)
1	43	4,431
0.5	95	2,059
0.3	270	—
0.1	405	—

TABLE 2
Baker method: Piperack Only—Ethylene with Some Hydrogen (high Reactivity); 2D; and a High Value for Obstacles

Cloud volume = 21,000 m³; Ground reflection; Flammable mass = 1,650 kg 50 percent of flammable mass contributes to percussion		
Flame speed = 1.765 mach		
Over-pressure (bar-g)	Distance (m)	Impulse (kPa ms)
5	20	5,027
1	30	2,793
0.5	66	1,396
0.17	187	—
0.1	280	—

Application of TNO's multi-energy method (see [2]) gave:

TABLE 3
TNO Multi-Energy Method: Piperack and Adjacent Equipment Data Generated by PHAST™

Distance (m)	Confined source 1 = piperack (21,000 m³) Strength: 9		Confined source 2 = equipment adjacent to piperack (42,000 m³) Strength: 6	
	Over-pressure (bar-g)	Pulse duration (ms)	Over-pressure (bar-g)	Pulse duration (ms)
25	2.6	32	—	—
50	1.4	49	0.5	126
75	0.6	62	0.4	126
100	0.4	71	0.4	130
200	0.1	89	0.2	137

The results were viewed with a certain skepticism and with trepidation. In order to rationalize these rather abstract results with a "physical" picture we considered a number of scenarios and their ramifications, including:

1. A flammable ethylene cloud is ignited by the fired heaters at the west end of the plant, creating a flame front that accelerates eastwards through the plant causing high over-pressure at the east end of the piperack. Such an incident could potentially bring down E2 and E3 or more than 70 percent of the future Joffre ethylene production; and
2. Inherent safety dictates that occupied buildings should be *away* from process hazards but where exposed to those hazards staffing levels should be minimized. *Smaller inventories* are *safer* and *cheaper*, this applies not only hazardous materials but also to people. These reduced staffing levels are not always possible for reasons such as organizational effectiveness, etc. Where occupied buildings need to be close to a process plant they need to be designed for the risks at hand (see [3]). Under these circumstances we like to apply considerations such as:

- buildings should be upwind of flammable release sources, i.e., in the west or in the north-west corners of E2 and E3,
- buildings should be in open areas where a flamefront will decelerate,
- buildings should be outside the hazard zone of fired heaters (firebox explosion),
- buildings should be separated from hydrocarbon containing process units by the likeliest ignition sources, i.e., E2's and E3's fired heaters in order that they will:
  —shield buildings from a westbound flamefront, and
  —in case they ignite a cloud themselves, create a flamefront that will move away from the building(s).

It is virtually impossible to substantiate the given scenarios/hypotheses with the gathered data. We need therefore a mathematical tool that:

- accounts for layout, obstacles and location of ignition source;
- calculates results as a function of actual *location* and *time*.

Computational fluid dynamics or CFD explosion modeling satisfies these requirements (see [2] and [4]). CFD modeling would also provide a more fundamental answer to the questions concerning building siting, building design requirements and RR yard layout. CFD modeling is coming of age in the chemical industry (see [5]) and is used for applications as diverse as room ventilation (see [6]) and explosion modeling (see [2], [4], and [7]).

CFD Modeling

GENERAL

Conceptual CFD modeling was done with AutoReaGas (see [8]), which performs three-dimensional explosion modeling. Input files for all conceptual design AutoReaGas simulations were prepared with the aid of digital images and drawings of E2, cooling towers (CT), RR tanker cars and RR hopper cars. Engineering consultants were retained to create the 3D CAD model and perform the simulation work using an AlphaStation 600$^{5/266}$ workstation and other computational resources. Final representation of the different geometries, used as a database, was approved by NOVA.

PLANT LAYOUT, COOLING TOWERS, AND BUILDINGS

The main objective was to study the effects of an explosion in E2 and its effects on an E2 building and cooling towers, which had been simulated east of E2. Results of this study were considered applicable to:

- the effects of a major vapour cloud explosion (VCE) in E2 on E3's (future) cooling towers;
- the effects of a major VCE in E3 on the existing E2 cooling towers; and
- building siting, i.e., is the open area near fired heaters appropriate (?), which is intended to be the basis for future and more detailed building design calculations.

The following items were defined:

- a numerical mesh consisting of 234,000 cells to represent the ethylene plant and cooling tower area (see Figure 2);

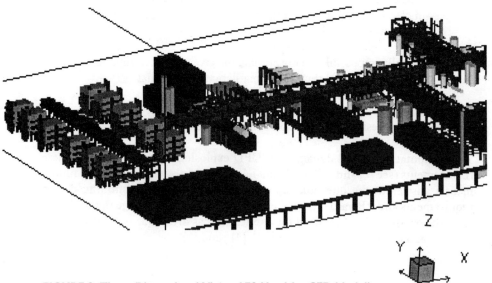

FIGURE 2. Three-Dimensional View of E2 Used for CFD Modeling

FIGURE 3. Progress of Eastbound Pressure Wave

- a (relative coarse) resolution based on 5 m × 5 m × 5 m cell size in the E2 and CT area was used for these conceptual design studies;
- gauge points at 1 m above the ground to record the time history of over-pressure;
- a rectangular stoichiometric gas cloud of 300 m × 135 m × 5 m.

The effort for model setup and analysis took approximately one week. Two simulations were performed, one with the ignition point in the fired heater area, which took approximately 20 CPU hours, and one with the ignition on the road just east of E2 area, which took approximately 26 CPU hours. Results are given in Figure 3 for the eastbound pressure wave and in Table 4. CPU time increases dramatically with better resolution calculations, e.g., a 1 m × 1 m × 1 m cell size would require about one year CPU time.

TABLE 4
Results of Low-Resolution CFD Calculations

Ignition location	Peak reflected over-pressure at Cooling Towers (bar-g); *at approximately 200m east of piperack*	Peak reflected over-pressure at Building 15 (see Figure 1) (bar-g)
Fired heaters	0.04	0.15 *(a more detailed calculation, described in section 4.0, gave 0.37)*
East end of piperack at edge of flammable cloud	0.009 negligible	0.14

RAIL ROAD YARD

The objective of these simulations was to establish if there are potential VCE problems with a RR shunting/storage yard filled with either RR tanker or RR hopper cars (see Figure 4), near an ethylene derivative plant. The "worst case" scenario was defined by filling the yard with RR cars; cars on alternate tracks were offset by half a car length. A stoichiometric ethylene cloud with a height of 6m was postulated to have been released by the derivative plant just south of the yard. This cloud blanketed the RR yard and was ignited in the center; cell size (resolution) was 1.5m × 1.5m × 1.5m. A 135m × 55.5m section of the RR yard (rather than the complete yard) was carried through the calculation for the purpose of CPU time reduction.

The analyses showed that very high *over-pressures* (> 1 bar-g) can be generated in a crowded RR storage yard and that hopper cars tend to generate higher pressures than tank cars. They also showed that the maximum over-pressure drops off sharply once the flamefront exits from between the RR cars.

p.198

* Don Ketchum - Baker

NOVA problem   p.191   * he confirms gan's calcs. on V-cloud

1650 kg ethylene

$$\sim \frac{48 \text{ MJ}}{\text{kg}} \rightarrow 7.9 \times 10^{10} \text{ J}$$

$$\times (.7376)(12) = 7.0 \times 10^{11} \text{ in-lb}$$
$$\times \tfrac{1}{2} = 3.5 \times 10^{11} \text{ in-lb}$$

2D - high

$\times 3.28$ ft/m

$$R = 200 + \frac{274}{2} \Rightarrow \underline{1105 \text{ ft}}$$

2 psi
66 psi·ms

Flame Speed	Distance (m)	P (psi)	i	P (bar)
2D High $M_w = 1.765$	20 (65.6 ft)	88.4	644	6
	30 (98.4)	37.1	388	2.5
	66			
	187			
	280 (918 ft)	2	50	.1
2D Med $M_w = 1.029$	20	42	659	3
	30	29	394	2
	280	1.9	55	.1

- says they would've used high for ethylene

- says side on wall point source o/p is what is calc from Baker Strehlow

P | (curves labeled 20%, 50%, 100%) — for brick Bldg
   i

P | (curves) — for steel structures
   i

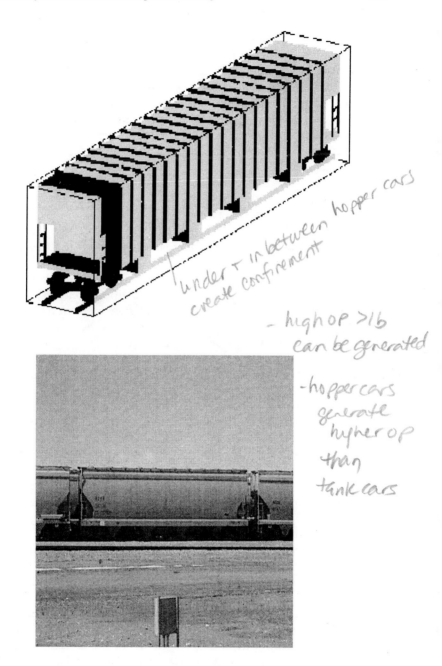

FIGURE 4 Simulated hopper car and original photograph for CFD calculations

Discussion

Plant simulation results with the ignition source near the fired heaters show that there is an initial pressure wave of approximately 0.1 bar-g peak reflected over-pressure, which drops to 0.01 bar-g peak reflected over-pressure over the next 250 ms. During the next 250 ms the pressure rises sharply as the east-bound flamefront reaches a compressor building where local peak reflected over-pressure attain values in excess of 1 bar-g (see Figure 3); from thereon the pressure subsides as the flamefront travels further through the plant.

A westbound flamefront, on the other hand (ignition at the east end of the piperack), did not generate those sudden sharp pressure increases. This westbound flamefront created local peak reflected over-pressure values of 0.23 bar-g in the compressor building area.

The sudden pressure surge associated with the eastbound flamefront were coined "Red-zones" after the sudden red blot showing up in time history graphics. "Red-zones" are of great concern, they show that excessive high over-pressures can be generated when only a fraction of the flammable mass has been consumed. While, so far we have discussed significant releases, which are very unlikely we are finding now that smaller and more frequent releases can potentially do as much explosion harm as a large release. A decision was made to do a further more detailed analysis of this "Red-zone."

The low-resolution simulations did not support the concerns we had with the cooling towers, in terms of location and domino effects since the calculated peak reflected pressures were too low to damage cooling towers. The circular pressure wave expansion pattern (see Figure 3) and the location of the cooling towers (see Figure 1) also indicate that the E2 and E3 cooling towers will be subjected to similar forces in case of a major incident. This means that there are no process safety reasons, neither personnel nor economic, to "swap" cooling towers or provide a crossover between the two cooling water systems. The "T" shape's linear layout appears to be congruous with "good" venting of over-pressures. Simulations also support our pre-contention that if buildings have to be near process plants they need to be in an open area near the likeliest ignition sources albeit outside the area of exposure of these ignition sources.

The RR yard simulation shows that off-site areas with obstructions can create high over-pressures and that they need to be considered with the same care as the process units themselves, in particular if they are close to process facilities.

As mentioned before, from an inherent safety perspective it makes sense to expedite process safety decisions as much as possible. We found it useful to move consequence modeling, including (CFD) up to the conceptual and process design stage. At this stage enough data is often available to derive at not necessarily a perfect decision but at least at a clear direction for decision-making, especially when done by knowledgeable people. In case of multiple projects, which typically impact each other, it is essential to get a grip on as many issues as early as possible in to the design. For example, in our case it was not intuitively clear that the cooling tower concerns were unfounded; by tackling it early on we were able to dismiss it (*for now*). Similarly, quantifying the RR yard's process safety concerns adds credibility to the cause of process safety *during* the process and RR yard design.

Red-Zone Analysis

The "Red-zone" area was defined with a numerical mesh of 243,000 of 1.66 m cubic cells, it included the fired heaters, building 15, the compressor building and the first north south lateral sub-piperack. The area around the compressor building was refined with the aid of detailed digital photographs. Sixteen pressure history recording locations were defined each with recording points at 1m and 3m. Cloud height and compositions as well as the ignition location were kept the same for consistency purposes. Cloud shape and volume were revised to reflect a smaller release. It took an AlphaStation $600^{5/266}$ workstation approximately 75 CPU hours to perform the calculations. Over-pressures in excess of several bar-g were calculated. The increase in flame speed and hence overpressure is consistent with the increase in congestion between the different gauge points. Congestion is enhanced by the fact that compressors in Alberta are typically housed in buildings, which is done to facilitate maintenance during the winter time when temperatures can drop to –40C. The solid concrete floor of the building and the number and density of objects, in particular the building columns and the large diameter piping which will protrude into a slumping ethylene vapour cloud, probably magnify the explosion forces (see Figure 5).

The peak reflected pressure at the north east corner of building 15 was determined to be 0.37 bar-g under the conditions given, during the earlier "scanning" type calculations a value of 0.15 bar-g was found.

Results/Conclusions

Multi-energy, Baker and CFD explosion modeling was carried out. Multi-energy and Baker indicated potential problems; however, only the CFD method was able to provide results where a detailed physical picture "fitted" the outcome of the mathematical modeling of a "stretched" confined area. A conceptual, low resolution, type analysis was found useful when determining potential problem zones; however, results from such an analysis cannot be used as end results, as was shown with the building 15 data. For a quantitative interpretation a more detailed analysis is required. Main findings were:

FIGURE 5. Example of Confinement and Obstacles Around One of The Compressor Buildings Causing A "red-zone"

FACING NORTH BELOW COMPRESSOR DECK

1. CFD is a valuable tool for explosion modeling.
2. No evidence was found that the E2 and E3 cooling towers are at an unreasonable risk.
3. Hunt for CFD's "Red-zones" and perform more detailed calculations.
4. Railroad storage yards can create substantial explosion hazards.
5. Buildings when located near process areas should be in open areas separated near the most likely ignition source(s) and away from the most likely hydrocarbon release sources.
6. A linear ethylene plant arrangement ("T-shaped piperack") appears to halt excessive flame acceleration.
7. Try to locate large diameter piping above slumping ethylene clouds, i.e., locate them at the first piperack level (when possible), otherwise consider mitigation and/or risk reduction measures.
8. Make process safety decision as early as possible; base the input on knowledge which can explain actual physical phenomena and where this is not possible try to find tools that remove uncertainty.

ACKNOWLEDGMENTS

The author likes to acknowledge the comments and insights provided by Greg Fairlie of Century Dynamics, which performed the AutoReaGas calculations, and Bert van den Berg of TNO Prins Maurits Laboratory.

References

[1] Q.A. Baker, M.J. Tang, E. Scheier and G.J. Silva, "Vapor Cloud Explosion Analysis." Prepared for Presentation at the American Institute of Chemical Engineers (AIChE)'s 28th Annual Loss Prevention Symposium, April 17–21, 1994. Industrial Explosions—Session 10.

[2] A.C. van den Berg, H.G. The, W.P.M. Mercx, Y. Mouilleau, C.J. Hayhurst, "AutoReaGas—A CFD-Tool for Gas Explosion Hazard Analysis," 8th Int. Symp. On Loss Prevention and Safety Promotion in the Process Industries" Antwerp, Belgium, Vol. 1, pp. 349–364, June 6–9, 1995.

[3] Center for Chemical Process Safety of the American Institute of Chemical Engineers. Guidelines for evaluating process buildings for external explosions and fires, 1996.

[4] K. van Wingerden, H-C Salvesen and R. Perbal, "Simulation of an accidental vapor cloud explosion," Process Safety Progress (Vol. 14, No. 3), pp. 173–181.

[5] N. Hamill, "CFD comes of age in the CPI," *Chemical Engineering* / December 1996, pp. 68–72.

[6] J.A. Denev, F. Durst and B. Mohr, "Room ventilation and its influence on the performance of fume cupboards: A parametric numerical study," *Ind. Eng. Chem. Res.* 1997, 36, pp 458–466.

[7] H. Ozog, G.A. Melhem, B. van den Berg and P. Mercx, "Facility siting: Case study demonstrating benefit of analyzing blast dynamics," International Conference and Workshop on process Safety Management and Inherently Safer Processes, Center for Chemical Process Safety of the American Institute of Chemical Engineers, October 8-11, 1996, Orlando, Florida, pp. 293-315.
[8] Century Dynamics/TNO, "AutoReaGas User Documentation," Version 1.2, September 1995.

Rhone polenc asked if we would buy it re AutoRegas

Consequence Modeling for the Insurance Industry— New Philosophies and Methods

Doug Scott
J&H Marsh & McLennan, Inc., London, England

Sanjeev Mohindra
Arthur D. Little, Inc., Cambridge, Massachusetts

ABSTRACT

The fire, explosion and other property damage incidents which occur in the process industries often result in large insurance claims – in some instances they may be large enough to threaten the survival of the companies concerned. Because of this, insurance underwriters must understand the loss potential of the sites they insure. Due to the large number of scenarios possible and the number of sites insured, the tools used must be capable of quick and easy use. As a result, usually only a limited number of scenarios are explored, sometimes resulting in a sacrifice of accuracy to speed and simplicity which may, in turn, result in non-optimal insurance arrangements.

The insurance industry has been relatively slow to develop the computer tools necessary to allow hazard assessment in terms of financial loss. Whilst a few are now available, they tend to be based on a single scenario or use simple methodologies. All still require the use of significant engineering judgement to accurately interpret the output. The development of a new set of software models designed to meet the needs of the more sophisticated insurance underwriters is described.

These models differ from those already existing by the use of improved calculation methods allowing a detail of calculation that is impractical without the aid of a computer. In addition, some scenarios that were previously considered too complex for analysis can now be examined. These methods allow greater accuracy without a significant increase in the information provided by the user, and also reduce the degree of individual interpretation allowing greater consistency – and consequently, the provision of optimal insurance solutions. These tools will also be useful to plant designers by allowing them to look in more detail at the potential financial consequences of plant layout decisions.

Introduction/Background

The insurance industry exists to provide a means of financial "security" in the event of fires, explosions, earthquakes or other unwanted events occurring. The origins of insurance lie in merchant shipping, where ship owners and merchants paid a premium per voyage with the intention that "the misfortunes of the few should be shared among the many." This philosophy has basically continued over many years, but the insurance market has become increasingly sophisticated and specialized. With the increase in industrialization throughout the world and the development of expensive, high technology industries, the financial risks have increased and the insurance industry can now be considered as a financial "fly wheel" for the industries protected.

An insurance underwriter (who will be liable for the cost of any claim) will essentially be trying to answer two questions: how big? and how often? The "how often?" is measured in qualitative ways (the claims record, the perceived quality of procedures, maintenance, etc. at the site).

Equally, or more important, is the "how big?" question. A prudent underwriter will obviously not want to commit himself to a liability that he or she may not be able to finance. For this reason, estimates of the potential loss (called by a variety of names, but usually the Estimated Maximum Loss – EML) form an essential part of insurance submissions. Prior to the explosion at the Nipro factory at Flixborough in the UK, estimates of loss were generally performed by non-specialist engineers covering all types of property insurance. However, Flixborough resulted in significant changes within the industry and most of the large broking and underwriting companies employed specialist engineers with a chemical engineering background, to survey and assess the risks posed by the oil and petrochemical industries. In particular, methods were developed to assess the potential damage from vapor cloud explosions, commonly seen as the most destructive type of events likely to occur.

Existing Methods

Over the years a number of variations on the modeling of vapor cloud explosions have been developed, these basically fall into two categories, both widely used. The first is the "IOI method"(1), this is an empirical method which has been revised on several occasions, based on the damage caused by a release of hydrocarbon with a defined epicentre. The other method is a TNT equivalent method(2). Both methods work on the basis of plotting circles over areas of known value on a plot plan and estimating percentage damage based on overpressure, commonly used figures are given in Table 1. The models both make

TABLE I
Property Damage Related to Overpressure

Percentage Damage	Overpressure (mbar)	Overpressure (psi)
80%	355.00	5.00
40%	142.00	2.00
5%	73.00	1.00

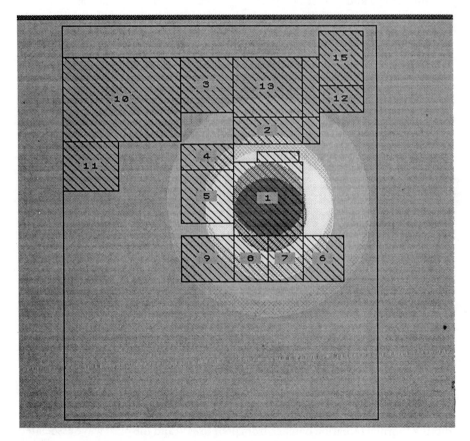

FIGURE 1. Typical Insurance Overpressure Contours for a VCE

centre of the explosion to move anywhere within the 80% damage overpressure circle. Figure 1 shows a typical plotplan with overpressure contours.

Both the IOI and TNT models follow a similar procedure for the estimation of explosion damage, which essentially takes three stages:

- Estimate the size of the release;
- Calculate the explosion potential;
- Estimate the location of the overpressure contours and associated damage.

An insurance risk assessor will have to analyze a wide variety of locations/scenarios relatively quickly, there is a requirement for methods to be simple, easy to use and robust in output. This latter point is important, as a method that produces major differences in damage estimation with relatively minor changes to the input parameters will not be regarded as credible.

These methods, in one guise or another, have been in widespread use since the mid-1970s and whilst being broadly accurate, they have not always accurately estimated potential damage. It is now commonly agreed that these methods have a number of limitations:

- No account is taken of the topology of the area. When the methods were first developed a common expression was Unconfined Vapor Cloud Explosion (UVCE), a term not now commonly used. Research over a period of time has shown that for combustion processes to develop into an explosive overpressure, a degree of confinement is usually necessary to allow turbulence and hence flame velocities to increase.
- The size of the release has usually been based on the total contents of the vessel involved. No allowance has been made for material which is above the upper flammable limit or below the lower flammable limit, which may result in the mass capable of supporting combustion being considerably smaller than the mass released. Similarly, no allowance has been made for restrictions in flow rate caused by flashing flow when liquefied gases are released. There is also some variation between the different methods in the extent of entrainment (or lack of entrainment) of aerosol liquids within the gas cloud.
- The extent of damage is also relatively arbitrary and is based largely on historical experience and often on theoretical studies of the explosion potential of condensed phased, or nuclear, explosions, which have very different overpressure profiles.[3] Note that the damage following a vapor cloud explosion will not only include direct blast damage, but also the effects of fire from materials released as a result of the blast.

The largest focus within the insurance industry has been on the modeling of vapor cloud events. There are, however, a few relatively simple models for predicting thermal radiation damage from pool fires [1] and the blast wave associated with high pressure vessel rupture.[2]

New Methods

In many cases the level of analysis is relatively basic and the complexity is well below that commonly used for safety analysis. The methods developed here are an attempt to instill a more scientific approach into the financial calculations of insurance loss. As with the existing methods, the aim is to provide simple, easy to use methods with, in this case, an improved theoretical basis and greater accuracy. The model is known as COINS (COnsequense analysis for INSurance).

Many of the parameters that are defined individually for complex models are set to standard defaults within COINS and there is also access to a large database of chemicals allowing the properties of mixtures to be modeled accurately.

Explosion

The explosion model is based on joint work between J&H Marsh & McLennan, Inc. and Strathclyde University,[4] which takes its roots from earlier research by British Gas and the Shell Thornton Research Station. These studies show that the effect of confinement increases flame velocities by inducing turbulence with the ultimate result that overpressures increase. Pressures fall as the turbulence inducers are removed, but may rise again if unburned gas is located in two adjacent, but separated, partially confined areas. In this case the overpressure builds up as the flame front accelerates through the gas in the obstructed area and then falls as the flame proceeds through an open area. Finally the flame front reaches the second area of partial confinement and flame velocities and overpressures, again begin to increase. The DACALS model is not unique in this approach, but appears to offer advantages over other methods by the relatively simple approach to determining the degree of confinement. Analysis of a number of vapor cloud explosions has shown good agreement between calculated overpressures and those estimated after the real events.

Taking the three key stages in the calculation process, as indicated above, the new program:

- *Estimate the size of the release:* The program will allow the normal calculation of release rate for a single-phase fluid but also allow two phase calculations, giving a more accurate estimation of gas within the cloud. The use of simple Gaussian gas dispersion modeling in any direction will also allow an improved analysis of the extent of drift.
- *Calculate the explosion potential:* The new DACALS model for VCE calculations will allow a better, more scientific analysis of the calculation making due allowance for confinement without expending significantly greater effort in site investigation or inputting data into the model.
- *Estimate the extent of damage:* As yet this area has not been explored but would be a fruitful area of future work.

A common problem when estimating the business interruption (BI) potential from an event is determining the speed with which key items of equipment can be replaced. Very often these key items have high resistance to overpressures that common items, such as piping, structural steelwork, atmospheric vessels, etc. do not. Similarly items with low blast resistance, such as cooling towers, can be readily identified. The new model has the ability to indicate these key items within the plot area and hence their likely extent of damage can be more easily assessed and more accurate assumptions made on the requirement to replace or repair items. This has important consequences for BI analysis as major items having replacement times of up to 12 to 24 months.

The overpressure contours for the VCE at Flixborough using both the TNT and DACALS methods are shown in Figures 2 and 3.

Fire

At present, there is only one, relatively simple, pool fire damage estimation method in the public domain within the insurance industry.[1] This makes many assumptions and does not allow for the effect of high winds or differences between the properties of different fuels that might be in the fire. It does, however, differentiate between the extent of damage in fireproofed and non-fireproofed areas.

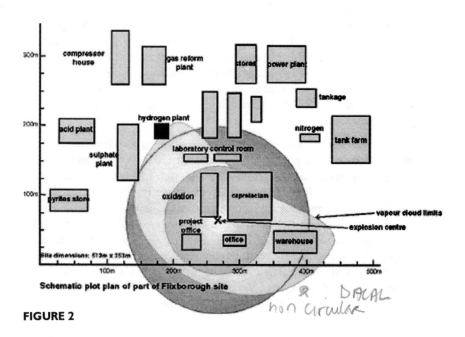

FIGURE 2

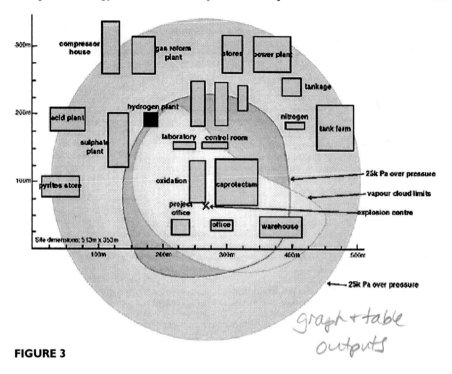

FIGURE 3

Pool fire modeling is relatively well established methodology. However, in theoretical terms, most pool fires will be short lived unless there is a long standing leak or a deep pool of confined liquid, which will burn. The model in use follows standard modeling practice and incorporates features allowing for flame tilt, drag, etc. Analysis of results by comparison with the existing method have shown good agreement when the arbitrary nature of the assumptions have been considered.

A separate pool fire scenario for tank farms has also been developed, with the intention of investigating how a fire originating in one tank might propagate to other tanks in both the upwind and the downwind direction. Again, this involves the use of standard methods, but considers both the potential for downwind tanks to be affected by thermal radiation or direct flame impingement, and upwind tanks to be affected by the release of flammable gases which then ignite and flashback to the tanks concerned.

A sophisticated algorithm has been developed to allow searching of all tanks for all potential wind directions within a practical period of time. At present it is common practice within the insurance world to consider that a fire may spread to all the tanks in a common dike area, this model might demonstrate that this is an over cautious approach.

Vessel Rupture

A number of separate vessel rupture models have been developed, but only one has been used, and only in a limited sense, in the insurance industry.[2] This makes the assumption that all of the energy in a ruptured vessel, which bursts at its hydraulic test pressure, is converted into a blast wave. This is, in fact, an oversimplification, as part of the energy it used to accelerate portions of the vessel structure, sometimes to considerable distances. The method used is based on a standard overpressure rupture model developed by Baker.[5] Overpressure contours are shown and the potential areas where projectiles are likely to land are also indicated. While this is of limited use in estimating EMLs, it will give an indication of the potential for secondary damage from a projectile and may allow some loss scenarios to be discounted.

The Future

The program represents a level of sophistication not currently available in the insurance market and includes a variety of models allowing a relatively complete analysis of common scenarios to be performed. Its development has allowed multiple scenarios for the same site to be combined easily, together with a calculation of the worst case events. It is also designed to be easy to update (for example, when a new unit has been constructed, or insurance values have changed). There are a number of potential improvements that could be incorporated in the program and these are currently under consideration.

References

1. Anon. *Estimated Maximum Loss from Explosion and/or Fire*. International Oil Insurers, London, 1992
2. Niggli, S. *Extool Theory and User Manual*. The Swiss Reinsurance Company, Zurich, 1993.
3. Glasstone S. *The Effects of Nuclear Weapons*. US Atomic Energy Commission, Washington, DC, 1962.
4. Crawley F. K., Scott D. S., Scilly N. F. "DACALS—A Simple Verified VCE Model." Proceedings of the Eighth International Conference on Loss Prevention and Safety: Promotion in the Process Industries, Antwerp, Belgium, 1995.
5. Baker W. E. *Explosions in Air*. University of Texas Press, Austin, Texas, 1973.

PSM Verification of Relief Device Sizing in the Case of Reactive Chemical Service

G. Bradley Chadwell
Battelle, 505 King Avenue, Columbus, OH 43201

ABSTRACT

The Occupational Safety and Health Administration (OSHA) Process Safety Management (PSM) Standard requires compilation of process safety information, including information on the relief system design and design basis. Much more than assembling the appropriate equipment data and original design calculations is required for relief systems. Improvements in design methodologies and changes in process operating conditions since the original relief system specification could invalidate the original relief system design and design basis. Therefore, relief system verification is necessary and includes: (1) reassessing the relief system design basis against current operating conditions; and (2) confirming that the existing relief system design is adequate for the revalidated design basis.

This paper presents a stepwise approach for establishing relief system design basis requirements in the case of reactive chemical service and analyzing relief systems against this design basis. Our approach addresses two-phase flow and explicitly accounts for internal and external heat sources (e.g., heat generation from exothermic reactions and fire exposure).

Introduction

The Process Safety Information element of the Occupational Safety and Health Administration (OSHA) Process Safety Management (PSM) standard, 29 Code of Federal Regulations 1910.119(d),[1] requires employers to compile information pertaining to relief system design and design basis. The regulation further requires employers to document that equipment complies with "recognized and generally accepted good engineering practices." For equipment designed and constructed to out-of-date codes, standards, or practices, the employer also must "determine and document that the equipment is designed ... and operating in a safe manner."

To satisfy the PSM standard and to ensure adequate protection from vessel overpressures, the employer must examine the relief system design in light of current relief device sizing methodology and compare the original design basis with existing operating conditions. Recent advancements in sizing technologies include modeling two-phase relief conditions and characterizing exothermic reactions as an internal heat source. Two-phase flow has a lower flow capacity compared to single phase flow. Heat input from exothermic reaction generally is large in comparison with external fire. For these reasons, devices placed in service before the availability of such sizing methods warrant revalidation. Likewise, current-operating conditions must be examined to ensure conformance with the original design basis.

This paper presents an approach for verifying relief device sizing in light of the issues discussed above. Required activities in the verification process include:

- establishing a vessel list
- determining a design basis
- gathering the information required for the analysis
- performing sizing calculations
- assessing existing device adequacy
- documenting the analysis.

Each step in the series is examined separately, followed by an example illustrating design basis determination.

Establishing a Vessel List

Obtaining a copy of the most recent relief device list for a process is not the first step in verifying relief device sizing. Starting with the list of in-service devices overlooks the possible existence of vessels needing overpressure protection, but with no relief devices.

Therefore, the ultimate concern of relief device verification is not whether all relief devices are properly sized, but rather whether properly sized relief devices are in place to protect against all credible overpressure scenarios. Defining appropriate scenarios is discussed in a subsequent section. To this end, the verification process should begin with an examination of the equipment list for a process. Each piece of equipment involved in one or more potential overpressure scenarios should be included in the analysis.

Once compiled, a comprehensive list of equipment requiring relief devices may be compared with the existing relief device list, if one exists. Relief device sizing for currently unprotected vessels may be performed directly during the verification process. This approach ensures consistency in applying design basis criteria and sizing methodology to both new and existing relief devices.

Determining a Design Basis

Choosing the appropriate design basis is a critical step in sizing a relief device. For each piece of equipment requiring overpressure protection, a credible worst case scenario should be defined. For a given vessel, several plausible scenarios may exist—from external fire to various operating contingencies, such as overfill or vessel swell conditions. Several sources of standard sizing methods and requirements are available for guidance in selecting design basis scenarios. These include the American Society of Mechanical Engineers (ASME) *Boiler and Pressure Vessel Code*,[2] American Petroleum Institute (API) Recommended Practice (RP) 520, *Sizing, Selection, and Installation of Pressure-Relieving Devices in Refineries*,[3] as well as guidelines from AIChE's Design Institute for Emergency Relief Systems.[4]

The original design basis for an existing device may serve as a useful starting point, but should be examined carefully against existing operating conditions. For example, chemical service, operating temperature and pressure ranges, vessel pressure rating, or any combination thereof could have changed since a vessel began operating. If any such change has occurred, then the original design basis is potentially invalid and should be set aside accordingly. Another issue is whether the original design basis accounted for internal heating effects due to runaway reaction. If internal heat sources were ignored in the original design basis, the relief device may be undersized, and the original design basis should not be used in the verification process.

The discussion in this paper focuses on sizing in the case of reactive chemical service, in which external fire and internal heating due to reaction may occur simultaneously. Assuming that a new design basis is to be defined, the following parameters need to be established:

- vessel inventory
- vessel contents
- operating temperature
- operating pressure
- relief device set pressure
- external heat sources
- internal heat sources
- mitigating features.

The first four items are taken at their respective maximum values, that is, at the upper operating limits with the composition that will result in the greatest heat evolution due to reaction. The relief device set pressure upper limit is the maximum allowable working pressure (MAWP) of the vessel. External heat sources include fire exposure and heating jackets. Exothermic reaction is the primary source of internal heating. Characterization of heat sources is discussed in a subsequent section. Mitigating features may be taken into consideration once reliability criteria have been established. Insulation will reduce the heat input rate of fire exposure, given that the fire duration is less than the time required to reach thermal equilibrium across the insulation layer. Another fire exposure mitigation feature to be considered in some cases is a water deluge system. Unlike insulation, a water deluge system is an active mitigation device requiring a reliability assessment.

The issue of reliability gives rise to a difficult question in establishing a design basis—when is a worst case scenario credible? It is possible, for example, to postulate a runaway reaction scenario in which a continuous reaction vessel is engulfed in fire, the cooling system has failed, all inlet flow control valves have failed wide open, while all outlet flow valves have failed closed. In addition, the reactant concentrations in the feeds are well above operating limits, and catalyst concentration exceeds the normal level. Although possible, the simultaneous failures of multiple engineering and administrative controls may be highly improbable, unless due to a common underlying cause.

One approach in dealing with this issue is simply to decide qualitatively on a credible worst case for the design basis. A second approach is to perform sizing calculations on several alternative scenarios and take the most conservative results. This approach is useful, although costly, in situations where it is difficult to determine which relief conditions would require the largest relief devices. A third technique is to apply risk-based decision making. In this method, the probability of the chosen credible worst case scenario is considered together with the severity of the consequences of the overpressure scenario exceeding the relief device capacity. Active mitigation devices, such as water deluge systems, can be factored into the scenario using this approach, since failure probabilities are addressed directly. Using a risk-based approach will result in reduced relief capacity requirements compared to absolute worst case conditions, but does require a value judgment in establishing a tolerable level of risk.

Gathering Data

A fair amount of process and equipment data is required to determine the design basis. The majority of the information should be available in the process safety information (PSI) files compiled for the process hazard analyses covering the equipment under analysis. Once the design basis is established, additional information beyond that typically included in the PSI files is required.

The dataset required to perform relief device sizing calculations is quite extensive. First, the equipment dimensions and physical properties must be assembled. Modeling heat flow across the equipment surface requires knowledge of the vessel material's heat capacity, thermal conductivity, and density (if vessel mass is determined indirectly from vessel dimensions and wall thickness). The vessel geometry—vertical or horizontal cylinder, spherical, etc.—is a necessary for calculating the wetted surface area, where the vessel contents contact vessel walls.

Second, the properties of the vessel contents must be quantified. This includes density, heat capacity, viscosity, and thermal conductivity. Values of each parameter are required for both liquid and vapor phases. Boiling points, vapor pressure, and thermal expansion coefficient values also are required. Ideally, the properties will be expressed as functions of temperature, pressure, and composition of the fluid. This allows for time-dependent modeling of the relieving fluid. If functional expressions are not available, values should be reported at relief conditions.

Next, expressions describing relevant reaction rates and heat of reaction must be developed. The rate expressions should relate either reactant consumption rate or heat evolution rate to reaction concentrations and temperature. Typically, an Arrhenius expression is used for this purpose. If rate data are not readily available, advanced calorimetry apparatus are available for collection and analysis of reaction data.

Finally, heat flux rates must be established for fire cases. API RP 520 Section D.5, Heat Absorption Equations, provides guidance in this area. In addition, data should be gathered for any mitigating devices or systems included in the overpressure scenario. For crediting insulation effects, an environmental factor is used. Again, API RP 520 describes the proper usage of these factors. If a water deluge system is modeled, the relevant parameters are spray rate and distribution details, water film physical properties, and heat transfer model assumptions and quantities for the associated variables.

Calculating Relief Device Size Requirements

After all data gathering is completed, relief device sizing calculations can be performed. Even if the original design basis still applies, the sizing calculations should be performed. This is because the original design calculations, while applied to the correct design conditions, may have been based on sizing technology that is no longer considered adequate or appropriate. This could be the case with either single-phase or two-phase flow, but is especially important with two-phase relief flow.

Because two-phase flow generally has a decreased flow capacity compared to single-phase flow, a greater relief orifice area often is required for two-phase flow. The lower flow capacity is attributable, in part, to frictional losses due to phase slippage in the fluid. Another cause is that flashing of the liquid component induces backpressure that, in turn, decreases flow rate and increases required flow area. However, oversizing a relief device with two-phase flow can have dangerous consequences. Excessive fluid flashing on the downstream side of an oversized relief device can cause the backpressure buildup to the point that the relief device function is impaired. The result could be a catastrophic vessel failure.

Internal heating from reaction also should be included even with external fire heating. The basic tenet of relief device sizing is to provide a relief area large enough to permit pressure-relieving flow at a rate that is at least equal to the maximum rate of pressure generation. Heat of reaction contributes to vessel overpressure and must be included with all other heat and overpressure sources.

The area through which the fluid passes limits the relief flow rate. A maximum mass flux—mass flow rate per unit cross-sectional flow area—exists at critical, or choked, flow conditions. The limit depends on the fluid properties. In short, the relief device area is calculated such that at the maximum mass flux, the mass flow rate is sufficient to prevent vessel pressure from rising beyond acceptable limits. Normally, the required area is reported as the next largest standard orifice or pipe size. The calculations can become quite involved and are solved best by iterative computer codes or in some cases by advanced spreadsheet applications.

Assessing Existing Device Adequacy

The process of identifying inadequate relief devices is straightforward once the required sizes are calculated. Any existing device smaller than the required size obviously is inadequate. The only difficulty is in determining if any devices are extremely oversized. In single-phase relief cases with extremely oversized valves, excessive valve chattering may be a problem. In scenarios where two-phase flow is predicted, relief flow should be modeled given the existing relief area to determine if backpressure due to flashing is a problem.

Documenting the Analysis

Finally, to satisfy PSM PSI requirements, the analysis must be documented properly. Three categories of information are recorded: the design basis, the design calculations, and the results. The entire design basis definition should be included for each relief device in the process. This includes operating conditions, relief set pressure, heat source term characterizations, and all assumptions for the credible worst case scenario. Any alternative scenarios analyzed should be recorded as well. The objective is to document why the particular design basis is the correct choice.

Recording design calculations may be difficult if proprietary programs were used. At the least, the input and output data should be placed in the PSI files along with a summary of the program or computer code used including software version number, where appropriate. The key is reproducibility—enough information being provided that if the valve sizing calculations are revisited in the future, the engineer can establish easily the original calculation method and design variable values.

The last pieces of information to document are the actual results. This includes the relief device size, setpoint pressure and temperature, and required flowrate. Enough information should be recorded that a specification sheet, suitable for submission to a vendor for a price quotation, can be generated directly from the PSI file.

Other Considerations

This paper addresses sizing individual relief devices. Therefore, several issues pertaining to relief systems were not covered. They are mentioned here for completeness. All of these issues may be addressed by analyzing piping losses using standard fluid dynamics principles or pipe flow analysis packages.

One issue is relief system piping, both inlet and outlet. Relief device inlet piping—the piping between the protected vessel and the relief device—is subject to certain restrictions by ASME code. Design constraints are placed on inlet piping concerning diameter and configuration that prevent excessive pressure drop of relieving fluid before entering the relief device.

Outlet piping or relief vent systems are constrained by the relieving flow conditions. Excessive pressure drop in the line due to frictional losses can result in built-up backpressure greater than acceptable operating limits for the associated relief device. Although a relief device may be sized properly, flow may stagnate if the outlet piping equivalent length is too great. The result would be the same as an undersized relief device.

Another possible source of backpressure is from simultaneous releases through the same vent system. In a multiple-relief scenario, such as a pool fire in a tank farm dike, flow from one vessel to a common vent header will impose backpressure on all other vessels relieving through that vent header. It is possible that the operation of one or more relief devices will be impaired to the point that relief capacity is reduced significantly.

Example Case

To illustrate how alternative relief scenarios may arise from a single process vessel, the case of a polymerization reactor is examined. Various approaches to choosing a design basis are explored.

The example situation is a batch polymerization of styrene monomer in toluene solvent. The reaction vessel is adjacent to several identical reactors, all located in a process building with an adequate floor drain system. The specific parameters are

- vessel inventory—1000 gallon reactor, operating at 80% capacity
- vessel contents—40% by mass styrene monomer in toluene with catalyst

- operating temperature—100°F at initial conditions
- operating pressure—60 psig at initial conditions
- relief set pressure—250 psig (vessel MAWP)
- external heat sources—external fire (e.g., spill from adjacent vessel, pool fire); heated vessel jacket
- internal heat sources—exothermic polymerization reaction
- mitigating features—insulation (fire case), good drainage (fire case).

Potential design basis scenarios include:

1. external fire, with good drainage; credit taken for insulation
2. external fire, with good drainage; no credit taken for insulation
3. heating jacket temperature controller fail high
4. heating jacket temperature controller fail high AND high catalyst charge
5. heating jacket temperature controller fail high AND high catalyst charge AND high monomer concentration in batch.

Each of the above conditions could precipitate a runaway polymerization reaction. In the fire cases, the heat of reaction would add to the external fire heat source term. The overall rate of pressure increase may be different in each scenario, depending on the rate of reaction and the total heat input rate at relief conditions. Thus, the calculated required relief device size could be different for each potential design basis. For the two fire cases, the scenario crediting insulation should have a lower heat input rate that may result in relatively smaller relief device size requirement. The two scenario sizing results would differ only if the design basis fire duration is significantly less than the time required for the vessel wall heat transfer conditions to reach equilibrium. Otherwise, only the time to reach runaway reaction onset would be different; the overpressure severity would be the same. For the last three scenarios, runaway reaction and heat input from the jacket are the two heat sources. Assuming typical polymerization kinetics, the scenario 5 reaction would be more vigorous than the scenario 4 reaction, which, in turn, should be more vigorous than the one in scenario 3. Conversely, the probability of the set of initiating events occurring simultaneously may decrease significantly from scenario 3 through scenario 5.

The question remains: which scenario is the "credible worst case scenario?" As discussed previously, one approach is to decide qualitatively which scenario is conservative—yet not too improbable—and perform the sizing calculations on that design basis. The approach of performing sizing calculations on multiple design bases, although time consuming and costly, may be useful in comparing equally probable scenarios where a priori determination of the most conservative case is difficult. For example, both case 2 and case 4 may be judged as equally credible, but it is not readily apparent which design basis will require the greater relief device area. Performing sizing calculations for both scenarios and taking the larger, more conservative, device size requirement will result in a relief device that will provide sufficient capacity for either design basis.

A third approach is to make a risk-based decision on design basis choice. This method entails performing a reliability assessment of each piece of equipment and procedural steps involved in the process to determine the expected failure rate of each one. The individual rates combine to give an overall likelihood, or frequency, of scenario occurrence. For example, scenario 3 entails a single hardware failure. Scenario 4 could result from one hardware failure (temperature controller) and an administrative control failure (too much catalyst added by operator), making scenario 4 more unlikely than scenario 3. Even more unlikely is scenario 5, where three failures are necessary—temperature controller failure, too much catalyst added, and an incorrect monomer–solvent ratio charged to vessel, unless any combination of these can be caused by a common mode failure. Once quantitative, or semi-quantitative, probabilities are calculated for each potential design basis, a tolerable frequency is chosen based on the consequence severity of an overpressure exceeding the relief device capacity. The range of consequences should include impacts to personnel, equipment damage, business interruption losses, and environmental impact from inventory release. A risk-based method requires risk tolerability criteria to be established, but does provide a decision-making framework for addressing complex systems with numerous relief scenarios.

Conclusions

Verification of relief device sizing is a complex multi-step process that involves much more than assembling existing documentation on the relief devices in service. Each piece of equipment in a process should be evaluated for potential overpressure scenarios. After the complete set of vessels requiring relief devices has been developed, an appropriate design basis must be established for each vessel. Choosing a design basis requires assessing alternative scenarios to find the credible worst case scenario. The choice may be made qualitatively or by using a risk-based approach. The design basis then is used to calculate the required relief device size. The sizing calculations should use the most current design methods incorporating such issues as two-phase flow and reaction heat sources.

Once the calculations are performed, the existing device sizes are verified for adequacy against the calculated size requirements. Inlet and outlet piping systems are analyzed for potential backpressure problems. Finally, design basis data, design calculations, and results are documented. Following this stepwise approach and employing the most recent sizing methods will ensure that the PSM verification of relief device sizing is conducted properly.

References

1. Department of Labor, Occupational Safety and Health Administration, "Process Safety Management of Highly Hazardous Chemicals," 29 CFR 1910.119 (February 24, 1992).
2. American Society of Mechanical Engineers, ASME Boiler and Pressure Vessel Code, Section VIII, "Pressure Vessels," Division I, ASME, New York (1995).
3. American Petroleum Institute, API Recommended Practice 520, Sizing, Selection, and Installation of Pressure-Relieving Devices in Refineries, Part I, "Sizing and Selection," 6th ed., API, Washington D.C. (March 1993).
4. Fisher, H.G., et al., Emergency Relief System Design Using DIERS Technology, AIChE's Design Institute for Emergency Relief Systems, DIERS, AIChE, New York (1992).

Accounting for Common Cause Failures When Assessing the Effectiveness of Safeguards

Henrique Paula and Emma Daggett[1]
JBF Associates, Inc., 1000 Technology Drive, Knoxville, TN 37932-3353,

ABSTRACT

Traditionally, companies in the chemical process industry (CPI) have provided multiple layers of protection (multiple safeguards) to help ensure adequate protection against process hazards. Safeguards include both engineering and administrative controls that help prevent or mitigate process upsets (e.g., releases) that can threaten employees, the public, the environment, equipment, and/or facilities. Examples of safeguards include process alarms, shutdown interlocks, relief systems, hydrocarbon detectors, fire protection systems, and plant process safety policies and procedures. Using multiple safeguards often reduces risk. However, the very high reliability theoretically achievable through the use of multiple safeguards, particularly through the use of redundant components, can sometimes be compromised by single events that can fail multiple safeguards (e.g., functional failure of all temperature sensors in an emergency shutdown system attributable to a miscalibration error during maintenance activities). The events that defeat multiple safeguards and are attributed to a single cause of failure are often called common cause failures (CCFs), and they have consistently been shown to be important contributors to risk. Despite the fact that the frequency of accident scenarios may be grossly underestimated if CCFs affecting multiple safeguards are not taken into account, many qualitative and quantitative approaches for analyzing safeguards do not consider CCFs. This paper presents a method to account for CCFs when assessing the effectiveness of safeguards in a CPI facility.

Introduction

Traditionally, companies in the chemical process industry (CPI) have provided multiple layers of protection (multiple safeguards) to help ensure adequate protection against process hazards. Safeguards include both engineering and administrative controls that help prevent or mitigate process upsets (e.g., releases) that can threaten employees, the public, the environment, equipment, and/or facilities. Examples of safeguards include process alarms, shutdown interlocks, relief systems, hydrocarbon detectors, fire protection systems, and

---

1 Speaker. Copyright © 1997 by Henrique Paula and Emma Daggett.

plant process safety policies and procedures. The types and number of safeguards for each process hazard depend on the nature of the hazard, but the number of safeguards generally increases with the frequency and/or consequences associated with a potential accident scenario. For example, redundant temperature sensors (as opposed to a single sensor) are often used in the emergency shutdown (ESD) systems in chemical processes that involve highly exothermic reactions.

When multiple safeguards are used to help ensure adequate protection against process hazards, accidents cannot occur unless multiple failures occur. This could happen as the result of the independent failure of each safeguard; however, operational experience shows that multiple independent failures are rare. This is easily understood with the following simple, numerical example. Consider a shutdown interlock that consists of three redundant temperature switches (A, B, and C), each designed to individually shut down the system upon high temperature. Also, assume that the probabilities that the switches will fail on demand ($P\{A\}$, $P\{B\}$, and $P\{C\}$) are constant and equal to 0.01 (1 in 100 demands). If it is further assumed that failures of these three switches are *independent*, then the total probability that all switches will fail, $P\{S\}$, is given by

$$P\{S\} = P\{A\} \times P\{B\} \times P\{C\} = (0.01)^3 = 10^{-6}/\text{demand}$$

That is, the system is expected to fail *once in every one million demands*. Further, if we had assumed that the probabilities of switch failure on demand ($P\{A\}$, $P\{B\}$, and $P\{C\}$) were equal to 0.001 (1 in 1,000 demands), which may be difficult but not impossible to obtain in practical applications, then the system would be expected to fail *once in every one trillion demands*. These are rather unbelievable numbers because, as exemplified later in this paper, systems with two, three, four, or even higher levels of redundancy have failed several times in commercial and industrial applications, including CPI facilities, aircraft, and nuclear power plants. That is, the assumption of *independence* among redundant safeguards results in unrealistic, very low estimates for the probability of loss of all safeguards; it gives too much credit for multiple safeguards, thereby potentially causing gross underestimation of risk.

But what makes the simple probabilistic evaluations presented above unrealistic? As more complex designs evolved in the 1950s, engineers and reliability specialists discovered that multiple safeguards can also fail as a result of a single event (a dependent failure event).[1-3] In the previous example, all three switches in the high temperature shutdown interlock could be miscalibrated during maintenance, resulting in the functional unavailability of the entire system. Because they are attributable to a single cause of failure, these dependent failure events are often called common cause failure (CCF) events. Many authors have used different terminology to describe this class of events, including "cross-linked failure," "systematic failure," "common disaster," and "common mode failure."[4-6] This paper defines and exemplifies CCFs, and it provides guidance and quantitative data to account for CCFs when assessing the effectiveness of safeguards in CPI facilities.

CCF Definition and CCF Coupling Factors

For CPI applications, a CCF event is defined as multiple safeguards (e.g., temperature switches) failing or otherwise being disabled simultaneously, or within a short period of time, from the same cause of failure (e.g., maintenance error or design deficiency). Thus, three important conditions for an actual CCF are that (1) multiple safeguards must be *failed or disabled* (not simply degraded), (2) the failures must be *simultaneous* (or nearly simultaneous as discussed next), and (3) the cause of the failure for each safeguard must be the *same*.

Within this definition, multiple failures occurring "simultaneously" (or nearly simultaneously) does not necessarily mean occurring at the same instant in time but rather sufficiently close in time to result in failure to perform the safety function required of the multiple safeguards (i.e., preventing and/or mitigating the consequences of an accident). For instance, if emergency cooling water is required from one of two (continuously running) redundant pumps for 2 hours to safely shut down a reactor, "nearly simultaneous" means "within 2 hours." That is, both pumps (safeguards) failing any time within the 2-hour mission constitutes a CCF. For interlock systems that use redundancy (e.g., the high temperature shutdown interlock discussed earlier), "nearly simultaneous" often means "within the time between testing of the redundant equipment" (assuming that once they occur, failures are detected and correct during the next test).

Note that the essence of a CCF event *is not* the cause of failure, which could be equipment failure, human error, or external damage (e.g., fire or external impact). In fact, the available literature shows that the causes of CCF events are generally no different from the causes of single, independent failures _ except for the existence of coupling factors that are responsible for the occurrence of multiple instead of single failures.[6-9] For example, the spurious operation of a deluge system can result in the (single) failure of an electronic component (A) in a certain location of the CPI facility. The same deluge system failure would probably have resulted in the failure of both redundant components (A and B) if they were in the same location. The cause of component failure (water damage to electronic equipment) is the same in both cases; coupling (same location in this example) is what separates CCF events from single failure events. Other coupling factors include common support system, common hardware, equipment similarity, common internal environment, and common operating/maintenance staff and procedures.

Thus, the essence of a CCF event is the *coupling* in the failure times of multiple safeguards. This is illustrated in Figure 1, which shows the failure times for three redundant safeguards over a period of about 20 years. In case (a), each safeguard has failed four times, and the times of failure are random (not linked or coupled). The pattern in Figure 1(a) should be expected if no CCF coupling factors exist. Figure 1(b) shows the failure times for three other safeguards. Just like the safeguards in case (a), each safeguard in Figure 1(b) has failed four times over about 20 years. However, the failure times are completely coupled in time (i.e., the safeguards always fail at the same time). The pattern in Figure 1(b) is hypothetical because complete coupling in the failure times does not occur even if all CCF couplings exist, but Figure 1(b) does illustrate the essence of a CCF event. There are six CCF coupling factors that act alone or (more often) in combination to create a CCF event. Each coupling factor is discussed and exemplified in the following paragraphs.

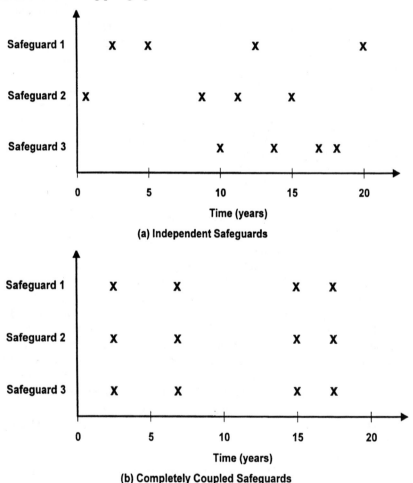

FIGURE 1 Failure Times for (a) Independent and (b) Completely Coupled Safeguards

CCF Coupling Factor 1: Common Support System

Several types of safeguards have a functional dependency on support systems, including control systems (distributed control system [DCS], programmable logic controller [PLC], etc.) and utilities (instrument air, electric power, steam, etc.). Although safeguards are often designed to "fail safe" upon loss of support systems (e.g., isolation valve closing upon loss of control signal or loss of instrument air), there are applications in which, intentionally or unintentionally, loss or degradation of support systems will defeat safeguards. This can be a source of coupling if the support systems are common to multiple safeguards.

For example, if two electric-driven firewater pumps are supplied electric power from the same motor control center (MCC), they will both be disabled if the MCC fails. Also, even when the safeguards rely on separate support systems (e.g., pump A gets electric power from MCC A, and pump B gets electric power from MCC B), it is possible that there are couplings factors involving the separate support systems (e.g., common offsite electric feeder to both MCCs A and B). Therefore, it is not enough to provide separate support systems for multiple safeguards; it must be ensured that CCF couplings within these separate support systems have also been eliminated or reduced.

Note that the **common support system** coupling factor refers to coupling that results from safeguards being disabled because of loss or degradation of the support system. It is also possible that the support system will malfunction in a way that damages the safeguards. For example, a power surge in the electric supply to the two firewater pumps A and B could damage the electric motor on each pump. This type of coupling is considered with the **common internal environment** coupling factor, and thus is excluded from the **common support system** coupling factor.

CCF Coupling Factor 2: Common Hardware

This coupling factor is similar to the **common support system** coupling factor, but the coupling here is the potential failure of hardware that is common (shared) by multiple safeguards. A typical example of multiple safeguards with common hardware are two (or more) firewater pumps that take suction from a common header. All pumps would fail if the header were inadvertently blocked, plugged, or ruptured. As another example, several pumps were used to help ensure an adequate and continuous supply of feedwater to steam boilers at the powerhouse for an oil refinery. However, the inadvertent operation of a single low-level switch in the feedwater surge tank caused simultaneous tripping of all boiler feedwater pumps.

The **common hardware** coupling factor has also been observed among redundant instrumentation, control/data acquisition equipment, and (to a lesser degree) protection systems. For example, Reference 10 discusses four one-out-of-two redundant systems that failed a total of 23 times because of hardware failures in shared equipment (bus, bus switches, wiring, etc.). In fact, based on the failure experience of redundant computer systems, Reference 10 concludes that failures within common or shared equipment (e.g., output modules) are one of the most important contributors to the frequency of failure of fault-tolerant DCSs typically used in CPI facilities.

CCF Coupling Factor 3: Equipment Similarity

Most CCFs observed in several industries have involved similar equipment. This is primarily due to similar equipment being affected by common design and manufacturing processes, the same installation and commissioning procedures, the same operating policies and procedures, and the same maintenance programs. These commonalities allow for multiple failures that are due to systematically repeated human errors or other deficiencies. For example, two redundant circuit breakers in the reactor protection system at a nuclear power plant in Germany failed to open. Investigation of the event revealed that, because of a deficiency in the manufacturing of the breaker contacts, the coating on the contacts melted during reactor operation and fused the contacts together. Both redundant breakers were manufactured following the same process and procedures, and obviously they were both susceptible to (and affected by) the same deficiency.

Equipment similarity has also been an important contributing factor to maintenance-related failure events. For example, during routine maintenance of a commercial aircraft, a maintenance mechanic failed to install an O-ring seal in each of the three jet engines. Shortly after takeoff, all three engines shut down after the lubricating oil was consumed because of the missing seal. Fortunately, one engine was able to be restarted, allowing the pilot to perform an emergency landing. The cause of this incident was that, unknown to the mechanic, the storeroom had changed the normal stocking procedure (because of a packaging change from the parts manufacturer) and now stocked the O-ring seal separate from the other components in the lube oil seal replacement kit. The similarity of the piece-parts (and maintenance procedures) resulted in the mistake being systematically made on all three engines.

CCF Coupling Factor 4: Common Location

Equipment in the same location may be susceptible to failure from the same external environmental conditions, including sudden, energetic events (earthquake, fire, flood, missile impact, etc.) and abnormal environments (excessive dust, vibration, high temperature, moisture, etc.). For example, redundant electronic equipment in a room could fail because of a fire in that location or from high temperature if the air conditioning system for that room fails.

Regarding sudden, energetic events, Reference 11 discusses two unrelated air tragedies (a Japan Air Lines Boeing 747 and a United Airlines DC-10) that resulted from the loss of redundant hydraulic systems caused by damage to the redundant hydraulic lines in the rudder of each aircraft; in both cases, all hydraulic lines were close together (common location). According to documents from the National Transportation Safety Board and the Federal Aviation Administration, the DC-10 accident resulted in 111 fatalities and numerous injuries when the plane crashed during an emergency landing in Sioux Gateway Airport, Iowa. It was caused by catastrophic failure of the tail-mounted engine during cruise flight. The separation, fragmentation, and forceful discharge of the stage 1 fan rotor assembly led to severing or loosening the hydraulic lines in the rudder of the aircraft, which in turn disabled all three redundant hydraulic systems that powered the flight controls.

Regarding abnormal environments, operational experience in CPI and other industrial facilities indicates that the **common location** coupling factor is often strengthened by the **equipment similarity** coupling factor because similar equipment has the same (or similar) stress-resisting capacity (strength) to environmental causes. Thus, similar components are more likely to fail simultaneously if the environmental-induced stress exceeds the strength of the components. Dissimilar components generally have different strengths regarding environmental causes, and the weakest component is likely to fail first, allowing the operating/maintenance staff to detect and correct the problem before additional failures occur.

CCF Coupling Factor 5: Common Internal Environment

The internal environment (air in an instrument air system, electric current in an electrical distribution system, water in an emergency cooling system, fluid in a hydraulic system, etc.) sometimes causes or contributes to safeguard failures. These events can fail multiple safeguards if the internal environment is the same or similar for these safeguards. An example mentioned earlier is a power surge in the electric supply to two firewater pumps A and B that could damage the electric motor on each pump. A more common example in CPI facilities is grass and other debris causing strainers in river water pumps to plug, resulting in loss of suction to redundant pumps. Redundant river water pumps have also failed because of accelerated internal erosion from abnormally high concentrations of sand in the water. Another typical example of this coupling factor in CPI facilities is contamination (e.g., moisture) in the air supply to pneumatically operated valves in safety-related applications, which has caused multiple failures on several occasions.

CCF Coupling Factor 6: Common Operating/maintenance Staff And Procedure

Some catastrophic accidents were the result of human or procedural errors such as misoperation, misalignment, and miscalibration of multiple safeguards. Theoretically, all safeguards (similar or dissimilar) operated or maintained by the same staff or addressed by the same procedure (written or otherwise) are susceptible to failure from a CCF. In the well-publicized accident at the Three Mile Island (TMI) nuclear power plant in the United States, the plant operators (acting on inadequate and misleading information) actually shut down the redundant trains of the emergency core cooling system (ECCS). The ECCS had started automatically to respond to a small loss of coolant event, and the operator action eventually led to uncovering the reactor core and core damage.

Operational experience indicates that when misalignment, miscalibration, and some other types of staff/procedural errors result in multiple failures, they often involve similar equipment. That is, this coupling factor is often strengthened by the **equipment similarity** coupling factor (or vice-versa). This is understandable because multiple misalignment and miscalibration errors are more likely to occur when the equipment involved is similar. For example, the likelihood of inadvertently closing a redundant set of valves A and B while attempting to close another set of valves C and D is much higher if these two sets of valves look the same. Also, the **common location** and **equipment similarity** coupling factors together can strengthen the **common operating/maintenance staff and procedure** coupling factor. For example, if an operator misaligns valves in one train of equipment, there is an increased likelihood of misaligning the valves on the redundant equipment if the redundant equipment is similar and is in the same location; the operator could rely on the incorrect alignment of one train to align the other train.

As another example of this coupling factor, on April 26, 1986, the worst accident in the nuclear power industry occurred at Chernobyl Unit 4 (Chernobyl-4) during a test designed to assess the reactor's safety margin in a particular set of circumstances. Descriptions of the details of the incident are somewhat inconsistent, but it has been established that the automatic trip systems on the steam separators were *deactivated* by the operators to allow the test. That is, multiple safeguards were disabled by the operators. (The ECCS was also isolated prior to the test, but experts now believe this had little impact on the outcome of the accident.) Because this type of reactor has a positive void coefficient (i.e., water turning into steam in the core *increases* the reaction rate [and power generation]), controlling pressure and temperature in the core is particularly critical; the misoperation of safeguards (deactivation of the trip systems) disabled the protection against inadvertent steam generation in the core. Subsequent actions by the operators while conducting the test resulted in an uncontrolled generation of steam in the core, causing the reactor power to peak about 100 times above the design power.

Identifying and Quantifying CCFs in CPI Facilities

As part of process hazard analyses (PHAs) required by government regulations,[1] CPI facilities use a number of methodologies (what-if, checklist, hazard and operability analysis, failure mode and effects analysis [FMEA], etc.) to identify process hazards and to reduce the risk that these hazards pose to employees, the public, and the environment. As specified in the regulations, PHAs must identify engineering and administrative controls (safeguards) applicable to the hazards as well as the consequences of failure of these safeguards. The higher the frequency of specific accident scenarios and the more severe the consequences of these scenarios, the more the PHA team focuses on the need for additional and/or more reliable safeguards. Because they can compromise the reliability of multiple safeguards, CCFs should be considered in the PHA team's evaluations.

The coupling factors previously defined provide the basis for identifying CCF potential among multiple safeguards; every set of multiple safeguards applicable to a potential accident scenario must be reviewed for the existence of CCF coupling. In our experience, PHA teams quickly learn the basic concepts associated with CCF coupling, and it is typically easy to incorporate CCF considerations in the PHA evaluations without significant impact on time and budget to perform these studies. The next two sections discuss the identification of potential CCFs and defenses against CCF coupling.

In addition to identifying potential CCFs, sometimes it is useful to estimate the probability of occurrence of CCFs. This quantitative information can improve the PHA team's understanding of risk, thereby helping develop PHA recommendations. Quantitative information is also useful if additional analysis (e.g., quantitative risk assessment [QRA]) is performed.[12,13] (See Quantifying CCFs section for further discussion.)

Identifying CCFs

Table 1 summarizes key points in the identification of CCFs. The coupling factors **common support system** and **common hardware** are usually apparent on piping and instrumentation diagrams (P&IDs), logic diagrams for interlock and shutdown systems, and other process safety information (PSI) documents. PHA teams generally review these diagrams and PSI documents as part of the PHA, and the review should reveal these types of dependencies.

However, PHA teams do not necessarily review all of this information in sufficient detail to identify subtle support system dependencies or hardware dependencies. For example, a detailed FMEA of the fault-tolerant (F-T) DCS (a Honeywell TDC 3000), which controls a fluidized catalytic cracking (FCC) unit

---

[1] Occupational Safety and Health Administration regulation 29 CFR 1910.119, *Process Safety Management of Highly Hazardous Chemicals*, and Environmental Protection Agency regulation 40 CFR 68, *Risk Management Programs for Chemical Accidental Release Prevention*.

in a large refinery, was performed as part of a special study to supplement the PHA for the FCC unit. The FMEA team included instrumentation and control (I&C) specialists and a technician from Honeywell—the DCS manufacturer. (Although they are often available on an "as-needed basis," these specialists rarely participate extensively in PHAs.) The FMEA involved an in-depth review of the DCS logic diagrams and associated instrumentation, and it revealed some shared instrumentation for interlock systems that had not been identified during the PHA for the unit. Also not identified during the PHA were a few shutdown interlocks that were not "fail safe." In addition, some of the redundant equipment in the F-T DCS was in the same location, thereby being susceptible to failure caused by loss of the heating, ventilating, and air conditioning system.

TABLE I
Key Points in the Identification and Quantification of CCFs for CPI Facilities

Coupling Factor	CCF Identification	CCF Quantification
Common support system	Support system dependencies and common hardware dependencies are usually not of interest if the safeguards "fail safe" upon loss of the support system or common hardware	These coupling factors are highly plant-specific, and plant personnel usually know the frequency of loss of support systems such as electric power and other utility systems. Plant data should be used to evaluate the probability of loss of multiple safeguards resulting from the unavailability of a common support system
Common hardware	These CCF couplings are identified by reviewing P&IDs, logic diagrams for interlock and shutdown systems, and other PSI documents associated with the set of multiple safeguards. Although PHA teams generally review these diagrams and PSI documents as part of the PHA, additional reviews may be required with specialists on each support system (e.g., DCS specialists, including a representative from the manufacturer)	Standard QRA techniques (e.g., fault tree analysis) and generic failure rate data can be used when plant data are not available (e.g., to evaluate the probability of failure of common hardware)
Equipment similarity	Any set of similar safeguards or safeguards that have similar piece-parts is susceptible to this coupling factor	Parametric models (based on empirical data) provide an estimate of the probability of CCF events resulting from this coupling factor. This estimate typically includes the contribution from this coupling factor as well as contributions from at least some of the causes considered in the coupling factors **common location, common internal environment,** and **common operating/maintenance staff and procedure**

Coupling Factor	CCF Identification	CCF Quantification
Common location	Any set of safeguards that are physically in the same location is susceptible to this coupling factor	All CCFs caused by sudden, energetic events (earthquake, fire, flood, hurricane, tornado, etc.) should be analyzed using techniques specially designed for the analysis of each type of event
		Parametric models are used to analyze CCF events resulting from the other causes (abnormal environments) associated with this coupling factor, including excessive dust, vibration, high temperature, moisture, etc.
Common internal environment	Any set of safeguards that have the same or similar internal environment is susceptible to this coupling factor	Parametric models are used to analyze CCF events associated with this coupling factor
Common operating/ maintenance staff and procedure	Any set of safeguards (similar or dissimilar) operated or maintained by the same staff or addressed by the same procedure (written or otherwise) is susceptible to this coupling factor	Operator errors during accidents (i.e., misoperation actions) should be analyzed using human reliability analysis techniques
		Parametric models are used to analyze CCF events resulting from the other causes (misalignment and miscalibration) associated with this coupling factor

Any set of similar safeguards (e.g., three identical temperature switches) is susceptible to the coupling factor **equipment similarity**. However, this coupling factor is not limited to identical, redundant components. Some "dissimilar" equipment (e.g., two pumps from different manufacturers) may have piece-parts (e.g., motor starter and I&C devices) that are similar, thereby being susceptible to this coupling factor. Also, any set of safeguards that are physically in the same location, have the same or similar internal environment, or are operated or maintained by the same staff or addressed by the same procedure (written or otherwise), is susceptible to the following coupling factors: **common location**, **common internal environment**, and **common operating/maintenance staff and procedure**, respectively.

Identifying Defenses against CCF Coupling

An important consideration in the identification of CCFs is the existence of (or lack of) defenses against coupling factors. It is obvious from the previous discussion of coupling factors that a search for coupling factors is primarily a search for similarities in the design, manufacture, construction, installation, commissioning, maintenance, operation, environment, and location of multiple safeguards. A search for defenses against coupling, on the other hand, is primarily a search for dissimilarities among safeguards, including differences in the safeguards themselves (diversity); differences in the way they are installed, operated, and maintained; and differences in their environment and location.

References 8 (pages 21–26) and 14 (Appendix A) discuss defenses against CCFs in more detail. For example, excellent defenses against the **equipment similarity** coupling factor include functional diversity (the use of totally different approaches to achieve roughly the same result) and equipment diversity (the use of different types of equipment to perform the same function). Spatial separation and physical protection (e.g., barriers) are often used to reduce the susceptibility of multiple safeguards to the **common location** coupling factor.

As another example of defense against CCF coupling, *staggering* test and maintenance activities offer some advantages over performing these activities simultaneously or sequentially. First, it reduces the coupling associated with certain human-related failures—those that are introduced during test and maintenance activities. The probability that an operator or technician repeats an incorrect action is lower when test or maintenance activities are performed months, weeks, or even days apart than when they are performed a few minutes or a few hours apart. A second potential advantage of staggering test and maintenance activities relates to the maximum exposure time for CCF events. If multiple safeguards are indeed failed because of a CCF event, and if this type of failure is detectable by testing and inspecting, then evenly staggering these activities reduces the maximum time that the multiple safeguards would be failed because of that CCF event.

Quantifying CCFs

Table 1 also summarizes key points in the *quantification* of CCFs. There are two ways to quantify CCF events:

- Use QRA techniques specially designed for the analysis of the specific causes of interest
- Use a parametric model (e.g., the beta factor or the multiple Greek letter model)[9]

The first two coupling factors in Table 1 (**common support system** and **common hardware**) are highly plant-specific, and they can be quantified using standard QRA techniques specially designed for the analysis of the specific causes of interest. These techniques include generic failure rate data and fault tree analysis.[12,15] However, plant personnel usually know the frequency of loss of support systems (instrument air, steam, etc.), and this information should be used to evaluate the probability of loss of multiple safeguards resulting from the unavailability of a common support system.

Selected causes associated with the coupling factor **common location** should also be quantified using standard QRA techniques specially designed for the analysis of these causes. Specifically, all CCFs caused by sudden, energetic events (earthquake, fire, flood, etc.) should be analyzed using techniques specially designed for the analysis of each type of event. The reason for considering these causes individually is because the techniques that are best suited for one type of event (e.g., estimating the frequency of an earthquake) are generally different from the techniques that are best suited for the other types of events (e.g., estimating the frequency of a hurricane or tornado). In Section 3.3.3, External Event Analysis, Reference 12 presents these techniques in some detail and provides additional references.

Selected causes associated with the coupling factor **common operating/maintenance staff and procedure** should also be quantified using standard QRA techniques specially designed for the analysis of these causes. Specifically, operator errors during accidents (i.e., misoperation actions) should be analyzed using human reliability analysis techniques. This type of human error includes the actions taken during the TMI and Chernobyl-4 accidents previously discussed. In Section 3.3.2, Human Reliability Analysis, Reference 12 presents these techniques in some detail and provides additional references.

Parametric models use empirical data, and they are used to quantify the remaining coupling factors (and the causes associated with a coupling factor that are not analyzed using standard QRA techniques). Specifically, parametric models are used to quantify (1) all causes (inadequate design, manufacturing deficiencies, installation and commissioning errors, environmental stresses, etc.) associated with the coupling factors **equipment similarity** and **common internal environment**, (2) the causes related to abnormal environments (excessive dust, vibration, high temperature, moisture, etc.) associated with the **common location** coupling factor, and (3) the causes related to misalignment and miscalibration associated with the **common operating/maintenance staff and procedure** coupling factor. In Section 3.3.1, Common Cause Failure Analysis, Reference 12 presents parametric models in some detail and provides additional references.

Quantifying CCFs Using a Simple Method

The quantification procedures available for CCFs have been briefly described. These methods are often used as part of QRAs for CPI facilities. However, the detailed and complete quantification of CCF events is not always required or cost-effective; there are some applications in which approximate numbers are adequate to support decisions about safeguards. This section presents a simple method that provides probabilities estimates in the right "ballpark." The simple method consists of a three-step procedure, which is performed separately for each set of multiple safeguards:

- **Step 1:** Review the set of multiple safeguards to identify the coupling factors and the defenses that are in place against coupling. Previous discussions in this paper provide guidance for this identification step, and Table 1 shows the key points in the identification of coupling factors
- **Step 2:** Establish the "strength" of the CCF coupling as **High, Moderate to High, Low to Moderate,** or **Low.** Table 2 provides guidelines for establishing the coupling strength as a function of the coupling factors and defenses identified in Step 1
- **Step 3:** Evaluate the probability of failure for the set of multiple safeguards using Table 3. This probability is a function of the level of redundancy and success logic (one-out-of-two, one-out-of-three, etc.), the probability of failure on demand (PFOD) for a single safeguard, the testing/maintenance strategy (staggered versus nonstaggered), and the coupling strength (**High, Moderate to High, Low to Moderate,** or **Low**)

TABLE 2
Guidelines for Determining the Coupling Strength

High
If one or more of the following coupling factors exist: **common support system, common hardware,** and **common location** (sudden, energetic events only) *AND* the probability of occurrence of the event (support system failure, failure of common hardware, or occurrence of a sudden, energetic event) is in the same order of magnitude as the probability of failure for a single safeguard
[Note: If the probability of support system failures is higher than the probability of single safeguard failures, the support system failures dominate and safeguard redundancy is irrelevant]
Moderate to High
If the **equipment similarity** and **common internal environment** coupling factors exist
Low to Moderate
If one or more of the following coupling factors exist: **common support system, common hardware,** and **common location** (sudden, energetic events only) *AND* the probability of occurrence of the event (support system failure, failure of common hardware, or occurrence of a sudden, energetic event) is about one order of magnitude lower than the probability of failure for a single safeguard
OR
If the **equipment similarity** and either **common location** (abnormal events only) or **common operating/maintenance staff and procedure** coupling factors exist
Low
If none of the conditions for **High, Moderate to High,** or **Low to Moderate** apply

TABLE 3
Probability of Failure for Multiple Safeguards

Level of Redundancy and Success Logic	Testing/Maintenance Strategy							
	Nonstaggered				Staggered			
	Coupling Strength							
	High	Moderate to High	Low to Moderate	Low	High	Moderate to High	Low to Moderate	Low
One-out-of-two								
PFOD = 0.1	5e-02*	3e-02	2e-02	1e-02	4e-02	2e-02	1e-02	1e-02
PFOD = 0.03	2e-02	7e-03	4e-03	9e-04	1e-02	4e-03	3e-03	9e-04
PFOD = 0.01	5e-03	2e-03	1e-03	1e-04	3e-03	1e-03	7e-04	1e-04
PFOD = 0.003	2e-03	6e-04	3e-04	9e-06	1e-03	4e-04	2e-04	9e-06
PFOD = 0.001	5e-04	2e-04	1e-04	1e-06	3e-04	1e-04	6e-05	1e-06
PFOD = 0.0003	2e-04	6e-05	3e-05	<1e-06	1e-04	4e-05	2e-05	<1e-06
PFOD = 0.0001	5e-05	2e-05	1e-05	<1e-06	3e-05	1e-05	6e-06	<1e-06
One-out-of-three								
PFOD = 0.1	5e-02	2e-02	1e-02	1e-03	3e-02	1e-02	5e-03	1e-03
PFOD = 0.03	2e-02	6e-03	3e-03	3e-05	1e-02	3e-03	1e-03	3e-05
PFOD = 0.01	5e-03	2e-03	9e-04	1e-06	3e-03	9e-04	3e-04	1e-06
PFOD = 0.003	2e-03	6e-04	3e-04	<1e-06	1e-03	3e-04	1e-04	<1e-06
PFOD = 0.001	5e-04	2e-04	9e-05	<1e-06	3e-04	9e-05	3e-05	<1e-06
PFOD = 0.0003	2e-04	6e-05	3e-05	<1e-06	1e-04	3e-05	1e-05	<1e-06
PFOD = 0.0001	5e-05	2e-05	9e-06	<1e-06	3e-05	9e-06	3e-06	<1e-06
Two-out-of-three								
PFOD = 0.1	6e-02	5e-02	4e-02	3e-02	5e-02	4e-02	3e-02	3e-02
PFOD = 0.03	2e-02	1e-02	7e-03	3e-03	1e-02	7e-03	5e-03	3e-03
PFOD = 0.01	5e-03	3e-03	2e-03	3e-04	3e-03	2e-03	1e-03	3e-04
PFOD = 0.003	2e-03	9e-04	5e-04	3e-05	1e-03	5e-04	3e-04	3e-05
PFOD = 0.001	5e-04	3e-04	2e-04	3e-06	3e-04	2e-04	8e-05	3e-06
PFOD = 0.0003	2e-04	9e-05	5e-05	<1e-06	1e-04	5e-05	2e-05	<1e-06
PFOD = 0.0001	5e-05	3e-05	2e-05	<1e-06	3e-05	2e-05	8e-06	<1e-06

Level of Redundancy and Success Logic	Testing/Maintenance Strategy							
	Nonstaggered				Staggered			
	Coupling Strength							
	High	Moderate to High	Low to Moderate	Low	High	Moderate to High	Low to Moderate	Low
One-out-of-four								
PFOD = 0.1	5e-02	2e-02	9e-03	1e-04	3e-02	8e-03	3e-03	1e-04
PFOD = 0.03	2e-02	7e-03	3e-03	<1e-06	1e-02	2e-03	8e-04	<1e-06
PFOD = 0.01	5e-03	2e-03	9e-04	<1e-06	3e-03	7e-04	3e-04	<1e-06
PFOD = 0.003	2e-03	6e-04	3e-04	<1e-06	1e-03	2e-04	8e-05	<1e-06
PFOD = 0.001	5e-04	2e-04	9e-05	<1e-06	3e-04	7e-05	2e-05	<1e-06
PFOD = 0.0003	2e-04	6e-05	3e-05	<1e-06	1e-04	2e-05	7e-06	<1e-06
PFOD = 0.0001	5e-05	2e-05	9e-06	<1e-06	3e-05	7e-06	2e-06	<1e-06
Two-out-of-four								
PFOD = 0.1	5e-02	3e-02	2e-02	4e-03	3e-02	1e-02	8e-03	4e-03
PFOD = 0.03	2e-02	8e-03	4e-03	1e-04	1e-02	3e-03	1e-03	1e-04
PFOD = 0.01	5e-03	3e-03	1e-03	4e-06	3e-03	1e-03	4e-04	4e-06
PFOD = 0.003	2e-03	8e-04	4e-04	<1e-06	1e-03	3e-04	1e-04	<1e-06
PFOD = 0.001	5e-04	3e-04	1e-04	<1e-06	3e-04	1e-04	4e-05	<1e-06
PFOD = 0.0003	2e-04	8e-05	4e-05	<1e-06	1e-04	3e-05	1e-05	<1e-06
PFOD = 0.0001	5e-05	3e-05	1e-05	<1e-06	3e-05	1e-05	4e-06	<1e-06
Three-out-of-four								
PFOD = 0.1	7e-02	7e-02	6e-02	6e-02	6e-02	6e-02	6e-02	6e-02
PFOD = 0.03	2e-02	1e-02	1e-02	5e-03	1e-02	9e-03	7e-03	5e-03
PFOD = 0.01	5e-03	4e-03	3e-03	6e-04	4e-03	2e-03	1e-03	6e-04
PFOD = 0.003	2e-03	1e-03	7e-04	5e-05	1e-03	6e-04	3e-04	5e-05
PFOD = 0.001	5e-04	4e-04	2e-04	6e-06	3e-04	2e-04	9e-05	6e-06
PFOD = 0.0003	2e-04	1e-04	6e-05	<1e-06	1e-04	5e-05	3e-05	1e-06
PFOD = 0.0001	5e-05	4e-05	2e-05	<1e-06	3e-05	2e-05	9e-06	<1e-06

*Scientific notation: 5e-02 = 5×10^{-2} = 0.05.

For example, if PFOD = 0.01, the coupling strength is **Moderate to High** for a set of three safeguards configured as two-out-of-three success logic, and safeguards are tested/maintained on a nonstaggered basis, the probability of at least two-out-of-three safeguards failing on demand is 0.003. That is, the probability that at least two safeguards would fail on demand is about one-third of the probability of failure for a single safeguard.

Conclusion

Using multiple safeguards in CPI facilities often reduces risk. However, the very high reliability theoretically achievable through the use of multiple safeguards, particularly through the use of redundant components, can sometimes be compromised by CCF events. CCF events have consistently been shown to be important contributors to risk, and the frequency of accident scenarios in CPI facilities may be grossly underestimated if CCFs affecting multiple safeguards are not taken into account. This paper presented a method to account for CCFs when assessing the effectiveness of safeguards in a CPI facility.

An important observation regarding CCFs is that the potential for the occurrence of CCFs does not imply that there is no value in using multiple safeguards. On the contrary, the use of multiple safeguards has been shown to reduce risks, and the information presented in this paper supports this contention. However, it is important to recognize that CCFs may limit the theoretical benefits achievable through the use of multiple safeguards, particularly through the use of redundant components. A good understanding of CCFs provides a more realistic appreciation of risk in CPI facilities, thereby allowing better decisions to be made about the use of safeguards.

References

1. E. Siddall, "Reliable Reactor Protection," *Nucleonics*, Vol. 15, No. 6, 1957.
2. G. C. Laurence, "Reactor Safety in Canada," *Nucleonics*, Vol. 18, No. 10, 1960.
3. E. P. Epler, "Common Mode Failure Considerations in the Design of Systems for Protection and Control," *Nuclear Safety*, Vol. 10, No. 1, January-February 1969.
4. G. T. Edwards and I. A. Watson, *A Study of Common Mode Failures*, U. K. Atomic Energy Authority, Safety and Reliability Directorate, SRD R146, July 1979.
5. I. A. Watson and G. T. Edwards, "Common Mode Failures in Redundancy Systems," *Nuclear Technology*, Vol. 46, December 1979.
6. H. Paula, "Technical Note: On the Definition of Common-Cause Failures," *Nuclear Safety*, Vol. 36, No. 1, January-June 1995.
7. H. M. Paula, D. J. Campbell, and D. M. Rasmuson, "Qualitative Cause-Defense Matrices: Engineering Tools to Support the Analysis and Prevention of Common

Cause Failures," *Reliability Engineering and System Safety*, Vol. 34, No. 3, 1991.
8. H. M. Paula and G. W. Parry, *A Cause-Defense Approach to the Understanding and Analysis of Common Cause Failures*, NUREG/CR-5460, Sandia National Laboratories, Albuquerque, NM, March 1990.
9. A. Mosleh, K. N. Fleming, G. W. Parry, H. M. Paula, D. H. Worledge, and D. M. Rasmuson, *Procedure for Treating Common Cause Failures in Safety and Reliability Studies: Procedural Framework and Examples*, Vol. 1, NUREG/CR-4780, U.S. Nuclear Regulatory Commission, Washington, DC, January 1988.
10. H. M. Paula, M. W. Roberts, and R. E. Battle, "Operational Failure Experience of Fault-Tolerant Digital Control Systems," *Reliability Engineering and System Safety*, Vol. 39, 1993.
11. J. Stephenson, *System Safety 2000: A Practical Guide for Planning, Managing, and Conducting System Safety Programs*, Van Nostrand Reinhold, New York, NY, 1991.
12. *Guidelines for Chemical Process Quantitative Risk Assessment*, Center for Chemical Process Safety of the American Institute of Chemical Engineers, New York, NY, 1989.
13. A. M. Huff and R. L. Montgomery, "A Risk Assessment Methodology for Evaluating the Effectiveness of Safeguards and Determining Safety Instrumented System Requirements," accepted for presentation at the *CCPS/AIChE International Conference and Workshop on Risk Analysis in Process Safety*, Atlanta, GA, October 1997.
14. H. Paula, E. Daggett, and V. Guthrie, *A Methodology and Software for Explicit Modeling of Organizational Performance in Probabilistic Risk Assessment (Report for Task 1 _ Methodology Development)*, JBFA-261.01R-94, Revision 2 (Draft), JBF Associates, Inc., Knoxville, TN, April 1997.
15. *Guidelines for Process Equipment Reliability Data, with Data Tables*, Center for Chemical Process Safety of the American Institute of Chemical Engineers, New York, NY, 1989.

Risk Acceptance Criteria and Risk Judgment

Chairs **Arthur Dowell, III**
Rohm and Haas Company
William G. Bridges
JBF Associates, Inc.

Benefits of Quantifying Process Hazards Analyses

Thomas I. McSweeney
Batelle Memorial Institute

ABSTRACT

Companies are being faced with an ever-increasing number of requirements to assess the hazards associated with their processes. All processes covered by the U. S. Occupational Safety and Health Administration's Process Safety Management Standard (29 CFR 1910.119) were to be analyzed by May 1997. Most of these same processes must be brought into compliance with the U. S. Environmental Protection Agency Risk Management Plan Rule (40 CFR 68) within 3 years. There are two approaches to meeting these requirements; one is to do the minimum. The other approach embraces the quote found in Chapter 1 of the American Institute of Chemical Engineer's Center for Chemical Process Safety *Guidelines for Chemical Processes Quantitative Risk Analysis* (Ref 1). The quote from Admiral Hymen Rickover:

> We must accept the inexorably rising standards of technology, and we must relinquish comfortable routines and practices rendered obsolete because they may no longer meet the new standards.

To embrace this quote is to recognize that there are benefits in doing more than the law currently requires. Current U. S. laws do not require quantification of process hazards analyses (PHAs). However, as the above quote states, the standards will continue to rise. This paper points out some of the real benefits of going beyond the U. S. Occupational Safety and Health Administration (OSHA) and U. S. Environmental Protection Agency (EPA) requirements for process hazards analyses by quantifying those hazards.

Introduction

Many hazards assessments performed today are qualitative or semi-quantitative. They look systematically at a process and determine that all accidents have been analyzed and that safeguards are in place for preventing and mitigating any of the severe accidents. While much of the focus of the PHA is on accidents, PHAs provide benefits in areas such as product quality and process availability (Ref 2). These benefits arise from the structured analysis and the team approach used in performing analyses. Qualitative analyses can be turned into quantitative analyses by doing at least one or all of the following: using models to estimate the consequences, estimating the frequency of the deviation, or estimating the risk by combining the consequence and the frequency estimate. In this paper, the focus will be on estimating the risk.

The OSHA PSM standard requires that facilities with greater than the threshold quantity of listed chemicals have documented process hazards analyses. The EPA RMP regulation requires regulated sources to conduct hazard assessments, including offsite consequence analyses and, thereby, goes beyond the PSM standard. This is an example of the "inexorably rising standards" of performance and also of a partial quantification of process hazards analyses.

Many companies do not attempt to quantify PHAs because it is not required by regulations. Others consider the quantification not worth the effort. When the quantification of process scenarios is suggested, most people have visions of the Reactor Safety Studies performed for the U. S. Nuclear Regulatory Commission, which were multi-year studies funded at millions of dollars per year. No chemical company has shown the willingness to allocate these kinds of resources toward quantification. However, some chemical companies have shown a willingness to quantify PHAs using order of magnitude estimates of consequences and frequencies. Such quantification is an inexpensive process if the company is already committed to performing PHAs. This paper focuses on the affordable quantification approaches rather than the detailed fault-tree/event-tree formulations performed for nuclear power reactors.

Even though worker safety and the concern for the offsite public seems to be driving many of the regulatory initiatives that have been imposed on the chemical industry, this paper will present the spin-off benefits as well.

Summary of Benefits

If both the consequences and likelihood of a deviation are estimated, the risk associated with that deviation can be calculated. By knowing both the consequences and frequency for a deviation, a worst-case accident required by the EPA RMP requirement can be compared with accidents with a similar consequence but with a very different frequency of occurrence. One of the worked examples quantifies the likelihood of the EMP RMP worst case accident. For the scenario considered, the likelihood of experiencing the worst case accident was less that once in a million years. This perspective is very valuable when discussing the scenarios with the general public. The airplane industry carries the analysis one step further when they compare the risk of flying with other activities. They will frequently state that "the most dangerous part of your trip is the ground transportation segment." Such statements would not be possible if the hazard scenarios had not been quantified.

If the decision is made to quantify the risk of a process, the risks for each deviation can be summed and expressed as a single best-estimate value. If the consequences are then converted to dollars, investment decisions can be made based on comparable risk levels. Two types of comparisons are possible: (1) Ranking processes on a relative basis and improving the process with the greater risk first, and (2) considering the risk reduction made by process improvements and comparing that to the cost of the investment. Those with a high benefit/cost ratio are candidates for funding. Without quantification, such analyses are not possible.

Given the numerous accident sequences, the risk can be expressed as a distribution by plotting the accident consequence measure on the ordinate and the likelihood of an accident with greater consequence on the abscissa. Such a risk spectrum can be used to place the worst case accident required by the EPA RMP in perspective. That worst-case accident becomes a point on a curve, showing the domination of less severe accidents in the risk equation. Experience in the nuclear industry has shown that the general public responds very poorly to worst-case accidents if they are not compared to other accidents. They envision them happening at least once every year. By placing that risk value as a point on a risk spectrum curve, their fears have been somewhat addressed. They can see that the facility has already experienced the dominant accidents and the very severe accidents are much less likely to occur than many natural phenomena hazards. Without quantification such a perspective would not be possible.

An essential element of the quantification is estimating the effectiveness of safety features. One of the worked examples demonstrates the relative importance of safety features. This information can be used in several ways. First, the most-relied-upon safety features can be designated as important and the surveillances of those features can be documented in auditable records. In addition, it is possible to model the change in effectiveness of the safety feature following process changes or upgrades. If there are several different processes with different safety systems, the relative importance of the safety systems is an important fact when allocating resources for surveillances. Lastly, given the estimated availability of safety systems, surveillance data can be used to determine the frequency of the inspections. Once again the use of valuable resources can be focused on enhancing safety.

In conclusion, quantifying PHAs can be thought of a section of the line drawn in Figure 1. At the bottom of the curve, are the requirements to comply with OSHA PSM and EPA RMP regulations. That brings the knowledge base to the steepest part of the curve. Quantifying process hazards analyses is the next logical step in expanding a company's safety knowledge. With a little expenditure in effort, e.g., quantifying the process hazards analyses, the curve becomes almost vertical. Your knowledge and understanding of the process and its controlling parameters greatly expands with only a small additional expenditure in effort. Indeed, it has been our experience that the quantification can easily be

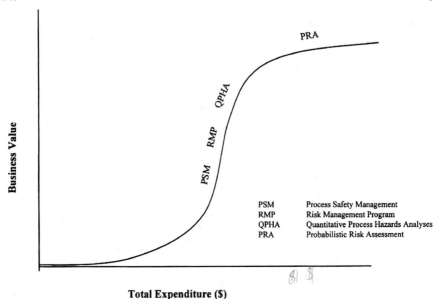

FIGURE 1. Value of Performing Process Hazards Analyses

done while performing more qualitative assessments, e.g., Hazard and Operability (HAZOPs) Studies. At this stage, quantification maximizes the benefits and minimizes the incremental cost.

Worked Examples and Their Benefits

In the following sections, five worked examples will be presented. The benefits shown cannot be realized without determining the risk that is associated with the process being evaluated. The examples are quantifying an RMP example, quantifying a PSM example, prioritizing critical safety components, evaluating a proposed design change, and justifying funding for a process improvement.

Quantifying a RMP Example

The owner or operator of a stationary source that has more than the threshold quantity of a regulated substance, as specified in 40 CFR 68.115, must present to EPA the results of a worst-case accidental release scenario for each covered process. Such a release scenario would be identified as part of the required process hazards analysis. This example will look at the benefits of quantifying just the worst-case example.

In preparing the worst-case scenario, the owner or operator is required to use many conservative factors. For the toxic scenario, the analysis must use a wind speed of 1.5 m/s and atmospheric stability class F unless it can be demonstrated that, during the last 3 years, dispersion parameters have always been more favorable. The release is to be at ground level and no credit can be taken for active mitigation measures. The affected population is determined by drawing a circle around the release point with the radius set at the toxic end point.

The EPA does not require the owner or operator to estimate the frequency of occurrence of this worst-case scenario. What benefits would be derived from the quantification of this analysis? How difficult would it be to estimate the frequency of the worst-case accident? This evaluation will only quantify the conservatism in some of the terms.

Assume the initiating event for the worst-case accident being quantified would occur with a frequency of 1/10 years if there were no preventive safety features present. If preventive safety features are effective, the accident sequence is never initiated. It should be quite easy to show that the frequency of the initiating event is not greater than 1/10 years. Such a high value implies that some of the preventive safety features might be tested over a 10-year operating period. A high-pressure alarm on a vessel might be considered a preventive safety feature if it would be possible for an operator to lower the pressure before the release occurs. The worst-case analysis cannot consider such active safety features. Most systems have some and it should be possible to show that these systems would be effective 999 times out of 1000. Thus the probability they would not be effective is 10^{-3}. The worst-case meteorology will occur less than 5% of the time. In addition, the entire population around the release point will not be exposed to each release. As shown in Figure 2, the plume typically covers less than 10% of the

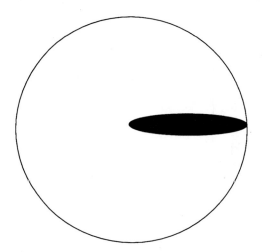

FIGURE 2. Actual Isopleth Area versus RMP Worst-Case Area

area within the circle. This is particularly true for highly stable plumes, e.g., F Stability and 1.5 m/s. Thus, just taking these conservatisms into account, the likelihood of experiencing the worst-case scenario is:

$$F = 10^{-1} \times 10^{-3} \times 0.05 \times 0.1 = 5 \times 10^{-7}/\text{year}$$

Other factors could also be considered. Perhaps only 50% of the releases would be at ground level. Elevated releases may disperse significantly before exposing any of the potentially affected population. Of course the more detailed quantification can show increased probability of exposure as well. If the population around the site is not uniformly distributed, the prevailing wind might be toward the population most of the time. This might require that the 0.1 factor in the above equation be removed.

Regulatory bodies like worst-case analyses because they are valuable tools for evaluating the effectiveness of protective and mitigative safety features. Properly used, they can be valuable to the company as well. However, by themselves, they provide a poor estimate of the overall risk. As shown in Figure 3, the worst-case accident represents just one point, denoted by the "x" on the risk spectrum curve. Companies that quantify their PHAs can make their process improvement decisions based on the entire risk curve not just the "x" point on the curve. Since the "x" represents the RMP worst-case analysis, the accident scenario represented by point "x" considers only passive safety features. As a general rule, passive safety features are more desirable than active features. However, when the risks are thought to be too high, using the entire curve enables the decision-maker to evaluate process improvements that consider both active and passive safety features.

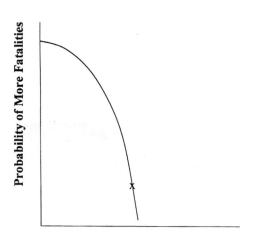

Number of Fatalities
FIGURE 3. Risk Spectrum

Quantifying a PSM Example

HAZOPs are frequently performed to satisfy the OSHA PSM standard. The columns in a typical HAZOP are: GUIDE WORD, DEVIATION, CAUSE, CONSEQUENCES, SAFEGUARD, SCENARIO #, and ACTIONS/COMMENTS. Each individual line of the HAZOP defines an accident scenario. For the consequences to be realized, the deviation or cause must occur and all the protections must be ineffective. While it is at the discretion of the HAZOP team, no modeling is commonly performed to estimate the consequences. The consequences statements might range from "No Significant Consequence" to "Potential for Injuries or Fatalities." The focus of the analysis is on the existence or absence of effective "Safeguards." The word "protection" is being used to cover several types of safety systems. A safety system can "Prevent" the initiation of a deviation; it can provide "Protection" by stopping the propagation of the accident sequence; or it can provide "Mitigation" of the accident sequence should the deviation propagate to a release. The adequacy of the "Protection" is a judgment call made by the HAZOP team.

A well-functioning and experienced HAZOP team can make accurate calls of the severity of the consequences and the adequacy of the protection. By quantifying the deviations listed in the HAZOP, it is possible to document the accuracy of these judgments. This procedure is described in the following paragraphs.

The example presented was taken from a HAZOP performed on a chlorinator used for a potable water supply for a large industrial complex (Ref. 3). The inventory is sufficient to fall under the OSHA PSM standard. A schematic of the chlorinator is shown in Figure 4. For the HAZOP, it was broken up into four

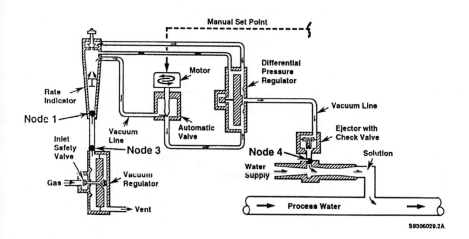

FIGURE 4. Chlorination Process

nodes and two procedural steps. The procedural steps are removing an empty chlorine cylinder and installing a full replacement chlorine cylinder. In the example, they are designated as Node 1 through 6. One page of the HAZOP, for Node 2, is shown in Table 1.

The quantified HAZOP, with two additional columns, is shown in Table 2. One column shows the likelihood of the deviation and the other is the effectiveness of the protection system, titled "EFF" for "effectiveness." The numbers shown are order of magnitude estimates. Additional details on the use of order of magnitude estimates are provided in Reference 4. The key to this method is estimating the likelihood and effectiveness terms. The most difficult term to estimate is the likelihood term because it must be estimated without any of the protections. Since this facility had been operating for many years, members of the HAZOP team often found it easier to estimate the likelihood of the deviation with some or all the protections present and then work backward to the likelihood of the deviation. There are some standard references that can be used as a guide. In the area of human error, Reference 5 provides data on the probability that an individual will make an error, such as a skipped surveillance. Effectiveness typically ranges from "–1" to "–2," meaning that even well-trained operators will miss a surveillance item at least once in a hundred times. For equipment such as heaters that fail on, there are data in Reference 6 can be used. The mean value for the probability of a heater failing on is given as 10^{-7}/hr which converts 10^{-3}/yr or a "–3" for the "order of magnitude" likelihood estimate shown in Table 2.

If there are two operators, there is the possibility that the second operator will detect a deviation when the first operator did not. The second operator would be listed as a "protection." If the operator was trained in the process, the probability that the second operator would detect the deviation might typically be assigned a "–1." This implies that the second operator would detect the first operator's error nine out of ten times. Equipment reliability is typically higher, if the equipment is regularly tested and inspected.

Since these are order of magnitude estimates, the availability of some systems can be inferred from the known availability of other systems. Based on Reference 7, the availability of fusible link actuated water sprinkler systems is "–3." HALON systems, on the other hand, have a reliability of "–2." Such reliabilities could probably be extended to related systems such as water curtains or deluge systems. The recent AIChE CCPS initiative, (Ref. 8), when further along, will collect failure rate data more directly applicable to the chemical industry.

In the chlorinator example, a total of 22 protective features were identified. Some provide safety against serious injury of death. Others provide assurance that no one becomes sick from inadequately chlorinated water. If there are multiple outcomes for the scenario, an order of magnitude estimate of the differences in consequences must be estimated to determine the importance of the safety features. In this example, the severity of a single fatality was assigned a "6" and the severity of an illness from inadequate chlorination was assigned a "3." If multiple

TABLE 1. Portion of a HAZOP for a Chrorinator Facility

PLANT OPERATION	Water Treatment Facility / Chlorination Process		REVIEW DATE:	
LINE/VESSEL/NODE	Node 2		DRAWING NO:	Chlorination Process Flow Diagram
DESIGN INTENTION	Storage cylinder provides Cl_2 gas to regulator at 65 °F to ambient, approximately 75 psig to 150 psig (at 110 °F); 20 – 60 lbs Cl_2/day		REVIEW TEAM:	

GUIDE WORD	DEVIATION	CAUSE	CONSEQUENCES	PROTECTION	SCE-NARIO	ACTION/COMMENTS
No	No Cl_2 Provided	Tank is empty	Decreased Cl_2 residual in water. Violates state code (W-AC 246-290); takes 1 – 2 hours to occur. If continued undetected, a bacterial problem could result in illness across the site (within a day)	1) Automatic Switch over if valves G-1 and G-5 are closed 2) High Vacuum Alarm; the operator diagnoses and restores the system if possible 3) Weight Check 4) Low Cl_2 Residual during surveillance a) The filter plant is checked every 2 hours b) The tour operator checks around the grid (at 12 points/shift) Mitigation: Restrict usage of potable water when low Cl_2 is detected	2-1	Sufficient Protection
		Tank valve is closed	Same as # 2-1	Same as # 2-1	2-2	Same as # 2-1
		Internal tank tubes are plugged/defective (blocked)	Same as # 2-1	Same as # 2-1	2-3	Same as # 2-1
More	More Cl_2 Provided	No Causes Identified				
Less	Less Cl_2 Provided	Valve partially closed Internal tubes partially plugged	Same as # 2-2 and # 2-3 except it takes longer to occur	Same as # 2-2 and # 2-3	2-4	Same as # 2-2 and # 2-3
More	High Temperature	Heater fails "on" during summer heat	Temperature greater than 160 °F, the fusible link may release, resulting in a Cl_2 release Potential for injuries and fatalities near the 315 building and neighboring buildings	Tour operator notices high temperature in room during 2-hour check Mitigation: Cylinder repair kit to reduce size of release	2-5	Calculate temperature based on heat input versus heat loss for this scenario; base further recommendation items on the results.

TABLE 2. Portion of the Chlorinator HAZOP Showing Quantification of the Devitions

PLANT OPERATION	Water Treatment Facility / Chlorination Process	REVIEW DATE:	
LINE/VESSEL/NODE	Node 2	DRAWING NO:	Chlorination Process Flow Diagram
DESIGN INTENTION	Storage cylinder provides Cl₂ gas to regulator at 65 °F to ambient, approximately 75 psig to 150 psig (at 110 °F); 20 – 60 lbs Cl₂/day	REVIEW TEAM:	

GUIDE WORD	DEVIATION	CAUSE	LIKE-LIHOOD	CONSEQUENCES	PROTECTION	EFF	SCE-NARIO	ACTION/COMMENTS
No	No Cl₂ provided	Tank is empty	1	Decreased Cl₂ residual in water; Violates state code (WAC 246-290); takes 1 – 2 hours to occur If continued undetected, a bacterial problem could result in illness across the site (within a day)	1) Automatic Switch over if valves G-1 and G-5 are closed 2) High Vacuum Alarm; the operator diagnoses and restores the system if possible 3) Weight Check 4) Low Cl₂ Residual during surveillance c) The filter plant is checked every 2 hours d) The tour operator checks around the grid (at 12 points/shift). Mitigation: Restrict usage of potable water when low Cl₂ is detected	-3 -2 -1 -2	2-1	Sufficient Protection
		Tank valve is closed	0	Same as # 2-1	Same as # 2-1		2-2	Same as # 2-1
		Internal tank tubes are plugged/defective (blocked)	-1	Same as # 2-1	Same as # 2-1		2-3	Same as # 2-1
More	More Cl₂ Provided	No Causes						
Less	Less Cl₂ Provided	Valve Partially Closed Internal tubes partially plugged	0	Same as # 2-2 and # 2-3 except it takes longer to occur	Same as # 2-2 and # 2-3		2-4	Same as # 2-2 and # 2-3
More	High Temperature	Heater fails "on" during summer heat	-3	Temperature greater than 160 °F, the fusible link may release, resulting in a Cl₂ release Potential for injuries and fatalities near the 315 building and neighboring buildings	Tour operator notices high temperature in room during 2-hour check Mitigation: Cylinder repair kit to reduce size of release	-2 -1	2-5	

fatalities or injuries were postulated, the consequence measure was increased an order of magnitude. As discussed in Reference 4, the relative differences in consequences is more important than the absolute numbers.

Table 3 shows a summary of the deviation sequences using the order of magnitude approach. The scenario number is in the first column and the next column is the estimate of the number of operations per year. For nodes 5 and 6, the designation "1" is used because about 10 exchanges of cylinders are made in a year. Using order of magnitude estimates, 10 is equivalent to 101 which is then designated as a "1" in the table. For Nodes 1 through 4, the number of operations per year was set at 0, meaning that this term was not used as part of the risk equation. Any effect of this term was included in the likelihood term, shown in the third column in the table. This is the likelihood of the initiating the event occurring. The fourth, fifth, and sixth columns are the estimated consequences, sum of the effectiveness for the protection systems and the risk estimate respectively.

TABLE 3
"Order of Magnitude" Estimates of the Risk Parameters

Deviation	Operations/Yr	Likelihood	Sum Of Effectiveness	Severity	Deviation Risk
1-1	0	1	-7	4	-2
1-4	0	-1	-9	4	-6
1-9	0	-1	-6	4	-3
1-13	0	-2	-5	7	0
2-1	0	0	-8	4	-4
2-5	0	-3	-3	7	1
2-8	0	-4	-2	7	1
3-1	0	-1	-6	4	-3
3-2	0	-3	-5	6	-2
4-1	0	-2	-7	4	-5
4-5	0	-3	-7	4	-6
4-9	0	-2	-6	4	-4
4-12	0	-4	-4	7	-1
4-13	0	0	-7	7	0
5-4	1	-3	-6	7	-1
5-6	1	-3	-6	3	-5
5-9	1	-3	-3	3	-2
5-11	1	-3	-1	3	0
5-15	1	-3	-3	3	-2
5-24	1	-2	-5	7	1
5-30	1	-4	-2	3	-2
6-2	1	5	-3	7	0
6-8	1	-4	-2	7	2
6-13	1	-3	-3	4	-1
6-19	1	-2	-4	6	1
6-20	1	-2	-4	6	1
6-21	1	-2	-4	3	-2
6-28	1	1	-5	4	1

Since all the numbers are order of magnitude values, the risk estimate is the sum of the previous columns.

In developing this risk estimate, active safety features that provide both prevention and mitigation have been included in the effectiveness term. Mitigation could affect both the consequences and the probability term. By including both types of protection systems, the assumption is being made that if the mitigation system is effective, there are no consequences from the release. If the active mitigation system only reduces the consequences, then separate deviations would have to be used to include the consequences with the mitigation system functioning properly.

Given these consequence measures, the importance of a safety feature can be determined by taking the safety feature away and looking at the change in overall risk. Such a comparison is shown in Table 4. From an overall risk standpoint, the two vacuum alarms and the rigging used to hoist the cylinders were found to be the most important safety features. In other words, the overall risk went up the greatest when these safety features were removed. Many of the remaining safety features were orders of magnitude lower in terms of importance

This method has been used for several purposes. Experience has shown that this method is able to detect those systems that are being heavily relied upon for safety. In a couple of cases, it was decided that much reliance was being placed on a single administrative control. Additional safety features were needed to provide the desired level of risk. On occasion, formal studies were performed to document that the assigned effectiveness could be relied upon to control risk. Such assessments would not be possible without quantifying process hazards analyses.

In this example, the HAZOP was quantified by placing the consequences, the likelihood of the initiating event, and the effectiveness of safety features on a numeric scale. In the example it was assumed that failure rate data had not been collected and, therefore, order of magnitude estimates for all the terms would be used in the quantification. It was then shown that by using this approach, the relative importance of safety features can be determined. These estimates should not be considered to be validated estimates of the actual risk. However, if the decision was made to collect incident and reliability data in the database, it would be a straightforward process to both validate the estimate and to develop the estimate into a probabilistic risk assessment. At any point in the assessment, the decision can be made to carry the quantification no further.

Specifying Surveillance Intervals for Critical Safety Components

If the process hazards analysis has been quantified, it is possible to estimate the importance of safety features, as shown in the PSM example. The next logical step is to keep records of the surveillances and begin to estimate the effectiveness of the safety feature. There are several measures of effectiveness that can be defined. Only two will be considered here: availability and loss of accuracy.

TABLE 4
Relative Importance if Safety Features

Safety Feature	Risk Increase Without Safety Feature
Low Vacuum Alarm	1.02E+04
High Vacuum Alarm	1.01E+04
Certified rigging used to hoist cylinders	1.00E+04
Personal Protective Equipment	2.01E+03
Leak Check	2.00E+03
2 hr Walkthroughs	2.00E+03
Cylinder Weight Check	1.00E+03
Regulator shutoff on loss of vacuum	1.00E+03
Regulator safety valve	1.00E+03
Ammonia Check when opening connections	1.00E+03
Other crew members	1.00E+03
Remote Chlorine Alarm	1.21E+02
Cylinder Repair Kit	1.00E+02
Serviceman notes position of cylinder requires correction	1.00E+02
Local Chlorine Alarm	1.12E+01
Automatic Switch Over if G-1 and G-5 closed	1.01E+01
Low Chlorine Levels during Surveillance	2.10E-01
Tubing Replaced every 2 years to minimize likelihood of leaks	2.10E-02
2 hr Rotometer Check	1.00E-02
Rotometer Check before Disconnecting Tank	1.00E-03
Emergency Procedures	1.00E-03
Check Valve on Ejector	1.00E-04

Availability is defined as the probability that a safety system can provide its design function when needed. Typically, system checks are made based on factory recommended inspection intervals. If this interval is once every quarter, then over a period of a few years, it is possible to develop data on the number of times the system was in the failed state when checked. If the required surveillance is just to check and not maintain the equipment, then the mean time to failure based on one failure is, at most, one quarter out of 12. Thus, the unavailability of the system is no greater than 3 months out of 36. Using the order of magnitude approach, the availability could be assigned a "–1," implying that it is not available 10% of the time. Most safety systems have availabilities much greater than 90%. This means that to get good data, the failure rate data on the many similar systems must be included in the database. This is where a national database, such as being proposed by CCPS, would be valuable.

There are many places in the chemical industry where temperature control is very important. If the control temperature drifts high, a reactor vessel could explode from a runaway reaction. While the set point would have to drift several tens of degrees to get such a violent reaction, the calibration of one controller showed that if it was calibrated once every year, the standard deviation of the trip point was 10°C. If it was calibrated every 6 months, the standard deviation of the results was about 3°C. If it was calibrated every 3 months the standard deviation was about 2°C. Given the standard deviation and assuming the deviations in the set point are normally distributed, it is easy to estimate that the probability of the set point drifting 20°C is 2.3% for the annual inspections, but is less than 10^{-6} for the other cases. Clearly, performing the surveillances more frequently than once a year would greatly improve the reliability of the safety system. If no effort had been made to quantify the importance of safety systems and then attempt to validate their assumed reliability, the safety of the system would not have been adequate.

Evaluating Proposed Change in Total Process Risk

This worked example is actually a situation that would occur during the design phase. It is during this time that "Inherently Safer Chemical Processes " (Ref. 8), can be evaluated. If two processes are being considered, a preliminary hazards analysis is performed on both. Major deviations can be identified and quantified using order of magnitude estimates of risks and impacts. Given the level of detail present in the designs, the assessment probably would not support anything more formal than the order of magnitude estimates. In performing such an assessment, it is valuable to consider more than just public and worker injury or fatality. In one recent assessment, performed at a Department of Energy facility, five additional risks were considered. These were ecological costs; compliance costs; mission impact; mortgage reduction; and social, economic, and cultural costs. The consequences were estimated in each of the categories for each deviation. It was then possible to identify which alternative had the least overall risk. Of course, risk is not the only factor in making an investment decision. The problem is that in the past, it hasn't been enough a part of the investment decision. Given the increasingly heavy burden posed by civil suits and regulatory fines, it is becoming more important to include risk in the decision making process.

Justifying Funding for Process Improvements

In the first example, it is possible to improve the effectiveness of any safety feature and determine the change in overall risk. If the change is significant, then one could justify the cost of making such an improvement. For example, the most important safety feature in the chlorine example was the "automatic shutoff on loss of vacuum." It was given an effectiveness of "–3." It is a straightforward process to change all the effectivenesses to a "–4" and determine the change in risk. The overall risk decreases about less than 1%. Thus, making such a change is probably not worthwhile.

Some of the value of such a sensitivity analysis is lost when only one process is evaluated. A typical plant is made up of numerous processes, all with different risk levels. It makes sense to focus on the process posing the greatest risk. One could look at the safety features associated with that process and evaluate the effect of additional surveillances, which would then increase the reliability of the safety feature. Alternatively, one could change to a more reliable instrument and look at the effect on the overall risk.

If the overall risk is expressed in dollars, then the change in risk resulting from a change in a protection can also be expressed in dollars. The savings can be divided by the estimated cost of making the change to obtain the benefit to cost ratio. A change that is much greater than one is a cost-effective change.

Such a calculation cannot be made without quantifying the process hazards analysis. Once quantified, it becomes possible to justify process improvements on a cost-effective basis. The process works equally well for a process change. If the change results in a larger release, typically the consequences change unless changes in the protection systems are made as well. Since these are also part of the quantified risk equation, they can also be addressed.

Conclusions

The worked examples shown above cannot be performed without quantifying the process hazards analysis. Once quantified, all the risk comparisons shown above can be performed with relative ease. The above examples show the benefits of quantification to any degree, with systematic improvement in safety as the most important benefit.

The examples all use order of magnitude estimates of risk. While more formal probabilistic risk assessments could be performed, for initial risk assessments, the "order of magnitude" estimates provide most of the benefits of the more detailed assessments at significantly less cost. Use of such methods does not preclude using more detailed assessment methods in the future. The order of magnitude estimates are a good springboard to the more detailed assessments. Indeed; the whole premise of quantifying risk is that the risk level will be monitored for change. Are the initiating events or the deviations more frequent than estimated? Are the protection systems less reliable than estimated? By making such continuing assessments, the "order of magnitude" risk assessments gradually turn into more refined quantitative risk assessments. This fits well into the initial quote from Admiral Rickover. The expectation is that, over time, facilities will be expected to meet higher standards of performance.

References

1. Center for Chemical Process Safety. Guidelines for Chemical Process Quantitative Risk Analysis. American Institute of Chemical Engineers, New York, 1989.
2. Jones, D. W., "Lessons from HAZOP Experiences," Hydrocarbon Processing, April 1992, p.77.
3. "Example Process Hazards Analysis of a Department of Energy Water Chlorination Process," DOE/EH-0340, U.S. Department of Energy, Washington, D. C., September, 1993.
4. Johnson, R. W., McSweeney, T. I., and Yokum, J. S., "Resource Optimization by Risk Mapping," Proceedings of the International Conference and Workshop on Risk Analysis in Process Safety, American Institute of Chemical Engineers, Atlanta, GA, October, 1997.
5. A. D. Swain and H. E. Guttmann, "Handbook of Human Reliability Analysis with Emphasis on Nuclear Power Plant Applications," NUREG/CR-1278, August 1983.
6. "IEEE Guide to the Collection and Presentation of Electrical, Electronic, Sensing Component and Mechanical Equipment Reliability Data for Nuclear-Power Generating Stations," IEEE Std 500-1984, Institute of Electrical and Electronic Engineers, New York, New York, 1983.
7. "Fire Sprinkler Facts," National Fire Sprinkler Association, Patterson, New York, 1989.
8. AIChE CCPS Initiative on Collection of Reliability Data, 1997.
9. Bollinger, R. E. et al., "Inherently Safer Chemical Processes," Center for Chemical Process Safety, American Institute of Chemical Engineers, New York, New York, 1996.

An Integrated Quantitative Decision Approach for Risk Management Problem Solving

Paul C. Chrostowski and Sarah A. Foster
The Weinberg Group Inc., 1220 Nineteenth St., N.W., Washington, D.C. 20036

ABSTRACT

A variety of quantitative techniques are available and widely used by risk managers to assess hazard, risk and consequences, including chemical process quantitative risk analysis, human health and ecological risk assessments, product life cycle analysis and operations research. A comprehensive framework that forms linkages between these techniques has not, however, been developed. In this paper, we introduce an integrated quantitative decision approach (IQDA) that encompasses an array of techniques and can vastly improve the usefulness of risk analysis tools in decision-making. Several initial applications of integrated quantitative approaches are presented and discussed with respect to their ability to fit in the IQDA process. The case studies focus on the selection of pollution control equipment for a hazardous waste combustor, identification of remedial alternatives for soils at a chemically contaminated site, and evaluation of chemical emissions from a manufacturing facility. Initial principles for an integrative framework are presented, drawing on the disciplines of decision analysis and operations research, and areas requiring additional work are identified.

Introduction

Risk managers are presented with a plethora of techniques for assessing hazard, risk and consequences. In general, these techniques can be categorized as probabilistic oriented, consequence oriented, or management oriented. Probabilistic-oriented techniques are those that either qualitatively lay the groundwork for, or culminate in, calculating the probability or the probability distribution of an event. Consequence-oriented techniques focus on the outcome rather than just the probabilities under which it may occur. Management-oriented techniques focus on the use of probabilistic and consequence information in the larger context of decision making.

Most engineers practicing in risk management are familiar with a variety of these techniques. The *Guidelines for Hazard Evaluation Procedures*, for example, lists 12 common hazard evaluation techniques, most of which are probabilistic oriented. Many of these techniques are capable of extension into chemical process quantitative risk analysis (CPQRA) through a variety of both determi-

process quantitative risk analysis (CPQRA) through a variety of both deterministic and stochastic methods. CPQRA logically leads to consequences analyses which predict both the probability and outcomes of an event. In addition to its wide use in the chemical industry, variants of CPQRA are starting to be used in analyzing a wider variety of problems from food safety (Osborne and Chrostowski 1995) to medical device failure.

Consequence-oriented techniques include the traditional human health and ecological risk assessments that are performed to satisfy regulatory requirements of USEPA, OSHA, NRC, and/or the FDA. As noted above, consequence-oriented techniques may be conducted sequentially from probabilistic analyses. Conditional consequence risk assessments are also performed that evaluate only outcomes without regard for the probability of the underlying causal event. For example, USEPA's worst case analysis is conditionally predicated on a worst case occurring. Health risk assessments performed under other regulatory programs such as Superfund and the Clean Air Act are also conditional in this sense. A relatively new group of consequence-oriented analytical tools is emerging that is being used by scientists and engineers in product development areas, namely product life cycle analysis (LCA). LCA is a formal process that evaluates the environmental or societal burdens associated with a product, process, or activity and identifies opportunities for affecting environmental improvements or societal benefits associated with the activity being analyzed. Conceptually and operationally, the tools used in LCA are different than those applied in the techniques mentioned above; however, there is a substantial overlap in many of their components and LCA is also amenable to both stochastic and deterministic procedures. Since full-scale LCAs are often difficult to implement, recourse is often made to LCAs that focus on particular portions of a problem. These are often known as limited life cycle analyses or LLCAs. Hubal and Overcash (1993) have published a form of LLCA known as a net waste reduction analysis that focuses on a life cycle inventory based on energy use and waste produced.

Many risk managers, especially those who deal with the financial aspects of risk, are familiar with another series of techniques for evaluating risks and their consequences, the management-oriented techniques. For example, operations research is useful for solving many risk analytical problems using optimization, decision theory, and utility theory tools. Although historically there has been little quantitative linkage between the results of a CPQRA and risk management, it is relatively simple to envision using the output of a stochastic CPQRA (for example from a Monte Carlo simulation) as the input to a decision tree problem that is approached using Bayes' rule.

Figure 1 is a summary of the risk analytical tools that we are considering in this paper. Although these tools are widely applied, the linkages among them are weak and often non-existent. Several authors have discussed individual techniques and their linkages. For example, Davoudian et al. (1994) discuss the linkages between work process analysis and probabilistic safety analysis. Balagopal

Probabilistic-Oriented Methods
Analytical process models
Numerical process models
Maximum entropy analysis (2)
Chaos theory
Fault tree/event tree (5)
Cause-consequence analysis (5)
Failure mode effects analysis (5)

Consequence-Oriented Methods
Life cycle analysis (1)
Health risk assessment
Ecological risk assessment
Comprehensive risk assessment (4)
Risk inventories

Management-Oriented Methods
Decision theory (3)
Operations research
Present value (cost/benefit)
Expected value of including uncertainty (6)
Expected value of perfect information (6)
Life cycle cost analysis

Sources:
(1) SETAC 1993, USEPA 1993
(2) Jaynes 1968
(3) Brooks and Borison 1996, Raiffa 1968
(4) Allenby 1996
(5) CCPS 1992, Schlechter 1996
(6) Dakins et al. 1994

FIGURE 1
Quantitative Methods Considered for Framework

(1989) integrated risk as calculated by an event tree analysis, with cost-benefit analysis. Kolluru and Brooks (1996) developed a conceptual framework for integrating risk assessment with business decisions; however, they focused on risk management rather than risk analysis aspects of the problem. It is apparent from the literature that no one has worked to develop and use a comprehensive framework. Due to this, a substantial amount of information is lost when a risk manager attempts to make strategic decisions using information that is provided from the application of risk analysis tools.

In order to facilitate the formation of linkages and full use of information, we have been in the process of developing an Integrated Quantitative Decision Approach (IQDA). The first application of the IQDA was serendipitous in that we encountered a problem that required the application of several risk analytical techniques for its solution. This solution used stochastic conditional health risk assessment in conjunction with a limited life cycle analysis and engineering reliability analysis to optimize the selection of a complex environmental control strategy. Other simpler applications involve integrating toxic release inventory (TRI) emissions, risk management plan (RMP) analyses, and refined consequence analysis into a single package. Since these initial applications, we have been working to develop an overall framework that encompasses the risk assessment methods discussed above. This paper will focus on the development process of the IQDA.

An Initial Foray into Integrated Approaches: A Pollution Control Problem

Our initial foray into integrated approaches was prompted by a problem that involved an industrial boiler that was under consideration for a permit to burn hazardous waste. Two important facts in this case were the USEPA requirement for a multiple pathway health risk assessment (HRA) and indecision on the part of the applicant as to which types of equipment should be used for particulate emission control. Normally USEPA uses the results of multiple pathway HRAs for comparison to a bright line criterion that determines the adequacy of emission controls. In this case, we decided that the information in the HRA could also be used to help to decide between the pollution control alternatives. The method we used is shown in Figure 2. Four pollution control alternatives were posited in the technology proposal stage (fabric filter, electrostatic precipitator, dry sorbant injection, and scrubber). A net waste reduction analysis in the form of a limited life cycle analysis was performed on the alternatives singly and in combination which identified fabric filters and electrostatic precipitators used singly as the most effective alternatives. A Monte Carlo multiple pathway HRA was performed for each of these alternatives and the outcome was incorporated into an engineering reliability analysis (Figure 3) which integrated capital costs and costs for potential replacement with the probability of environmental compliance. The results showed that there was no significant difference between the alternatives and lead to the recommendation that factors other than cost and environmental compliance may be used in making the final selection. This project showed that more information could be drawn from risk assessments than is commonly assumed, however, there is no adequate decision framework for doing so.

Life Cycle Analysis and Risk Assessment

Risk assessments often fail to address problems whose impacts are spatially or temporally deferred. The classic situation is evaluation of three alternatives for

FIGURE 2. Overview of Integrated Risk Management Application

municipal waste management. Incineration's impacts are immediate (emissions occur during the process) and local. Landfilling's impacts are often deferred in time (until the landfill fails). Recycling impacts are often deferred in space (emissions occur from a processing plant located far from the point of waste generation). These impacts can be addressed by life cycle analysis, however, it is generally unknown how the two techniques of health risk assessment and life cycle analysis compare or which may be most appropriate in a particular circumstance. As a first step in addressing these questions, it was decided to perform a risk assessment and a life cycle analysis in parallel. The case study involved four alternatives for remediating soils at a chemically contaminated site. The risk assessment used worker safety, worker health, and off-site health impacts as risk metrics. The life cycle analysis involved a traditional inventory, followed by scoring using environmental and worker standards. Both procedures were performed for the four alternatives, and the alternatives were ranked in order of diminishing impact. Interestingly enough, the life cycle analysis yielded different results from the risk assessment. In general, the life cycle analysis empha-

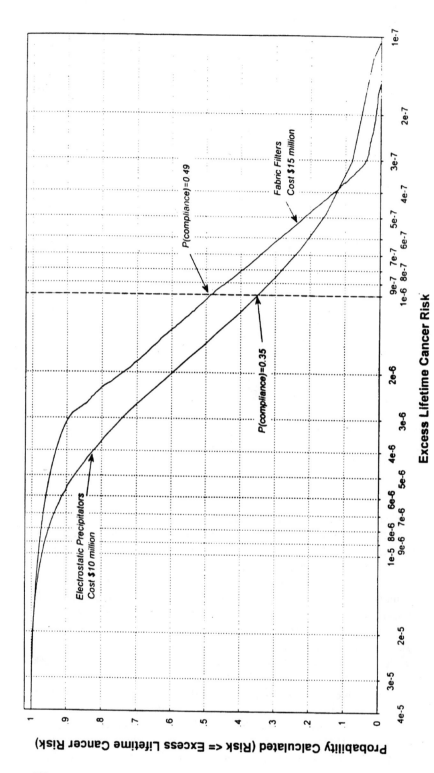

FIGURE 3. Cumulative Density Function. Electrostatic Precipitators and Fabric Filters

sized impacts associated with energy production (e.g., sulfur dioxide from coal combustion), whereas the risk assessment emphasized local impacts of toxic substances. This project showed that life cycle analyses and risk assessments could both be used to evaluate sets of alternatives; however, they may yield different results. These differences may be real or artifacts of the process.

TRI/RMP

USEPA's TRI and RMP programs may be used to illustrate the applicability of several different tools in an IQDA process. TRI reporting has expanded to include almost all industries that emit chemicals. Unfortunately, most of the reporting is done on a mass-balance basis without considering actual chemical releases and without evaluating the consequences of a release. The RMP has turned out to be a worst-case risk analysis which conveys little information to the industry, regulators, or the local community. In some cases, information reported in the TRI has contradicted information used to satisfy the RMP requirements. We overcame some of the shortcomings of these programs by constructing a comprehensive framework that dealt with all emissions whether routine or emergency from a facility. This was accomplished using fault tree analysis to identify the probabilities of emergency releases and refined engineering calculations to identify the rates of both routine and emergency releases. The outcomes of these analyses were then incorporated into a series of air dispersion models and health and safety risk assessments to provide a detailed analysis of consequences. The outcome showed that, with the exception of a low probability emergency release of a single acutely toxic substance, consequences to the neighborhood associated with emissions from this facility were likely to be negligible. The facility managers were then able to direct financial resources into designing "fail safe" measures to protect against release of the single significant substance.

Steps Toward an Integrative Framework

One thing that became apparent after conducting several of these exercises was that the information in most risk assessments was under utilized and that most risk assessments were being conducted outside of a larger risk management context and only used to a limited extent in decision making. For example, performing a risk assessment of TRI releases or industrial boiler emissions will only result in a probability of a consequence. If the risk assessment is performed within the larger context, say of solvent substitution or pollution prevention, it is of substantially greater utility. Thus, the first objective of any risk assessment should be its use as a decision-making tool. It also became apparent that there was a need for an integrated framework to assist decision makers in selecting and using various risk analytical tools.

As a first step, we adopted some of the basic principles of decision analysis as our ground rules (Brooks and Borison 1996):

- In making good decisions, you should combine in logical fashion what you can do (alternatives), what you know (information) and what you want (objectives).
- Alternatives should constitute courses of action, not worries. Unless there is more than one course of action, you don't have a decision problem.
- Information, or lack thereof, should be expressed precisely using probabilities. In a world of uncertainty, making a good decision does not necessarily result in a good outcome.
- Objectives should be expressed precisely using relative preferences. In a world of multiple objectives, making a good decision does not necessarily result in achieving all objectives fully.

As a second step, we adopted a loose framework for analysis and planning from Patton and Sawicki (1993) (Figure 4). Inspection of this framework certainly reveals many analogies to the generally accepted framework for operations research (Figure 5). For example, we can look at evaluation criteria from either a policy analysis or operations research standpoint:

Policy Analysis Criteria	Operations Research Criteria
Regulatory compliance	Dominated action
Technical feasibility	Maximin
Economics	Maximax
Political viability	Minimax regret
Administrative operability	Expected Value

Thus, we decided to incorporate some concepts from operations research into our approach.

More importantly, however, we were interested in how risk analytical tools could be used in the overall framework. The first place that risk analysis appears to fit into the overall framework is in the establishment of evaluation criteria. From a risk standpoint, criteria can be the rank of an alternative compared to all others, a bright line value, a probability distribution, a cost, or the avoidance of a liability. Obviously each of these should be reviewed for its place in the overall framework. The evaluation of alternatives is another obvious place where risk analytical tools can be employed. Figures 6 and 7 illustrate how the examples given above fit (or fail to fit) into the framework. The pollution control problem fit the framework quite well and went on to implementation. The LCA/HRA problem fit the framework reasonably well; however, it failed to produce a con-

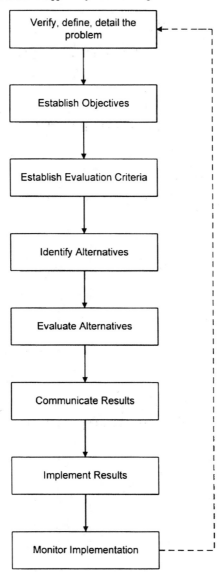

FIGURE 4. Framework for Analysis and Planning

sistent outcome due to conflicts between two of the evaluation criteria. In these situations, one may either alter the criteria or select another alternative. Since, in this case, the criteria were imposed by a regulatory authority, the only recourse was to develop additional remedial alternatives. In the TRI case, the problem

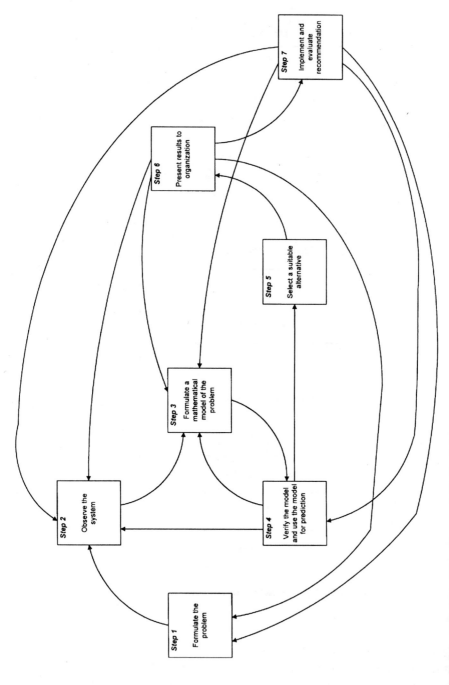

FIGURE 5. The Operations Research Methodology

An Integrated Quantitative Decision Approach for Risk Management Problem Solving

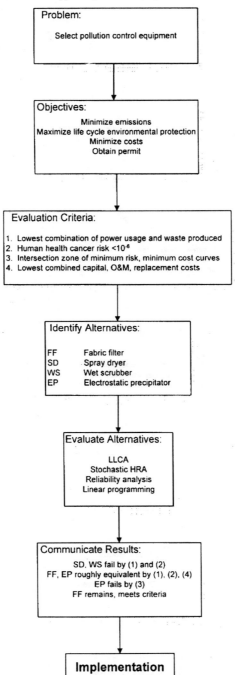

FIGURE 6. The Pollution Control Problem

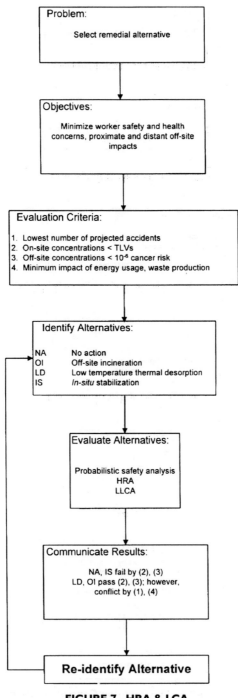

FIGURE 7. HRA & LCA

failed to fit the framework, primarily because of the lack of process management alternatives. On further examination, it was found that an alternative leading to solvent substitution would have been more attractive than one leading to a "fail safe" design; however, alternative analysis was not addressed from the beginning. In this case, use of the framework alone could have resulted in considerable savings of both time and money.

Conclusions

The use of the framework described here formalizes risk assessment/risk management problems. As such, it aids in selecting effective solutions by utilizing a comprehensive and well thought out approach. Once items are placed into an initial framework for planning purposes, it is then necessary to select the appropriate quantitative tools for evaluation purposes. This is the current emphasis of our on-going work. It is obvious that a substantial amount of work needs to be done before this framework has general utility. Work regarding combining dissimilar risk metrics, valuation, and sensitivity analysis of disparate elements is also currently underway. We believe that such a framework, when fully developed, will enable risk analysis to fulfill its promise as a true decision making tool rather than merely a box to be checked in fulfillment of a regulatory program.

References

Allenby, B.R. 1996. Industrial ecology and comprehensive risk assessment. In Kolluru, R., Bartell, S., Pitblado, R., and Stricoff, S. (eds.). *Risk Assessment and Management Handbook for Environmental, Health, and Safety Professionals.* New York: McGraw-Hill, Inc. Pp. 17.3–17.16.

Balagopal, V. 1989. Total probable risk analysis: A technique for quantitative risk evaluation of hazardous waste disposal options. *Hazardous Waste and Hazardous Materials* 6(3):315–325.

Brooks, D.G. and Borison, A. 1996. Risk-based decision making: Integrating risk management into business planning. In Kolluru, R., Bartell, S., Pitblado, R., and Stricoff, S. (eds.). *Risk Assessment and Management Handbook for Environmental, Health, and Safety Professionals.* New York: McGraw-Hill, Inc. Pp. 18.1–18.14.

Center for Chemical Process Safety (CCPS). 1992. *Guidelines for Hazard Evaluation Procedures.* Second Edition with Worked Examples. New York: American Institute of Chemical Engineers.

Dakins, M.E., Toll, J.E., and Small, M.J. 1994. Risk-based environmental remediation: Decision framework and role of uncertainty. *Environ. Toxicol. Chemistry* 13(12):1907–1915.

Davoudian, K., Wu, J.-S., and Apostolakis, G. 1994. Incorporating organizational factors into risk assessment through the analysis of work processes. *Reliability Engineering and System Safety* 45:85–105.

Hubal, E.A.C. and Overcash, M.R. 1993. Net waste reduction analysis applied to air pollution control technologies. Air & Waste 43:1449-1454.

Jaynes, E.T. 1968. Prior probabilities. *IEEE Tran. on Sys. Sci. and Cybernetics*, SSC-4(3). As cited in Englehardt 1997.

Kolluru, R. and Brooks, D.G. 1996. Integrated risk assessment and strategic management. In Kolluru, R., Bartell, S., Pitblado, R., and Stricoff, S. (eds.). *Risk Assessment and Management Handbook for Environmental, Health, and Safety Professionals*. New York: McGraw-Hill, Inc. Pp. 2.1–2.23.

Osborne, C.G. and Chrostowski, P.C. 1995. Quantifying the tree. Food Quality. June/July. Pp. 32-38.

Patton, C.V. and Sawicki, D.S. 1993. *Basic Methods of Policy Analysis and Planning*. Second Edition. Englewood Cliffs, NJ: Prentice Hall.

Raiffa, H. 1968. *Decision Analysis: Introductory Lectures on Choices Under Uncertainty*. Reading, MA: Addison-Wesley. As cited in Dakins et al. 1994.

Society of Environmental Toxicology and Chemistry (SETAC). 1993. *A Conceptual Framework for Life-Cycle Impact Assessment*. Pensacola: SETAC and SETAC Foundation for Environmental Education, Inc.

Schlechter, W.P.G. 1996. Facility risk review as a means to addressing existing risks during the life cycle of a process unit, operation or facility. Int. J. Pres. Ves. & Piping 66:387–402.

U.S. Environmental Protection Agency (USEPA). 1993. *Life-Cycle Assessment: Inventory Guidelines and Principles*. Washington, DC: Office of Research and Development. EPA/66/R-92/245.

CCPS Equipment Reliability Database: A Solid Foundation

Harold W Thomas
Air Products and Chemicals, Inc., Allentown, PA.
© Air Products and Chemicals, Inc.

ABSTRACT

This paper details the current CCPS equipment reliability database effort, as well as the database design, including the overall data structure and relationships that exist. In addition, various examples illustrate how the various data can be used to the benefit of participating companies. Potential analyses include estimation of equipment MTTFs for repairable and non-repairable systems, plant availability and/or reliability, optimization of maintenance schedules/strategies, and modeling of equipment failure distributions.

Introduction

In 1989, the CCPS *Guidelines for Equipment Reliability Data* was published with the intent to support chemical process quantitative risk assessments. At the conclusion of this project it was apparent that better data was needed. To obtain such data, it was further determined that the emphasis had to change from risk assessment to reliability and tools to improve maintenance decisions, as these were the disciplines who were in control of the maintenance management systems in the plants where the data resided. By doing this, companies would be able to improve their operating performance and risk analysts would have the best, highest quality data ever imagined. Over time, the composition of the committee was modified to include the additional disciplines necessary to broaden the effort. In January of 1996 all the pieces fell into place, allowing the current project to formally begin.

Current Mission

The mission of the current project is to begin operation of an equipment reliability database, making available high quality, valid, and useful data pertaining to the hydrocarbon and chemical process industries enabling analyses to support availability, reliability, and equipment design improvements, maintenance strategies, and life cycle cost determination.

To accomplish its objective, the CCPS has formed a group of sponsor companies to support and direct this effort. The sponsor companies, through representatives on a Steering Committee, are actively participating in the design and implementation of the database.

Benefits

Sponsors, through participation, gain the knowledge that enables them to develop meaningful plant management systems that provide a foundation for data collection and analysis. Learning comes through project development and hands on involvement. The successful implementation of this project is expected to yield several benefits of increasing value as our knowledge and tools evolve over time. These include:

- The guidelines book, quality assurance protocols, and software that is developed will reduce the engineering effort at individual companies with respect to the design of new maintenance management systems or revisions to existing ones.
- Cost of obtaining quality data will be reduced as companies improve the compatibility of their plant information management systems with the CCPS software data fields and definitions, allowing for electronic transfer of data from participating companies to CCPS without an additional data input step as seen in many other database efforts. This in turn will encourage an increasing number of companies to become data contributors.
- Quality data will allow an increasing level of sophisticated analysis. These include but are not limited to plant/unit op availability, MTTF and MTTR for plants, unit operations, and equipment systems, predict plant reliability, support of risk based inspection decisions, quantitative safety analysis input, and allow life cycle cost analyses.
- Consistent protocols and compatible data fields will enable sharing of data industry wide.
- Better analysis and shared data allows bench marking ones performance against other company plants or against industry, improve maintenance optimization, and provide insight as to improving the design of equipment and a better understanding of the mechanisms of equipment failure.

Organization/Ground Rules

There are 25 sponsors from around the world as of May 1997. They are:

Air Products	AMOCO	ARCO
BP Oil	Caltex Services	Chevron Research & Technology
DOW Chemical	DuPont	Eastman Chemical
Exxon	Factory Mutual Research	Fluor Daniel
GE Plastics	Hartford Steam Boiler	Hercules
Hoechst Celanese	ICI, UK	Intevep S.A. (Venezuela)
Mitsubishi Chemical	Philips Petroleum	Rohm and Haas
Shell	Syncrude Canada	Texaco
Westinghouse Savannah River (DOE)		

These sponsors have come together under the aegis of CCPS which is a nonprofit organization dedicated to technical advancement and knowledge. CCPS provides a forum to facilitate development and sharing of technical information. A contractor with both practical and theoretical experience in equipment reliability databases, Det Norske Veritas (DNV), has been retained to facilitate development of our new guidelines book as well as to produce the software tool needed to retain the industry data which is to be merged.

The committee membership has been set up to be as inclusive as possible, while still providing the incentives for companies to participate and contribute data. Cost of participation has purposely been kept low to further encourage participation and put maximum benefit and control of expenditures necessary to supply data information in the hands of the sponsors. The scope of the CCPS effort is to provide a technically sound foundation with supporting software which will assist analysis of plant and equipment reliability data. The company specific maintenance management systems which are the major source of information, are better left in the hands of the individual sponsors. Only they know best how to spend and allocate their resources for their own systems. They can determine what parts of the CCPS database are useful to them and determine the most cost effective way to populate their copy of the CCPS software which can then be used to contribute data to the CCPS industry merged data set.

Not all of the companies wishing to participate are in a position to contribute data. This may be due to not owning the data they possess, as in the case of insurance companies, or as in the case of engineering A&E's, they may not be in a position to gather data, which requires operating a plant. Finally, some operating companies may not initially have a maintenance management system which makes it cost effective to provide data. As a result of these realities, certain ground rules were established which allows continued participation, yet encourages sponsors to contribute to the fullest extent. In summary these ground rules for sponsors in good standing are as follows:

- All sponsors have voting rights
- All sponsors receive free copies of all technical information developed
- All sponsors receive a copy of the software tool being developed minus data
- Only those sponsors contributing data meeting the minimum requirements as set forth by the sponsors themselves will receive the full merged and anonymized data sets as they come available
- CCPS will provide for a fee, specific data requested by non data contributing sponsors to assist in ad hoc analyses. The fee will fully cover all costs associated with providing the data. The CCPS database may be referenced as the source of the data.
- All sponsors may participate on any of the sub committees that have been formed to further the development of the technical information and tools necessary to collect and analyze meaningful data.
- Sponsors always retain ownership of their own data.
- CCPS owns the merged data as anonymized
- DNV retains ownership of the software source code.

During the course of the project, several decisions need to be made. Every effort is made to achieve a consensus, however, if this is not possible, a simple majority of a quorum present at officially scheduled meetings rule.

CCPS is not the only organization trying to make headway in this area. Several other concerns have systems in place or projects underway. Some of these related activities include:

- ISO/WD 14 224 *Collection of reliability and Maintenance Data for Equipment*
- OREDA
- API Risk Based Inspection Committee
- Solomon Associates—SA Onstream
- Institute of Nuclear Power Operations
- Strategic Power Systems
- AEA Technology

Each of these have their own niche which make them unique. As can be expected, there is a certain amount of overlap such that each can learn from the others. As part of our project, which is predicated on technical advancement, we have developed an on going technical sharing arrangement with some of these groups. In fact several of our sponsors participant in some of these activities as well as the CCPS effort.

Concepts

Prior to beginning the project, the question of who the database was to be targeted was asked. To answer this question, another question was asked, which was; who needs the data? The answer came back as:

- Operating and Maintenance Staff
- Reliability Engineers
- Design Engineers
- Risk Analysts

With this in mind, the fundamental concepts were developed in order to provide a structure that would ultimately allow these different functional disciplines to achieve their goals if companies were willing to expend the necessary resources. The data must be of high quality, so that it is "trusted". To make this happen, quality systems must be in place at the plants with an auditable way to verify compatibility with the CCPS database. In addition, the level of detail has to be applicable to the level of detail that is required by the specific analysis being performed. To make these things happen, it is necessary to:

- *Define* what you need/want to know
- Provide for both a numerator *and* a denominator
- Reduce the effort required by field personnel
- Improve the management and information processing tools

It was recognized that this could be an expensive proposition for companies, so the project is structured in a way that minimizes their up front cost commitment and allows an evolutionary process whereby the companies themselves decide how fast to proceed and into what depth to get involved. In essence, the project recognizes that companies will invest only in those areas where it is demonstrated that a value added return will result from the investment.

By utilizing basic industry management systems which exist at every facility, it is quite possible to leverage them in such a way to provide data in a cost effective manner for further value added analysis and subsequent decisions, such as best equipment for application, optimal test intervals, knowledge of system or component wear out, optimal maintenance strategy or improved risk analysis. In turn these provide the foundation for achieving improved plant availability and/or reliability. In order to support these analyses, data is required which includes population information, maintenance information and failure event data as shown in Figure 1.

DATA AGGREGATION AND SHARING

It is the wish of CCPS to provide the foundation that allows industry to establish a standard. This is important as it allows the data for different systems to become more compatible, increasing the value of industry data. Benefits accrue from being able to aggregate data without degrading its quality. Figure 2 shows the CCPS concept, with CCPS acting as an industry clearing house.

This concept allows individual plants the ability to measure their performance and to seek improvement in their designs and maintenance strategies. At the company level, it allows the performance comparison of its plants.

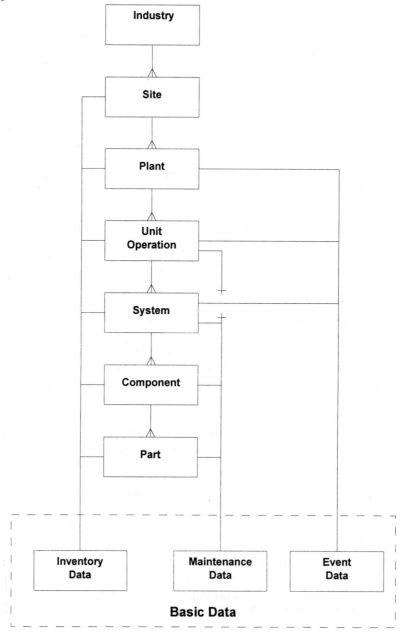

FIGURE 1. CCPS Taxonomy

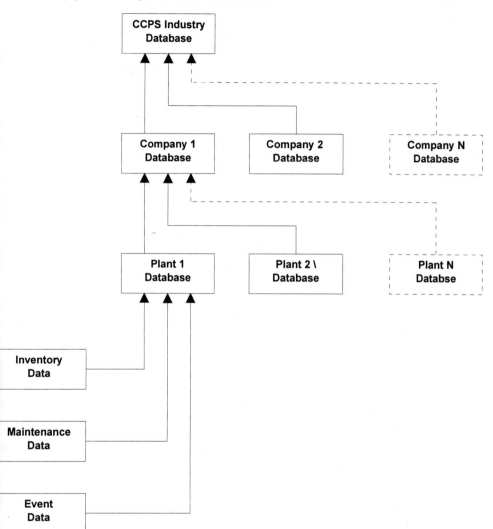

FIGURE 2. CCPS Data Aggregation

Sharing information through CCPS also allows companies to benchmark themselves against industry. It is further a resource of information in areas where they are less expert. The operating concept of CCPS acting as an industry clearing house is shown in Figure 3.

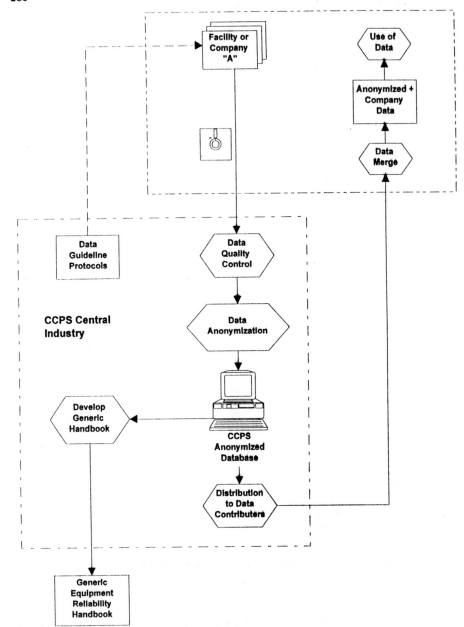

FIGURE 3. CCPS Equipment Reliability Databases Operating Concept

Data Structure

The database consists of a series of tables that allow collection of data for analysis. The data base program enables data analysis and reporting according to certain specifications. The Equipment Reliability database currently consists of the following tables:

- Subscriber Information
- Site Information
- Plant Information
- Unit Operations Information
- System Inventory
- Component Register
- Part Register
- Plant Failures
- Unit Operations Failures
- System Failures
- Component Failures
- Part Failures
- Loss of Containment Events
- System Maintenance Events
- Component Maintenance Events
- Part Maintenance Events

To make the most effective use of data requires that its organization provides for a numerator (maintenance plus events) and a denominator (inventory). These comprise the basic data. In order to facilitate analyses, a data structure (taxonomy) is necessary. Refer to figure 1 for an overview. Within this data structure, established relationships make up the actual taxonomy. These relationships reflect the features and functions that identify and characterize a specific system including the permutations that result from different entries into data fields for the same equipment type. Please note that the taxonomy is not an explicit part of the database structure.

At the higher levels of the data structure, categorization relates to industries and plant operations regardless of the equipment systems involved. For these levels, the user inputs the appropriate choices from pick lists for each equipment system being tracked. In this way, development of the actual taxonomy for the specific application occurs. Figure 4 illustrates the types of relationships existing at the higher levels.

It is not practical to display the entire taxonomy on paper. Basic relationships are evident within the software, supplemented by standard pick list tables and failure mode definitions. It is possible for the user to query the relational database using filters, to essentially define a unique taxonomy with attendant population and event data for the purpose of analyzing a specific issue.

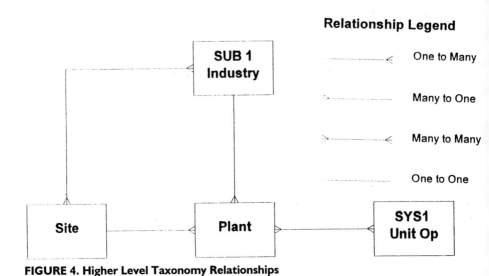

FIGURE 4. Higher Level Taxonomy Relationships

The last three levels are specific to equipment systems. Some equipment systems such as heat exchangers, motors, or control valves can function as an independent system in some applications; or as a support system in other applications. When supporting a larger system, the supporting systems function more like components of the larger system, rather than systems themselves. These supporting systems are considered *subordinate systems* when they function as components for a larger system. This is also referred to as a parent child relationship. An example of this parent child relationship is an interstage heat exchanger that is part of a compressor system; the compressor being the parent, and the heat exchanger being the child. Both the compressor and the heat exchanger are equipment systems. The heat exchanger is subordinate in this case. Use of the relational database software allows recording of this relationship and allows the user to specify equipment system populations that are of specific interest to the analyst.

Development Work Process

The Equipment Reliability Database intends to collect data for many diverse types of equipment. The diversity of equipment in the database demat equipment-specific fields exist to collect the appropriate data. The development of these equipment-specific data fields for all possible equipment is an arduous task. To facilitate this effort and to maintain a consistent approach, DNV developed a procedure applicable to all equipment types. The procedure provides a step-by-step approach for developing the taxonomy and data field specifications for unique equipment types. It also provides an understanding of the role of the taxonomy and data field specifications in the database.

The general steps contained in the procedure to establish the data field specifications for a system being added to the Equipment Reliability Database are as follows:

Step 1.   Draw the boundary diagram
Step 2.   Verify and revise pick list for higher level (above system level) taxonomy tree
Step 3.   Identify the functions and failure modes/definitions of the new system
Step 4.   Revise the system boundary diagram
Step 5.   Revise the pick lists for general inventory data
Step 6.   Define the new data fields and pick lists for the specific system inventory data
Step 7.   Define any new data fields and pick lists for the specific system failure data
Step 8.   Define any new data fields and pick lists for the loss of containment data

The database has a predefined higher level taxonomy that is applicable for all equipment systems. The taxonomy development process determines the make up of the taxonomy for the equipment system level and below.

Project Deliverables (3 items)

As part of phase 1, several deliverables have been defined so as to enable the startup of an on going industry equipment reliability database. The first is the guidelines book that provides the technical foundation for the project. The book has three major thrusts. The first is to document and explain the theory of failure rate data and the data structure employed by CCPS to accomplish its goals.

The second is to demonstrate the potential use of the data with actual worked out examples. After all, why spend money collecting data unless its use leads to something that adds value. If a particular analysis is important, then the data required for that analysis is defined by the specific analysis. The book will help to illustrate this point as well as helping understand how one actually performs the analysis.

The third thrust is to provide an overview and performance criteria for the necessary quality assurance that must be maintained by both the plants submitting data as well as the CCPS contractor maintaining and operating the database in order to obtain and maintain a high quality and integrity of data. It is important that data from different plants that exist in different companies use the same definitions in order to allow merging of the data. It is also important so that misinterpretation of the failure modes of interest does not occur by field personnel and/or analysts.

Another major deliverable is the software that is intended as an off line tool which will facilitate analysis. The software is a run time application made available to every sponsor for use anywhere within their company for their own internal benefit. The database software initially comes without any data. It is up to the sponsors to determine how to best and most cost effectively populate the software with their own data. With their database populated according to the quality assurance requirements, they then are in a position to contribute to the CCPS industry database. Contributing this data qualifies them to receive the fully merged and anonymized data set.

The last deliverable consists of the actual audit and quality assurance procedural protocols. These procedures enable the plant certification process and determine whether suitable data ready to contribute exists, as well as the necessary procedures to merge and anonymize the data. This will allow CCPS to actually start up and begin operation of the database on an initial limited pilot basis, setting the stage for long term operation evolving in a continually improving environment.

Conclusion/Future Path Forward

The CCPS database is laying a solid foundation that should prove beneficial for many years to come. The database is due to begin operation in early January 1998 with the equipment systems rigorously developed. It is anticipated that only some small percentage of information will be available initially as compared to all of the data fields that exist within the database. The database design allows and supports continuous improvement. Once high quality data proves to allow value added analysis with the minimums required to contribute, it will inevitably lead to a desire for increasingly in depth analyses. Rigorously designed, the database should prove up to the task, supporting the level of analysis required by the individual sponsor companies. With the large diversity of equipment types in our plants, the task of developing rigorously specific taxonomies will be continuing for some time.

References

1. Draft CCPS *Guidelines for Improving Plant Reliability Through Data Collection and Analysis*, May 1997.
2. OREDA-1992, *Offshore Reliability Data*, DNV Technica, Høvik, Norway, 1992
3. Rausand M. & Øien, K., *The basic concepts of failure analysis*, In *Reliability Engineering and System Safety* 53 (1996), pp. 73-83.
4. Wolford, A. J., *The Role of "Good" Data in Reliability Assessments*, Proceedings of Process Plant Reliability Conference, 1995.
5. *CCPS Guidelines for Process Equipment Reliability Data*, AICHE, 1989.

Using Quantitative Risk Analysis in Decision Making: An Example from Phenol-Formaldehyde Resin Manufacturing

Don Schaechtel
Borden Chemical, Inc., 520 112th Ave. NE, Bellevue, WA 98004

David Moore
EQE International, 1411 4th Avenue Building, Suite 500, Seattle, WA 98101

ABSTRACT

An out of control phenol-formaldehyde resin reaction challenges traditional pressure relief systems. To prevent overpressure and the potential for a catastrophic failure of the reactor, Borden Chemical has followed strict manufacturing safety rules since 1974, when a reactor ruptured and killed two people.

With the advent OSHA's process safety management standard, process hazard analyses were conducted to further improve reactor safety. As part of this work, a method was needed to assess operator safety as required by the "facility siting" clause in the standard. Since moving or reinforcing control rooms was not considered feasible, a quantitative risk assessment was used to assess operator safety. This approach included:

- Developing risk tolerance criteria for the phenol-formaldehyde resin manufacturing industry
- Determining the individual risk associated with the manufacturing process
- Identifying scenarios which could lead to reactor overpressure
- Conducting a risk assessment that included quantitative fault tree analysis
- Comparing the calculated risk to the pre-developed risk tolerance criteria
- Implementing risk reduction measures suggested by the fault tree analysis

This paper will highlight the art and strategy behind developing and using risk tolerance criteria for an industry that was not accustomed to making explicit risk-based decisions. Managers endorsed these techniques, as they allowed them to allocate capital and engineering resources where they could most effectively reduce risk.

Introduction

Borden Chemical is a leading supplier of phenol-formaldehyde (PF) resins used in the forest products industry. These resins are made in batch reactors by reacting phenol and formaldehyde together in the presence of an alkaline catalyst. This is a highly exothermic reaction. The heat of reaction is removed by a large condenser, which condenses the water vapor boiled from the reactor contents and returns it as reflux. The reactor typically holds 10,000 or more gallons and contains a motor driven agitator. The process follows these steps:

1. Water and phenol are charged into the reactor.
2. Vacuum is established in the reactor
3. Catalyst is charged into the reactor.
4. Formaldehyde is slowly charged to the reactor. As the reaction begins, the heat of reaction brings the reactor contents to a boil at a temperature determined by the absolute pressure in the reactor.
5. Reactor temperature is controlled by adjusting the rate of formaldehyde addition or the reactor vacuum (which changes the boiling temperature). The heat of reaction is removed by the condenser, with reflux returned to the reactor.
6. After all of the formaldehyde has been added, the batch continues to reflux until the desired polymer characteristics are obtained. This condensation reaction is not as exothermic as the initial methylolation reaction.
7. The batch is then cooled, which essentially stops the reaction.

Successful reactor control depends on the condenser's ability to remove heat faster than it is generated. Under normal circumstances this is easily achieved. However, control is quickly lost if prompt action is not taken upon failure of the cooling system due to power failure, mechanical failure, or other operating problems.

Borden Chemical has a healthy respect for the energy released by an out-of-control exotherm. In 1974, two men were killed when a reactor ruptured catastrophically after reaction control was lost. The shock wave destroyed everything within 100 feet of the reactor, including the control room. The men, one the reactor operator and the other a supervisor, were inside the control room when the blast occurred.

Following this incident, Borden established a Reactor Safety Committee, which researched safety measures for batch PF resin operations and developed a set of reactor safety rules. These included requirements for more pressure relief capacity, rules for batch sizing, and criteria for applying deluge water to quench an out-of-control reaction. As the committee continued its work they commissioned studies of the reaction kinetics using the Reactive System Screening Tool (RSST) and vent sizing tools that took two phase flow into account. The results posed a new challenge: Given the most reactive combination of phenol, formaldehyde, and catalyst; a traditional pressure relief system, with no other safeguards, cannot adequately prevent overpressure. This led to further refinement of the reactor safety rules to include additional safeguards for preventing runaway reactions.

When OSHA's Process Safety Management standard became effective, process hazard analyses were conducted to further improve safety. A qualitative fault tree analysis was performed on one reactor, and using the results of this study, a checklist review was developed for evaluating over twenty Borden PF resin reactors. The goal was to standardize the level of safeguards installed on the reactors. As a result of this work, significant capital was approved to improve reactor safety beyond the level required by the reactor safety rules.

Addressing Facility Siting

An issue not addressed specifically in the qualitative study was facility siting As the accident in 1974 demonstrated, control room siting, and the safety of those inside the control room, is an important concern. OSHA's recommended approach to locating control rooms is to determine the radius around a process area that could be adversely impacted by a fire, explosion, or release. Ideally, the control room is built outside of this radius. If that is not feasible, the control room is built using blast resistant construction.

Borden Chemical, like many other companies, struggled with how to apply this approach to at least 20 existing control rooms in almost as many locations. Using the Dow Fire and Explosion Index as a screening tool, an exposure radius of over 100 feet was calculated. This is consistent with the radius of destruction seen in 1974. While the control rooms could be moved away from the reactors, that creates two significant problems. First, the operators collect numerous samples during production, so it is desirable to be close to the reactor. Second, moving the control rooms is an expensive proposition. While it would protect an operator who was inside, it offers no protection to anyone outside. And if a reactor were to rupture catastrophically, the damage to surrounding equipment would be costly to repair and create a significant business interruption. It was clear to management that investment in preventing a reactor rupture in the first place was by far preferable to spending money to mitigate the effects. Reinforcing existing control rooms was not considered feasible because the energy released by a pressure-volume rupture is high enough to damage even reinforced buildings. This was another clear reason to explore preventive measures.

This approach, one of risk management, is supported by API RP 752, "Management of Hazards Associated with Location of Process Plant Buildings." By conducting a risk assessment, and comparing the risk posed by an operation to pre-established risk acceptance criteria, a company can make decisions on where resources can be spent most effectively to reduce risk. Risk reduction measures

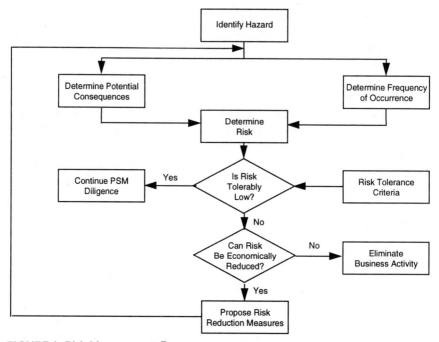

FIGURE 1. Risk Management Process
From: *Guidelines for Evaluating Process Plant Buildings for External Explosions and Fires*, CCPS, 1996, page 94.

that prevent the incident are most desirable. This process is shown in Figure 1. The focal point of this process is comparing the risk posed by the process to risk tolerance criteria. To address the facility siting issue, Borden Chemical commissioned EQE International to assist with establishing appropriate risk tolerance criteria and then conduct a quantitative risk assessment. Results were to be compared and risk reduction efforts taken as needed.

Selecting Risk Tolerance Criteria

In the chemical process industry, and virtually all other human endeavors, there is no such thing as zero risk. Therefore, when making reasoned decisions about issues such as facility siting, whether qualitative or quantitative, there must be some criteria for deciding whether a risk is tolerable or not. Reasonable risk tolerance criteria will be:

- not so high that plant personnel are subjected to inordinate risk
- not so low that significant cost expenditures are required to meet criteria
- supportable by regulatory or industry precedent
- defensible, even under adverse scrutiny

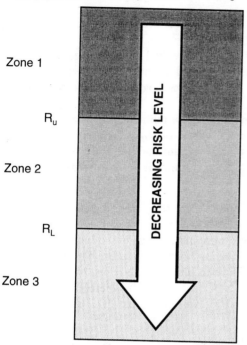

FIGURE 2. Zone of Risk Tolerance.
From: *Guidelines for Evaluating Process Plant Buildings for External Explosions and Fires*, CCPS, 1996, page 48.

Because of the uncertainties involved with risk assessment, one effective approach to developing risk tolerance criteria is to consider a range of risks, as shown in Figure 2. The conceptual basis is that there is an upper level of risk tolerance, R_U, above which the risk is too high and risk reduction measures should be taken. At the same time, there is a lower level of risk, R_L, below which it does not make economic sense to continue to spend more money to reduce the risk even further. In fact, efforts to reduce these already low risks may be counterproductive because they divert resources from other higher risks. In between R_U and R_L there is a region of risk where risk reduction measures should be considered, as appropriate, on a cost–benefit basis.

One of the first considerations in selecting risk tolerance criteria is to determine which risk parameter (or parameters) will be used. For this study, two parameters were used, *maximum individual risk* and *average individual risk*. Individual risk expresses the risk to a person in the vicinity of a hazard. It is normally calculated as the frequency of serious or fatal injuries per year (fatalities/year).

Maximum individual risk is defined as the risk to the most exposed individual in an exposed population. For process-plant buildings, this is the person who spends the most time in the building under study. For PF resin operations, this is the reactor operator. Maximum individual risk is used when calculating the impacts from a reactor overpressure event.

Average individual risk is defined as the individual risk averaged over the population that is exposed to the risk. This would include everyone who works in the plant. Average individual risk is used for comparison to other industry average risks.

In the United States, the government has not established specific risk tolerance criteria. Guidance for the numerical values of R_U and R_L comes from a number of sources, including historical data on industrial risk and risk tolerance criteria from other countries. For PF resin operations, the risk tolerance criteria were developed following these steps:

1. *Average individual risk was determined for companies manufacturing PF resin. This number was compared with other industry data.*

Industry experience data and worker populations were obtained from five PF resin manufacturing companies, including Borden Chemical. The data show that there were three fatalities in the PF resin industry in the last twenty years, which represents 16,000 employee-years of operation. The average individual risk is therefore 1.9E-4 fatalities/year for PF resin operations. Table 1 shows how

TABLE I
Average Individual Risk for Various US Industries

Industry (Total Population)	Death Rate[1]	Fatal Accident Rates (FAR)[2]	Average Individual Risk (Total Population)[2] (Fatalities/Year)
Manufacturing	3	1.5	3×10^{-5}
Services	3	1.5	3×10^{-5}
Trade	4	2.0	4×10^{-5}
All Industry	7	3.5	7×10^{-5}
Government	9	4.5	9×10^{-5}
Refining Direct Hires (All Refining)	14.3	7.2	1.4×10^{-4}
PF Resin Reactor Operations	19	9.5	1.9×10^{-4}
Trans., Pub., Utilities	20	10.0	2×10^{-4}
Refining Contractors (All Refining)	21.7	10.8	2.2×10^{-4}
Construction	22	11.0	2.2×10^{-4}
Mining	29	14.5	3×10^{-4}
Agriculture	37	18.5	4×10^{-4}

1 Deaths/100,000 workers per year
2 Based on 2,000 hours worked per year per worker
From:*Guidelines for Evaluating Process Plant Buildings for External Explosions and Fires*, CCPS, 1996, page 51 (PF Resin Reactor Operations added by authors).

this compares with other industries in the United States. The table shows that the risk posed by PF resin operations falls between direct hires in the refining industry and refining contractors, and is about the same as the average individual risk for transportation and public utility employees. Based on this comparison, the historical average individual risk for PF resin operations was neither exceptionally good nor exceptionally bad. Because Borden is interested in improving safety, an average individual risk of 1E-4 was chosen as reasonable. This also exercises some care with the risk number for PF resin operations, which is based on limited data.

2. *Average individual risk was converted to maximum individual risk.*

Among a plant population, some people will be exposed to more risk than others. Operators and mechanics will be exposed to higher risks by the virtue of the tasks they perform. The API document, *Serious Incidents in the U.S. Petroleum Refining Industry, 1985–1989* (M. Rusin and L. Hoffman, 1991) found that 88% of the fatalities occurred to individuals having the most exposed occupations. Other people, such as office workers, will be exposed to less risk. This range of risk is depicted in Figure 3, which is adapted from previous EQE projects and training material. It shows that the operator's risk can be estimated as ten times greater than the average individual risk. Note that this is a combined occupational risk, and not just risk from a single event such as a reactor overpressure and rupture.

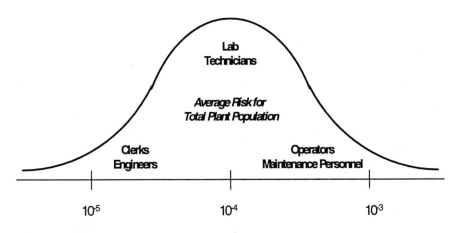

FIGURE 3. Distributions of Risks by Occupation.
From: *Developing Risk Tolerance Criteria for Facility Siting,* Course Material, EQE International, 1997.

3. *Maximum individual risk was adjusted to account for PF resin overpressure events only, rather than all occupational events.*

In determining the risk tolerance levels for an individual exposed to a single event, such as a PF resin reactor rupture, the tolerance criteria levels should be lower, since that single event represents only one risk to which the individual is exposed. For example, death could also result from a fall, vehicle accident, phenol exposure, or other workplace hazard. An approach used by the UK Safety and Health Executive is to assume the risk from a single hazard represents about one-tenth of the total occupational risk exposure, depending upon the other activities included in the person's work assignment. For resin reactor operators, this should be conservative, as reactor rupture should be one of the more significant hazards to which the operator is exposed.

The following equation was developed to establish maximum individual risk tolerance levels for PF resin operators:

$$MIR_{SE} = (AIR_{TP})(F_{MIR})(F_{SE})$$

where

MIR_{SE} is the maximum individual risk for a single occupational event
$AIR_{TP} = 1E-4$ fatalities/year (the average individual risk for all occupational events; Step 1)
$F_{MIR} = 10$ (the factor relating average individual risk to maximum individual risk; Step 2)
$F_{SE} = 0.1$ (the factor relating overall occupational risk to single event risk; Step 3)

Applying the equation:

$$MIR_{SE} = (1E-4)(10)(0.1) = 1E-4 \text{ fatalities/year}$$

This number was chosen as the maximum individual risk tolerance level for reactor rupture.

4. *Risk tolerance goals were set, with both R_U and R_L established for continuous improvement.*

When the above analysis was finished, the results were reviewed with Borden operations managers. The managers approved the work and agreed to risk tolerance criteria of:

$R_U = 1E-4$ fatalities/year (above this level, risk reduction measures are necessary)
$R_L = 1E-6$ fatalities/year (below this level, further risk reduction is not necessary)

Establishing risk tolerance criteria was new for the managers, as they were not accustomed to making explicit risk-based decisions. However, this approach was well received for several reasons. First, the managers had a lot of confidence in the PF resin safety measures that had been instituted through the years. They had confidence that the quantitative risk assessment would not reveal any insurmountable hazards. Second, the upper risk level was consistent with published regulatory data and the risk tolerance criteria of the UK Health and Safety Executive (the maximum individual risk for all occupational exposures, R_U, is 1E-3 for both). This gave the managers confidence that the criteria were supportable and could be defended if necessary.

Quantitative Risk Assessment

Once risk tolerance criteria were established, the next step was to develop a quantitative model to assess the probability of reactor overpressure and rupture. Fault tree analysis was chosen for this model. A qualitative fault tree analysis had already been developed, and this was used as the starting point. A representative reactor, which met the checklist review criteria and was operated using a programmable logic controller (PLC) and computerized operator interface, was chosen for study.

The benefits of quantifying the fault tree were quickly realized. In the qualitative fault tree, the logic was modeled as shown in Figure 4. To reach the top event of reactor overpressure, three events had to occur. The heat of reaction had

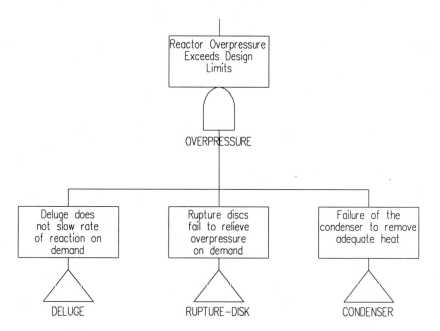

FIGURE 4. Top Event Logic: Initial Qualitative Fault Tree Analyis

to exceed the cooling capability of the reactor, *and* the deluge (quench) system had to fail to cool the reaction sufficiently to regain control, *and* the installed pressure relief devices had to fail to relieve the pressure in the reactor. Modeling these quantitatively was challenging because each can fail to some degree and still not cause overpressure sufficient to rupture the reactor. To refine the fault tree model to account for this an event tree was constructed and success criteria were developed.

An unexpected result of developing success criteria was the development of scenarios where the pressure relief devices could fail to relieve overpressure even if the deluge system operated as designed. These scenarios were termed "inadvertent bulk charges." Normally reaction rate control is maintained by metering the formaldehyde charge rate and stopping flow if necessary until control is regained. With an inadvertent bulk charge, the entire formaldehyde charge gets into the reactor with little or no reaction occurring. This could happen in several ways, although two are most probable. One way is if, by mistake, the catalyst is not charged. The formaldehyde charge follows and no reaction occurs because there is no catalyst. Once all of the formaldehyde is charged, the operator realizes he failed to charge the catalyst and then charges the catalyst too quickly. The heat generated by the reaction quickly overwhelms the cooling capability of the condenser. Studies of the reaction kinetics show that the rate of temperature rise may be so fast that deluge cannot quench the reaction and the pressure relief capacity will be marginal. The other likely case is where formaldehyde is added at a batch temperature too low to initiate the methylolation reaction. The operator continues, in error, to charge formaldehyde. When the reaction finally begins, the heat generated by the reaction again overwhelms the cooling capability of the condenser. In this case deluge may not quench the reaction and the pressure relief device must be successful to prevent overpressure. To account for these cases, the top fault tree logic was revised as shown in Figure 5.

What is significant about the inadvertent bulk charges is the reactor control system under study did not include a means to prevent either of these conditions (though checkpoints were included in operating procedures). This became apparent when the fault tree was quantified. The frequency of reactor rupture was 4E-3, which is above the desired R_U value of 1E-4 (for the purposes of this study it was assumed that reactor rupture would cause on average one fatality). Analysis showed that the higher risk was attributable to the risk posed by inadvertent bulk charges.

Working with process engineers, an enhanced control logic was proposed that would detect the inadvertent bulk charge. This logic would stop a formaldehyde charge if the reaction was not initiated within 10 minutes. Reactor temperature and condenser heat load, which were already being measured, were used as indicators. This change alone reduced the risk by a factor of 28 to 1.5E-4. Although this risk is slightly above R_U, it is certainly within the accuracy bounds of the quantitative assessment process. When taken in context with the conserva-

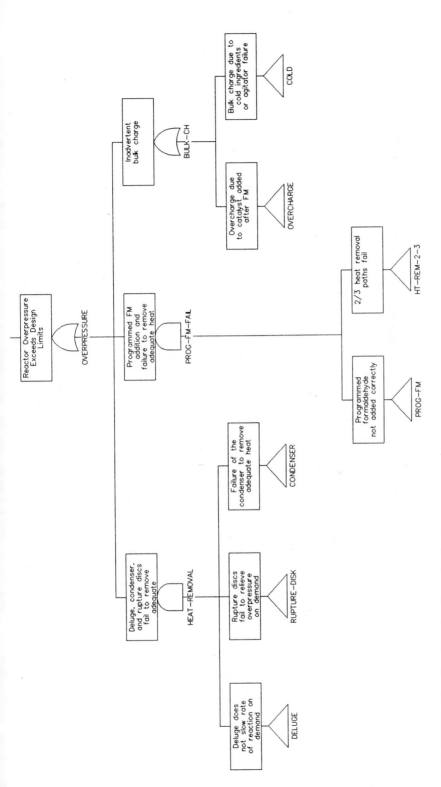

FIGURE 5. Top Event Logic: Quantitative Fault Tree

tive assumptions of the analysis, this level of risk is judged to meet the suggested risk tolerance criterion.

Analyzing the fault tree for the enhanced control case suggested another possibility for risk reduction. Because many of the control functions are based on reactor temperature, any problem with temperature measurement or the associated control action could compromise the safety system. Adding diversity to the parameters that control formaldehyde addition, especially reactor temperature, will reduce the risk of reactor overpressure. While this effect was not modeled, it was estimated to further reduce risk by a factor of 1.6. This would put the overall risk into the range between R_U and R_L. Ways to add this diversity are under study. It is believed that it can be done with little expense, making this risk reduction effort well worth consideration.

Conclusions

When the study was finished, the results of the risk assessment were reviewed with the operations managers. They supported the conclusion that operating a PF resin reactor with an enhanced control system reduces the risk of reactor overpressure below the agreed-upon upper risk tolerance criterion. As a result they developed an action plan for upgrading reactor control systems throughout the company. It is interesting to note that the managers' intuition was correct regarding the safety of resins produced with proper charging of formaldehyde. The risk of overpressure in this case is quite low. What was not expected was the impact of an inadvertent bulk charge.

The study also demonstrated the importance of process safety management diligence. Compromising any of the safeguards modeled in the fault tree analysis will impact risk. These safeguards include maintaining the integrity of the control system, performing preventive maintenance on schedule, reviewing the safety impacts of revisions to resin formulas, and training operators in reactor safety concepts and procedures.

Benefits of This Approach

Compared to the complexity of the PF resin reaction system, the risk reduction measures recommended by this study are relatively easy to implement. As a result, a significant increase in safety will be made with a minimal expenditure of resources. While advanced control systems were already being installed on newer reactors, demonstrating the safety benefits quantitatively provided an excellent case for upgrading the control systems on all reactors.

By applying preventive measures, not only are personnel in the plant safer, the risk of extensive property damage and business interruption is greatly reduced. And while the focus of this study was on preventing catastrophic failures, the preventive measures also reduce the likelihood of activating the pressure relief system and subsequent venting. This provides environmental benefits. In the same way, the preventive measures can prevent operability and quality problems. Mitigative measures, which were not considered feasible in existing plants, could not offer these same benefits.

Finally, from a regulatory perspective, the basis for these risk-based decisions is sound and defensible.

The Synergy of the Business Planning Process, Safety Management, and the Acceptance of Risk

John Kimball
Safety InSites, Clifton, Virginia

ABSTRACT

Too often, risk-based decisions on Process Safety and Emergency Management are not done in concert with the business management organization. The process engineering, business operation, and emergency planning and management functions have much in common but too often have little interaction. The issues of regulatory compliance, measurement of risk, risk mitigation and downsizing will be addressed in this presentation. Discussion will also include methods to quantify "business/emergency" risk as well as the need to incorporate emergency plans into one "Integrated Contingency Plan."

Introduction

One all-pervading issue in the process industries is the compliance with regulatory drivers such as Process Safety Management, Risk Management Planning, and HAZWOPER. Business decisions or strategic planning that will impact emergency planning and response operations are seldom, if ever, made with regulatory compliance or emergencies in mind. While the business drives the activities that support it, an operation must never lose sight of those unplanned events that could threaten the ability to continue operation.

Opportunities have been lost for facilities to maximize their compliance by failing to use the regulatory mechanism as a vehicle to implement changes in processes or other improvements to enhance safety and production. An example of this is the HAZWOPER response which may exist in isolation from after-action investigation reporting and follow-up. While the HAZWOPER rule (29 CFR 1910.120) requires an orderly termination of an incident, unless the emergency falls under the PSM guidance for near-miss or incident investigation, the cause of a spill, fire, or release may go undocumented and unremedied. How many times have response teams gone out to a fire or release in the same location or process without the root cause being fixed or eliminated?

Another issue is the efficiency of response forces, both on and off-site. A facility may comply with all SARA, HAZWOPER, PSM, and RMP rules, but may be unable to quantify just how efficient this response effort is. This efficiency, or lack thereof, translates to lost dollars in direct and indirect costs from an incident. How do you measure "efficiency?" Is it time required to deploy? The ability of a team to assemble and respond to scripted scenarios? The ability of a team to resolve an incident requiring hazard and risk assessment and judgment calls? By what yardstick do you measure safety in response? Sheer numbers? Hours of training? Training and equipment budget? Years of experience of responders? None of these factors alone can measure just how effective is your response effort, but some idea of what the operation expects of its responders and what level they achieve should be measured. In addition to the unquestionable safety issues, the hard business facts remain that an inefficient response results in longer downtime, greater potential lost to capital property, and the extreme possibility of the threat to continued operation .

The business priorities should be at least communicated to emergency planners. The business impact of the loss of a particular product line, if that product is "irreplaceable," and the impact of interruption of products and services should be part of emergency planning in order for planners to prioritize the allocation of resources to prevent and mitigate emergencies.

The value of various products to customers, whether the customers are end users or the recipients of feedstocks should be somewhat quantified to planners and responders. At some point the facility must determine what is the "acceptable loss." Again, excepting the unquestionable issue of life, safety, and the environment, the level of loss able to be absorbed must be considered. Does your operation know what the loss is in units or dollars per hour of down time? At what point does this loss increase exponentially due to the need for movement of raw materials in or finished product out? Many other issues can affect this outage outside of the realm of process areas: communications, power failure from outside the fences, failure in computing systems controlling the ordering and processing flow. Something as simple as a local weather condition keeping the data processing people from their work stations could idle a chemical plant or a portion of it for a time. What contingency plans are in place to mitigate those non-engineering causes of production failure?

The business issue of Downsizing, whether it is termed "rightsizing," "reinvention" or "reengineering," is one that can directly impact emergency planning. Some of these effects, such as losses due to incidents are direct and clear and others are indirect and less easily identified.

One example is the increase in the average age of emergency responders due to reduction in force of the less senior people. While this may be a benefit because the most experienced operators, technicians and engineers remain in highly strenuous response activities, the age factor is a liability. When the AVERAGE age of a four-person Level A entry and backup team is 58 years old, I believe the line has been crossed from "risk management" to "risk taking." Even with people in superb physical shape on a diet and exercise program, a case could be made that the company is expecting too much at the expense of the safety of its responders.

Conversely, with early-out retirement programs and buy-outs of senior employees, a plant can be left with operators and technicians lacking the experience and savvy of the seasoned veterans.

Other effects in cutting costs can be very direct. For example, specific actions in an emergency plan can be assigned to work positions in the facility. If that position is eliminated by business necessity, the emergency function still remains. These emergency functions can be very important and critical to life safety; various notifications, evacuation, control activities, or direct intervention must be completed early in the incident. While the reality of downsizing continues, the reality of emergency response task remains as well. I want to emphasize these are not just figures, such as numbers of persons on a hose line, but very specific response activities that may be critical to safety and property damage prevention.

In the planning phase, downsizing may increase the number of differing tasks for which a person may be responsible. A common example is the "inheriting" of PSM responsibilities for emergency planning responsibilities thrown in with other PSM work. A logical extension is to include RMP duties as well. In the case of the response technicians, the most experienced person has just walked out the door with their wealth of corporate knowledge; the person replacing them not only has this new task, but their former primary job as well, and probably some new ones in addition to PSM, RMP, SARA, and RCRA.

How can we deal with this harsh reality? First, we must acknowledge it for the reality it is. Downsizing is a fact of life and it is by no means over for any industry. "Doing more with less" is not just a motivational slogan any more.

A very plausible decision is to alter the role and expectations of the response group. Falling back to a purely defensive posture is sometimes the only answer if the response organization has been reduced below the level required to operate safely or according to HAZWOPER. The risk of shifting this response planning strategy must be understood by all levels of management. The business arm of the organization must understand that the downtime or production outage due to an incident may increase due to the longer time needed to mitigate an incident. The decision to cut on-site response capability and rely on off-site response efforts should not be made without assurance of the level of off-site response units. More fire departments, for example, are falling back to defensive hazardous materials response teams due to the expense of field technician level offensive posture teams. Regional response organization are supplanting or replacing many of the local fire departments who formerly fielded a technician-level HAZMAT team.

More controlled systems will help in incident prevention as well as mitigation. Better operator training is always going to pay dividends in helping to cope with abnormal operating parameters. A well-trained operator can mean the difference between an anomaly and an explosion. Automated systems that accurately map or predict release data in real time conditions as well as "expert" or "smart" systems that actually prompt operators through the early stages of an incident will help. Realistic expert-driven interactive training programs also can enhance operator training; virtual reality training is the next step to give trainees emergency "experience" on the technical level of flight or combat simulators.

The definition of an "Emergency" must also be addressed in order to deal with them. The regulations such as PSM, RMP, HAZWOPER, OPA '90 and others bring to mind fires, spills, catastrophic releases, and/or explosions. While these are indeed emergencies by anyone's definition, other less dramatic events can also be classified as emergencies. Power failures, labor unrest, transportation accidents clogging vital traffic arteries; less dramatic—but no less serious—spills or fires, severe weather, even a product liability issue can be termed as an "Emergency." Any event that has an impact on normal operations of the business of the facility can be termed as an emergency. In this day of "integrated global supply chain," just-in-time inventory, and Economic Value Management, the flow of raw materials and finished goods is on a very tight time line with little allowance for slippage. Any event from a major fire to equipment failure to transportation interruption can adversely affect the business life blood of an operation.

The measurement of risk and the impact of an incident can be difficult to quantify. While direct costs, such as those to life, health, property, stock, equipment buildings can be very easily identified—indirect losses are much more difficult to pinpoint. Issues such as losses of market share, tarnished corporate reputation, and disgruntled customers are not handily tracked. In extreme cases, the eventual outcome of a tarnished corporate reputation may result in the loss of that company's ability to operate in a particular jurisdiction. If that plant produces a vital feedstock for the other facilities in the organization, the impact of an incident is far-reaching beyond the direct property losses.

Some type of risk ranking method is needed to quantify events as to their frequency impact and to assign a numeric value to the type and location of incident. The time-honored PSM hazard evaluation methods can be modified or tailored to meet the business aspects of planning for emergencies.

Several methods are especially suited for this. One example is the What-If/Checklist method which combines the brainstorming aspects of What-If with the structure of a systems checklist. In the preparation phase, the boundaries can be defined to include specific business aspects of the checklist.

Failure Mode and Effects Analysis (FMEA), is another method that readily lends itself to this process. The effects of the event must include the business impact of the particular mode, the risks involved, and the costs associated with mitigating these risks. The drawbacks should be obvious. The time involved to prepare, conduct, and deliver one of these exercises is extensive. Typically, it is difficult enough to implement these exercises when the regulations require it. None of what I have described is remotely driven by external regulations and will be a tough sell to management.

Management may decide to accept the risks rather than to expend the resources to completely eliminate them. While this is understandable from a business perspective, the benefits of the Hazard Evaluation exercise come from the process of identifying potential response issues and what actions can be taken to lessen the impact of an incident.

Integrated Contingency Plan

The introduction of the Integrated Contingency Plan guidelines by the National Response Team in June of 1996 is a major step in consolidating emergency response plans into one document. It directs that all existing response planning documents can be combined into a single plan. The ICP consists of an Introductory section, the Core Plan, and Annexes.

The Introduction is just as the name implies, the highlights of the operation. The Core plan is probably the greatest advantage to the ICP as it is designed to be clear and concise, able to be placed in the glove box of a response vehicle. The annexes are merely the other existing plans such as OPA '90, RCRA, and others that call for response plan documents.

The advantage of the ICP is the compilation of existing plans in one user-friendly document that can be rapidly accessed and used in an incident.

The drawbacks are that the ICP is only geared to the incident types of fire, spills, and releases. Weather related problems, transportation incident not involving hazardous material, or other incidents damaging to the infrastructure are NOT included. Typically, the ICP is in the hands of the shift response leader whoever he or she may be in whatever capacity at an operation. All serious contingencies should be addressed albeit briefly and concisely in a true Integrated contingency Plan.

Risk Analysis and Risk Management

Chairs **Dennis Hendershot**
Rohm and Haas Company
Keith Cassidy
Health & Safety Executive, U.K.

How Does the Public Perceive Risks from Major Industrial Hazard Sites?

Ann Brazier
Health & Safety Executive

Gordon Walker
Staffordshire University

ABSTRACT

The UK Health and Safety Executive (HSE) provides expert technical advice to Local Planning Authorities on risks that would be posed by new hazardous installations and pipelines and risks from existing installations to proposed development in their vicinity. HSE uses quantified risk assessment (QRA) to set a consultation distance around each of the major hazard sites; the risks may be derived from dispersion models which estimate concentration levels and exposure times for a range of loss of containment accidents.

Forthcoming EC legislation will require information on safety measures and emergency planning to be made available to persons liable to be affected by a major accident originating from a major industrial hazard site. HSE has commissioned an interdisciplinary research project to inform the development of policies on the tolerability of risk, to aid in developing criteria for the siting of major hazards and the use of land within their vicinity and to guide the explanation of risk issues to local authorities and the public. The research is considering the level of public awareness of the nature and extent of these risks, the sources of this information and the factors which influence attitudes, such as economic benefit, the public image of a company, good neighbor policies, etc.

The research is based on comparative case studies of a cross-section of hazardous installations and pipelines which would present risks to people off-site in the event of a major incident. It uses complementary data collection and analytical methods, including archival research, census data, focus group discussions, personal interviews, Q-sort exercises, discourse analysis, computer assisted qualitative data analysis and factor analysis.

The methodology and emerging results of this major 3 year research project will be discussed and the paper will also describe how HSE has piloted a Geographical Information System to support the decision making process and assist in communicating consistent responses to Local Planning Authorities within statutory deadlines.

Introduction

The UK strategy for controlling industrial major hazards was first developed by the Advisory Committee on Major Hazards which was set up following the 1974 cyclohexane explosion at Flixborough which killed 28 employees and caused widespread damage offsite. The strategy was taken up by the European Commission to provide the framework for the Seveso directive which will be superseded by the Control of Major Accident Hazards (COMAH) directive.

UK Strategy for Controlling Major Hazards

The strategy for the control of major accident hazards is based on 3 principles:

- **Identification**—establishing where major accident hazards exist. This is done by legislation requiring notification.
- **Prevention and Control**—reducing the possibility or risk of an accident occurring by on site controls required under control legislation
- **Mitigation**—reducing the impact of an accident event by means of strategic land use planning to separate populations from major hazards and emergency planning

Identification

The Planning (Hazardous Substances) Act and Regulations require site operations to obtain consent for any use of storage of specified hazardous substances at or above certain threshold inventory levels. This consent is obtained from the hazardous substances authority (normally the local planning authority) and for new sites is applied for alongside normal planning permission. HSE is a statutory consultee on all applications for consent, ensuring that expert advice on safety is always obtained.

Prevention and Control

This includes the assessment of risks by the operator of the hazardous installation and the setting up of appropriate safety measures and controls on site which are monitored by HSE inspectors. These measures should reduce the risk of an accident to a low level but do not completely eliminate the hazard so there remains some "residual risk" of an accident taking place. It is this residual risk which HSE takes into account when advising Local Planning Authorities on new developments in the vicinity of major hazard sites.

Mitigation

Land use planning—Because of the residual risk it is prudent to exercise control over the relative locations of the sources of hazard and any population in the vicinity. If there are fewer people in the vicinity of an installation when a major accident happens, there will be fewer potential casualties or fatalities and emergency plans will be easier to implement effectively. In Bhopal, India, the high number of offsite fatalities was in part due to the very close proximity of a dense population to the site boundary. The objective of strategic land use planning is to minimize the number of people exposed by an informed choice of site for new notifiable installations and to control development within the consultation distance of the existing notifiable installations. This philosophy was developed by the Advisory Committee on Major Hazards (ACMH) which recommended legislation to control and reduce risks following the Flixborough disaster. The ACMH philosophy was to stabilize or reduce populations around notifiable installations by strategic land use planning.

Estimating Hazard and Risk

When acting as statutory consultees for planning applications within the consultation distance of hazardous installations HSE undertakes a technical assessment of the safety implications of the proposed development. This assessment has two closely related components:

- an estimation of the types of accident events which could take place at an installation, the scale of these accidents and how far away their impacts could be felt. This is termed an estimation of the scale of "hazard" or "consequence."
- an estimation of how *likely* it is that these events will take place, and therefore how *likely* it is that certain levels of damage could be experienced away from the installation.

The estimates of consequences and likelihoods can be combined to give an estimation of the scale of risk and may be either qualitative or quantitative in nature. HSE performs the QRA and interprets it for the Local Planning Authority so as to advise them that either

- the risk would not justify refusal of planning permission

or

- that the risk is sufficiently high to justify refusing planning permission on safety grounds.

The advice that planning permission should be refused is accompanied by an assurance that HSE would provide expert support in the event of an appeal against the Local Planning Authorities decision.

Criteria Against Which Judgments Are Made

HSE uses a suite of around 50 computer models to predict effects and estimate risk levels. For thermal and explosion hazards the assessments undertaken to formulate advice to Local Planning Authorities are mainly consequence or hazard based. A typical example would be the assessment of a liquefied petroleum gas installation. HSE would calculate the fire ball radius which is used as the inner zone and then the distance to 1000 thermal dose units for the middle zone and 500 thermal dose units for the outer zone or consultation distance.

For explosion risks HSE performs overpressure calculations and uses the decision criteria 600 mbar for the inner zone, 140 mbar for the middle zone and 70 mbar for the outer zone or consultation distance. 600 mbar is deemed to produce almost total demolition of the building and there is a high probability of death for the occupants. At 140 mbar some structural damage will occur which may lead to some fatalities among building occupants. At 70 mbar structural damage is unlikely although windows will be broken and other minor damage may occur. It is assumed that below 70 mbar there will be no fatalities even amongst vulnerable populations.

For the toxic substance chlorine the method considers plant failure, release of chlorine, its vaporization and mixing with air, travel downwind, dilution, inhalation by people and resulting injury. The essential elements of the modeling are source term specification, dispersion modeling to predict spatial and temporal variations in toxic dose then conversion of the dose into harm by mapping out zones in which at least the specified level of harm is realized.

Quantified risk assessment is also used to consider the benefits of risk reduction measures such as daytime only deliveries to minimize night time weather conditions which may give rise to poor ambient dispersion (i.e., low wind speeds and stable atmospheres), automatic shut off valves actuated by gas detectors to limit release durations or even mechanical ventilation and gas scrubbing. Implementing such measures can dramatically reduce the consultation distance.

HSE's specified level of harm or dangerous dose is that which causes severe distress to almost everyone, a substantial fraction would need medical attention, some would be seriously injured and require prolonged treatment and the highly susceptible might be killed. The QRA calculations result in contours of isorisk probabilities which are analogous to heights above sea level on a map. These contours are set as follows:

- the inner zone equates to 10 chances per million (cpm) (or 10^{-5}) per annum of receiving at least a dangerous dose of the toxic substance
- the middle zone is set at 1 cpm or 10^{-6}
- the outer zone is set at 0.3 cpm or 3×10^{-7}.

Within the 10 cpm inner zone the risk roughly equates to 1/10th the average chance of being killed in a road accident whereas the 1 cpm middle zone roughly equates to 10 times the chance of being killed by lightning. The risk levels used as criteria are based upon the Royal Study Group on risk assessment which concluded in 1983, that risks below 1 cpm were trivial when applied to a typical pattern of user behavior and in a development. For vulnerable people such as the elderly or children HSE uses the criterion of 0.3 cpm of receiving a dangerous dose or worse Because populations contain a proportion of highly vulnerable people the consultation distance is set at 0.3 cpm risk of receiving a dangerous dose or worse.

In the UK the public information zone is usually set to be coterminous with the consultation distance assigned for land use planning purposes.

Categorization of Developments

HSE uses differing criteria depending upon the level of sensitivity of the people likely to be present at a proposed development thus reflecting in some respects some elements of the public perception of what is tolerable for such groups. Factors which are taken into consideration include the following:

- Type and purpose (height, ventilation of building, etc.)
- Occupancy time (proportion of time spent by an individual in the development)
- Number of people
- Location of people (indoors or outdoors)
- Vulnerability/sensitivity (are they children, elderly, disabled, etc.)
- Effectiveness of emergency action (e.g., ease of evacuation)

The approach divides developments into four categories as follows:

A   Housing, hotel holiday accommodation
B   Workplaces for less than 100 people, car parking
C   Retail outlets, community and leisure
D   Educational establishment, institutional accommodation or more than 1000 people out of doors

Having determined the category into which a proposed development falls and the zone of the consultation distance in which it is to be located HSE advises the local planning authorities based upon Table 1

It is usually possible to allow small industrial developments or offices or warehouses near notifiable installations on the basis that their occupants may be readily evacuated in the event of an emergency but developments for sensitive people within the consultation distance are usually resisted. Where the decision is not obvious further calculations are performed.

TABLE I
Decision Matrix for HSE Land-Use Planning Decisions

DEVELOPMENT CATEGORY/TYPE	INNER ZONE	MIDDLE ZONE	OUTER ZONE
B (commence/industry[a])	ACCEPTABLE	ACCEPTABLE	ACCEPTABLE
C (community/leisure[b])	MAYBE/NOT	MAYBE/NOT	ACCEPTABLE
A (housing)	NOT	MAYBE/NOT	ACCEPTABLE
D (sensitive development)	NOT	MAYBE/NOT	MAYBE/NOT

(a) includes small housing developments (under review)
(b) ditto-but for cases which may set a precedent

Risk Perception Research

To prepare for the implementation of a the Control of Major Accident Hazards Directive, HSE has commissioned an interdisciplinary research project to help understand:

- the level of public comprehension of hazard and risks from major hazard sites, for example, awareness of the nature and extent of these risks, and the sources of this information.
- public perception of these risks, that is, of the perceived likelihood of a major accident, compared with other perceived risks, and how this perception compares with assessed risk levels.
- the level of risk the public is willing to tolerate from major hazards, and the factors which influence attitudes, such as economic benefit, the public image of a company good neighbor policies, etc.

Practical Application of the Research

The primary purpose of the research is to assist HSE and other Government Departments:

- to develop policies on the tolerability of risk
- to develop criteria for the siting of major hazards and the use of land within their vicinity
- to explain risk issues to local planning and hazardous substances authorities and to the public
- to evaluate the effectiveness and impact of legislation
- to provide a baseline of knowledge from which to measure future change.

The Research in Context

The research forms part of a wider HSE behavioral and social sciences research program and coincides with both the revision of the HSE publication on the Tolerability of Risks from Nuclear Power Stations which is being redrafted to encompass other industries and the revision of the 1989 HSE discussion document which suggested quantified risk criteria for land use planning within the vicinity of major industrial hazards and outlined the rationale underpinning these criteria. The revised document will deal with

- the siting of new hazardous installations
- the tolerable risks from existing installations
- the further development of land in the vicinity of such installations
- the tolerable risks to existing development in the vicinity of such installations
- the location of pipelines containing hazardous substances
- the control of developments (both new and existing) in the vicinity of such pipelines

Existing Risk Perception Research Base

Risk perception involves people's beliefs, attitudes, judgments, feelings, and values toward hazards and their benefits. "Hazard" (i.e., threats to people and the things they value), means different things to different people; different sections of society perceive risk differently and accommodating these differences requires both political and scientific judgments.

Chapter 5 of the 1992 Royal Society Study Group report on risk, Risk: Analysis, Perception and Management stated that risk assessment is subjective and entails judgment, e.g., in assessing consequences and uncertainties or balancing death against long-term injury and illness, or effects on humans versus the environment. Quantified Risk Assessment (QRA) assumes that a complete set of pathways to failure can be defined but some consequences, e.g., those involving organizational and management factors, may be difficult to forecast or control, particularly during the onset of an emergency.

Risk assessment is not totally objective but depends on the risk analyst's judgments and assumptions. There are uncertainties such as extrapolation of health effects in animals to humans, the limited amount of data available on past accidents necessitating predictions by calculation or mathematical modeling, differences in individual reactions and susceptibilities to the impact of a particular harmful effect (heat, toxic gas exposure or explosion blast), etc.

Several methods have been developed to cope with the effects of uncertainties in hazard and risk assessments. The two main approaches are "cautious" where any assumptions overestimate the risk; and *best estimate*, where efforts are made to ensure that all assumptions are as realistic as possible. HSE uses an approach which sits between these two positions termed *cautious best-estimate*. Every attempt is made to use realistic, best-estimate assumptions (whilst clearly defining the basis of the assumptions), but where there is difficulty in justifying an assumption, some overestimate is preferred. The "cautious best-estimate" approach helps to offset any uncertainty arising from the possibility of grossly abnormal human behavior and other unquantified causes of accidents.

The Royal Society Study Group report also reminded us that risk means more than just predicted number of fatalities. The work of Slovic, Fischhoff & Lichtenstein (1980) revealed three important factors:

- **"Dread" risk** involves uncontrollability, dread or fear, involuntariness of exposure, and inequitable distribution of risks. Hazards which rate high on this factor include nuclear weapons, nerve gas and crime, whereas home appliances and bicycles rate low.
- **"Unknown risk"** involves observability of risks, whether the effects are delayed, familiarity of the risk, and whether the risks are viewed as known to science or not. Hazards that rate high on this dimension include solar electric power, DNA research and satellites. Those that rate low include motor vehicles, fire fighting and mountain climbing.
- **"Number of people exposed."**

Slovic (1987) concluded that perceptions of risk are related to the position of an activity in the factor space. The dread risk factor is the most important because the higher a hazard scores on this factor (the further to the right it appears in the space in the diagram), the higher its perceived risk and the more people wish to see its current risks reduced. Slovic notes that 'expert' perceptions were more synonymous with assessments of numbers of expected fatalities and less influenced by the qualitative characteristics.

The following hierarchy of risks is referred to in Tolerability of Risks from Nuclear Power Stations report:

- voluntary versus involuntary exposure
- natural versus man-made risks
- perceptions of personal control
- familiarity
- perceptions of benefit or disbenefit
- the nature of the hazard or consequence
- the nature of the threat
- the special vulnerability of 'sensitive' groups
- public perception of the extent and type of risk
- perceptions of comparators
- the reversibility of effects

Major Hazards Risk Perception Research

Although major hazards have been recognized as a particular category of technological risk for over 20 years, there have been comparatively few studies which include major hazard risks amongst the list of risks to be considered. In the UK the Social and Community Planning Research (Prescott-Clarke 1980, 1982) undertook a study of public perceptions of a range of risks incorporating analysis of perceptions of major hazards, through a small-scale, qualitative feasibility study on Teesside and a larger nationally representative quantitative survey. The latter (Prescott-Clarke 1982) found that the nearer someone lived to a major industrial site the less likely they were to consider that such sites collectively posed a risk to the public, but the underlying reasoning behind this spatial trend is not explored. The discussion of the results of the focus group work on Teesside (Prescott-Clarke 1980) began to open up to dimensions of trust in regulators, economic dependency and industry–community relations.

Smith and Irwin (1984) undertook a survey of public perceptions of risk in the vicinity of major hazard industry in Halton in Cheshire, concluding that in neither of the two survey areas, "did the risks associated with factory accidents emerge as a major concern." However, they questioned the value of attitudinal survey evidence, stressing the dangers of overgeneralization and the need to recognize the diversity and contextual nature of risk attitudes. A number of studies have also considered public perceptions and responses in the context of information given out to the public under the CIMAH Regulations. These include research contributing to comparative European studies led by Brian Wynne (1987, 1990) and work undertaken by Jupp and Irwin (1989) around the Carrington complex in Manchester. Irwin (1995) discusses the results of questionnaire and semistructured interview work around plants in Eccles and Clayton/Beswick which found a generally high level of concern about factory accidents and pollution sitting alongside other concerns such as unemployment, crime and violence. The trust placed in various possible disseminators of information about hazards is explored, revealing a low level of trust in industry but also an overall pattern of skepticism and wariness about information sources. It is stressed that both the hazards and their perception are very much embedded in the nature of the locality and the lives of local people, so that they are "an intrinsic part of everyday social reality and the very identity of these areas."

The research team anticipates that the approach to be adopted in the present study, involving both multiple sites and a combination of psychological and sociocultural research methods, will provide some further insight into the issues that have been raised.

Approach to the Research

The approach adopted by the research team to study how people experience the risks associated with major hazard sites sees the various dimensions of aware'

installation (referred to as CIMAH sites in the UK), or as a lower inventory installation (referred to as NIHHS sites in the UK). CIMAH sites will have disseminated information to local publics, whilst there is no such requirement for NIHHS sites
- the physical size of the installation (ranging from a major complex to a warehouse)
- the length of time the installation has existed
- the history of accidents or publicized incidents at the plant
- the socioeconomic characteristics of the surrounding population (derived from census data)

The team needed to select sites with a reasonably substantial population living nearby and used census data to estimate how many people lived within the 'consultation distance' (CD) specified around each installation by HSE. So far field work has been completed in five of the case study areas and details of these are shown in Table 2. Further case study sites are to include a warehouse holding ammonium nitrate and a stretch of "major hazard" pipeline.

Contextual Research

The initial stage in the work in each case study area is to carry out contextual research involving a search of information sources held in local libraries and archives. These include local press reports, local histories, industrial and company histories, local development plans and ordnance survey maps of the area. This research is supplemented by interviews with key local actors including site management at the company, the HSE Inspector responsible for the site, emergency planners, land use planners and the chair of the local liaison committee, together with other people from the local community such as head teachers of schools near to the site or local councillors.

Focus Groups

The main method for finding out about how people perceive, understand and feel about risk is through focus group discussions which allow people to interact with one another, creating an opportunity to observe how their views are formulated, expressed and challenged. Six groups of 6–8 local people in each area are recruited (*without* informing them of the principal themes of the research). Each group meets twice. Each meeting lasts 90 minutes and the two meetings take place about three weeks apart. In the first meeting a discussion guide is used and participants are led through a list of topics, beginning with their views about the area and gradually focusing in on the major hazard site. At the end of the meeting the Q-sort exercises (see below) are demonstrated to participants and they complete these in the period between the two meetings.

TABLE 2
Case Study Sites

Company	Site	Hazard Type	No. of Employees	Size	Year Established	Extent of CD
Albright and Wilson	Langley, Sandwell, West Midlands	CIMAH Toxic	600	55 acres	1851	750–1000 m
Allied Colloids	Low Moor, Bradford, West Yorkshire	CIMAH Toxic	2000	42 acres	1953	300 m
Rohm and Haas	Jarrow, South Tyneside	NIHHS Toxic	220	14 acres	1952	400 m
BOC	Brinsworth, Rotherham, S Yorkshire	NIHHS Flamm	150	10 ha	1952	1000 m
Esso/Exxon	Fawley Southampton Hampshire	CIMAH multiple	1500	450 acres	1920	500–1000 m

The second meeting begins with a discussion of three stimulus sheets containing quotations which express different points of view about (a) the provision of public information about industrial hazards, (b) the regulation of industry and (c) who should be responsible for setting safety standards. In the second half of the meeting two sets of materials are employed which use planning and facility siting scenarios to stimulate discussion of a variety of risk issues, including questions of risk assessment criteria and societal risk. The first of these scenarios involves deciding about the location of new housing in the vicinity of an existing hazardous installation in an already built up area. The second relates to a choice between two possible locations for a hazardous plant extension.

Q-Method *to find sensitive locations - geographical risk contours*

The Q-method is a pattern analytic technique developed to study subjectivity. In this project two Q-sort exercises are undertaken by all the people involved in the focus groups, one concerned with people's sense of place, the second concerned with their view of the local major hazard site.

Using Q-method first involves collecting statements about major hazard sites from a range of different sources. A sample representing the range of statements is created and from this a set of cards is produced, each containing a single statement related to the local major hazard site. People then sort the cards into a predetermined pattern according to the strength of their agreement or disagree'

being presented in Amsterdam Oct 1997 by HSE

on perceptions of a site, this collective memory sometimes extending back a long time and passing from generation to generation
- the importance of experience of incidents in shaping local discussion and perceptions and relationships between the company and the local community
- the way in which inferences about risk are drawn from sensory "evidence" such as smells, visible plumes, the general appearance of the site and the quality of local nature
- the way in which people reason about risk, in relation to probability and consequence arguments, trade-offs between risk and other concerns and the scope for individual choice in risk-taking
- the stigmatization that can result from the presence of a hazardous site and the implications this has for local identity
- the variety of images that people hold of the company and the range of ways in which initiatives by the company, such as open days, community newsletters, and liaison groups are responded to
- the strength of views on regulation and regulators and the expectations of and trust in regulatory practice that are expressed

Q-Sort Data

Initial analysis of the Q-sort data has found that in each area two clearly differentiated, orthogonal factors, or points of view, emerged. In each case the strongest of these was a point of view which was in general distrustful of the company and had little confidence in the regulators or in the emergency services. This point of view was also characterized by concern about the risks associated with the site.

The second contrary point of view was more trusting in the competence of the company, the regulatory authorities and the emergency services and more tolerant of the risks. Significantly, very few of the participants in these case studies appeared to hold such a robust view. A comparative analysis of the factors identified for all five ? sites found them to correspond very closely, confirming that the researchers were finding very similar patterns of response at all four sites. The most significant differences between the factors can be accounted for by specific differences in the local contexts.

Communication

HSE inspectors are well versed at dealing with industry and have regular contact with planning authorities and locally elected councilors. They also become involved with local residents and the media, particularly after major incidents such as the fire at Allied Colloids. One of the major aims of this research is to help us to explain risk issues most effectively to local authorities and the public in terms that allow them to weigh these risks against potential economic benefits such as increased employment. Employer organizations such as the Chemical Industries Association with their Responsible Care program and the Confederation of British Industry are showing great interest in the emerging results of the research as it clearly has implications for the way in which they approach local residents.

An important factor in the equation is the degree of trust awarded to different players in the game. Recent UK research by Marris, Langford and O'Riordan (1996) has shown that family and friends are seen as the most reliable sources of information, followed closely by environmental groups which people see as independent of government and commercial biases. Doctors and scientists were also seen as quite trustworthy, but trades unions, religious organizations, the media, companies and government all scored badly.

As the research is not complete I will draw on some general good practice principles for risk communication:

- choose your communicator carefully, a local person is more likely to be well received than a stranger from head office
- identify and understand your audience and their concerns to establish a dialogue
- create the right environment for successful communication
- know your goals then plan, implement, monitor and replan following feedback
- tailor your message, use technical information carefully and do not ignore uncertainty
- demonstrate your commitment

At IChemE's 75th Jubilee Lecture Howard Newby put forward five rules for turning an environmental risk into a public relations disaster which are worth sharing:

- Experts know best
- Be wise only after the event
- Blame someone who is powerless to fight back
- Only manage what you can measure
- Che sera sera, anything else is too expensive

HSE is also piloting a Geographic Information System which might become a further tool for communication.

Geographic Information Systems

HSE is consulted on around 7,000 planning decisions a year of which roughly 10% are referred to the major hazards assessment unit in circumstances where the advice is not obvious from the codified guidance which has been given to HSE's local offices. It is vital that consistent advice is given to local planning authorities and that the advice can withstand the scrutiny of a planning inquiry if an appeal is lodged against a decision where health and safety formed the grounds for refusal. A national building company which had had planning permission refused on safety grounds in one part of the country must be clear it would be turned down in similar circumstances elsewhere.

Many planning authorities, emergency planning authorities and pipeline operators are using Geographical Information Systems (GIS) in a range of tasks and HSE now has a pilot system to assist in decision making and to demonstrate consultation distances and the consistency of our advice. Some emergency planners are using GIS to investigate the characteristics of the area which might be affected in the event of a major incident, e.g. to decide how many private cars may be available to aid evacuation. There would be advantages in modeling the progression of an incident in real time, e.g. to investigate the relative benefits of sheltering or evacuating the exposed population or to decide whether to let a fire burn itself out while spreading combustion products downwind versus using large quantities of water to extinguish the fire thus risking groundwater pollution from the runoff. To be of real benefit, however, the system would have to model events several times faster than the incident progresses and to date few, if any, have proved sufficiently fast.

While GIS undoubtedly has many benefits there are downsides not least of which is the start up cost of purchasing background mapping data and adding data to the system.

Conclusion

Although there are uncertainties in the modeling of the consequences and risks from loss of containment the UK HSE approach ensures consistent advice is given to local planning authorities although the final decision rests with them and is subject to political and economic influences. It is important to develop a better understanding of public risk perceptions in the settings and contexts in which these are experienced and to recognize that living with risk is very much part of everyday life for many people. Research which allows people to express points of view in their own terms is needed if we are to develop a deeper understanding of *why* particular perceptions of risk are held and the stability and certainty of particular points of view. The use of focus groups allied with q-method provides an innovative and effective approach to the formulation of recommendations for the development of major hazard policy and regulation in the UK. The use of a Geographic Information System, albeit only yet at a pilot stage, could assist HSE to communicate the results of its deliberations and calculations and to demonstrate the consistency of that advice.

It is vital to avoid arrogance in the way we communicate. In conclusion I turn again to Howard Newby's IChemE's 75th Jubilee Lecture. The declension "I am an expert, you are ignorant, they are irrational" might well provoke the response "You may know more but do you know better?"

References

Beck U (1992) *Risk Society: Towards a New Modernity*, Sage, London, UK.
British Royal Society (1992) *Risk; Analysis Perception and Management*, London, England.
Cassidy, K (1997) Developments in HSE criteria *Risk 2000 acceptability to core technology*, London, UK
Commission of the European Communities (1994) *Proposal for a COUNCIL DIRECTIVE on the control of major accident hazards involving dangerous substances (COMAH)*. COM(94) 4 final: Brussels 26.01.94
Erikson, K. (1990) Toxic Reckoning: Business Faces a New Kind of Fear, *Harvard Business Review*, Vol. 90:1, pp. 118-126.
Health and Safety Executive (1989) *Risk Criteria for Land use planning in the vicinity of Major Industrial Hazards*, HMSO, London, UK.
Health and Safety Executive (1992) *The Tolerability of Risk from Nuclear Power Stations*, HMSO, London
Irwin, A. (1995) *Citizen Science: A study of people, expertise and sustainable development*, Routledge, London, UK.
Jupp, A., and Irwin, A. (1989) Emergency Response and the Provision of Public Information under CIMAH, *Disaster Management*, 1:4.
Marris, C., Langford, I., and O'Riordan, T. (1996) *Integrating sociological and psychological approaches to public perceptions of environmental risks; detailed results from a questionaire study*. Centre for Social and Economic Research on the Global Environment (CSERGE working paper GEC 96-07), University of East Anglia, Norwich.
Newby, H (1997) Risk analysis and risk perception, the social limits of technological change *Transactions of the Institution of Chemical Engineers, Part B Process Safety and Environmental Protection* vol 75, no B3
Nussey, C., Carter, D. A., Cassidy, K. (1995) The application of consequence models in risk assessment: a regulator's view. *Proceedings of the International Conference and Workshop on Modelling and Mitigating the Consequences of Accidental Releases of Hazardous Materials*. American Institution of Chemical Engineers, New York.
Prescott-Clarke, P. (1980) *Public Attitudes Towards the Acceptability of Risks*, Social and Community Planning Research, London, UK.
Prescott-Clarke, P. (1982) *Public Attitudes Towards Industrial, Work Related and Other Risks*, Social and Community Planning Research, London, UK.

Slovic, P., Fischhoff, B. and Lichtenstein, S. (1980) Facts and Fears; understanding perceived risk. In *Societal Risk Assessment: How Safe Is Safe Enough* (R.C Schwing and WA Albers, eds.), New York: Praeger

Slovic, P. (1987) Perception of risk. *Science* 236, 280–285

Smith, D. and Irwin, A. (1984) Public Attitudes to Technological Risk: the Contribution of Survey Data to Public Policy Making, *Transactions of the Institute of British Geographers*, 9.

Stallen, P.J.M. and Tomas, A. (1988) Public Concern about Industrial Hazards, *Risk Analysis*, 8:2, pp. 237-245.

Van der Pligt, J. (1992) *Nuclear Energy and the Public*, Blackwell, Oxford, UK.

Vlek, C. and Stallen, P.J. (1981) Judging Risks in the Small and the Large, *Organizational Behaviour and Human Performance*, 28, pp. 235-271.

Walker, G. P. (1995) Land use planning, industrial hazards and the "COMAH" Directive, *Land Use Policy*, vol. 12, no. 3, pp. 187–191.

Wiegman, O., Gutteling, J.M. and Boer, H. (1991) Verification of Information through Direct Experiences with an Industrial Hazard, **12:3**, pp. 325–339.

Wynne B. (1987) *Implementation of Article 8 of the EC Seveso Directive: A Study of Public Information*, Report to the European Commission, DG XI.

Wynne B. (1990) *Empirical Evaluation of Public Information on Major Industrial Accident Hazards*, Report to EC Joint Research Centre, ISPRA

Wynne B., et al. (1993) *Public Perceptions and the Nuclear Industry in West Cumbria*, Centre for the Study of Environmental Change, University of Lancaster, Lancaster, UK.

The General Duty Clause under the Clean Air Act: Issues of Implementation

John Ferris
Chemical Emergency Preparedness, and Prevention Office, U.S. Environmental Protection, Agency, 401 M Street, SW (5104), Washington, DC 20460

ABSTRACT

Sections 112 and 113 of the Clean Air Act provide EPA with the authority to implement and enforce the general duty clause. However, these sections do not require EPA to write regulations clarifying it. The challenge to EPA is to develop a proactive implementation strategy for the implementation of the general duty clause to foster owners and operators of facilities to know their responsibilities, to prevent accidents and minimize the consequences of accidents that occur. This paper will discuss the issues the EPA is facing regarding accident prevention and possible strategies for EPA and the chemical processing industry to implement the general duty clause.

Introduction

The goal of EPA's Chemical Emergency Preparedness and Prevention Office is to reduce or eliminate accidental chemical releases. The Clean Air Act Amendments of 1990 provided EPA with a powerful new tool to accomplish this goal: The general duty clause. The general duty clause (GDC) has untapped potential for accident prevention, and not merely as an enforcement tool. Simply put, if you have an extremely hazardous substance, you are obligated as the caretaker of that substance, to use any feasible means to eliminate or substantially reduce the hazards posed by that substance.

In the Clean Air Act Amendments of 1990, Congress added subsection (r) to section 112 of the Clean Air Act for the prevention of chemical accidents. Section 112(r)(1), known as the general duty clause, states:

> It shall be the objective of the regulations and programs authorized under this subsection to prevent the accidental release and to minimize the consequences of any such release of any substance listed pursuant to paragraph (3) or any other extremely hazardous substance. The owners and operators of stationary sources producing, processing, handling or storing such substances have a general duty in the same manner and to the same extent as section 654, title 29 of the United

States Code, to identify hazards which may result from such releases using appropriate hazard assessment techniques, to design and maintain a safe facility taking such steps as are necessary to prevent releases, and to minimize the consequences of accidental releases which do occur. For purposes of this paragraph, the provisions of section 304 shall not be available to any person or otherwise be construed to be applicable to this paragraph. Nothing in this section shall be interpreted, construed, implied or applied to create any liability or basis for suit for compensation for bodily injury or any other injury or property damages to any person which may result from accidental releases of such substances.

The general duty clause provides unique statutory language to be implemented by EPA. It does not require EPA to develop clarification regulations nor guidance. It's brevity provides flexibility for persons subject to it, yet it is clear that these persons must perform their duty to prevent accidents and minimize the consequences of accidents that do occur. This paper is not meant to be a definitive legal analysis of the clause, but it is meant to foster a discussion with those who are subject to the clause of what it means to be performing their general duty. The use of legislative history in this paper is meant to provide guidance on the issues associated with the various parts of the clause.

What Chemicals Are Involved?

The general duty clause (GDC) is not specific as to the universe of chemicals that are covered by the general duty. The Act states "any substance listed pursuant to paragraph (3) or any other extremely hazardous substance." Paragraph (3) of section 112(r) is the statutory basis for the list of regulated substances for the Risk Management Plan regulations. This list was published in the *Federal Register* on January 31, 1994 (59 FR 4478). Paragraph (3) states "the Administrator shall use, but is not limited to, the list of extremely hazardous substances published under the Emergency Planning and Community Right-to-Know Act of 1986 [EPCRA.]" While the Act does not specifically define extremely hazardous substances, it is reasonable to assume that Congress intended to have the GDC apply to substances beyond the listed substances under 112(r)(3) and EPCRA. It is not clear exactly what the scope of "any other extremely hazardous substance" is. Using the language of paragraph (3) it is likely that substances "which, in the case of an accidental release, are known to cause or may reasonably be anticipated to cause death, injury, or serious adverse effects to human health or the environment" can be considered extremely hazardous under the GDC.

The Senate Report[1] also references the extremely hazardous substances list under EPCRA, but it further adds

---

1 Senate Report No. 228, 101st Congress, 1st Session (1989).

Extremely hazardous substances would also include other agents which may or may not be listed or otherwise identified by any Government agency currently which may as the result [of] short-term exposures associated with releases to the air cause death, injury or property damage due to their toxicity, reactivity, flammability, volatility or corrosivity. The release of any substance which causes death or serious injury because of its acute toxic effect or as the result of explosion or fire or which causes substantial property damage by blast, fire, corrosion or other reaction would create a presumption that such substance is extremely hazardous.

It is impossible for EPA, or any other organization, to develop a complete list of all of the chemicals that could be considered "extremely hazardous." Many factors, including the unique circumstances by which the chemical is present at the facility and possible release scenarios, may be involved in determining the hazards posed by a chemical. It is unreasonable for anyone to expect an organization can develop a complete list and state that chemicals not on the list would never be "extremely hazardous." The owner or operator needs to determine which chemicals at their facility could cause the impacts described in the Law and Senate Report.

Who Is Subject?

The Act places the requirement to comply with the GDC on "owners and operators of stationary sources producing, processing, handling or storing [an extremely hazardous substance]." Further, the Senate Report states

> The responsibility to prevent releases of these extremely hazardous substances is the general duty of the owners and operators of facilities producing, processing, handling or storing such substances, whether or not explicit requirements have been imposed under this section or other authorities.

The Senate Report states that the phrase "producing, processing, handling or storing" should be interpreted to include the broadest set of activities associated with the manufacture and use of chemical substances including, but is not limited to, transportation and the use of the chemical in commercial, agriculture, mining and industrial activities.

Stationary sources are defined in the Act as "any building, structures, equipment, installations, or substance emitting stationary activities. . . ." Using this definition, virtually any stationary structure that could have an accidental release can be considered a stationary source.

Therefore, the owners and operators of these stationary sources (facilities) that have an extremely hazardous substance are required to perform their general duty.

What Is Meant by the Reference to OSHA's General Duty Clause?

The GDC uses the phrase "in the same manner and to the same extent as section 654, title 29 of the United States Code[.]" This phrase refers to the general duty clause that is implemented and enforced by the Occupational Safety and Health Administration (OSHA).

The Occupational Safety and Health Act of 1970 includes a similar general duty clause which provides that:

> Each employer... shall furnish to each of his employees employment and a place of employment which are free from recognized hazards that are causing or are likely to cause death or serious physical harm to his employees. [As reported in the Senate Report]

The Occupational Safety and Health Administration cites this clause for enforcement purposes when a recognized hazard is found in the workplace for which there is no specific OSHA regulation or standard.

The Senate Report states that the standard for applying the general duty clause has been described (for the OSHA statute) in the case *Secretary of Labor* v. *Duriron Company, Inc.* 11 OSHC 1405 (upheld by the courts on appeal). This decision is cited to indicate that a "similar standard is an appropriate application of the general duty to operate a facility free from accidents[.]" Thus, the Duriron decision indicates what is necessary for OSHA to establish a violation of their general duty clause:

> In order to establish a... violation, the Secretary must prove: (1) the employer failed to render its workplace free of a hazard, (2) the hazard was recognized either by the cited employer or generally within the employer's industry, (3) the hazard was causing or was likely to cause death or serious harm, and (4) there was a feasible means by which the employer could have eliminated or materially reduced the hazard. [*Secretary of Labor* v. *Duriron Company, Inc., 11 OSHC at 1407*]

What is not clear in Section 112(r)(1) is what is meant by "in the same manner and to the same extent." OSHA's general duty clause applies to a much wider scope of hazards in a workplace than hazards associated with chemicals (which seems to be the scope of EPA's GDC). Also, OSHA's general duty clause deals with the relationship between employer and employee on-site. EPA's GDC does not seem to be limited to workers exposed to hazards because EPA's programs are to protect human health and the environment off-site.

The meaning of the phrase "in the same manner and to the same extent" may have been clarified by the following sentences from the Senate Report.

OSHA frequently found that its health and safety regulations (which are designed to protect against continuous exposures to routine releases) were not well-suited to prevent or mitigate sudden, accidental releases and used the general duty clause frequently to justify citations for unsafe conditions requiring corrective action which were identified by [OSHA's Chemical Special Emphasis Program] audits. The general duty clause . . . may be used by EPA in similar ways and for similar purposes.

The Senate Report also adds the following paragraphs to clarify what is necessary to document an employer's awareness of a hazard and what is meant by feasible:

The employer's awareness of the hazard may be documented by the existence of an industry code or consensus standard (such as the "Boiler and Pressure Vessel Code" or the "Compressed Gas Standard") which is used by others operating similar facilities within the industry.

Feasible means of preventing death, serious injury or substantial property damage outside of the boundary of the facility would include not only the prevention and mitigation measures employed to reduce the likelihood or extent of a hazard moving across the boundary, but also public warning (alert) and emergency response systems which may reduce exposure to the hazard if it reaches the community. Facility owners and operators would need to implement all feasible means to reduce the threat of death, serious injury or substantial property damage to satisfy the requirements of the general duty clause.

Even though the Law does not specify exactly how OSHA's general duty clause applies to EPA's, *Duriron* and other references to OSHA's GDC provides some understanding of what is meant by the GDC under the Clean Air Act.

What Needs to Be Done?

Owners and operators of sources subject to the GDC are required to perform the following tasks:

1. Identify hazards which may result from [accidental] releases using appropriate hazards assessment techniques;
2. Design and maintain a safe facility taking such steps as are necessary to prevent releases, and;
3. Minimize the consequences of accidental releases which do occur.

The term "appropriate hazard assessment technique" is not defined in the Act. However, the legislative history of the Act provides some details. The

Senate Report defines hazards assessments as "an engineering analysis of a facility or process in which an extremely hazardous substance is used to determine the potential for any accidental release and the consequences (death, injury or property damage) which would likely occur in the event of an accidental release." Also, the Conference Report[2] states that a hazards assessment includes:

(1) basic data on the source, units at the source facility which contain or process regulated substances (including the longitude and latitude of such units), operating procedures, population of nearby communities, and the meteorology of the area where the source is located;
(2) an identification of the potential points of accidental releases from the source of regulated substances;
(3) an identification of any previous accidental releases from the source including the amounts released, frequencies, and durations;
(4) an identification of a range (including worst case events) of potential releases from the source, including an estimate of the size, concentration, and duration of such potential releases and a correlation of such factors with the distance from the source of the release;
(5) a determination of potential exposure (including the concentration and duration of exposure) for all persons who may be put at risk as the result of a release from the source;
(6) information on the toxicity of the regulated substances present at the facility; and
(7) a review of the efficacy of various release prevention and control measures, including process changes or substitution of materials.

It is not clear in the Conference Report whether this definition applies to the GDC or the Risk Management Provisions under section 112(r)(7) of the Clean Air Act (or both). However, this definition provides some guidance as to the scope of the hazard assessment under the GDC.

In the chemical industry, the term hazards assessment is used in conjunction with such terms as hazards identification, hazards analysis, and risk assessment. All of these terms are considered "terms of art," having specific and unique meanings. Hazards assessment typically means using such techniques as HazOps, What ifs, and fault tree analysis. Using the term "hazards assessment" implies going beyond knowing what the hazards are, to an assessment of the hazards as well. While owners and operators are not expressly held to these definitions, they are expressly required to use "appropriate hazards assessment techniques," a term which signifies going beyond the identification of the hazards.

---

2   Conference Report of the Clean Air Act Amendments of 1990, Report number 101–952. (October 26, 1990).

In the Senate Report, it states that "The facility owner or operator is obligated to take all feasible actions that are available to reduce hazards which are known to exist at that particular facility or which have been identified for similar facilities in the same industrial group."

Industry has a multitude of standards which may be employed in performance of the general duty, ranging from design specifications for such equipment as tanks and pressure vessels to procedures defined in CCPS's Guidelines for Technical Management, API's Recommended Practices and portions of the Chemical Manufacturer's Association's Responsible Care. These private standards were developed to provide industry with what needs to be done to prevent accidents and may also be applied to meet GDC requirements..

Each facility subject to the GDC has generic and unique hazard features. The generic hazards are those commonly found at other similar facilities in the same industry sector. The unique hazard features are those that arise from specific activities and other specific conditions that are unique to each facility. To meet the standard of care set by EPA's GDC, the owner/operator must address *both* types of hazards at its facility by:

- using appropriate hazard assessment techniques,
- assuring safe facility design and maintenance, and
- using effective accident prevention and mitigation procedures.

Letting Experts Be Experts

The owner/operator should have the best expertise for preventing/mitigating accidents at its own facility, and the GDC mandates that this expertise be fully used to address any recognized hazards which have not been eliminated by the owner/operator's compliance with formal regulations, permits and standards. Thus, under the GDC, the owner/operator needs to look beyond regulatory compliance and determine what more needs to be done to prevent/mitigate accidents. This is a continuing obligation. Organizations like CCPS can help owners/operators in several ways, for example:

- developing and improving hazard assessment methods,
- identifying and defining recognized hazards,
- recommending best practices for accident prevention/mitigation, and
- fostering inter-company sharing of technical expertise.

EPA's Implementation Issues

Thus, the GDC imposes responsibilities for safety management which go beyond regulatory requirements, and spurs continuous improvement in safety management among companies. Although EPA can use the GDC to punish companies for failing to address recognized hazards (before and after an accident), the Agency's primary goal is to have companies understand their GDC obligations and take proactive steps to prevent accidents before they occur. As a result, EPA hopes to enlist the support of CCPS, CMA, API and other organizations which have to capacity to help owner/operators meet the GDC standard of care.

The chemical industry presents several scenarios that may warrant different degrees of EPA involvement pursuant to the general duty clause. Some facilities may benefit from voluntary audit programs of EPA, states and companies under product stewardship and process safety programs. Findings from these audit programs may help businesses improve their accident prevention programs and obviate the need for GDC enforcement actions..

Other facilities may require stronger actions though compliance inspections and enforcement to improve their prevention programs. For example, EPA may use injunctive relief or administrative orders to assure that a facility improves their accident prevention program.

Finally, EPA may assess penalties to the owner and operator of a company for failure to comply with the GDC. This not only to penalizes the business but has the broader effect of demonstrating that EPA is serious about having businesses comply with the GDC.

What Do We Do Now?

The chemical process industry has taken great strides to recognize the hazards posed by extremely hazardous substances as well as to learn what needs to be done to prevent and mitigate accidents. Now it is time to ensure that all facilities are doing what we know needs to be done.

Industry sector programs, voluntary standards, and government regulations will prevent many accidents. But, each owner and operator needs to look at the unique aspects of their facilities, and do an even better job, as the GDC mandates. However, these owners and operators are not in it alone. The concept of the general duty clause is for these owners and operators to work with industry groups and agencies **before** accidents for assistance in what needs to be done. Once they know, it is then up to the owners and operators to ensure that it gets done.

Disclaimer: The views expressed in this document are the opinions of the author and may not represent official agency positions.

Development of Integrated Fixed Facility and Transportation Risk Criteria

Lisa M. Bendixen
Arthur D. Little, Inc., Cambridge, Massachusetts

ABSTRACT
While a limited number of chemical companies and regulatory bodies are using some type of risk criteria for fixed facilities, very few have tackled criteria for transportation risks. Even fewer recognize that there are several ways in which to make risk criteria consistent for both transportation and fixed facilities. Having a common basis for evaluating different sources of risks encourages the full consideration of risks from all aspects and life cycle stages of a project or an activity. In turn, this full consideration of risks resonates better with decision-makers, and increases both the actual and the perceived value of the risk management function. This paper explores the basics of developing, implementing and communicating risk criteria for decision-making in the framework of addressing both transportation and fixed facility risks.

As the use of hazard and risk assessments increases, so does the need for and importance of an overall risk management system. Such a system allows organizations to prioritize both their efforts to identify and understand risks, and their allocation of resources to manage and control risks. Integral to many successful risk management systems are risk criteria for use in decision-making. Over the last 15 years, such criteria have evolved in their presentation and communication from risk *acceptability* criteria to risk *tolerability* criteria to risk *prioritization* criteria. This reflects both more accurate representations of the purpose of such criteria and an increased sensitivity to the dynamics between industry and the public.

The purpose of having risk prioritization criteria is to provide consistent, reliable guidance to personnel conducting and interpreting risk studies as to how the resulting risk levels are viewed by an organization, without having to present all results to senior management. This allows:

- consistent treatment worldwide of all facilities and transportation operations and their surrounding populations;
- prompt recognition of problematic risk levels for early resolution by the affected business;
- a formal treatment of risk in key decision-making situations; and
- the selection of effective risk reduction measures.

Developing Criteria

Acute risk may be evaluated in terms of either individual and/or societal risk; when considering and incorporating existing approaches, both types are important. Societal risk represents the full set of potential impacts on a given population from a given source of risk—such as a specific unit, site, transportation activity, or pipeline. It is generally displayed as a F/N curve. Individual risk gives the risk level at a specific location and therefore looks at the risk to a particular individual (at that location). It is generally expressed as a risk level per year and may be shown as a risk contour. Criteria can be developed for these risk measures and/or for average outcomes (expected values) or maximum consequences.

Risk prioritization criteria for a particular organization may consider both individual and societal risk and are generally developed by reviewing:

- the approaches set forth publicly by regulatory authorities (both local and national) in a number of countries;
- those approaches used or under consideration by other chemical companies;
- recommendations by academicians and scientists based on public perceptions and attitudes as well as historical (or background) risk levels; and
- the way that the criteria would be implemented by the organization.

This last point is critical, because it demonstrates the potential need for slightly different criteria for different organizations due to variations in their methodologies and the degree of conservatism of the results of risk assessments for each organization. If the criteria are to be utilized on both a site-wide and a single-unit basis (as is common in most process hazard analyses), separate criteria may also need to be developed for each case.

When following this approach, it is best to first develop the criteria to be used in quantitative risk assessments and then to translate the selected criteria for screening approaches or semi-quantitative analyses based on an evaluation of the differences in methodologies for the different approaches. Generally one would also start with site-wide criteria prior to developing criteria for single units, as site-wide criteria most directly correspond to the existing background or historical risk data. For similar reasons, fixed facility criteria should be developed before those for pipelines or transportation activities.

At the most detailed level of analysis, with appropriate definitions, it is possible to have criteria that are the same for fixed facilities, pipelines and transportation activities. Such consistency in criteria can be viewed as implying that an organization does not want to pose higher risks from one type of operation than it does from any other.

Fixed Facility Criteria

Major U.S. and European chemical companies, as well as government agencies, are evolving philosophies or criteria for evaluating the tolerability or prioritization of risks. A very brief summary of these positions and the general approach used to derive them is given below.

Risk criteria are generally related to background or historic risk levels. There are several reasons for this:

- *a need to understand what risk levels actually are.* The general public (and others) often assume that any risk they can imagine is a credible risk. This creates problems when analysts provide comprehensive lists of identified risks and then eliminate some of them from further consideration on the basis of low probabilities of occurrence. Worst-case analyses reinforce this perception. By linking criteria with historic risk levels, there is at least some foundation for the criteria.
- *a desire not to increase risks (for individuals) significantly.* By understanding background risk levels for a number of age groups, a facility/project/operation can assure that it is not increasing the risk to any population group in any significant way (e.g., less than a 1% increase in risk).

In addition to developing individual risk criteria (usually for an entire site) linked to background risk levels, many organizations are finding it useful to have explicit criteria for multiple-fatality events. These may be strict thresholds on the maximum number of potential fatalities to reflect the overall risk attitudes of the organization or the communities in which they operate, or the criteria may be combinations of consequences and frequencies—such as in an F/N curve. Such criteria can easily be applied for all operations of concern: fixed facilities, transportation activities, and pipelines.

The developed criteria can be selected so as to be internally consistent for individual and societal risk, based on the mathematical relationships in the risk calculations and experience with the typical size of exposed populations. This generally implies that if the societal risk prioritization criteria are satisfied, then the individual risk prioritization criteria are also satisfied. This often implies a sequence in which the criteria are applied to the results of a risk assessment.

There are a number of factors which may affect the scaling of the selected values for site-wide risk criteria to single-unit operations. For instance, a given unit will generally have its maximum contribution to off-site individual risk levels at a point on the fence line that is close to the unit, while other units will have their maximum contributions at other locations. Thus, a maximum individual risk criterion would not need to be scaled significantly in going from site-wide operations to unit operations. However, average individual risk levels do receive contributions from numerous units and on a per-unit basis the associated criterion should be reduced (i.e., made more restrictive).

With regard to societal risk, the selected risk prioritization levels are generally reduced (i.e., the frequencies are made more restrictive) when they are to be applied to a single unit, as most sites include multiple units. However, experience has shown that it is quite unusual for more than one or two units on-site to pose the potential for the maximum consequence event, so the scaling for frequency at the low frequency/high consequence end of the F/N curve is different than it is over the rest of the F/N curve where many units may contribute to societal risk.

Both fixed facilities and transportation activities have more aggregated levels of risk for which risk prioritization criteria are typically not derived. This is largely due to the fact that many different companies may be involved, not just the organization for which the criteria are being developed. For example, there may be a number of sites in one complex or area, or multiple materials may be moved along the same route, or rail lines may be adjacent to highways. This has precedence with regard to looking at background risk, if one looks at rates for different causes of death rather than the overall death rate.

Transportation Criteria

The first step in developing risk prioritization criteria for transportation is determining the logical equivalence between certain levels of activity and fixed facility operations. Overall site risk levels could be considered as on a par with the risks associated with a major transportation activity *within a community*, such as all shipments of a given material via a selected mode from a particular origin. The single unit level could then be considered to be on a par with a smaller-scale transportation activity *within a community*, such as from the origin to a major customer or a set of customers in one geographic area (which may be beyond the community of interest but which require all the shipments to follow one route through that community).

When considering transportation risks, the definition of "community" is a critical part of linking the risk criteria with those for fixed facility operations. Doing so allows the same basic risk criteria to be used for fixed facilities and transportation activities. Individual risk criteria (average and maximum) for transportation can be established just as those for fixed facilities, whether full or partial routes are examined. (Full routes are origin-to-destination, partial routes are a segment or portion of a full route.) However, societal risks from a full route can be many orders of magnitude higher than are typically associated with a fixed facility. This is a direct result of the accident rates, spill probabilities, route lengths, and number of shipments made each year. The relatively high societal risk levels and lack of good comparative or background data for transportation risks suggest the development of another approach, such as community risk levels.

In general, the definition of community must be narrow enough to ensure that the population covered would all relate (that is they would take it as a local threat that could have affected themselves) to a specific incident, and yet broad enough so that events outside of the "community" are either not of concern or are of much less concern. A definition of community as being represented by a 15 to 20 mile segment of a transportation corridor is therefore suggested. Such a definition is:

- broad enough to cover bordering towns in suburban/rural areas where an incident in a neighboring town may be considered just as critical as one in one's own town;
- broad enough to cover a full city for most small to moderate cities where the citizens are likely to feel somewhat at risk from any event that occurs in that city;
- narrow enough to retain a sense of community in large cities, where so many different events take place that the population tends to ignore all but those from which they feel directly threatened; and
- broad enough to account for circuitous routing in certain cities.

Given such a definition of community, the same criteria that are used for a site can now be used for major transportation activities. The societal and average individual risk criteria can then be reduced (by the same factor as for fixed facilities) for use with smaller-scale transportation activities. However, since all activities involving a given material and mode are expected to contribute to the maximum individual risk level, this value is fully scaled for transportation unlike fixed facilities. Similarly, any activity for a given material can yield a maximum consequence event, and the associated frequency value (if there is one) should also be fully scaled (reduced) for use with smaller-scale transportation activities. This might apply to either the low frequency/high consequence portion of an F/N curve, or to an independent criterion for maximum consequences.

This approach to establishing societal risk criteria can also guide the risk assessment process. If it is possible to determine which 15 to 20 mile segment is at greatest risk via a screening process, a more detailed risk analysis may only be required for the chosen segment. Even if the highest risk portion of the route cannot be accurately identified through screening techniques, it may still be possible to reduce the amount of the route to be analyzed in detail to a much smaller fraction of the total.

This approach can limit one's knowledge of the total societal risk level associated with a transportation activity. However, since the societal risks for every transportation activity are so different and are so dependent on route length and number of shipments, decision-makers do not seem to be able to process or effectively use information on total risk levels. Comparisons to other transportation activities will often be made *after* normalizing for route length and the number of shipments, which also removes the focus from total risk. This latter approach also does not offer the internal consistency between fixed facility and transportation criteria, and means that the criteria are based on what other organizations are actually doing, without a good sense of or basis for the priority of the associated risk level.

Pipeline Criteria

For that portion of a pipeline which is located beyond the fence line (as opposed to the property line since the public may be present anywhere beyond the fence line), the approach to risk prioritization can also be based on a 15 to 20 mile segment representing a community. If a pipeline is shorter than the chosen segment length, its actual length should be used. For a single pipeline on its own or for joint evaluation of all of the hazardous material pipelines in a pipeline corridor, the appropriate criteria could be the same as for a site or major transportation activity. If only one pipeline out of a set is being evaluated, the criteria could be scaled down, usually by a different factor than that used for a smaller-scale transportation activities. This results directly from the typical number of pipelines of concern in any given pipeline corridor.

Adjusting Criteria for Simplified Assessments

There are a number of important differences in the methodologies and assumptions used in detailed quantitative risk assessments versus more simplified, process hazard analyses. The sections below highlight some of the more influential differences and summarize their expected impact on the results for fixed facilities, transportation activities and pipelines. This discussion focuses on general trends only, as the actual methodologies used by different organizations may vary significantly, which will influence their individual adjustments.

Fixed Facility Assessments

Factors which are characteristic of many simplified analyses and which need to be considered in order to properly adjust the risk criteria include:

- not identifying secondary (or knock-on or domino) effects or other interactions with other units and utilities in the hazard identification process;
- taking event frequencies from standardized databases that do not account for a variety of site- and equipment-specific factors—and which are generally more conservative than those associated with detailed analyses;
- simplifying the treatment of ignition probabilities for flammable releases—such as using one probability rather than considering early versus delayed ignition and the location of specific ignition sources;
- assuming all exposures to a hazardous situation result in the impact of concern;

- not addressing the potential for intervening after a release;
- using less sophisticated consequence models and simplifying assumptions on spill area, entrainment of aerosols, etc.—which can produce either over or under stated hazard areas;
- not considering variations in population distributions; and
- using fewer wind directions, speeds and stability classes.

Depending on which of these factors are relevant for a particular simplified assessment methodology, it is possible to under or over state the risk levels. In general though, most simplified methodologies are designed to err on the side of conservatism.

Since average individual risk is the average of all the individual risk levels among those exposed to risk from the fixed facility, pipeline or transportation movement, it can actually be most meaningful in simplified analyses where the population distribution is not analyzed in detail, etc.

Transportation Assessments

There are differences between detailed and simplified analyses relating to: accident rates, conditional spill probabilities, potential consequences, ignition potential, modeling conditions, consequence models, likelihood of fatality, assumed characteristics of the route, and population density. In general, simplified analyses use conservative generic accident rates and spill probabilities, do not fully account for certain limiting factors such as early ignition or less than 100 percent likelihood of fatality, and use conservative simplifying assumptions for the consequences of concern and the average population density.

Virtually all of these items make simplified analyses more conservative than detailed ones. The modeling conditions and types of consequence models used could somewhat under predict the largest hazard area, but the combined effect of all the items listed above is usually an overstatement of risk levels compared to the results of a detailed study.

Pipeline Assessments

Many simplified pipeline analyses use the same basic failure rates and release sizes as in a detailed analysis. However, they also tend to use average population densities along a segment as opposed to location-specific population distributions—which generally means that the potential consequences farther away from the pipelines may be underestimated. Wind direction probabilities and likelihoods of fatalities are generally not considered. It is typical to consider both immediate ignition (e.g., jet fires) and worst case hazards, but not intermediate events such as smaller vapor cloud fires. As a result, risk levels are likely to be somewhat over stated.

Application of Risk Criteria

Many organizations use both individual and societal (F/N curves) risk criteria and assign risk priority levels based on whichever measure produces the highest estimate of risk. For some organizations it will be important to also have employee risk criteria (usually only for fixed facilities and expressed as individual risk levels) that are used in conjunction with the other measures.

Typically there are three priority categories, equivalent to: *high* (do something), *medium* (use judgment) and *low* (no action required at this time). Priority categories may need to change over time in order to sustain their usefulness in resource allocation and decision-making by discriminating amongst a set of risks. As an organization reduces the risks associated with its operations, it is foreseeable that it may be necessary to subdivide the judgment category *(medium)* to indicate which risks should be reduced next. Similarly, the "no action at this time" level might change.

Practical difficulties can arise when the societal risk estimates are sufficiently high that they are considered a top priority, but the individual risk level can be shown to be very low. This is not an uncommon outcome for storage tank failures, pipeline incidents, and transportation accidents. The credibility of the use of risk criteria and the degree of support from line-of-business managers can be sorely tested in such situations, where one organization could be directed to a very different risk priority (and risk reduction expectation) than an organization that relied solely on individual risk criteria.

To avoid this type of situation, a hybrid approach to risk prioritization could be used. This hybrid approach would allow low individual risk levels to take precedence over societal risks when establishing priorities – under certain circumstances. These circumstances would need to be defined in advance, and might consider whether or not the risks of concern arise from systems that meet industry standards and practices and if further practical mitigation is available to reduce the risk levels. Under the appropriate circumstances, slightly higher societal risks could then be offset by low individual risk levels. If the specified conditions were not met (i.e., practical mitigation were still available, the system did not meet industry standards and practices, or individual risks were also high), the priority would be established based on the higher level of societal risk.

This hybrid approach is meant to illustrate that the application of risk criteria can be realistic and flexible, but that the need for such flexibility should be considered in advance and designed into the system. If it is created on an as-needed basis, the risk management system loses its robustness and objectivity. This is not to say that the system cannot be changed with time as more experience is gained, particularly to account for increased knowledge in the area of the degrees of conservatism and nonconservatism associated with the use of simplified methodologies.

Communication of Risk Prioritization Criteria

Another issue of concern is how the use of the criteria is described, particularly to senior and line-of-business managers. The successful strategy must take into account the nuances of an organization's culture. Should the criteria (and the associated responses) be described as being absolute versus just a part of the decision-making process? Is the use of the criteria suggested or required? Do the criteria apply to completed studies or only those conducted after a certain date? Are future projects and activities included or only ongoing operations? The answers to these questions can also affect workloads, particularly in the short-term if completed studies must be revised.

The scope of the criteria must be stated clearly and be restated fairly frequently. If the criteria, and therefore the prioritization, only address risk to the public and not risk to employees, management needs to be aware of this.

Most importantly, it must be recognized and communicated that the application of risk criteria helps reduce and control risks across an organization, but does not eliminate the potential for some undesired event to occur. In such communications it is also important to stress that the risk criteria are used to prioritize the risks for action—be it risk reduction or further study, and that such criteria do not imply that any particular level of risk is acceptable to the organization, its employees, its stockholders, the public or regulatory agencies. This holds true whether communicating internally or externally. This is why the low priority category should be described as "no action required at this time," rather than "no action required."

The use of risk criteria to rank risks and direct resource allocation brings the risk manager into closer alignment with the objectives of line-of-business managers within an organization. This is of increasing importance as risk management activities become more integrated with other business processes, and their benefits need to be made explicit. Risk criteria provide a means of continuous improvement of an organization's risk posture. The use of linked and internally consistent criteria also facilitates the communication of risk management concepts and help assure all involved that all risks are being treated equitably. It also provides a basis for aggregating fixed facility and transportation risks for new ventures, or for comparing the full set of risks for two or more alternatives.

The RMP as a Tool for Risk-Based Decision-Making

Craig Matthiessen and Lyse Helsing
Chemical Emergency Preparedness and Prevention Office, US Environmental Protection Agency, 401 M Street, SW (5104), Washington, DC 20460

ABSTRACT
The Risk Management Program (RMP) Rule required by section 112(r) of the Clean Air Act Amendments of 1990 was promulgated June 20, 1996. The RMP rule is designed to reduce the potential for catastrophic chemical accidents at certain facilities through the development of risk management programs and sharing information about these programs in a risk management plan. Once this information is shared, industry, government and the local community can work together toward reducing risks to public health and the environment. The elements in the RMP rule requirements and supporting guidance and tools have broader application to the assessment of, and decisions about, chemical and process hazards, worker and public safety, and environmental protection even before the risk management plan is prepared, and in an ongoing way afterward. Consequently, integration of the RMP elements into company safety, health and environmental programs can help companies adopt and implement risk-based decision-making approaches.

The purpose of this paper is to briefly describe the final RMP rule and to discuss some of the questions and interpretations generated since. The status of some of the many tools (guidance and models) that support the RMP rule will also be provided followed by a brief description of the uses of the rule elements and tools that can support and stimulate risk-based decision-making.

Introduction

In the past 10 to 15 years, a number of major explosions, fires and toxic chemical vapor cloud releases have occurred causing worker deaths and injuries, injuries and evacuations in the communities surrounding these facilities, and environmental damage. As a result of these ongoing incidents, the US Congress enacted legislation in the Clean Air Act of 1990 that calls for the promulgation of regulations by the Occupational Safety and Health Administration (OSHA) and the Environmental Protection Agency (EPA) for the prevention of accidental chemical releases that harm workers, the public and the environment. Consequently, OSHA promulgated the Process Safety Management of Highly Hazardous Chemicals (the PSM Standard) in February, 1992, and EPA promulgated the Accidental Release Prevention Requirements: Risk Management Programs under the Clean Air Act Section 112(r) on June 20, 1996. The PSM Standard contains specific, performance-based actions and recordkeeping requirements designed to protect workers from accidental releases while the RMP rule contains requirements that build on the PSM standard for the protection of the public and environment surrounding facilities handling certain chemical substances.

We could discuss in detail the merits and value of many of the specific requirements in these two rules for the prevention of accidental releases. However, in this paper we want to step back and take a broader look at the use of these requirements as tools for risk assessment and risk-based decision-making. Although the primary focus will be on the RMP rule and the protection of the public and the environment, the PSM standard is a main building block for the prevention program requirements in the RMP; consequently, the discussion here can also benefit the protection of workers. Further, many companies have recognized the significant value in integrating safety, health and environmental programs and requirements into their management systems. The discussion here, and the use of the risk management program and plan approach, could also be applied to efforts on pollution prevention, worker safety and health and environmental protection. This paper does not discuss risk assessment in detail; rather opportunities for use of the information gathered under the RMP rule requirements for risk assessments and risk-based decision-making are raised.

The RMP Rule

The purpose of the RMP rule is to reduce the potential for catastrophic chemical accidents and to reduce the severity of those that do occur. Facilities covered by the RMP rule must develop and implement a risk management program that includes a hazard assessment, a prevention program and an emergency response program. The risk management program must be described in a risk management plan that is registered with EPA, submitted to state and local authorities, and made available to the public. The major features of the rule requirements and some of the issues currently under analysis are briefly described below followed by a brief discussion of their application to risk-based decision-making.

As stated in the preamble to the final RMP rule, EPA recognizes that regulatory requirements, by themselves, will not guarantee safety. Instead, if information about the hazards present at a facility along with the safeguards is communicated by the facility to workers, first responders, and the public, then the workers, first responders, and the public can work together to better understand these hazards and the safeguards to ensure that potential risks are addressed, managed and reduced. Attention is focused at the local level where the risks are found. Consequently, the RMP rule requirements are largely information driven.

Although many processes and facilities are similar in design, layout, and operation, there are unique circumstances surrounding each process. Further, there are often a wide variety of safeguards available to address a variety of hazards. New ways to manage these problems are continuously evolving. Therefore, the RMP rule requirements are flexible and performance based; the burden is on company management and company safety experts to thoroughly understand the hazards and the best approaches to manage them and to communicate this to workers, the public first responders, emergency planners and all levels of government through the risk management plan and other using other communications tools.

EPA recognizes that the RMP rule is largely hazard and consequence based rather than risk based. While the rule contains specific requirements for hazard and consequence assessments, there are no specific requirements to implement any particular prevention or emergency response actions based on the result of these hazard and consequence assessments. EPA believes it is up to the company, along with dialog between workers, first responders, and the public, to decide how best to tackle these problems. This is where comprehensive risk assessment, risk-based decision-making and risk communication can play a role as described below.

The RMP rule applies to facilities that have processes that handle more than a threshold quantity of certain toxic and flammable substances. The processes subject to the rule are divided into three tiers, labeled Programs 1, 2, and 3, that reflect the hazards and consequences present. Program 1 has the least requirements and is available to any process that has not had an accidental release with offsite consequences in the five years prior to submission of the risk management plan and has no public receptors within the distance to a specified toxic or flammable endpoint associated with a worst case release scenario. Program 3 has the most requirements and it applies to processes already covered by the OSHA PSM standard and to processes in certain standard industry classification (SIC) codes. The SIC codes identify processes that have had a number of incidents in the past that generated considerable consequences. Finally, Program 2 contains streamlined requirements and it applies to all other covered processes.

The major elements of the RMP rule include the hazard assessment, the prevention program, the emergency response program, the management system and the risk management plan. The purpose and objectives of these elements are briefly described; a detailed description and the specific requirements can be found in the rule itself.

The Hazard Assessment

The hazard assessment portion of the RMP rule primarily includes an assessment of worst case and alternative scenarios and a compilation of a 5 year accident history. In some respects it also includes information normally collected and assessed under the process safety information section of process safety management or in the prevention program (see below). The purpose of the hazard assessment portion is to thoroughly understand what hazards are present, where the vulnerabilities are, what could go wrong, what are the consequences of something going wrong and what has gone wrong in the past.

The Prevention Program

The prevention program portion of the RMP rule applies to Program 2 and 3 processes only as described above. As in the PSM standard, the purpose of the prevention program is to bring together all of the elements necessary to operate a process safely day-after-day, the right way, every time. The Program 3 prevention requirements are nearly identical to the OSHA PSM standard while the Program 2 is a streamlined version more suitable to less complex processes. Regardless, the key features of the prevention program requirements are the performance-based flexibility and the need for interconnectivity of the elements. The owner or operator of the process has the flexibility to establish and tailor the prevention program to best suit needs, the complexity of the process, hazards present and site-specific conditions. The formal evaluation of the process, its possible failures and its safeguards, or process hazards analysis (PHA), serves as a foundation for all the other elements necessary to operate the process safely and within design and technology limits. The owner and operator has the responsibility and performance-based flexibility to make sure that these other elements function together; for example, the process information must reflect the equipment design and technology present, the operating procedures must reflect the actual equipment and operating limits, operators need to be trained on current procedures, and so on. In addition, the owner or operator has an obligation to constantly monitor the performance of the prevention program through mechanisms such as audits, updates and revalidation to ensure that all the elements are functioning together as intended.

The Emergency Response Program

Owners and operators of processes covered by the RMP rule must ensure that existing or newly developed emergency response plans implemented under the RMP requirements are coordinated at the local level. As above, owners and operators have the performance-based flexibility to develop the emergency response plans that best suit their needs and to reflect the potential needs of local first responders and the community.

The Management System

As described under the Prevention program, successful prevention programs hinge on the integration of all the elements into a system that ensures that the facility is safely operated and maintained the right way, every day. EPA also believes that the entire risk management program -the hazard assessment, prevention program, emergency response and risk management plan -should also be integrated and managed, not just the prevention program. Consequently, the RMP rule contains a flexible performance-based requirement that owners and operators establish a management system to make sure accountable responsibility is established, assigned and the risk management program system is maintained and monitored.

The Risk Management Plan

The risk management program is described in a risk management plan that must be registered with EPA, submitted to state and local authorities, and made available to the public. EPA is working on mechanisms for non-paper submission and availability of the information and information will likely be made public by the time you read this. The purpose of the plan is to communicate key features of the actions being taken by the owners and operators to successfully manage the hazards and risks present at the facility. It should form the foundation of interactive dialog between workers, first responders, the public, and all levels of government about the hazards and risks. EPA believes that the owners and operators of processes subject to the rule will want to fully understand, implement and take ownership of the risk management program and the information in the risk management plan before placing the plan under the bright lights of public, first responder, worker and government scrutiny.

Issues and Status of the RMP Requirements

Since promulgation of the final RMP rule, EPA has been working with OSHA and EPA Regional Offices and states on implementation, oversight, and enforcement of the requirements. In addition, as mentioned above, EPA will soon publish plans for electronic submission of the risk management plans. Several questions and issues have been raised with respect to the relationship of the RMP requirements to transportation, applicability of the rule to certain processes and substances, and worst case and alternative case scenario and offsite consequence assessment technical questions. EPA is also working on peer review of technical guidance, development of general guidance, peer review of model risk management programs and plans and development of new models. We suggest that you regularly contact our information service hotline at 800-424-9346 or visit the Chemical Emergency Preparedness and Prevention Office (CEPPO) homepage at www.epa.gov/swercepp or www.epa.gov/ceppo for updates on these issues. A question and answer database is available for downloading.

Regardless of the outcome of many of these issues and questions, we believe that the elements of the risk management program and risk management plan can serve as a valuable tool in risk assessments and risk-based decision-making.

The Role of the RMP in Risk-Based Decision-Making

Many companies endeavor to assess the risks associated with chemical processes in quantitative, or at least, qualitative ways. Often the purpose of such assessments is to gain confidence that the risks confronting the operation are being properly managed or to assess new opportunities for risk reduction or risk management from a cost and benefit perspective. In addition, decisionmakers often face the difficult task of evaluating and choosing safeguards, alternative processing techniques or process modifications that are designed to manage or reduce risks. These evaluations or choices should not be based solely on existing hazards and potential consequences but on a sound understanding of the existdecision-making for a particular process must also reflect the current capability of the existing process design, safety management systems, and safeguards in comparison to alternative process designs, safety systems and safeguards. Collection of the information necessary to support a risk assessment or risk-based decision can present a daunting task, especially at facilities with multiple and varied processes.

The RMP program and plan fosters the collection of process-specific detailed information, potentially into one location, critical to risk assessments and risk-based decisions regarding:

- the process technology, design bases, chemical and process hazards;
- screening and detailed quantitative consequence assessments of a variety of worst case and alternative case accident release scenarios;
- historical performance information and findings of incident investigations;
- process safeguards and release mitigation and response; and
- the status of the management system, process safety integration and results of audits.

One outcome of the collection and evaluation of this information might be an evaluation of whether the elements of process safety are properly integrated and linked into a dynamic system to sufficiently manage the hazards present in the process to ensure safe operation all day, every day. The process of implementing a risk management program, collecting this information, and preparing the risk management plan may reveal data gaps that could be critical to a risk assessment. In addition, preparation of the risk management plan and initiating the process of risk communication and dialogue with workers, the public, first responders and local planners may also reveal areas that need strengthening that can also factor into the risk-based decision-making process. Many companies in a particular local area are collaborating on their risk management plans and risk communication efforts. This collaboration and information sharing can also foster greater understanding of hazards, consequences, safeguards and risks. There is also considerable value in learning from others. The risk management plans prepared by other companies with similar processes may contain useful information on hazards, consequences, past accidents and safeguards to foster greater understanding of chemical and process hazards. Finally, the risk management plans of other companies can reveal opportunities for risk management, risk reduction, and alternative technologies to make processes intrinsically safer.

Conclusion

Today there is greater focus on safety and cost performance of chemical and petrochemical processes than ever before. This focus and demand for performance places more value on sound risk assessments and risk-based decision-making. Although the RMP final rule requirements are mostly hazard and consequence based and information driven, the process of hazard and consequence information collection and assessment, the development of prevention andecision-making. In addition, the process of dialogue with workers, the public, first responders and local emergency planners regarding the hazard and consequence information in the risk management plan can reveal local risk concerns, areas where the company may need to focus attention, and key information data gaps critical to the company risk assessment or risk decisions; and gaining feedback on communication of the results of comprehensive risk assessments and risk decisions at the local level. The final RMP rule does not specifically address these considerations and companies are certainly under no obligation to make use of the RMP effort this way. However, the risk management program and plan information provides another tool available to the risk and safety manager toward greater accidental release prevention beyond mere rule compliance.

Disclaimer: The views expressed in this document are the opinions of the authors and may not represent official agency positions.

Risk Management and Hazard Reduction Practices in the Ammonia Refrigeration Industry

D. R. Kuespert and M. K. Anderson

International Institute of Ammonia Refrigeration, 1200 19th St., NW, Suite 300, Washington, DC 20036-2422

Modern society depends upon the ability to remove heat from materials through refrigeration, and since its earliest days, mechanical refrigeration has depended upon anhydrous ammonia refrigerant. In 1773, Priestly prepared pure ammonia vapor, and in 1834, the vapor was liquefied for the first time by Faraday.

Mechanical refrigeration started with Jacob Perkins' invention in 1834 of a vapor-compression machine which used diethyl ether to manufacture ice. By the 1860s, an ammonia absorption-cycle ice machine was in production (the prototype was used in New Orleans while it was besieged during the American Civil War), and in 1872 the first patent was granted on an ammonia compressor. Refrigerants such as ether, ammonia, and sulfur dioxide dominated the refrigeration industry until the advent of chlorofluorocarbons in the 1920s. These compounds eliminated the market for ether and SO_2 and somewhat reduced ammonia's market penetration; their suitability for use with small hermetically-sealed motors brought about refrigeration into the home. Ammonia continued as the dominant industrial refrigerant.

Although they were originally touted as "safety refrigerants," chlorofluorocarbons were found to have undesirable environmental safety properties, and the Montreal Protocol established a schedule for the elimination of such ozone-depleting substances. Hydrofluorocarbon replacement refrigerants bear the burden of possible global warming impact and various technical differences from the original chlorofluorocarbons that affect their suitability as refrigerants. As a result, the natural refrigerants such as ammonia, propane, and isobutane are enjoying a renaissance. Table 1 identifies some of today's common refrigerants; fluorocarbon refrigerants are usually binary or ternary mixtures of the pure components listed.

Uses of Ammonia Refrigeration

Ammonia refrigeration is employed in many applications. Most industrial refrigeration facilities in the United States use anhydrous ammonia refrigerant, with a slightly smaller proportion in Europe. The most familiar ammonia refrig-

TABLE I
Common Refrigerants[12]

Designation	Chemical Name	Atmospheric Boiling Point	
		°F	°C
R-22	chlorodifluoromethane	-41	-41
R-32	difluoromethane	-62	-52
R-125	pentafluoroethane	-56	-49
R-134a	1,1,1,2-tetrafluoroethane	-15	-26
R-143a	1,1,1-trifluoroethane	-53	-47
R-290	propane	-44	-42
R-717	ammonia	-28	-33

eration systems operate in the food industry, both for production and storage of perishable foodstuffs. At any given time, 10–25% of global food production is being processed or stored somewhere in the "cold chain"[1] Refrigerated warehouses dot the landscape in most industrialized countries, storing food near the point of supply and throughout the processing and distribution chain. Food processing plants also employ ammonia refrigeration for cooling of such diverse products as margarine, meat, ice cream, and bread dough. Harvest crews deploy commercial field-cooling systems to remove field heat from vegetables to retard spoilage.

Why Ammonia Refrigeration?

Several properties make ammonia the refrigerant of choice for many industrial and commercial applications.[2] Ammonia is usually the most efficient refrigerant available. Thermodynamic considerations give ammonia a 2–10% theoretical advantage in energy consumption over fluorocarbon refrigerants. Although ammonia is incompatible with copper heat transfer equipment (steel, aluminum, or titanium are used), convective heat transfer coefficients are usually 1.6–4 times greater than fluorocarbons. Ammonia's high latent heat of vaporization means that less ammonia must be circulated to move the same amount of heat. Thus, compressors and piping may both be smaller and capital costs are reduced. The $0.10–0.25/lb ($0.20–0.50/kg) price of refrigeration-grade ammonia is also important when compared with $5–10/lb ($10–20/kg) for fluorocarbons.

Ammonia is also a prime refrigerant choice for safety reasons. Improved environmental safety is an obvious benefit of ammonia, as the compound has a very short atmospheric lifetime. Ammonia has neither ozone depletion potential nor global warming potential. The energy-efficiency advantages of ammonia translate to a very low total warming equivalent impact, another measure of potential climate change in refrigeration. Operational safety is also good in ammonia-refrigerated facilities (see below).

Characteristics of the Industry

Refrigeration's ubiquitous nature shapes the industry. Commonly, refrigeration is considered a utility rather than a process in and of itself: it is a closed-cycle operation where no easily identifiable product is produced nor is a feedstock routinely consumed. The community of users and facility neighbors tends toward the unsophisticated. A facility manager is rarely familiar with the details of refrigeration system design and operation, being concerned with the operation of his plant and the production of its product. Tellingly, cold storage firms now describe themselves as "logistics providers" for their customers rather than refrigerated warehouses used to store product.

Technological development evolves at a stately pace in the ammonia refrigeration industry. Full-automatic control is common on new plants, but many older facilities still employ teams of experienced operators who control the plant manually. Retrofits to incorporate modern control and detection systems depend on the ability to justify capital expenditures (and plant shutdowns) in very competitive and cost-driven industries. The most successful approaches tie process improvements to energy efficiency as a cost-reduction strategy.

The population surrounding a large refrigerated facility is frequently not aware of the presence of ammonia and is not accustomed to fairly assessing and discussing chemical risks. As a result, refrigeration systems are silent neighbors. A large meat-processing plant may contain 250,000 lb (114,000 kg) of anhydrous ammonia in its refrigeration systems. Despite this, the local population (and often the local emergency personnel) may be completely unaware either of the presence of a hazardous material or of its implications—unless an accidental release occurs. This obliviousness applies also to systems using other refrigerants, all of which are hazardous in some way.

Not surprisingly, these factors drastically affect the approach to risk communication which must be taken, both within the facility and with the community. In addition to discussing the specific risks from the facility, the public and plant personnel must be educated about risk in general, more thoroughly than in areas where chemical process industries dominate (such as the Kanawha Valley in West Virginia, USA).

Hazardous Properties of Ammonia

Flammability

Ammonia can form flammable mixtures at concentrations of approximately 15–28%, although the exact limits depend on the test conditions. Because the lower flammable limit is very high, anhydrous ammonia is transported as a nonflammable gas (UN Division 2.2). The accepted refrigeration safety classification schemes[3] treat ammonia as a "lower flammability" material as opposed to higher flammability refrigerants such as propane (Table 2) or "nonflammable" materials such as R-22 (chlorodifluoromethane). It is important to note that flammability is a matter of degree: R-22 explosions have occurred several times. Entrainment of compressor lubricating oil in released refrigerant expands the flammability envelope as well.

Ammonia is not a good fuel. When burned, it releases insufficient energy to support its own combustion. This affects the design of treatment systems for released ammonia, as any flare system employed requires a supplementary fuel for operation.

Ammonia explosions can occur in confined spaces. Deflagration of an ammonia atmosphere is a low-energy event, as flame fronts move very slowly and produce low overpressures compared to hydrocarbons. When heated to around 842°F (450°C), ammonia can dissociate slowly to form nitrogen and hydrogen, creating a more serious explosion hazard. Dissociation may be responsible for the lower ignition energy required for ammonia-air mixtures ignited by inductive discharge (Table 2).

Ammonia's flammability becomes a concern at far higher concentrations than its toxicity. It is common practice in the ammonia refrigeration industry to search for small leaks using a lighted sulfur candle. The oxides of sulfur in the candle smoke react immediately with any ammonia nearby to produce a visible ammonium sulfate precipitate. This generally appears as a white stream emanat-

TABLE 2
Flammability Data For[13]

	Ammonia	Propane
Flammable Limits	15–28 vol%	2.1–9.5 vol%
Ignition Temperature	1201°F/651°C	842°F/450°C
Minimum Ignition Energy	680 mJ	0.25 mJ
Min. Igniting Current Ratio[14]	6.85	0.82
Max. Expt'l Safe Gap[15]	3.17 mm	0.97 mm

ing from the smallest pinhole leak, greatly assisting the maintenance technician. As a general rule, if the ammonia concentration is such that personnel do not require air-supplied respiratory protection, no hazard results from the practice; even at much higher concentrations, use of "sulfur sticks" is safe.

Toxicity

Ammonia's moderate toxicity has been known and studied since the nineteenth century. Although much of the literature is old and thus not conformant with modern toxicological methods, the toxic behavior of ammonia is not subtle and thus older studies generally agree qualitatively with more recent ones. German companies touted ammonia as a possible chemical weapon during the First World War, but it was found to be useless for this purpose, requiring far too high a dose for militarily-useful toxicity. (Chlorine and phosgene, among other common industrial chemicals, found widespread use in this application. These compounds exhibit toxic effects at hundred or thousands of times lower doses.)

Ammonia has no remarkable long-term effects; acute toxicity rather than chronic toxicity concerns dominate its use. In refrigeration practice, where ammonia is contained within a closed piping system, fatalities are not common. Occupational death rates in halocarbon-based refrigeration systems are comparable or higher to those in ammonia.[4]

The toxic properties of pure ammonia stem from its corrosivity. At about 10,000 ppm in air, ammonia will burn moist exposed skin. Ocular exposure can cause severe injury; liquid ammonia directly striking the eyes can result in blindness.

Respiratory injury occurs at lower levels. The United States Permissible Exposure Limit is set at 50 ppm for ammonia (35 ppm in some areas); Britain uses a threshold limit value of 25 ppm (with a 35 ppm short-term limit). Actual potential for injury begins at around 1,000 ppm, and death may occur for exposures of approximately 5,000 ppm over one-half hour. As with most chemicals, longer exposures result in injury at lower levels; the 1-hr (rat) LC_{50} for ammonia is approximately 1,700 ppm. Other exposure levels for ammonia are shown in Table 3.

Important Properties of Ammonia

One of the most useful properties of ammonia for hazard reduction is its odor. The compound is usually described as "pungent," which in practice means that it stinks unbearably. The odor threshold for ammonia varies with the individual but is usually around 5–50 ppm. Experienced technicians and engineers sometimes report that their detection and irritation thresholds are increased; this acclimation is verified by literature reports. Sensitive individuals can detect but not identify ammonia fumes at low part-per-million levels.

TABLE 3
Toxicity Data for Ammonia[16]

Exposure limits	
US-OSHA PEL	35 ppm
US-OSHA IDLH	500 ppm
AIHA ERPG-3	1000 ppm
AIHA ERPG-2	200 ppm
UK TLV	25 ppm
Physical effects	
Throat Irritation	400 ppm
Cough	1,200 ppm
Threat to Life (30 min exposure)	2,400 ppm
LC_{50} (rat, 60 min)	17,401 ppm
Immediate Threat of Death/Injury	5,000–10,000 ppm

The normal human response to 300–500 ppm ammonia atmospheres is to flee immediately. This *self-alarming property* confers an important safety advantage over halocarbon refrigerants, which are either odorless or have very faint odors, as no human will remain willingly in an ammonia atmosphere which presents a hazard. Accidents have occurred in halocarbon refrigeration plants due to suffocation or cardiac arrhythmias caused by refrigerant overexposure, which can easily occur in low areas such as pits. Despite the high (10,000–100,000 ppm) refrigerant levels required for such effects, the absence of an identifiable warning odor can be fatal to refrigeration technicians. This may explain the relative accident rates of the two classes of refrigerant.

IIAR and the Ammonia Refrigeration Industry

In 1971, a group of six ammonia refrigeration specialists met in Philadelphia and formed the International Institute of Ammonia Refrigeration (IIAR) "to promote education, information, and standards for the safe use of ammonia as a refrigerant." Over the ensuing 25 years, IIAR grew to 1,200 members encompassing a broad cross-section of the ammonia refrigeration industry. IIAR includes manufacturers, users, engineers, contractors, and others involved in the use of anhydrous ammonia refrigerant, unlike many trade associations, which tend to include only manufacturer interests.

IIAR works to promote the proper and safe use of ammonia refrigeration in many ways. Activities range from participation in codes and standards development as subject matter experts to providing training materials for safe operation of ammonia refrigeration systems. The IIAR Annual Meeting includes both a trade show and many technical sessions; IIAR Technical Papers receive more complete technical and production editorial service than most peer-reviewed journals. IIAR is regularly involved in regulatory issues through participation in negotiated rulemakings and presentations to regulatory groups (such as process safety management inspectors).

The presence of an effective trade association confers several advantages to the industry. The scope of IIAR operations is industrywide rather than company-focused; this allows IIAR to create ANSI-approved standards and other advice using a broad base of input and with broad applicability. In an industry made up mostly of small businesses or small operating units, IIAR provides economies of scale, making accessible technical resources and speaking for the industry as a monolithic entity.

Explosion Prevention in Ammonia Refrigeration

Prevention of fire and explosion in ammonia refrigeration focuses on two major areas. Following the "fire triangle" approach, preventing the formation of an explosive atmosphere is the first concern, since air cannot practically be excluded from the rooms in which a refrigeration system operates. Most of the equipment in a refrigeration system is isolated in a *machinery room* of special design. General practice is to provide engineering controls to maintain a machinery room concentration of ammonia below 25% of the lower flammable limit; for ammonia, this equates to 4% in air.

This reduced-concentration atmosphere is attained by traditional methods. System integrity practices are applied to prevent generation of an ignitable ammonia–air mixture during normal operations. Vessels are constructed to the ASME Boiler and Pressure Vessel Code, and welded construction of piping systems is common practice, although some flanged connections to parts requiring frequent service (strainers, etc.) remain. Brazed connections are not permitted. Advances in shaft seals and valve packings originally developed for fugitive emissions control in the chemical industry have reduced ammonia concentrations in a modern machinery room to barely detectable levels. (In years past, a typical ammonia machinery room might exceed 50 ppm ammonia during normal operation.)

In addition to point source control, machinery rooms are usually equipped with engineered ventilation systems to further reduce concentrations. Minimum design criteria for machinery room ventilation are specified in mechanical codes in the US and in various European and International Standards rather than

TABLE 4
Refrigeration Machinery Room Ventilation Requirements (all refrigerants)

Exhaust system must achieve all of:
a. Continuously maintain -0.05 in H_2O (-12.4 Pa) relative to adjacent spaces
b. Continuously provide 0.5 ft^3/min/ft^2 floor area (152 l/s/m^2) airflow
c. Limit temperature rise to 104°F (40°C) max.
d. Provide emergency purge of escaping refrigerant of
 100 ft^3/min/$lb^{1/2}$ x (refrigerant charge)$^{1/2}$ (70 l/s/$kg^{1/2}$)

Uniform Mechanical Code, 1997 (ICBO)

being left to the design engineer's discretion (Table 4); for smaller rooms, IIAR recommendations increase the ventilation rate considerably above code requirements. In addition to normal room ventilation suitable for control of nuisance ammonia releases (and the heat generated by high-power electrical equipment), emergency ventilation is usually provided and/or required by code. Often triggered from a room ammonia detection system, the emergency system removes ammonia quickly from the room in the event of a large accidental release. This illustrates a tradeoff in the use of prescriptive codes to control engineering design decisions: While these systems can control most credible releases, it is impractical to engineer a system to immediately vent a catastrophic release, and so the code authority must decide what level of performance constitutes an acceptable risk.

Steps are taken to eliminate possible ignition sources, the second side of the fire triangle. Ignition by flames or hot surfaces is relatively easy to prevent. Permanent open flames are banned in machinery rooms. In ammonia and hydrocarbon systems, the ignition hazard is obvious. In halocarbon systems, the production of toxic and corrosive hydrogen halides by open flames or hot surfaces is a more insidious danger. A code prohibition against locating boilers in the same room with refrigeration systems has been widely ignored through large parts of the United States due to lax code enforcement. Smoking and the use of sulfur candle leak detectors (see above) are still permitted.

While open flames are prohibited, combustion engines and turbines are increasingly applied as prime movers for large refrigeration systems. An example is in district cooling systems, where combustion turbine, power generation equipment, and a refrigeration compressor may be coupled to the same shaft. Special precautions, including properly-equipped detection and shutdown systems, insulation of combustor and exhaust lines, and provision for 100% outside combustion air apply. Code variances (technically, "alternate method-and-material" equivalency findings) are usually required; after more experience is gained with these systems, variances may no longer be needed.

Properly ventilated ammonia refrigeration facilities are not normally Hazardous (Classified) areas under the National Electric Code[5] (NEC). Because the emergency ventilation system is intended to operate in a potentially explosive atmosphere, it should be constructed as a Class I, Division 2, Group D (or IEC Class I, Zone 1) system. To reduce the expense associated with this system, the motor and much of the ancillary equipment are often mounted outside the machinery room and out of the effluent airstream. When this is done, most of the emergency ventilation system is not within a hazardous (classified) area and normal equipment may be used. In special cases, such as the refrigeration section of an ammonia production plant, explosion-proof or other reduced-hazard electrical equipment is applied.

Where industry-standard precautions do not reduce the hazard sufficiently, additional measures can be taken. A common practice is to relocate the electrical switchgear outside of the machinery room. Additionally, the refinery technique of limiting the extent of hazardous (classified) areas has begun to penetrate the ammonia refrigeration industry. Hazardous (classified) areas usually exist for only a few feet around major leakpoints (flanges, seals, etc.), so the cost of special electrical equipment can be drastically reduced by a small relocation of lights, switches, etc.

A final practice, which provokes some controversy, is provision for automatic shutoff of electrical power to the machinery room on activation of the ammonia alarms. Required by code in some localities, the attraction of an automatic shutdown is obvious. Unfortunately, several problems pertain. Shutdown systems are frequently misapplied, creating hazards of their own (such as disabling the emergency ventilation and escape lighting). In addition, emergency electrical cutoffs can produce the very sparks and other discharges that it is meant to guard against as capacitors, coils, etc. power down.

The other type of explosion which can occur in facilities handling liquified gases is the catastrophic breach of a pressure vessel. Aside from the strict conformance with the ASME Boiler and Pressure Vessel Code (section VIII covers unfired pressure vessels), refrigeration vessels and piping adhere to high design safety factors. Piping generally employs a 4 : 1 safety factor and 5 : 1 is sometimes used on vessels. This may be compared with the 3 : 1 and 3.5 : 1 safety factors common for nuclear applications. While this difference might be taken as an increased attention to safety in refrigeration, it is actually indicative of a difference in philosophy. Nuclear vessels are subject to stricter quality control, both on the original materials and on radiography and heat-treatment of welds; refrigeration contractors are commonly sold pipe which meets normal ASTM steel specifications but would fail the more specific nuclear-application tests. The philosophy in refrigeration is to specify safer designs rather than attempt to exercise tight control over field-built equipment.

Pressure relief system design for ammonia refrigeration is rapidly changing. While longstanding specifications for required relief discharge capacity are embedded in mechanical safety codes and industry standards, the excellent work done by the Design Institute for Emergency Relief Systems is being used to update design practices in the ammonia refrigeration industry. The American Society of Heating, Refrigerating, and Air-conditioning Engineers (ASHRAE) is discussing new design specifications to be included in ASHRAE Standard 15, *Safety Code for Mechanical Refrigeration*.

Toxic Releases from Ammonia Refrigeration Facilities

A variety of accidents can occur in ammonia refrigeration systems, some more frequently than others. Most types of refrigeration accidents are common to all refrigerants, being more a function of system design and refrigeration fundamentals than the particular refrigerant in use. The properties of the refrigerant partially determine the severity of the incident, the appropriate response, and the methods used to prevent it. Many different credible scenarios could result in a release of ammonia with potential offsite consequences.

The most common release is a simple leak in valve packing, flange, or shaft seal. Refrigeration valves are a highly specialized product because they are designed for extremely low pressure drops. (Pressure drop corresponds to an increase in refrigerant saturation temperature and a consequent loss in system capacity.) They tend to employ traditional packing designs with some modifications for their low-temperature, low-pressure-drop use. Operating at –20°C or below has obvious effects on the pliability of packing materials, and small vapor or liquid leaks sometimes occur. Likewise, rotating shaft equipment such as compressors and pumps are generally of open design (the heat load from the motor of a hermetically-sealed unit again decreases capacity), and older seal designs can be subject to leaks, especially if not properly maintained. The development of better mechanical seals in the past decade has reduced the frequency of these sort of nuisance releases considerably. Such releases are not generally difficult to control and do not normally result in anything more than unscheduled maintenance activity.

To illustrate typical accidents and their effects, we will employ a fictitious facility. The facility is a large refrigerated warehouse with a single ammonia refrigeration system providing 1900 tons of refrigeration (6700 kW).[6] While modern packaged equipment can sometimes operate with around 0.25 lb/TR (0.032 kg/kW), a more reasonable estimate for the facility would be 35,000 lb (15,900 kg) ammonia charge distributed throughout the system. Of this 35,000 lb, 10,000 lb (4545 kg) might be contained in a high-pressure receiver vessel operating at 75°F and 140 psia (24°C, 965 kPa), while the remainder would be contained in various low-pressure components (evaporators and associated vessels).[7]

For purposes of discussion, the facility is assumed to be located at the residence of one author (DRK) in Columbia, MD. The location is within a kilometer of a hospital, 4 schools, 3 public recreation facilities, parks, major highways, and several shopping centers (Figure 1). While most industrial refrigeration facilities are not sited so near population centers, a community may grow around an existing plant. This site approximates an extreme example of this situation.

We will model the facility using the ALOHA computer program available from the National Safety Council in the United States, which is used by the local emergency planning authorities to plot release footprints over computerized maps of the locality. We will provide distances to toxic endpoints of 200 and 1000 ppm, representing the ERPG-2 and -3 planning guidelines. (Emergency Response Planning Guideline -2 is meant to represent an injury threshold, while the -3 guideline represents risk of death or serious injury. The ERPGs are set by the American Industrial Hygiene Association.) We will use actual meteorological conditions for a spring day in the area (May 14, 1997; Table 5).

The accident scenarios presented below represent credible releases occurring under credible conditions (except for the catastrophic pressure vessel failure). The authors believe that presenting putative "worst-case" scenarios as a tool in communication of chemical hazards to the public is of dubious value. Modern chemical dispersion models contain a multitude of engineering approximations designed to account for unknown or unpredictable variables. Furthermore, the majority of model development has been performed in the high-concentration zones of large-scale hazardous materials releases such as the Desert Tortoise series conducted at the U.S. Department of Energy Nevada Test Site and the Thorney Island series conducted by the U.K. Atomic Energy Authority. Extension of such results to the 200 ppm ERPG-2 is not reliable.

TABLE 5
Meteorological Conditions for Modeling[17]

Facility Location	Columbia, MD (USA) 39°13'3"N, 76°53'8"W
Temperature	52°F (13°C)
Relative Humidity	80%
Date/Time	0800 EDT 14 May 1997
Wind	West 5 miles/hr (2.2 m/s)
Cloud Cover	15% (by observation)
Barometric Pressure	29.87 in Hg (98.2 kPa)
Urban/forested location	

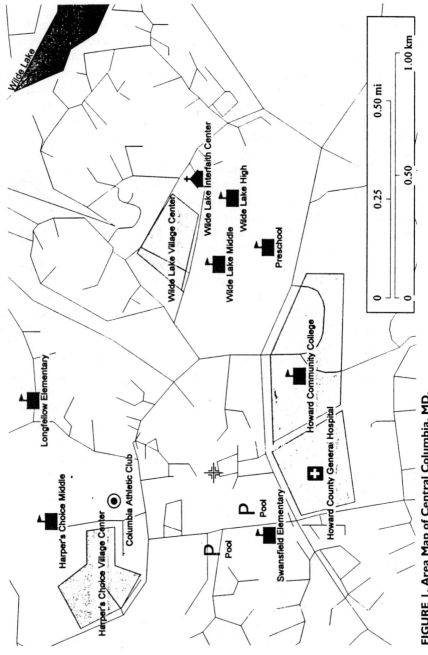

FIGURE 1. Area Map of Central Columbia, MD.

While chemical dispersion modeling is a useful tool for informed professionals to use in emergency planning, communication of model results to a public unprepared to judge the technical details and willing to seize on any available pseudo-quantitative results regardless of reliability is potentially disastrous for any industry. The recent Risk Management Program regulations in the United States provide an illustration of the pitfalls of this approach. To force the use of a standardized modeling technique and thereby provide some modicum of comparability between the risk measures of differing facilities, the US Environmental Protection Agency limited the hazard distance calculations to a single input: the size of the release. All other relevant parameters are specified extremely conservatively, to the point that any hypothetical "worst-case" release must take place under atmospheric stability conditions which only occur late at night but also at the highest temperature recorded for that locality. This meteorological condition does not occur. It is unreasonable to expect the general public to understand the approximations inherent in such calculations of "risk." Reporting a "hazard distance" under such conditions is merely a transformation of variables, that is, changing the units from kilograms of chemical released to kilometers allegedly threatened by the release (including no actual additional information).

Overconservatism in hazard assessment distorts any possible constructive dialogue with the public and with emergency planning authorities. It is a serious concern of IIAR that such distortions in public debate will result in poor choices in emergency planning, allocating scarce resources toward faraway populations rather than concentrating attention near the battery limits of the plant, where any threat from a release will be strongest. Use of conservative assumptions can pervert the results of relative risk ranking procedures often used to allocate scarce resources to safety improvements in a facility.[8]

Relief Valve Releases

Relief valves protecting ammonia pressure vessels are a common source of uncontrolled releases from refrigeration systems. In a typical recent accident, a solenoid valve controlling liquid flow into a roof-mounted surge tank stuck open, overfilling the vessel. As it is intended to do, the tank's relief valve opened to prevent a tank rupture, releasing ammonia liquid at 190 psi to the atmosphere. The sudden pressure drop in the tank caused the solenoid valve to reseat. The relief valve continued to release ammonia until its design blowdown pressure (usually about 60% of set pressure) was reached, terminating the release without any emergency responder intervention. The incident freed approximately 500 lbs (220 kg) of ammonia over 10 minutes, released 66 ft (20 m) above grade from a relief stack.

Using the ALOHA emergency response modeling program, which chose a Gaussian dispersion model for this release, we find an 82 yard (75 m) distance to the 1000 ppm ERPG-3 endpoint and 183 yards (167 m) to the 200 ppm ERPG-2. This is not sufficiently close to threaten any receptor marked on the reference map (Figure 1).

Catastrophic Vessel Failure

Catastrophic vessel failure is a rare occurrence in any industry. Nevertheless, this accident could be of interest as a "worst-case" release scenario for community risk management. The value of such scenarios is considerably lower for community risk communication purposes, as it is both low-probability and so dramatic as to draw attention and resources from more credible risks.

While rare, incidents of vessel failure are known. We will model the catastrophic failure of the 10,000 lb (4545 kg) high-pressure receiver in several different ways using ALOHA. Assuming the ammonia inventory is instantaneously released, the distance to the 1000 ppm endpoint is 810 yards (741 m), while 1,365 yards (1,248 m) is required to reduce the concentration to 200 ppm. This modeled release threatens several schools in the area. Figure 2 plots the projected cloud footprint and shows the substantial difference in the area encompassed by the 1000 ppm endpoint (which represents immediate hazard) and the 200 ppm endpoint (which represents a minor injury threshold). Sheltering-in-place would be appropriate for the facilities within the affected area.

If we instead model the release as a hole in a pressure vessel, drastically different results are obtained. Assuming a horizontal pressure vessel 6 ft (1.83 m) in diameter by 12 ft (3.67 m) long, a 4 inch (10 cm) hole in the bottom of the vessel would drain the entire 10,000 lb (4545 kg) charge in approximately 1 minute. Failure of a nozzle connecting a liquid line to the bottom of the vessel is an example of this sort of scenario. ALOHA's projected hazard distances for this incident are 1,118 yards (1022 m) to the 1000 ppm ERPG-3 and 1.1 miles (1770 m) to the 200 ppm ERPG-2. Note that the projected affected zones (Figure 3) are much larger than those found for an instantaneous release; this is not the expected result.

The reason for this anomaly is that ALOHA can vary its choice of release model depending on the situation. In this case, it chose to apply its internal heavy-gas model (based on the DEGADIS[9] program), calculating that the release would result in a cold aerosol cloud heavier than air. In actuality, the release jet would likely impinge upon the walls of the machinery room and much of the aerosol would deposit as a liquid pool. The danger inherent in blindly accepting the output of a computer model is obvious in this example, as is the potential for misunderstanding in the communication of such results. Modeling the accident using the mandated EPA worst-case scenario (Table 5 meteorology) of 1,000 lb/min (7.6 kg/s) for 10 minutes produces yet another prediction (368 yd/336 m to 1,000 ppm; 831 yd/760 m to 200 ppm); a great deal of engineering judgement is required to determine which number to regard as a reasonable calculation. (Table 6 summarizes this modeling data.)

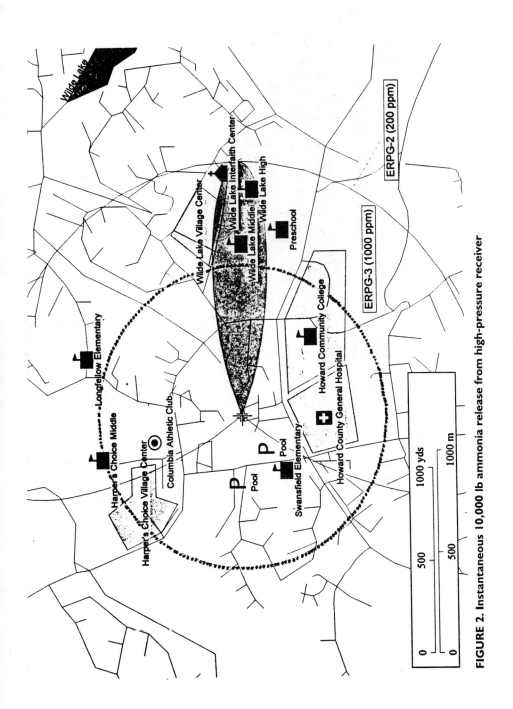

FIGURE 2. Instantaneous 10,000 lb ammonia release from high-pressure receiver

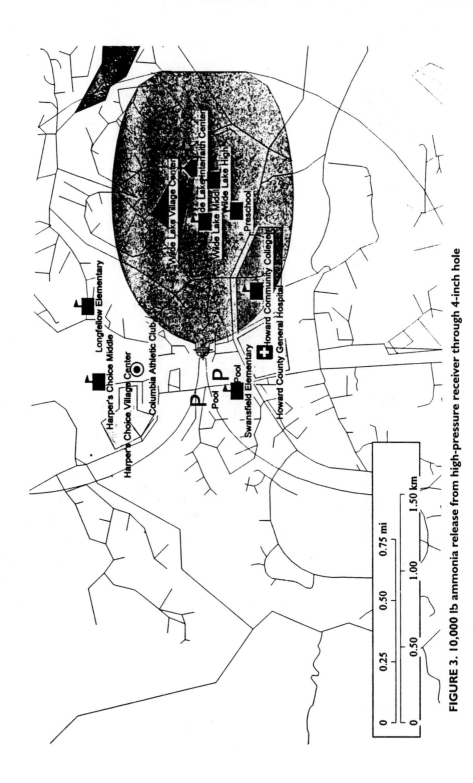

FIGURE 3. 10,000 lb ammonia release from high-pressure receiver through 4-inch hole

TABLE 6
Catastrophic Pressure Vessel Failure

Modeled As	Distance To	
	1000 ppm	200 ppm
Instantaneous Release (Gaussian model)	810 yards	1,365 yards
	741 m	1248 m
4" (10cm) Hole in Vessel (Heavy-gas model)	1,118 yards	1.1 miles
	1022 m	1770 m
US-EPA 10–min constant release (Gaussian)	368 yards	831 yards
	336 m	760 m

Vapor-Propelled Liquid Slugging

Piping systems handling saturated vapor and liquid can sometimes pick up and accelerate liquid slugs. Such slugs are driven by vapor pressure behind and/or by the condensation of the vapor pocket contained ahead of the slug. If a long enough pipe run is present, these slugs can reach tens or hundreds of meters per second; upon striking a fitting, valve, etc., a catastrophic event can occur. Accidents of this sort have killed workers in steam systems, notoriously at US Department of Energy facilities. Because flow rates and pipe sizes are not as large in ammonia refrigeration, the main risk is that of a rupture and consequent ammonia release rather than a large-scale explosive release of pressure. Liquid slugging (liquid hammer) incidents occur occasionally in refrigeration systems using any refrigerant. Since the physics governing slug generation is not well understood, operational and design changes aimed at eliminating these accidents have been only partially successful.

A potential liquid-hammer accident in our model facility would be a failure in the condenser coil, breaching it and releasing ammonia into the atmosphere. Assuming our hypothetical facility used three 600-ton (2100 kW) condensers, approximately 300 lb/min (2.3 kg/s) ammonia would be released. Stopping such a release is relatively easy, as the operators need only stop the compressors and close a single valve on the receiver vessel. Thus, we can assume that the leak will be terminated within 10 minutes. Ammonia vapor would be released at approximately 210°F (111°C).

Modeling this accident with ALOHA (which requires a 1-hour release time) produces projected hazard distances of 201 yards (184 m) to 1,000 ppm and 453 yards (414 m) to 200 ppm (Gaussian dispersion). This accident scenario illustrates a situation in which the release may or may not pose a threat to nearby

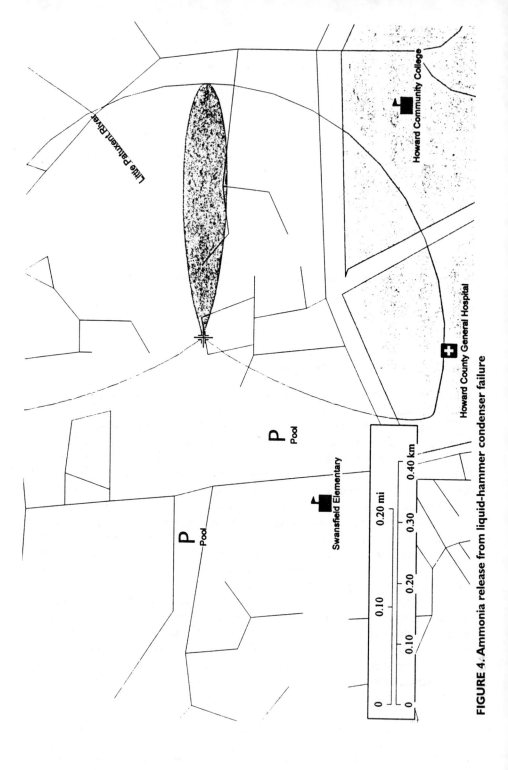

FIGURE 4. Ammonia release from liquid-hammer condenser failure

receptors. As Figure 4 shows, the cloud path may, depending on wind shifts, intercept the local hospital. Since this population is restricted in mobility, it might be appropriate for the facility and local emergency planners to analyze the hospital facility and the probability of the incident in more detail. The exact predictions of a modeling package such as ALOHA are not to be relied upon quantitatively, but they can be an excellent screening tool.

Process Safety and Risk Management Regulations

The ammonia refrigeration industry's experience with process safety management (PSM) programs has been mixed. The original US Occupational Safety and Health Administration PSM regulation was developed without substantial participation by the industry. This was primarily because the industry was unaware that the regulation would apply to anhydrous ammonia in a closed refrigeration cycle until after the regulation was adopted. The result of this was a confused, bureaucratic, and paperwork-intensive regulation oriented more toward complex chemical processes. While many aspects of OSHA PSM have caused compliance problems, the net result for the ammonia refrigeration industry is positive. Many facilities are using PSM as an opportunity to create current as-built drawings and equipment inventories, to develop standard operating procedures, and to establish routine surveillance of the system's condition; this results in improvements both in safety and in operating efficiency. The status of refrigeration as a utility rather than a core business has often meant neglect on the part of management in the past, and regulatory requirements are helping focus attention on proper design, use, and upkeep of refrigeration systems (and to make capital and operating funds available more readily).

With the recent extension of PSM concepts to community risk communication and management, the industry was determined to have a voice in shaping the program. During development of the US EPA Risk Management Program (RMP) regulations, the ammonia refrigeration industry, through the IIAR, offered public comments and participated actively in hearings. Much of the similarity between program elements in PSM and RMP is due to industry comments. Also, IIAR agreed to cooperate with EPA on the development of a model guidance document for ammonia refrigeration facilities as a compliance aid, especially for smaller facilities.

A fortunate aspect of the EPA regulations is the modular approach, where program elements are effectively shared between PSM and RMP. This approach was a key result of industrial participation in regulatory development, and we expect that it will reduce the industry's cost of compliance significantly. Consultant estimates for RMP preparation are around $10,000–$20,000 (for large companies with many plants), and Sacramento County, California (USA) plans to charge approximately $12,500 in fees per process for RMP review. As a result, we expect compliance costs for our industry to run into tens of thousands of dollars for each facility, while PSM drew fees of upwards of $100,000 per process.

In the end, though, the ammonia refrigeration industry gains little additional benefit from the new Risk Management Program. Facilities already covered by PSM are faced with additional paperwork, additional requirements for consultants, and a potentially significant community relations problems resulting from the disclosure of highly unrealistic "worst-case" information. Eventually, IIAR and many other trade associations filed lawsuits challenging the regulation. Although as of this writing a settlement is expected, the fact that this occurred at all should serve as a caution to those who expect high returns from "negotiated rulemaking" and other industry-government cooperative regulatory programs.

The results of our engagement and cooperation with regulatory authorities are mixed. Still, we believe that our industry benefits from being engaged and aware rather than disconnected and oblivious. The sole positive benefit gained from the RMP regulation is an increased impetus to improve cooperation between facilities and their local emergency response community. Thus, an upcoming issue is our industry's relationship with emergency response organizations and personnel. In order to prevent incidents such as the university ice rink release described below, the ammonia refrigeration industry must take the initiative to build a good working relationship so that incidents are handled promptly before escalation can take place.

Because they are either small businesses or small components of larger companies, few refrigeration-oriented businesses could afford to develop and implement comprehensive Process Safety Management programs from scratch. This created an opportunity to develop guidance for the industry, provide educational materials and give structure to the industry's voluntary efforts. In 1993, IIAR published the *Guide to the Implementation of Process Safety Management for Ammonia Refrigeration*. This document provides a model program for the industry, organizing the fourteen elements of PSM into a set of easy-to-use aids to compliance. The use of checklists for maintenance and hazard assessment, provision of forms for a facility's use, and simple, nonbureaucratic language has proven useful in implementing PSM throughout the ammonia refrigeration industry. In addition, IIAR offered PSM training specific to ammonia refrigeration and our guide. This was mainly a tool to raise awareness of PSM and was of necessity limited in scope: as a broad-based trade association, our membership includes consultants, design engineers, and contractors who offer PSM-related compliance services to refrigeration users, and as a general matter of policy, IIAR does not compete with its members.

The IIAR model PSM program for the ammonia refrigeration industry has been very well received, being used both directly by facilities and indirectly by consultants engaged to create facility-specific PSM programs. The detail and regulatory language of PSM confuses facilities in our industry. The chemical-process orientation of the regulations introduces complexity unnecessary in refrigeration and engenders further confusion. Providing forms and checklists proved useful and, in the ideal case, permits facilities to concentrate on actual safety aspects of their program rather than maintaining and creating records. In particular, the use of the What-If/Checklist method for hazard assessment is very useful in an industry where facilities are custom-engineered from fairly standard components and subsystems. Several detailed Hazard and Operability studies on actual refrigeration facilities were used to prepare checklists and "What-If?" question sets for use industrywide. Facilities found that this simplified approach assisted substantially in constructing an effective process safety management program.

An important aspect of developing such industry model programs is maintaining the guidance to keep abreast of developments in technology, regulations, and enforcement. We are updating the PSM *Guide* to include lessons learned from the first five years of the Process Safety Management program, and to concentrate attention on areas where compliance problems occur. Some means to ensure (or at least encourage) facilities actually to follow the requirements of the program should be incorporated, but this has proven difficult in practice. We have received reports of facilities which engaged consultants to create a PSM program, placed the consultant's report on the shelf, and did nothing further to implement it. In one case, a facility was found to have both "a PSM program" and drawings which showed 2,000 more valves than the facility actually possessed. Other industries implementing Process Safety Management or the Risk Management Program have experienced similar problems. We have also noticed that much PSM enforcement attention focuses on paperwork requirements rather than program performance.

Since RMP is largely an extension of PSM, we intend to add Risk Management to our existing *Guide*. A major focus of our future efforts will be risk communication. We intend to prepare a variety of materials to assist in communicating risk, risk reduction and control, and emergency response information to the public, elected officials, code enforcement personnel, and emergency planning and response organizations.

Risk management and process safety programs have created needs for training of many different audiences. Through our charter to promote the safe use of ammonia refrigerant through education, information, and standards, we provide a variety of training resources to the industry, alone and in conjunction with other organizations. Educational seminars are aimed at compliance officers (PSM/RMP), engineers (a school on system design at Kansas State University offered twice yearly), and operators of ammonia refrigeration systems. Operators are offered two weeklong schools, a twice-yearly lecture-oriented course presented with the Refrigerating Engineers and Technicians Association (RETA), and a hands-on course at Garden City Community College, Garden City, Kansas. The Garden City course is in a building containing four ammonia refrigeration systems, three of which are used exclusively for training and a fourth which provides comfort cooling for the building itself. IIAR, RETA, and many of our members provided funds, equipment, and encouragement for the construction of the training center.

Another method of providing training is through video materials. Seminars and schools provide good coverage and retention of material, but the cost can be high, limiting the potential audience. Videos allow the seminar to reach a very broad audience, since material can be covered in the workplace and in small lessons. Limited coverage (of material) is a major drawback, as is the lack of feedback and interaction with the student. Production of video training material is very capital-intensive: IIAR's first production of five 30–minute tapes ("Introduction to Ammonia Refrigeration") required an investment of $300,000. The Institute's advantage here is access to funds; we solicited donations from members as seed money. Proceeds from video sales (around $2,000/set) are used to fund further development, including the second set of videos (safety and emergency response) and the Spanish translation of the initial videos (for use in the US and Latin America).

Trends in Risk Management

Development of risk reduction practices continues constantly in the ammonia refrigeration industry. As with most industries, the twin goals for risk management are to increase safety (a major element of the IIAR charter) and to reduce costs and burden while doing so (which promotes the adoption of safety initiatives). IIAR has found that simpler practices tend to produce more effective results: increasing the safety factor on a vessel's design, for example, rather than providing complex field-erection and -testing procedures.

Design

The initial design of refrigeration systems provides many opportunities for incorporating inherent safety. An important recent trend in the industry is the drive to minimize the charge of ammonia contained within the system. Several innovations have appeared in this area.

The use of plate-and-frame heat exchangers has become more common. Plate units, which often provide a larger heat transfer surface with a smaller size and ammonia charge, are increasingly applied in fluid-cooling applications, both for condensing (rejecting heat, usually to atmosphere) and evaporating (absorbing heat from the material to be cooled) service. Crucial to the successful application of plate-and-frame units was the introduction of refrigeration-service semi-welded plate cassettes, which do not employ gaskets on the ammonia side. Past experience with gasketed plates was disappointing, since elastomeric gasket materials generally perform poorly in low-temperature service with aggressive chemicals.

Another innovation which reduces ammonia use is the sealed unitary chiller package. These units, similar to chlorofluorocarbon building water chillers used for years to provide air-conditioning, consist of a compressor, motor, and evaporator skid-mounted to chill water, glycol, brine, or other heat transfer fluid (a *secondary refrigerant*). In addition to the safety advantages of intensive engineering, testing, and factory quality control, the ammonia refrigerant is confined to a self-contained unit in the machinery room rather than being circulated into a process area in large quantity. Such chiller systems can exceed the energy efficiency and capacity of fluorocarbon-based refrigeration systems.[10]

Plate-and-frame heat exchangers are commonly used in package systems, and some units are *critically-charged*, having just enough ammonia refrigerant to transfer the required heat load. Applications of critically-charged systems are limited, since they perform best on constant heat loads (e.g., an ice rink), while many industrial systems have highly variable loads (e.g., a batch freezing station).

Treatment Systems

Development and application of treatment systems for discharged ammonia refrigerant is another area of innovation. Although the use of safety relief valves is essential to prevent vessel explosions, their operation inevitably results in refrigerant release. While releases from safety relief valves should properly be directed to the atmosphere in most cases, situations arise where this could present a public hazard or be unacceptable to code officials.

The dual necessities of preventing toxic releases and pressure vessel explosions sometimes result in conflicting solutions. Certain areas of the United States require that relief systems be diverted to a treatment facility; in practice, this is a large tank of water beneath whose surface the ammonia is discharged. With sufficient measures to ensure adequate gas–liquid contact, the ammonia can be readily absorbed and release (at least, *immediate* release) to the community is avoided. Unfortunately, these systems also provide opportunities to compromise the integrity of the relief system, causing the relief piping to rust shut, for example, or permitting water contamination of the system through backsiphoning. Thus, a mandated "safety" solution could carry with it the risk of *causing* a catastrophic pressure vessel failure. Also, problems such as provision for the sometimes-large water tanks and disposal of the resulting 30% or so aqua ammonia arise. IIAR's members are working to develop recommendations on the application and design of treatment systems for use by the ammonia refrigeration industry and by mechanical and fire code officials.

A variety of alternative treatment systems have been applied. These range from solid absorption/desorption systems to recover the ammonia to high-speed water mixers draining to a normally dry holding pond. Flare systems are also of interest but bring their own set of risks. As ammonia does not support its own combustion, auxiliary fuel must be supplied. The risks of storing and managing large quantities of propane or natural gas must be weighed against the benefits offered by treatment of ammonia discharges.

A related issue is mitigating the effect of large releases such as a punctured evaporator or a catastrophic pipe rupture. In other industries, such as chlor-alkali or propane, excess flow valves have been applied to good effect; however, these situations center around material handling and distribution: tankers, pipelines, barges, etc. The pressure drop induced by modern excess flow valves is unacceptable in a refrigeration system. Sectioning the system with fail-closed solenoid valves (actuated by ammonia detectors) is a viable alternative to excess flow valves. Such sectioned active safety systems have been installed in a number of facilities. The primary disincentives to such installations are added cost and maintenance, although with clever design, many normal control valves may be used as active mitigation.

Secondary Refrigeration Loops

As mentioned above, the use of ammonia with secondary refrigeration loops (employing ethylene or propylene glycol, calcium chloride brines, or a variety of other heat transfer fluids) is possible as well. The benefits of containing the ammonia within an engineered machinery room are obvious, and the necessity of doing so is just as strong with halocarbon refrigerants because of regulatory pressures to reduce environmentally damaging refrigerant emissions. As a result, there is substantial interest in increasing the use of secondary refrigerants throughout the entire refrigeration industry.

Secondary loops can be more practical for ammonia than for halocarbons, as there is an inherent loss in efficiency due to the additional heat transfer step. Having a higher-capacity/higher-efficiency refrigerant in the machinery room mitigates the effects of this loss. Ammonia chiller installations for building air-conditioning are increasing, and ammonia-based district cooling systems (such as the new McCormick Place convention center expansion in Chicago, IL, USA) are now operating.

Integrity of Containment

Desire to reduce the chance of refrigerant release from containment creates a need for high-integrity systems. Ammonia refrigeration systems traditionally employ welded construction rather than brazed. Systems are thoroughly leak-tested before commissioning. Interestingly, pneumatic proof testing is followed by a 100 psi ammonia test charge. The self-alarming characteristics of ammonia enable location of far smaller leaks than a pneumatic test will show.

Vessels intended for use in refrigeration systems of any type *must* be constructed and stamped in accordance with the applicable pressure-vessel code. (In the United States and Canada, this is the ASME Boiler and Pressure Vessel Code; in Britain, BS 5500, *Specification for Unfired Fusion-welding Pressure Vessels*, applies as well. Many Canadian provinces require design registration and approval for all refrigeration systems and components.) A high safety factor is employed (often 5 : 1). Many pressure vessel failures are traceable to code violations (or wholesale disregard of the code).

A specific feature of ammonia refrigeration systems is the need to drain oil from various locations from time to time. Conventional mineral oils are not soluble in ammonia and thus oil can be carried through the system, collecting at low points. Cold oil flows very slowly, and an operator waiting to drain a few gallons of oil through a small valve may become unwary. Once the oil is removed, ammonia can flow out from the drain valve under pressure, causing a release and exposing the operator to death or injury. The industry now requires secondary spring-loaded "deadman" valves for oil-draining points in new facilities to shut off the flow immediately and regain control in the event of an accident. This design has introduced problems of its own, including the reluctance of valve manufacturers to make the device (for liability reasons) and the tendency of plant personnel to lock the valves open, negating the benefit of the device. There is some disagreement within the industry about the efficacy of self-closing valves.

Ammonia refrigeration systems are usually designed and operated to flood the evaporators (the cold-side heat exchangers) with liquid ammonia. In older facilities, it is common for the evaporators to be mounted with little or no protection from impact. This can create a risk of damage from forklifts and other material handling equipment which generally operate in refrigerated spaces. A ruptured pipe or evaporator can release large quantities of ammonia in a short time. Modern design practice moves evaporators and associated piping out of harm's way; in a refrigerated warehouse, the evaporators are often housed in penthouses high above the space. Again, a tradeoff of risks is involved, as one is substituting the possibility of damage for the difficulty of maintaining equipment located in a confined space 40 feet (12 m) above the warehouse floor.

Emergency Response

An area needing further effort is emergency response to incidents in ammonia refrigeration plants. The relationship between refrigeration plant personnel and local emergency responders is often poor, and this severely hampers proper emergency response. An emergency in an ammonia refrigeration facility requires two sets of skills: expertise in hazardous materials emergency response and knowledge of the configuration and operation of refrigeration systems. Because these two skill sets usually reside in separate groups of people who may not communicate often (or well), accidental ammonia releases and other emergencies frequently have more severe consequences than the actual incident would justify. This should be contrasted to the common situation in chemical process plants, where larger facilities are likely to have a strong working relationship with local fire and emergency personnel or to have a captive emergency team.

A recent incident at a university ice rink illustrates the difficulties posed by a lack of planning and cooperation between refrigeration plant personnel and emergency response organizations. During a tiny release (1–2 lb) of ammonia, the local fire and environmental officials denied entrance to refrigeration technicians seeking to stop the leak, unnecessarily prolonging the incident and causing loss of service at the rink. The university did not have an adequate emergency response plan (including refrigeration technicians qualified to enter unknown ammonia atmospheres), while the fire and environmental personnel had exaggerated ideas of the toxicity hazards of ammonia. Encouraging plants and emergency personnel to jointly plan and drill for emergencies is an important area of emphasis. IIAR regularly collects examples of well-executed plans and drills1, and we are preparing materials to encourage cooperation.

A key IIAR strategy for improving cooperation with emergency responders is developing simple materials for use by emergency responders and planners. An example would be the promotional videotape, *Refrigerant of the Future*, which explains the reasons why ammonia refrigerant is used; this tape is aimed at a general audience. Part of IIAR's new guidance for implementing a risk management program for ammonia refrigeration will likely be a guide to creating and coordinating emergency response plans: we envision specifications for response equipment, critical plant information for fire and environmental use, model accident scenarios, and other useful information. Examples might include posters, videos, bulletins, or a laminated, ammonia-resistant emergency card intended to be placed on fire trucks

Conclusion

Ammonia refrigeration is an often silent and invisible part of many communities, but it is a technology central to modern life, touching the food and personal comfort of most citizens of industrialized countries. The community in and surrounding an ammonia refrigeration facility is frequently unaware of ammonia's presence and unprepared to evaluate risks posed by the use of refrigerants.

Ammonia is a moderately toxic and slightly flammable compound. Both types of hazards can be effectively handled in refrigeration applications through strong mechanical codes and good engineering practice. This design-side emphasis on safe practices is a major feature of the ammonia refrigeration industry.

Implementation of an effective process safety management program can be an important factor in the safe operation of a refrigeration facility. We recommend that other industries concentrate on the most important aspects of PSM first: adequate process safety information (particularly plant as-built drawings), simplified process hazard analyses, and establishing (and using!) good standard operating procedures. The paperwork requirements of PSM are of interest only to the extent that they must be produced to satisfy enforcement personnel; an effective program, not extensive files, is the key to safety. The ammonia refrigeration industry has benefitted from an industrywide approach to PSM programs that simplifies compliance.

Most members of the ammonia refrigeration industry are small businesses, either by virtue of their actual size or of their small utility-provider function within a large process plant. The presence of an association such as the IIAR improves risk management and hazard reduction in the industry in several ways. We act as a credible industry spokesman both within the industry and to the outside world. We make available shared resources that our members cannot afford alone, such as our technical committees and publications, training programs, and technical staff. Finally, we provide an understanding of the industry not available elsewhere, to influence regulatory initiatives such as RMP, and to simplify and customize those programs so that our industry can effectively comply with them.

Newer risk management regulations, although similar to PSM, offer little to our industry in additional benefits and much in additional burden. In particular, any attempts at rational communication of risks with the public are hampered by the mandatory disclosure of specious "worst-case" scenarios that create the unwarranted appearance of immense public hazard from facilities and invite public condemnation and unfounded lawsuits based on exaggerated "hazard distances" after minor incidents. Our response to RMP will be similar to that for PSM: create materials to educate the industry (and in this case, the community) and simplify compliance.

The toxicity and flammability of anhydrous ammonia can be effectively managed in a variety of refrigeration systems. Over a century of experience in systems ranging from large cold storage and food processing plants to small home comfort-cooling and hotel minibar refrigerators has shown that safe work practices, stringent mechanical design and construction codes and standards, and process safety management techniques can prevent significant accidental releases. The best evidence of this is ammonia's record as a refrigerant: although ammonia injuries and deaths among the public occur with some regularity in transport and agricultural use, we are not aware that any offsite ammonia fatalities have ever occurred from a refrigeration accident. Meanwhile, billions of people eat food kept healthful and safe from spoilage by ammonia refrigeration.

References

1. UNEP Technical Options Report on Refrigeration, Air Conditioning, and Heat Pumps. June 1989.
2. Stoecker, W.F. *Growing opportunities for ammonia refrigeration*. Proc. IIAR Ann. Mtg., Austin, TX, 1989.
3. ANSI/ASHRAE 34; ISO 5149.
4. Lunde, H., Lorentzen, G. *Accidents and Critical Situations Due to Unintentional Escape of Refrigerants*, in *New Applications of Natural Working Fluids in Refrigeration and Air Conditioning*, IIR/IIF (International Institute of Refrigeration/Institut International du Froid; not related to IIAR) Commission B2, Hannover, Germany 10-13 May 1994.
5. ANSI/NFPA 90, *National Electric Code*.
6. A ton of refrigeration—TR— is that rate of heat transfer which would melt 2000 lb (907 kg) of ice in 24 hours, or 12,000 Btu/hr (3.5 kW).
7. Facility sizing per Rudolf Nechay, Industrial Refrigeration Services, Inc., Baltimore, MD; personal communication.
8. Ferjencík, M. *The role of the two-phase scenarios concept in the matrix relative risk ranking procedure*, Process Safety Progress 16(2), 1997, p. 117-120.
9. Spicer, T.; Havens, J. *Users guide for the DEGADIS 2.1 dense gas dispersion model*. EPA-450/4/89-019, EPA/OAQPS, Research Triangle Park, NC, 1989.
10. Ritmann, J., Wiencke, B. *Safety aspects and installation experiences with packaged ammonia water chillers in public buildings*, Proc. IIAR 19th Annual Meeting, New Orleans, LA, March 23-26, 1997.
11. Kulp, C. *Organization and execution of a mock ammonia spill response*, Proc. IIAR Annual Meeting 1997, New Orleans, LA.
12. ANSI/ASHRAE 34-1992, with Addenda *a-o, q-x*.
13. ANSI/NFPA 497A, ANSI/NFPA 325.
14. MICR is relevant for inductive electrical ignition.
15. MESG is relevant for capacitive discharge ignition.
16. Refer to *Ammonia Data Book*, International Institute of Ammonia Refrigeration, 1992 for primary sources.
17. National Weather Service (USA) Baltimore report, May 14, 1997. Cloud cover by personal observation (DRK), facility location from United States Geological Survey map (Clarksville, MD quadrangle, 1979).

high level

Improving Safety, Environmental Protection, and Reliability through Integrated Operational Risk Management

Glenn B. DeWolf and Theresa M. Shires
Radian International LLC, Austin, Texas

Keith Leewis
Gas Research Institute, Chicago, Illinois

Introduction

Formal risk management programs have become a fundamental approach for process safety (1,2,3). Extending beyond the traditional concepts of loss prevention, risk management applications are expanding to encompass environmental protection, reliability, and other business risks. Both government regulatory approaches and fundamental business processes in an increasingly competitive global business environment have driven this movement.

This paper discusses an integrated approach to managing technical and business risks, where process safety is just one risk category. An understanding of the relationship between technical and business risks can be used to develop a comprehensive, operational risk management program that recognizes common risk factors in operations that can lead to multiple consequences; the basis for defining various risk categories. An operational risk management program can be used to analyze data and make informed decisions about risk control in multiple operational risk categories at the same time. This paper examines commonalities among various types of risk management programs such as the Occupational, Health, and Safety Administration (OSHA) Process Safety Management; U.S. Environmental Protection Agency (EPA) Risk Management Plans; U.S. Department of Transportation, Office of Pipeline Safety (DOT-OPS); risk management demonstration program pursuant to the Accountable Pipeline Safety and Partnership Act of 1996; and the ISO 4000 standard. It examines how these programs, though each with a different focus, lead naturally to an integrated risk management approach.

The paper discusses the role of performance measures and cost-benefit analysis in making resource allocations decisions and how risks can be monetized to place different risk categories on a common basis for comparison. It examines how an integrated risk management program can be developed and implemented within a facility or company. This paper derives from work originally done with the Gas Research Institute (GRI) related to pipeline risk management (4,5,6,7). The approach presented is based on experience in work performed within the context of the natural gas industry, blending risk management concepts and practices from that industry with those of the chemical industry.

What Is Risk?

Risk is defined differently in different settings, but fundamentally it is composed of two parts: a frequency or probability of an adverse event and the severity or consequences of that event. In mathematical terms this can be expressed as:

$$\text{risk} = \text{probability} \times \text{consequences}$$

Numerical values expressing both probability and consequences in absolute or relative terms can be used to evaluate a given risk either quantitatively or qualitatively. This allows different risks to be compared. If a common measurement unit for risks of different types can be established, risks can be compared on the same basis across different operational areas.

From a business viewpoint, risk is any adverse impact on business activity. This includes financial impacts and adverse societal impacts. Financial and societal impacts of activities are inextricably linked, as some firms have learned too late when their reputations have been tarnished by an operational flaw.

What Are Operational Risks?

Operational risks are adverse events or situations that occur in an operational area.

With this in mind, we believe that most of the operational risks can be grouped into the following primary risk categories:

1. *Public and worker health and safety*—Health and safety risks pertain to the loss of containment risks that can result in a fire, explosion, or release of a toxic substance, especially when such an incident can affect the community beyond the boundaries of the affected facility. It also includes risks to employees of a company.
2. *Environmental protection*—Actual environmental damage is covered by this risk category. This includes pollution of air, water, and land, and damage to flora and fauna.

3. *Reliability*—Reliability refers to risks associated with the production and delivery of an acceptable service or product to customers. Reliability also implies quality.
4. *Compliance*—Compliance risks refer to the risks of non-compliance with any mandated or other program that has formal obligations to an outside entity. In addition to government regulations, such regulations include formal commitments made to certification agencies [e.g., International Standards Organization (ISO)] and to insurers.
5. *Other Business Risks*—Some generic business risks include: failure of a capital project to produce acceptable return on investment, unplanned expenses, property losses, business interruption, liability, adverse public relations, cost increases, loss of sales, etc.

Each of the above risk categories corresponds to a fundamental operational area of a company and the associated organizational function. Operational procedures and practices within that function influence risk factors and individual risks in each of these categories. Table 1 shows an example of how risk categories, specific risks, risk control decisions, organizational functions, information, and data are connected. This specific example is for a pipeline application.

Starting from the left hand side of Table 1, the fundamental data column lists relevant data items that are collected for the various risk categories. Various functions in an organization use several processes (e.g., software tools, engineering judgement) to turn the data into useable information. This information is used by functions of the organization to make decisions about operational risks. The table illustrates some of the overlaps in information used for decision making and in the operational activities affected by those decisions. Understanding these relationships is fundamental to operational risk assessment.

What Are the Components of Risk Management?

Like risk, risk management is also defined in various ways depending on the application. The general definition we have selected was presented in a report published by GRI (4):

> *Risk management is the systematic application of procedures, policies, practices and finite resources to identify, assess, and reduce risk based on activities with the greatest potential for risk reduction.*

We selected this definition because it is comprehensive and explicitly includes the reference to "finite resources," a fundamental reality of risk management often overlooked in the zeal to focus only on pressing immediate issues or conversely to try to solve all problems at one time. It gets to the heart of why a risk management is so important—the ability to identify and compare risks.

TABLE I. Example of Linkage between Risk, Operational Decisions, Functions, Information, and Data

Risk Category	Data	Information	Organizational Functions	Operational Decisions	Risks
Pipeline Safety	Visual inspections CIS data Hydrostatic tests Smart pigs	Pipe condition Corrective actions schedule	Corrosion Engineering Compressor Services Design Engineering Codes and Standards	Routing Preventive measures Mitigation measures Suppliers	Leak—fire; explosion Rupture—fire; explosion
Reliability (Customer Satisfaction)	Visual inspections CIS data Hydrostatic tests Smart pigs	Pipe condition Corrective actions schedule	Corrosion Engineering Compressor Services Design Engineering Contracts Department		Leak—service interruption Rupture—service interruption Compressor outage Valve—closed failure Other
Environmental Impacts	Visual inspections CIS data Hydrostatic tests Smart pigs	Pipe condition Corrective actions schedule	Environmental, Health, and Safety Codes and Standards	Suppliers Cleaning residue disposal	Leak—environmental damage Rupture—environmental damage Cleaning residues
Non-compliance (regulations, other)	Compliance audit results	Status Corrective actions schedule	Environmental, Health, and Safety Codes and Standards	Procedures Training Supervision auditing content and frequency	Activity Documentation Reporting
Worker Health and Safety	Injury reports Inspections	Status Corrective actions schedule Training needs	Environmental, Health, and Safety District Supervisor Station Supervisor	Procedures Training Supervision auditing content and frequency	Physical injury Leak—chemical exposure Leak—fire; explosion Rupture—chemical exposure; fire; explosion

Based on the comparisons, control options are selected and prioritized so that resources are allocated to provide the highest value to the stakeholders according to the risk acceptability consensus of those stakeholders at any given time. It is also apparent that risk management requires the definition of goals or objectives for each category of risk addressed and requires performance measures to assess the success of the program in achieving these objectives.

The overall risk management process is a "cradle-to-grave" concept which considers the entire life cycle of the system: design, construction/implementation; operations; shutdown. Development of a risk management program requires setting up the means for carrying out the various specific activities which are defined by the risk management process. The process is defined conceptually in Figure 1. The three basic steps of this process are:

1. *Risk assessment*—identifies the causes, likelihoods, and consequences of adverse events;
2. *Risk control and decision making*—selects control measures and allocates resources for each category of risk to either accept, modify, or transfer the risk; and
3. *Performance monitoring, evaluation and modification*—determines the effectiveness of the program and evaluates modifications for improvement.

Risk management as a process applied appropriately, results in a comprehensive systematic structuring of the cause-effect relationship for activities, decision, technologies, and information in a way that ranks the options for the most cost-effective resource allocation decisions. Risk management critically and systematically examines a system to define failure modes and effects, makes decisions about options to prevent the failure modes and mitigate the effects, and makes those decisions based on a ranking that allows the most efficient allocation of resources (that is, people, material, and money) to achieve the best benefit-to-cost ratio. Effective risk management seeks the most efficient application of resources for controlling risks. Risk management involves the total management of all technical and non-technical activities needed to achieve and maintain an acceptable level of performance in any risk categories. It is apparent that risk management requires the definition of goals or objectives for risk control in each category. It requires performance measures to prove that the desired results are achieved.

The risk management process can be applied equally well to a physical system as well as to activities, procedures and practices within an organization. It ties together the relationships between the risks of failure and the decision processes and information in a system. By applying risk management in a comprehensive manner to operations, rather than piecemeal to individual aspects of operations, better accounting for interactions between different operational areas can be achieved. Previously unrecognized risks can be identified and a better relative ranking of all risks on a common basis can be achieved, leading to a better allocation of resources.

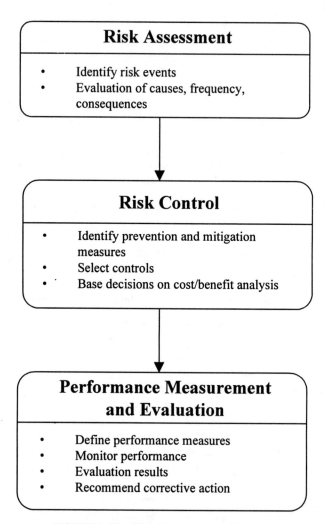

FIGURE 1. The Risk Management Process

With this in mind, operational risk management can be viewed as the application of the risk management process globally across all areas of operations in an integrated way. This is illustrated in Figure 2. Support for this concept can be seen if we examine the background for risk management as applied to process safety and environmental protection.

FIGURE 2. Operational Risk Management as Unifier of Multiple Risk Categories

Risk Assessment

As can be seen from the risk management process diagram (Figure 1), risk assessment is that part of the process that identifies and evaluates various risks. Risks can be ranked and priorities established for selecting appropriate risk controls.

Risk assessment begins with identifying the hazards associated with the physical system or activities, procedures or practices. Risk assessment answers the questions: What can go wrong? How likely is it? What are the impacts? Causes of failure and their likelihoods are determined. Consequences are identified and evaluated.

In process safety and reliability engineering, such formal techniques as the hazard and operability study (HAZOP); failure modes, effects, and criticality analysis (FMECA); fault tree analysis (FTA); and event tree analysis (ETA) are well known techniques commonly applied to physical systems. They are equally valid as disciplined systematic analytical techniques for activities, procedures, and practices. They can be used to analyze the failures and their effects leading to environmental impacts and business consequences just as much as process safety and equipment failure consequences.

In an operational management program, the risk assessment process would be more integrated. When an analysis is undertaken to examine, say, the process safety risks of a system, the environmental, reliability, and other risks would be recognized and examined in the same effort. By using common data and information sources on the system under examination and understanding the interrelationships of various risks, the resulting effort will be consistent and comprehensive.

Another area for integration is information management. Data and information on the fundamental characteristics of the system to be managed can be collected in a distributed effort using a coordinated and consistent plan. Data would be maintained in a central database for storage, and provided in distributed access for analysis by users. This ensures consistency of information upon which risk assessments and control decisions are based, and also offers direct cost savings by avoiding duplications of effort.

For all risk assessments, operational risk management requires that both the costs of controls and the costs of consequences be developed. This is to allow a cost / benefit analysis to be applied in the risk control decision process. The risk assessment must be set up to define events, the controls associated with preventing those events, and the consequences in equivalent terms, the most obvious of which is cost.

Risk Control and Cost / Benefit Analysis

Whereas risk assessment determines if all significant risks have been recognized, risk control analyzes existing prevention and mitigation efforts and determines what additional controls might be required. This iterative process between the risk assessment and risk control steps can lead to a sensitivity analysis where the effects of alternative controls on risk rankings can be evaluated. These results, combined with the costs of various options and weighed against the benefits, can be used to make informed and cost effective risk control decisions. From an *integrated operational* risk perspective, it means being able to convert various safety, environmental, and reliability risks into monetized rankings so that decisions can be made on a common basis.

The costs of controls and consequences must be developed to allow a cost / benefit analysis to be applied in the risk control decision process. The whole premise of risk management is based on more effective resource utilization. In an operational risk management mode, where all categories are considered in the final decision processes, there is a greater opportunity for more cost effective use of an organization's resources.

To determine the potential value of the risk management effort in strictly quantitative or semi-quantitative terms is not an easy task. Showing benefits to the organization requires the ability to monetize activities and changes in those activities. Ultimately the value of risk management includes assessing the benefit of avoiding future adverse events. Other factors to be considered in determining costs and in monetizing benefits include: cost of program implementation, cost of data collection and measuring program performance, present value of avoided capital costs and operating expenditures, and indirect effects such as improving relationships with regulators and the broader community.

As with risk assessment, risk control also offers opportunities for activity integration. An example of an opportunity for integration in the risk control area is in auditing. Safety, environmental, and operational audits can be combined into a single audit effort by suitable audit protocols and audit team composition. Combining audit activities avoids duplication of effort and results in cost savings.

Performance Measures

Determining the success of any risk management program depends on the ability to measure outcomes defined in terms of the objectives of a program. These performance measures can be direct outcomes of the risks being controlled, precursor events to these outcomes, or surrogate measures related to these. The validity of these performance measures then depends on the success in understanding and linking these practices and precursor events to the catastrophic outcomes.

Meaningful performance measures must enable the observation of statistically significant changes in risk in a relatively short time period. This requires first that progress can be measured over time, and second that a baseline exists from which to determine change. In addition, performance measures must be: Simple, Meaningful to the risk of concern, Attainable, Reliable, and Timely (SMART). For rare, catastrophic events, the direct measurement of outcomes (e.g., the accidental release of a hazardous substance) may not provide a statistically meaningful measure of the risk controls applied.

This leads to the practical requirement that performance measures other than direct outcomes be defined. Therefore, conformance to requirements in management systems and precursor or surrogate measures of performance are required as substitutes for the actual outcome events. For risk management programs such as OSHA PSM, and EPA RMP, the conformity of the organization to its own declared risk management plan and to its management system is the primary performance measure. The logic is that if the management system defines what must be done and these requirements are based on a linkage to risk control, then the effectiveness of risk control derives from the effectiveness with which the organization complies with the management system.

It is interesting that among the three government initiatives on risk, only the DOT-OPS program explicitly requires the definition of performance measures when the risk management program is declared. The others assume that outcomes will be the measure, and rely on periodic reviews and audits. For these programs, conformance with requirements is the basis for determining if the desired performance is achieved.

Business Goals and Operational Risk Management Benefits

What are we ultimately trying to achieve? Risk management, as it relates to business goals, can be related to strategic and tactical goals. The strategic goals are the fundamental reasons that the business exists. The tactical goals are goals that must be met to meet the strategic goals.

Using these definitions, strategic business goals include attaining and maintaining market share, revenues, and profitability. The strategic goals are achieved through tactical goals of cost control which includes both direct current costs and future liabilities. This is where the connection between risk management and business financial objectives occurs. Formal risk management controls short term and current costs by carrying out various risk management activities more cost effectively in the short term; it controls long term costs of future liabilities by better risk control.

An operational risk management program provides a company flexibility in selecting risk controls commensurate with differing levels of risk across operational areas. Resources can be applied to achieve the greatest degree of risk control for the resources expended.

- Better resource allocation, more cost effective spending, and allocation of resources on a risk basis, rather than a mandated compliance basis achieve equivalent or greater performance at less cost;
- Better understanding of the system, which includes identifying the causes of failures and potential impacts of failures;
- Short term line item cost reductions and possible long-term cost reductions that include the costs of future adverse events (e.g., business interruption, damages, and liabilities);
- Improved performance in managing risks through comprehensive identification of problems and putting resources where they do the most good; improved safety, more reliability, less liability; decisions made with better data and methodologies;
- Avoided costs of future regulations through reduction in perceived threats to the public;
- Increased system efficiencies and hence lower operating and maintenance costs through increased knowledge;
- Improved business climate through increased customer satisfaction and favorable perceptions; increased sales and revenues;
- Improved long-term profitability by improved business climate and deferral of capital expenditures through extension of equipment life and improved replace versus repair decisions;
- Improved public image through improved stakeholder communications and perceptions; and
- Reduced losses by more rapid recovery from unplanned outages and other events.

These benefits derive from changes made at the working level and are the result of applying a more systematic risk management process to operations.

Implementation of a Risk Management System

Motivated by the potential benefits, a number of firms are moving forward with some form of the concepts outlined. However, bringing the concepts of operational risk management to fruition requires a sound implementation plan. As with any program, clearly defined goals with a road map or plan directed toward these goals are required. To accomplish this a committed, interdisciplinary team or working group is needed. For the program to succeed, there has to be a genuine high level management commitment.

The practical issues that arise involve a mixture of organizational and technical issues.

The comprehensiveness of the operational risk management concept engenders some formidable implementation challenges. This is because operational risk management touches virtually every aspect of operations and aggressively crosses the boundaries between operating functions and departments. This requires securing and holding high level management commitment and building a team that ensures inclusion of all pertinent stakeholders. The overall implementation process is schematically represented in Figure 3.

Beginning with a comparison of current performance against goals, the current system is evaluated and modified as needed to introduce an operational risk management approach. We see the development of a program facilitated by a three-phase project structure:

1. Feasibility and Scoping Study
2. Program Design
3. Implementation

The feasibility and scoping study determines the requirements for a program. The first step is to evaluate current risk management practice against risk management goals and the structure of the risk management process described earlier. Although aspects of risk management are currently practiced by many companies, the scoping phase requires identifying gaps between desired and current practices. This includes reviewing current functional roles and responsibilities as well as current practices and policies.

The feasibility portion of this step links the results of the risk assessment to the decision process and estimates both the costs and benefits of changes. This is a significant area of effort for most programs. Initially, some of the cost/benefit analysis will be qualitative and intuitive. Keep in mind, however, that one of the reasons for proceeding with an operational risk management program in the first place is to improve the decision process for allocating resources more effectively.

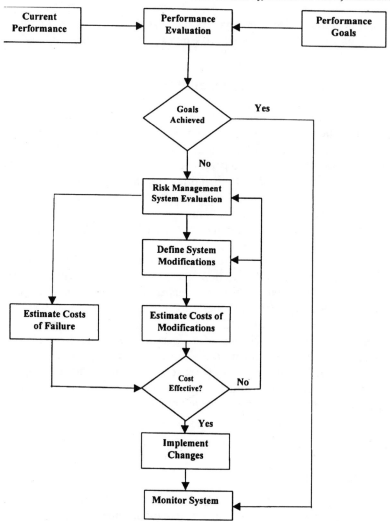

FIGURE 3. The Path to System Implementation

Therefore, the value of the program itself can only be estimated initially, and must be validated later. Improvements in this process result in a move toward a more quantitative approach to assessing the costs and benefits of those controls.

Program design requires design of both the overall structure of risk management as an organizational program and the technical details of the risk management process itself. Fitting the program into the existing organizations, requires evaluation of administrative and management issues. In the Program Standard developed by DOT-OPS and industry for the pipeline risk management demonstration program (9), the term *program elements* was used to describe these components of a risk management program. The program elements include: clearly defined roles and responsibilities, internal and external communication plans and practices, training specific to risk management, management of change, performance evaluation, and other processes as might be appropriate.

Design of the process elements, the technical core of the risk management program, requires formulating means to address the gaps identified in the scoping and feasibility study. In some cases, this will require introducing new methods, procedures, or tools. For example, a company may decide to introduce a new computer software tool for risk assessment and ranking. Program design will define such new requirements that will become part of standard procedures.

Because operational risk management is to be integrated into existing operations, there should be an effort to take advantage of other management systems in place that address risk management.

Recognizing similarities allows one to take advantage of synergies and to synthesize an operational risk management program that builds on previous experience and knowledge, both within and outside of the organization. Operational risk management attempts to unify risk management awareness, methods, and data sources across functional areas, resulting in new working relationships between functions such as safety, environmental and operations. Program design will define how these functions might interact in ways different from current practice.

Once the program design is completed and approved by management, the next step is program implementation. Implementation is likely to be the most difficult phase of the total effort because it requires actual changes to occur. If the earlier phases of the project have successfully involved the key stakeholders, many of the potential problems of implementation can be avoided. Some of the implementation difficulties commonly encountered include those listed in Table 3.

During implementation, some of the major activities that will take place include:

- Developing training materials and carrying out training;
- Writing new procedures;
- Implementing new procedures and practices;
- Acquiring, developing, testing and using new software tools and other technologies;
- Collecting, compiling, analyzing, interpreting, reporting, and storing new data and information.

Following implementation, the operational phase of the program evaluates and monitors performance and addresses changes. As with any management system, continuous improvement requires re-evaluating the process on a timely basis to monitor progress toward established goals.

TABLE 3
Some Commonly Encountered Implementation Difficulties

Lack of management commitment
Lack of clear performance goals
Inadequate resources
Conflicting interpretations
Integration with existing organization
Magnitude of documentation requirements
Diversity of training requirements
Difficulty in determining benefit/cost ratios
Concerns on disclosure, confidentiality, liability

The Evolution of Risk Management Systems

The concepts described in the preceding section are part of a natural evolution of risk management systems. In recent years, there has been an explicit emphasis on risk management as a discipline applied to process safety and environmental protection. These developments have been part of an evolution from loss prevention and earlier piece-meal approaches toward an integrated approach to both safety and environmental management. Table 4 shows the progression of risk management initiatives starting with the American Institute of Chemical Engineer's publication of the *Guidelines for Hazard Evaluation Procedures* in 1985.

The most recent industry directed risk management initiative is the risk management demonstration program issued by the Department of Transportation (DOT) under the authority of the Accountable Pipeline Safety and Partnership Act of 1996. A unique aspect of the pipeline industry program is the flexibility to substitute approved risk management programs for current prescriptive regulations. A four-year demonstration program will permit up to 10 pipeline operators (natural gas and liquid pipeline companies) to test the concept of risk management as an effective means of providing safety, environmental protection, and system reliability.

In addition to the initiatives listed in Table 4, there have been new and updated consensus standards by various standards setting organizations such as the American National Standard Institute (ANSI) and the International Standards Organization (ISO). Trade groups (e.g. American Gas Association, Interstate Natural Gas Association of America, etc.), research organizations (e.g., Gas Research Institute, Electric Power Research Institute, etc.) and professional associations (AIChE, American Society of Mechanical Engineers, etc.) have also supported the development of risk management programs in their respective industries.

TABLE 4
U.S. Risk Management Initiative Milestones

1985	American Institute of Chemical Engineers (AIChE), Center for Chemical Process Safety—*Guidelines for Hazard Evaluation Procedures*
1985	New Jersey—Toxic Catastrophe Prevention Act
1986	Environmental Protection Agency—Emergency Planning and Community Right-to-Know Act, (SARA Title III)
1986	California—Risk Management and Prevention Program
1988	Chemical Manufacturers Association—Responsible CARE®
1990	American Petroleum Institute—Recommended Practice No. 750
1992	Occupational Safety and Health Administration—Process Safety Management Rule
1996	Environmental Protection Agency—Accidental Release Prevention and Risk Management Plan Rule
1996	Department of Transportation—Pipeline Accountability and Partnership Act

Comparisons between Risk Management Programs and Other Management Systems

Each system listed above addresses a specific aspect of operations. We believe that the operational risk management is a unifying management system. Understanding the principle facilitator—risk management—enables risks from all operational areas to be examined on an equivalent basis.

If one takes the risk management process as a template it is easy to see how a number of different programs for safety, environmental protection, and reliability fit within the framework of the risk management process defined earlier in this paper. The fit occurs because of many common features to these systems:

- Each is a management systems sharing many elements;
- Each addresses one or more specific operational risks;
- Each promotes a systematic application of specific operational activities defined by the elements of the system;
- Each requires written documentation of procedures and written record-keeping; and
- Each requires audits, periodic reviews, and modification as needed.

Tables 5 and 6 compare some key elements from various management process. Table 5 compares program elements (administrative and management elements) and Table 6 compares risk management process (technical) elements.

Table 5. Comparison of Various Risk Management System Program Elements

Program Elements	Operational Risk Management	OSHA (Process Safety Management)	EPA (Risk Management Plan)	DOT (Program Standard)	ISO 14000 (Environmental Management Std)
Goals	Define goals	Broad goals stated in standard	Broad goals stated in standard	Broad goals stated in standard	Planning -Objectives and targets -Legal and other requirements -Environmental management program
Administration	-Roles and responsibilities -Communications -Training -Documentation	-Management systemI -Responsibility implicit in process elements (see Table 6) -Employee involvement -Employee training	-Management system -Responsibility implicit in process elements (see Table 6) -Employee participation -Training	-Roles and responsibility -Communication -Training -Documentation	Implementation and Operation -Structure and responsibility -Communication -Training and awareness -Documentation -Document control
Management of Change	Operating Procedures	Management of Change	Incorporates OSHA PSM requirements	Included within	Operational Control
Information Management	-Written documentation -Records -Audit	Written documentation	Written documentation internally and RM Plan Summary submission to agency.	Written RMP requirement through application for demonstration project	-Written EM system requirement -Environmental documentation -Document control -Records -Audit
Program Evaluation and Improvement	Explicit as part of program	Implicit through stated goals	-Implicit through stated goals -Compliance -Audits	-Explicit as part of program -Audits	-Explicit as part of program -Audits

Table 6. Comparison of Various Risk Management System Process Elements

Risk Management Process Steps	Operational Risk Management	OSHA (Process Safety Management)	EPA (Risk Management Plan)	DOT (Program Standard)	ISO 14000 (Environmental Management Std)
Hazard Identification	Process information Prestart impact reviews Incident investigation Management of change	Process safety information Prestart safety reviews Incident investigation Management of change Up-to-date procedures and drawings	Process safety information Prestart safety reviews Incident investigation Management of change Up-to-date procedures and drawings	Combined with frequency analysis	Environmental aspects of operations must be recognized
Hazard Evaluation	Process failure analysis Management of change	Process hazard analysis Management of change Up-to-date procedures and drawings	Process hazard analysis Management of change Up-to-date procedures and drawings	Frequency analysis Management of change	Environmental aspects of operations must be recognized
Consequence Analysis (of a release)	Consequence analysis with monetization	—	Hazard assessment with off-site impact modeling Up-to-date procedures and drawings	Consequence analysis	Environmental aspects of operations must be recognized
Risk Evaluation (ranking all risks)	Risk ranking procedure required	—	—	Determination of risk value	Impacts must be determined
Risk Control—Prevention Measures	Define prevention measures and relate to specific goals. Define decision processes based on defined decision criteria.	Mechanical integrity training Operating procedures Hot work permit Management of change Up-to-date procedures and drawings	Maintenance (mechanical integrity) Training Standard operating procedures Management of change Up-to-date procedures and drawings	Prevention measures must be defined in RMP. Design, construction, operations, and maintenance practices	Training, awareness, and competence Operational control
Risk Control—Mitigation Measures	Define mitigation measures and relate to specific goals. Define decision process based on defined decision criteria.	Emergency planning and response Operating procedures Emergency participation	Emergency response plan Standard operating procedures	Mitigation measures must be defined in RMP. Emergency planning and response and operating procedures implied.	Emergency preparedness and response

Because each of various existing programs already encompass risk management in some respect, integration of programs under the umbrella of operational risk management is an evolution; not a revolution. Areas of commonality offer the greatest opportunities for integration of operational activities and a more cost-effective approach to managing risks.

Three important aspects of operational risk management, which set it apart from other risk management programs developed through regulations or industry standards, are:

1. The integration of risk assessments and risk control decisions across functional areas;
2. The quantification of costs versus benefits as part of the decision process; and
3. The explicit measure of performance.

Indication of Movement toward Operational Risk Management Concepts

We believe that the integration of operational risk management with current practices is emerging in various areas. Recent industry experience suggests that the application of risk management principles to natural gas pipeline safety activities could reduce costs of operations, including maintenance costs and capital investment requirements, while increasing system efficiencies. Applied to pipeline safety, risk management is expected to improve safety by ensuring that resources are allocated based on risk, rather than on the basis of prescriptive requirements, so that the most serious concerns receive first priority.

There are already examples that recognize integration of various elements of risk management as a benefit. For example, the EPA and other federal agencies have sponsored guidelines for the "One-Plan" for Emergency Response. This allows consolidation of various emergency plan requirements of different agencies into a single plan that meets the requirements of all. Another example is the Accountable Pipeline Safety and Partnership Act of 1996 itself which explicitly addresses *safety, environmental protection, and reliability*. Several pipeline companies see an opportunity to not only apply more formal risk management to pipeline safety as part of the Risk Management Demonstration Program, but they are also considering the application of risk management principles to other operational areas of gas pipelines. In the end, the effort and cost of implementing an operational risk management program should be returned by the benefits achieved.

Conclusions

This paper examines the concepts of risk management in the context of programs initiated to address public safety risks. Risk management offers a process by which data can be used to make more effective decisions for resource allocation. Operational risk management applies the risk management techniques to all aspects of business operations such that risks from different processes or business operations can be compared on a common basis. This comparison across different kinds of risks and types of processes is the true value of the operational risk management approach. Operational risk management provides a mechanism for considering all of the complexities of a system and how they relate to the broader aspects of business operations. It moves beyond compliance with regulations to dealing with issues such as public safety, environmental protection, and business reliability in a proactive and systematic manner where decisions are based on a sound analysis and disciplined decision process.

ACKNOWLEDGMENTS

The authors of this paper deeply appreciate the support, encouragement, and other contributions of their colleagues both inside and outside of their own organizations who have and continue to support these concepts and the work deriving from them. Specifically the authors would like to thank Dr. Ted Willke of GRI; Bernie Selig of Hartford Steam Boiler Inspection and Insurance Company; Mark Hereth of Hartford Steam Boiler Inspection and Insurance Company; and Mike Cowgill and Matthew Harrison of Radian International. These individuals and others have contributed to the development of these ideas through projects, meetings, and many conversations.

References

1. 29 Code of Federal Regulations (CFR), Part 1910.119, Process Safety Management (OSHA PSM).
2. 49 Code of Federal Regulations (CFR), Part 68, Accidental Release Prevention (U.S. EPA RMP).
3. 49 U.S. Code (U.S.C.), Accountable Pipeline Safety and Partnership Act of 1996, Public Law 104-304, October 12, 1996.
4. Hartford Steam Boiler Inspection and Insurance Company and Radian Corporation. *Natural Gas Pipeline Risk Management, Volume I - Selected Technical Terminology, Final Report.* Prepared for Gas Research Institute, GRI-95/0228.1. Austin, Texas, October 1995.
5. Hartford Steam Boiler Inspection and Insurance Company and Radian Corporation. *Natural Gas Pipeline Risk Management Volume II - Search of Literature Worldwide on Risk Assessment/Risk Management for Loss of Containment, Final Report.* Prepared for Gas Research Institute, GRI-95/0228.2. Austin, Texas, October 1995.
6. Hartford Steam Boiler Inspection and Insurance Company and Radian Corporation. *Natural Gas Pipeline Risk Management, Volume III - Industry Practices Analysis, Final Report.* Prepared for Gas Research Institute, GRI-95/0228.3. Austin, Texas, October 1995.

7. Hartford Steam Boiler Inspection and Insurance Company and Radian Corporation. *Natural Gas Pipeline Risk Management, Volume IV – Identification of Risk Management Methodologies, Final Report*. Prepared for Gas Research Institute, GRI-95/0228.3. Austin, Texas, October 1995.
8. Center for Chemical Process Safety. *Guidelines for Hazard Evaluation Procedures, American Institute of Chemical Engineers*, New York, NY, 1992.
9. The Joint Risk Management Program Standard Team, *Risk Management Program Standard*, U.S. Department of Transportation, Office of Pipeline Safety, et al., January 17, 1997.

Maximizing the Economic Value of Process Safety Management

Joseph Fiksel and Kenneth Harrington
Battelle Memorial Institute

ABSTRACT

A properly implemented process safety management program will not only reduce the occurrence of adverse incidents at operating facilities, but will also deliver direct economic benefits to the business. Examples of such benefits include improved reliability, reduced downtime, improved yield, reduced operating and maintenance cost, and of course reduced liabilities and insurance costs. However, process safety performance measurement has traditionally focused on incident tracking and compliance issues. This paper describes a "value-based management" approach toward explicitly recognizing the contribution of process safety management toward achieving business objectives related to profitability, return on investment, and shareholder value.

The value-based management approach involves identification of key aspects of value added, selection of economic performance indicators, and definition of appropriate performance metrics and targets corresponding to specific process safety activities. Examples will be drawn from the experiences of several major chemical manufacturers in the U.S., who have recently implemented this type of approach. By linking process safety to business objectives, these companies are able to move beyond a compliance-oriented posture. They can develop a clear, strategic rationale for their process safety management programs and can make well-informed decisions about the appropriate risk-based criteria for their worldwide operations.

Process Safety Performance Measurement

The historic mission of process safety management has been to prevent the catastrophic release of hazardous materials that may have adverse impacts upon human health, property, or the environment. Therefore, it is not surprising that the metrics traditionally used to track process safety performance have been expressed in terms of the frequency and/or severity of such incidents. For example, in the past, the Chemical Manufacturers Association (CMA) adopted the following as a standard measure of process safety performance: annual number of incidents involving (a) fire or explosion exceeding $25,000 in damage, (b) release of a flammable chemical in excess of 5000 lb, (c) release of a reportable quantity of a SARA-listed "extremely hazardous substance" or (d) one or more human fatalities or serious injuries.

The shortcomings of this type "after-the-fact" empirical incidence measurement are well-known:

- It introduces a signifcant "lag time" in the recognition of process safety problems
- It does not address the underlying causal mechanisms that result in adverse incidents, and
- It does not address the management processes that can provide beneficial interventions.

In addition, incidence-based measurement of process safety performance fails to recognize the economic value to the firm associated with process safety improvement. This is an important shortcoming which tends to deprive process safety management of the top management support and capital availability that is essential for continuous improvement.

In view of these shortcomings, we contend that process safety performance measurement should be shifted toward a "value-based" approach that emphasizes cost-effective intervention and value contribution rather than incident tracking. To maximize the value contribution of process safety, a company should promote the following practices:

- Recognizing and eliminating the root causes of process safety incidents
- Reducing the costs of process safety compliance, risk management and incident management, without compromising the effectiveness of process safety management programs
- Improving the availability, reliability, efficiency, and profitability of process operations

By linking process safety to the underlying performance drivers, and by explicitly accounting for the financial benefits of process safety improvement, it is possible to develop a more pro-active, cost-effective management system that is responsive to corporate strategic goals.

Quantifying Economic Value

Recently, many companies have adopted a financial performance metric called Economic Value Added (EVA), which combines both profitability and cost of capital to develop a more meaningful measure of shareholder value. The EVA of a business unit or profit center is defined as follows:

$$EVA = \text{after-tax profit} - \text{attributable cost of capital}$$

The hidden costs of capital such as production facilities and equipment can dilute the value of short-term profits; conversely, improving capital utilization increases the competitive leverage associated with available capital.

For example, many chemical firms are beginning to integrate process safety considerations into the design of new facilities and processes, thus creating inherently safe plants. This can significantly improve their EVA in two ways:

- Capital investments are reduced due to the decreased complexity of the plant and the elimination of expensive containment and mitigation systems
- Profits are increased due to the reduced operating costs associated with process safety-related training, maintenance, monitoring, and documentation.

Even for retrofits or process modifications, process safety improvement will tend to enhance economic value added in similar ways.

Battelle's Value-Based Management Approach

Battelle has developed a framework called VM Advantage,™ which several companies have used to identify the value contributions of their EH&S activities and to support strategic resource allocation. The framework consists of a series of steps, combined with supporting methodologies and tools, that provide a logical and comprehensive assessment of how a collection of EH&S activities contributes value to the company. For example, aspects of value associated with process safety management may range from the traditional ones of compliance and risk reduction to competitive concerns such as business continuity and resource productivity. This framework can be used to:

- designate specific aspects of value added and determine how they relate to key business objectives, such as growth or market penetration,
- characterize, classify and evaluate their activities and business processes in terms of both their actual and potential value contributions,
- implement performance indicators and targets that encourage value creation, and can be used for continuous improvement tracking,
- establish priorities and allocate resources among multiple activities in a way that maximizes the value contribution of the entire group, and
- support day-to-day decisions about whether and how to address specific or investment opportunities or risks.

As an example, one major U.S.-based chemical manufacturing and distribution company used VM Advantage™ to assess the value contribution of its entire central EH&S department. Apart from facility compliance improvement, the aspects of value associated with their process safety management programs included (in order of importance):

- Liability reduction (corresponding to risk reduction)
- Business continuity enhancement (reduced downtime)
- Process performance enhancement (improved yield)
- Resource productivity enhancement (improved efficiency)
- Operating cost reduction (reduced overhead)
- Capital expense reduction (inherently safe design)

Thus, process safety engineers are able to identify linkages between their efforts and the strategic goals of the enterprise, which include increasing the return on investment on existing facilities, acquiring new facilities as part of their growth strategy, and designing new facilities that are cost-competitive. Moreover, the department managers are able to allocate resources in a way that maximizes value contribution. For example, it was found that process safety risk management was being pursued independently from other areas of EH&S risk management, such as product and occupational risk, leading to redundancy and conflicting terminology. The department is now working on unifying the various risk management initiatives into a comprehensive management system framework, which should both improve the productivity of EH&S staff and reduce the system complexity and training requirements for operating personnel.

As part of their annual strategic planning process, the EH&S department is using the results of the VM Advantage™ effort to develop quantitative performance indicators that reflect important value contributions. In fact, the key to creating value is finding the right indicators, or metrics, that provide the basis for employee performance incentives in specific dimensions of continuous improvement. The new indicators being developed by this company are pro-active, rather than retrospective, and reflect an understanding of economic value added rather than focusing purely on compliance results. Examples of this approach are provided below

Developing Pro-active Value-Based Indicators

A fundamental principle in the development of performance indicators is to strike an appropriate balance between process measurement and outcome measurement.

As illustrated in Figure 1, there is a general sequence of factors that influence the frequency and severity of adverse occurrences, including both acute and chronic risks. Performance metrics may be applied at each stage of this sequence:

- *Hazard* metrics address the inherent propensity for adverse occurrences, e.g., average lb. of flammable materials stored on-site
- *Prevention* metrics address the management systems designed to prevent such occurrences, e.g., frequency of fire and loss prevention training sessions

Maximizing the Economic Value of Process Safety Management 401

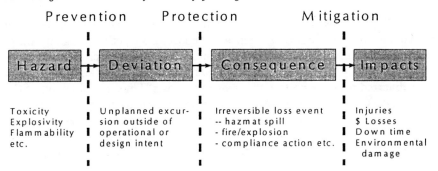

FIGURE 1. Hazard Impact Model

- *Deviation* metrics address the likelihood or frequency of unplanned excursions, e.g., number of exceptions noted in flammable material handling procedures
- *Protection* metrics address the operational systems designed to prevent adverse consequences, once a deviation has occurred, e.g., number of vapor sensors installed
- *Consequence* metrics address the actual occurrence of adverse events due to deviations, e.g., number of fires reported annually
- *Mitigation* metrics address the operational systems designed to minimize the magnitude of impacts, e.g., percent of facility workspaces with fireproof barriers
- *Impact* metrics address the ultimate severity or magnitude of the consequences, e.g., total annual dollar losses and liabilities due to fires

Finally, there are a set of influencing factors each stage of the above sequence that may tend to increase or decrease the degree of risk. These range from workforce behavioral characteristics that can influence the likelihood of a deviation to site and environmental characteristics that can influence the magnitude of the impacts. To the extent that these factors are controllable, they might also be selected for performance measurement; however, these are typically intrinsic conditions.

The above model can be used to support the selection of an appropriate set of process safety performance metrics, based on a strategy that targets specific activities in the cause-effect chain. It makes sense to focus process safety management resources on the areas of greatest opportunity or vulnerability—for example, if most fires are caused by equipment failures then training programs may have little beneficial impact. Ideally the set of metrics selected should include both outcome metrics and activity metrics that are believed to correlate with the outcomes. By tracking these metrics over time, managers can determine whether investments in enhancing the selected activities have indeed achieved the desired changes in outcomes.

Finally, the chosen set of metrics should be linked to economic value, using the types of financial indicators mentioned above. Each activity (e.g., training) has corresponding operating and/or capital costs, and each outcome has corresponding costs that include direct losses or liabilities, business interruption costs, diverted resource costs, and opportunity costs. Thus, the objective of the process safety management program is to deploy intervention activities in such a way that the net benefits in terms of economic consequences are maximized. To the exte tthat these economic consequences can be tracked, a process safety management business case can readily be developed to support the rationale for future investments.

In-Depth Case Study —Ashland Chemical

The experience of one Battelle client that utilized the above techniques provides a useful illustration. A gas processing facility had been monitoring deviations in flammable material handling procedures, and decided to perform a quantitative study of several alternative methods to reduce the associated risks. Battelle performed an analysis of the risk of fire and explosion hazards at the facility using a base case plant risk profile. The objective of this analysis was to determine the benefit of recommended improvements to the facility. The benefit to the facility was realized through a decrease in operating risk, which translates into a decrease in total liability. The proposed design changes were:

- Improve emergency shutdown system and install a facility blowdown system
- Reduce the process inventory of the product surge tank
- Replace the firewater system for the facility, and
- Provide and upgrade fireproofing in selected portions of the facility.

After completion of the base case profile, we analyzed the four potential design changes to identify how much they reduce the facility s operational risk. These design changes were evaluated based on their effectiveness to reduce the severity of consequences of fire and explosion hazard and/or reduce the frequency of an accident. The impacts of these design changes were compared to the base case and were translated into estimated monetary loss rate savings ($/year) or the annual benefit.

We used a risk ranking process to determine the impact of the proposed design changes. The risk ranking process calculates order-of-magnitude accident risks for each of the various process units. A process unit was considered to be a distinct major unit operation that performs a unique function. The facility was divided into 27 process units where data was collected on materials of construction, process operating conditions, loss prevention practices, process unit locations, and frequency of accidents based on historical data and/or engineering judgment.

The potential dollar loss rates considered in this analysis included property and equipment damage, business interruption due to lost production, and onsite population casualties. Although diverse types of risks may be calculated for each process unit, the approach was to assess the frequency of major accidental loss events in terms of incidents per year, and the severity of consequence of the loss events in terns of dollar loss or liability per incident. These assessments were then multiplied together to get a measure of risk in terms of dollar loss per year. If more than one loss mechanism existed for a given process unit, the loss rates from the various mechanisms were all added together for a combined total loss rate for each process unit.

Figure 2 shows the estimated annual loss rates for the base case and each proposed design change. Of the designs analyzed, the most effective at reducing operational risk is the proposed ESD and blowdown systems and the least effective is the fireproof coating.

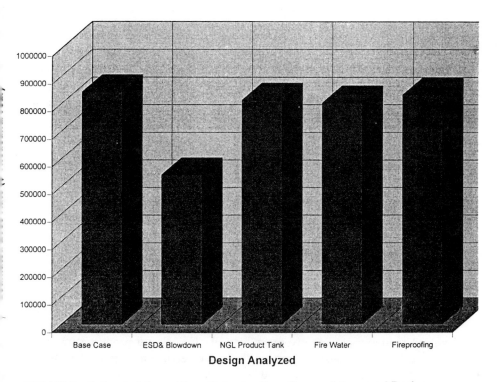

FIGURE 2. Estimated Annual Loss Rates for Base Case and Proposed Design Changes.

Base Case

Figure 3 shows the base case results for the current design of the facility. The results indicate that approximately 15 percent of the total estimated loss rate in attributed to the regeneration heater. This unit dominates the risk profile because the estimated accident frequency is three times higher than the other process unit accident frequencies. The regeneration heater accident frequency was estimated based on historical accidents associated with the operation of the unit.

Approximately 30% of the total risk is from the following group of process units located at the north end of the plant:

- Product surge tank, contractor and solvent separator
- Low Pressure cryo unit
- High pressure cryo unit
- Mol sieve beds

These procss units have close proximity to one another and they have significant impact on adjacent equipment items for a given process unit effect radius.

ESD and Blowdown System

The benefit from implementing the ESD and blowdown systems was estimated to be $256,000 per year in total loss rate liability savings. Of the 27 process units studied, 12 showed a consequential risk reduction as a result of the design change and 11 of the units showed an accident frequency risk reduction. The one process unit where the accident frequency did not decrease but showed a consequential risk reduction was in the product surge area. The consequential liability chang3ed as a result of being abe to isolate/block-in the product surge tank and reduce inventory of flammable/explosive chemicals from adjacent units in the event of an accident. The accident frequency in this area will not change since the product surge tank is isolated and will maintain its process inventory. If an accident event sequence involves process material loss of containment at the product surge tank, the inlet and outlet to the system can be closed; however, the 94,000 pounds of NGL will be released to the process area.

Process Inventory Reduction

The process inventory reduction of NGL from 94,000 pounds to 23,500 pounds involved reducing process inventory by replacing the old product surge tank with a new, smaller one. Compared to the base case, the estimated annual liability reduction is approximately $9,000 per year.

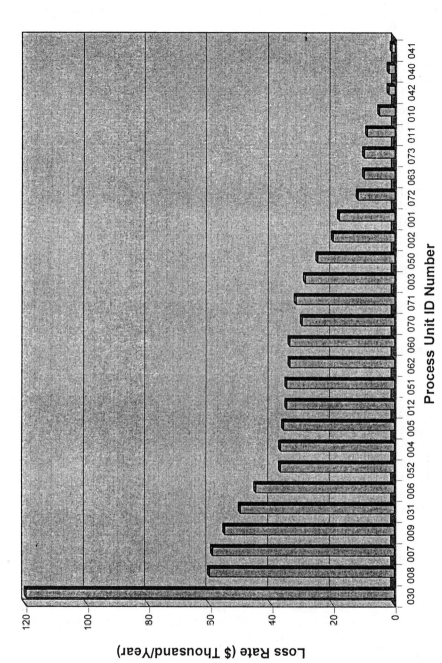

FIGURE 3. Base Case Annual Loss Rate for Each Process Unit

New Firewater System

The fire water design change involved the addition of a firewater system capable of delivering the maximum calculated demand of r at period of four hours to all process units, as well as, the implementation of a firewater deluge system for three process units. The annual benefit of implementing the firewater system design change is approximately $13,000 per year.

Fire Proofing Upgrade

Although several process units will receive a fireproof coating for the proposed design change, the study gave credit only for coatings applied to a height of greater than 15 feet. Only the high pressure cryo unit was proposed to receive coating greater than 15 feet. As part of the design change, metal sheets would be used to shield instruments and electrical cable trays in the process area. These metal sheets would also be coated with a fireproof coating. Six process units received credit for protective fire shielding. The annual benefit of this design change is approximately $3,000 per year.

Conclusions

The above case study illustrates how process safety management can combine traditional risk analysis mthods with economic benefit estimation techniques to determine the economic value associated with facility design improvements. It also demonstrates how pro-active performance indicators can be used to recognize process safety vulnerabilities and to initiate interventions prior to the occurrence of an incident. Through recommendation of design improvements that create measurable economic value, the process safety managers are able move beyond a compliance role and contribute explicitly to the business goals of the firm.

References

Chemical Manufacturers Association, Responsible Care® Progress Report: The Year in Review, 1995-96, Washington, DC, 1997, p. 8.

Fiksel, J., Competitive Advantage Through Environmental Excellence, *Journal of Corporate Environmental Strategy*, Summer 1997.

Integrating Quality Management Principles into the Risk Management Process

Kevin Mitchell and Jatin N. Shah
Four Elements, Inc., 355 East Campus View Blvd., Suite 250, Columbus, Ohio 43235

ABSTRACT

Effective risk management optimizes the allocation of limited resources to achieve maximum risk reduction. Presently financial risk assessment does provide a predictive tool to identify, predict, and control potential accidental release sources with the associated impact on people, property, and the environment. However, it is difficult to monitor and measure the actual performance of a facility against pre-defined risk targets since the catastrophic accidents occur very rarely. A means to identify and establish Key Performance Indicators (KPIs) is essential for effective monitoring of the risk management system.

When viewing the risk management process at a macroscopic level, it usually takes the form of an iterative procedure where low risk processes are separated from high risk processes with successive levels of refinement. Follow-up inevitably focuses on high risk factors and reducing these to acceptable levels. Key safeguards and their relation to the overall objective are identified and enumerated in the quantitative risk assessment process. Often failures that place demands on safety critical systems initially manifest as quality deficiencies, or excursions beyond the permitted environmental operating envelope. These operational targets form the basis of establishing KPIs to measure operational performance, thus ensuring that the risk management objectives are met.

An approach that incorporates features of a quality program within the risk management framework has been developed to identify KPIs and monitor the effectiveness of the risk management process. A case study is included to demonstrate the effectiveness of this approach.

Background

Many global companies within the process industries utilize risk management principles to control major liabilities in a consistent manner. Effective risk management optimizes the allocation of limited resources to target business risks and achieve maximum return on EHS investment. The goal of these systems is to control and minimize business risks by avoiding the occurrence of incidents that can adversely effect the corporate bottom line. Some of the most significant losses are incurred when rare, high impact accidents take place. Examples of these events include:

- Major toxic chemical release resulting in injury to workers and public;
- Plant explosion resulting in extended production outage; and,
- Major oil or chemical spill in sensitive environment setting leading to significant cleanup and natural resources restoration costs.

The present financial risk assessment models provide a predictive tool to identify and control the potential for an accidental release with the associated impact on people, property, and the environment. An integrated approach developed specifically for the process industries has been documented by Mudan et al. [1]. However, it can be problematic to monitor and measure the actual performance of a facility's risk management system against predefined risk targets since catastrophic events occur rarely, even within an entire industry. Further, the events leading to catastrophic losses are the ones the companies are trying to avoid with engineering and management controls. Therefore, how can a company demonstrate progress toward its own risk management objectives? There is an imperative need to establish Key Performance Indicators (KPIs) to measure, monitor and control the risk management system.

Many tangible benefits can be realized by implementing a monitoring program for the risk management process. These include:

- Ensuring that the risk management objectives are met;
- Demonstrating measurable progress toward achieving the objectives;
- Targeting limited resources for maximum benefit;
- Controlling expenditures while achieving real risk reduction;
- Reducing waste;
- Improving both customer and employee satisfaction; and,
- Increasing competitiveness, and improving financial performance.

Some of these objectives may be achieved without a monitoring program. In such cases, the progress is usually more difficult to identify and quantify. With a monitoring system in place, facility management has the feedback it needs to ensure that performance is targeted and is moving in the right direction, and a steady progress is being made toward reducing potential liabilities.

The challenge is significant, and it is further complicated by the fact that comprehensive risk management systems attempt to exert simultaneous control over many types of liabilities. Progress needs to be monitored on many diverse fronts, including:

- Public Safety,
- Worker Safety,
- Environmental Damage,
- Property Loss,
- Production Interruption/loss of market share,
- Public perception, etc.

FIGURE 1. Comprehensive Risk Management

KPIs have been developed to monitor and track performance of other management systems, including Environmental, Health, and Safety (EHS) systems. More companies are formulating a single risk management function that seeks to comprehensively control exposure from such a broad range of potential losses. This function is displayed in *Figure 1*.

Quality Principles

Integrating quality principles into management systems has been a topic that has received considerable attention. The Center for Chemical Process Safety (CCPS) has developed a "Guidelines Series" text concerning integrating quality management with EHS systems [2]. Similar principles apply when integrating quality principles into the risk management process.

Quality management depends upon industry type, and even within the Chemical Process Industry (CPI) varies from company to company. However, there are some fundamental principles that all quality management systems adhere to. These include:

- Management Commitment to Quality
- Quality Policy
- Quality Targets
- Quality Control/Quality Assurance
- Progress Monitoring, etc.

Quality management goals are also variable from company to company, but they are typically **quantified** to facilitate measurement and control. Quality itself can be measured and controlled by monitoring *deficiencies* or *defects* per *production unit* or per *opportunity*. For example, a product may have the quality target of no more than 3 defects per million units produced.

In general quality targets are applicable to almost any management system. For example, a quality goal for an EHS system may be no more than 1 lost time injury accident per 100 employees per year. In this example, the goal is not to deliver the customer a quality product; the beneficiary is the employee by virtue of a safer, higher-quality work environment and the facility with increased productivity.

The risk management process has many parallel functions, including:

- Management Commitment
- Corporate Risk Management Policy
- Risk Management Targets, etc.

Again, risk management goals are variable from company to company. Here the focus is not on the occurrence of product defects, but management system failures that could result in a major loss. The risk management objectives are more frequently taking the form of quantifiable targets that can be measured with statistical data. For example, investment in the potential economic benefits of a new production process may be authorized only if a facility can demonstrate through risk assessment that:

- average liability of $3 million per year or less;
- maximum liability of $500 million, and
- chance of exceeding a $100 million loss no more than 10^{-3} per year (i.e., 1 chance in 1000 per year).

Figure 2 compares the typical components of the two management philosophies, Quality Management and Risk Management. It shows that risk management principles closely parallel the quality system philosophy.

This comparison shows that risk management generally fits well within the quality management framework. In fact, quality management principles and techniques can be extended to address all required functions of the risk management system. Most companies will find that there is an opportunity to increase the value of the risk management process by ensuring that it fully complies with the company's commitment to overall quality management, and therefore engender world-class customer and employee satisfaction.

In reality, existing risk management usually lacks at least one critical function, monitoring and measuring system performance. It is important to distinguish the information derived from a *risk assessment* from information obtained from a *risk monitoring* program. Risk assessment does provide a predictive, forward-focusing tool to establish whether measurable risk targets can be met.

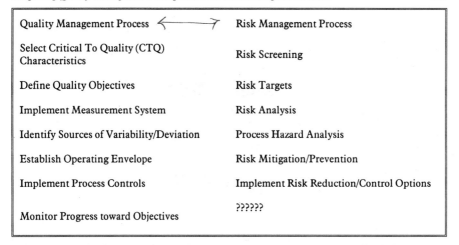

FIGURE 2. Comparison of Some Quality Management and Risk Management Program Elements

However, an equally important question is, "Are the risk targets being met on an ongoing basis?" Risk monitoring program provides essential feedback, and it closes the loop in the overall risk control process as shown in *Figure 3*.

The challenge to developing an effective risk monitoring activity is multifaceted. Each company will have individual needs that must be addressed. Some of the issues that must be commonly addressed include:

- How to integrate Key Performance Indicators or KPIs into the risk management process; and
- How to identify KPIs and establish performance targets consistent with the risk management objectives.

Integrating KPIs Into The Risk Management Process

One of the biggest challenges is determining how to monitor the chances of having a major accident, by definition a very low probability event. In searching for effective KPIs, the goal is to identify process parameters that can be tracked on a periodic basis (e.g., quarterly, annually, etc.) and that have some bearing on the likelihood of a catastrophic accident. Monitoring the number of catastrophic accidents is too late, and therefore unacceptable. Precursors to major accidents, however, can and should be monitored so that effective control of process risks can be maintained. In fact, most precursors to major accidents are failures that place demands on safety critical systems. These may initially manifest themselves as quality deficiencies, or excursions beyond the permitted operating

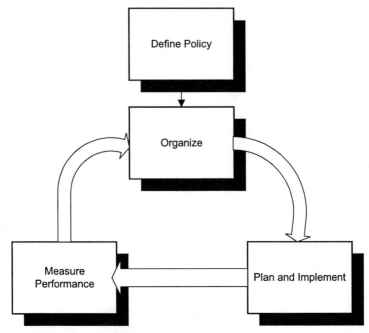

FIGURE 3. Risk Control Process

envelope. These operational targets form the basis of establishing KPIs to measure performance, thus ensuring that risk management objectives are met.

KPIs should necessarily be significant deviations from normal operating conditions, and precursors to near-miss accidents. Near-misses are, in turn, precursors to catastrophes. In between each layer are engineering and administrative safeguards preventing the incident from becoming elevated to the next level. As shown in Figure 4, the actual loss experience may only represent a small fraction of potential unrealized liabilities.

Therefore, while it is too late to monitor the number of major accidents, we can monitor the pre-cursors to them. These precursors form the accident pyramid or triangle as illustrated in *Figure 5*. KPIs are more-frequent deviations that can be observed, monitored, and controlled. They form the bottom layer of the accident pyramid. Other more serious events compose the intermediate layers, and can be monitored over a long period of time at a facility or within a company. At the top of the pyramid is the most serious accident category, an event that is generally the principal focus of the risk management system.

Statistical data within an individual company or across an industry should be used to quantify the relationship between accident precursors in the different monitoring categories. Other inductive techniques like Fault Tree Analysis (FTA) can also be used for specific applications. At the bottom level, KPIs

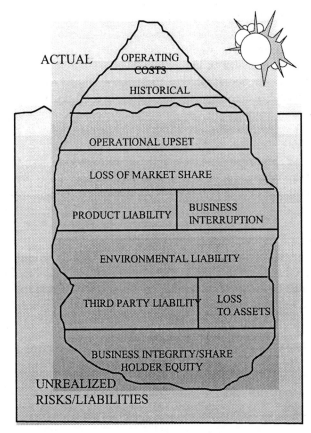

FIGURE 4. Realized and Unrealized Losses

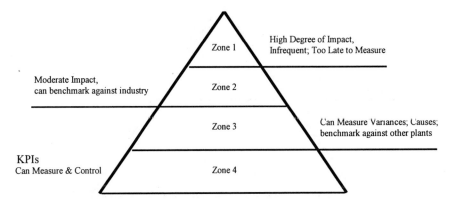

FIGURE 5. Generalized Accident Severity Categories

should be process specific so that the actual process parameters can be integrated into the risk management system, and subsequently monitored and controlled. An example is the relationship between accidents resulting in various degrees of environmental impact, as shown in Figure 6.

For a typical US chemical facility, this figure shows the relationship between the occurrence of non-reportable releases, federal reportable releases, and releases resulting in significant fines, cleanup, or restoration activities. The objective of the risk management process is primarily avoiding the more significant liability of "Zone 1" incidents. Progress can be monitored using this quantified relationship and observable data collected over a long period of time in Zone 2 or Zone 3. However, ongoing monitoring of KPIs in Zone 4 represents the most efficient way to measure the progress toward avoiding Zone 1 incidents.

Developing Key Performance Indicators

To implement effective monitoring and control over process risks, Key Performance Indicators must be identified and tracked. There are numerous parameters that could be tracked, but which ones should be tracked? Typical examples include:

- Excursions in critical operating conditions (temperature, pressure, composition),
- Demands placed on critical control or shutdown systems,
- Mechanical integrity data collected from on-line or field analysis, etc.

The process fundamentally involves two steps, (1) identify KPIs, (2) quantify the KPI targets.

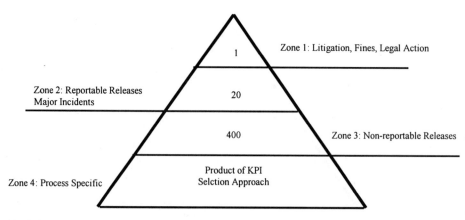

FIGURE 6. Sample Risk Monitoring for Environmental Impact

When developing KPIs, the objective is to identify significant deviations from normal operation and possible consequences that lead to pre-cursor events like reportable spills, worker loss-time accidents, or minor property losses. There is an obvious similarity here to Process Hazard Analyses (PHA), were plausible deviations are postulated and evaluated by an expert team.

PHA techniques have been used in the process industries for decades. In the US, the most PHAs conducted over the past 5 year period have been for the purpose of compliance with OSHA's Process Safety Management (PSM) standard 1910.119. This regulation focuses exclusively on worker safety. It requires that PHAs be conducted by a qualified team and revalidated on a periodic basis. The first round of validation has begun this year. Many companies are looking for ways to get more value out of the investment in the PHA team's time during the revalidation process. In fact, retooling the PHA process is also a regulatory requirement for many facilities with the advent of EPA's Risk Management Program (RMP) rule, which addresses significant off-site environment and public safety concerns, not just worker safety.

When modifying existing PHA techniques (e.g., Hazop, What-If, etc.) to achieve goals of monitoring parameter selection, a team of experts must consider what the goals of the company's risk management program and how to monitor progress. Additional guidewords, checklists, or what-if questions can be added to the process to identify critical pre-cursors to major incidents. Now, PHA becomes a value-added process when it is used as a technique to identify a broad range of business risks, including: environmental, public safety, product quality/liability, and loss prevention issues.

Case Study

A monitoring approach was developed within the overall framework of a corporate risk management system for a major US plastics manufacturer. Within this particular organization, risks need to be controlled for a broad variety of manufacturing processes, many including highly toxic and flammable substances. The company has established fully integrated risk management goals for all major liabilities. The company has also initiated an ambitious effort in the area of quality management that will effect almost every aspect of the business. The quality goals are sufficiently general so that they can be applied to almost every management activity. The objective is to be a world-class company in the areas of customer and employee satisfaction.

Initially, a pilot study was commissioned to evaluate the feasibility of integrating the risk management process into the company's quality system. A team of risk management and quality management experts was convened. This group studied the two management systems and identified areas where similarities and deficiencies existed. A comparison similar to Figure 2 was produced. Overall, there was a good fit. Significant opportunities existed to integrate the risk management system the quality system framework. One of the biggest challenges was implementing the monitoring activity.

The process of establishing monitoring parameters sought to link the relationship between a major accident that may result in significant impact to the corporate well-being to more frequent, less consequential incidents such as near-misses, and major process deviations. Company and industry data was evaluated to establish several monitoring categories in the areas of:

- Occupational Safety,
- Process Safety,
- Environmental Risk, and
- Financial Exposure (Property Loss and Business Interruption).

Relationships were quantified in each of these areas similar to that shown in Figure 6. KPIs were identified by applying a Process Hazards Analysis (PHA) technique. The What-If/Checklist approach was modified and extended to include questions that deal with aspects of the risk management process including environmental and loss prevention concerns. This resulted in a more value-added PHA process as compared to the initial, compliance oriented PHA activity that has been in place since 1992. Some of the sample What-If questions for identifying KPIs are shown in Figure 7.

Category	What-If Questions
Environment	Are carcinogenic materials used in the process?
	Can materials used in the process impact wastewater adversely?
	Can releases get to waste water system? Do we have any protection?
	What can we monitor? What do we monitor? What should we monitor?
Business Interruption	What can cause property damage and BI? What are we doing to prevent this from occurring?
	What can we do to minimize exposure?
	Can we replace unit production? If so how, where, how much, and at what cost?
	What can we monitor? What do we monitor? What should we monitor?
Quality	What are the requirements by the customer? Do they vary by customer?
	What can happen that would prevent us from meeting these requirement? Sources of contamination?
	What can we monitor? What do we monitor? What should we monitor?

FIGURE 7. Example PHA Issues for Implementing a Monitoring Program

Once KPIs have been identified, the relationship between the risk management objective and the observable KPI tolerances must be established. For example, consider a high pressure, high temperature process involving an oxidation reaction in a process furnace. Similar processes throughout industry have had major fires, resulting in extensive property loss and production interruption. A instrumented shutdown system is in place to prevent excursions beyond the safe operating envelope that may result in a near-miss (more-frequently), minor fire (credible), or major explosion and fire (less-frequently). This system is designed to a Safety Integrity Level (SIL) 2, meaning the probability of failure of the shutdown system is 0.01 to 0.001 per demand.

If the risk management objective is to control the likelihood of an incident exceeding $100 million in property loss to a frequency of 10^{-3} per year or less, how many demands can we tolerate for this safety critical system over the monitoring period of one year? Fault tree analysis shows that, given the design of the system, no more than 3 demands per year can be placed on the instrumented shutdown system to meet the risk management objective. Actual monitoring indicates that about 10 demands were placed on the system. This operating experience indicates that the objective may not be attained with the current system. At this point alternatives can be considered such as:

- Improving basic process control to reduce demands to less than 3 per year, or
- Improving shutdown system reliability to meet SIL 3 or 4 requirements.

A simple cost comparison between the two options can justify the preferred alternative.

After applying the KPI selection technique during the pilot study, the company was able to identify a list of KPIs. The procedure of setting limits on the performance parameters that are consistent with the risk management objective is ongoing. Performance of KPIs will be monitored over a period and losses will be compared to expected performance. As the process moves forward, the program will be re-evaluated to establish the value of performance monitoring within the risk management process. Already performance monitoring has yielded a revitalized PHA process that will produce additional benefits during the period in which PHAs will be revalidated.

Summary

Quality management principles can be successfully integrated into the risk management process. For organizations with a strong quality initiative, it is desirable to demonstrate that quality principles have been integrated into key management process. The procedure does involve close interaction with company experts in the areas of *quality* and *risk* management.

Important benefits can be achieved including demonstrating that measurable progress is being made toward satisfying the risk management objectives through the tracking the Key Performance Indicators. This provides critical feedback to close the control loop in the risk management process. Along with risk assessment, performance monitoring provides a strong tool to measure and control corporate liabilities.

References

[1] Mudan, K.S., Shah, J.N., and Myers, P.M., "Financial Risk Assessment A Uniform Approach to Manage Liabilities," 29th Loss Prevention Symposium, August, 1995, Boston, MA.
[2] Center for Chemical Process Safety (CCPS), *Guidelines for Integrating Process Safety Management, Environment, Safety, Health, and Quality*, 1996, American Institute of Chemical Engineers.

Case Study
- Reactor, FE hazard, SIS to SID system, SIL2

Team identified TI as an KPI
goals: $<10^{-3}$ freq of $100M loss
PTA should < 3 demands/y tolerated
Current operation is 10 demands/y

WORKSHOP A
Regulations (EPA and OSHA)

Chair **Craig Matthiessen**
U. S. Environmental Protection Agency

Importance of Scenario Selection in Preparing Hazard Assessments for EPA's RMP Rule

J. Ivor John
Arthur D. Little, Inc., Santa Barbara, California

Henry Ozog
Arthur D. Little, Inc., Cambridge, Massachusetts

Introduction

In a previous paper, it was demonstrated that the calculated distance to the endpoint for a toxic release scenario is highly dependent upon the scenario selected (John and Ozog, 1997). This was found to be a more important variable than the parameters that may be varied as part of the offsite consequence analysis (OCA). Exhibit 1 is an example showing how the scenario selection (hole size) dominates the hazard distance. Depending on the scenario selected, the hazard distance for ammonia refrigeration system varied from as little as 50 meters to as much as 1,500 meters.

For toxic releases, the RMP regulation states that the alternative scenario should be more likely to occur than the worst-case release scenario, and it should reach an endpoint offsite, unless no such scenario exists. Release scenarios that should be considered include:

- Transfer hose releases due to splits or sudden hose uncoupling;
- Process piping releases from failures at flanges, joints, welds, valves and valve seals, and drains or bleeds;
- Process vessel or pump releases due to cracks, seal failure, or drain, bleed, or plug failure;
- Vessel overfilling and spill, or overpressurization and venting through relief valves or rupture disks; and
- Shipping container mishandling and breakage or puncturing leading to a spill.

The EPA's OCA Guidance document reiterates that the selection process should consider those release scenarios that result in concentrations above endpoint at fenceline, and those with the potential to reach the public are of greatest

EXHIBIT 1. Results of a Sensitivity Analysis for Ammonia Release Calculations

	Peak Release Rate (kg/s) [1]	Ratio to Base Case [2]	Sensitivity Factor [3]
Hole size increased from 1/2 inch to 3 inches	35.6	36.0	36.0
Hole size decreased from 1/2 inch to 1/8 inch	0.1	0.1	15.7
Vapor release	0.2	0.2	6.2
System temperature decreased from 298K to 250K	0.4	0.4	2.6
Length of pipe added (10m)	0.4	0.4	2.5
Discharge coefficient increased from 0.62 to 0.98	1.6	1.6	1.6
Discharge coefficient decreased from 0.62 to 0.50	0.8	0.8	1.2
System temperature increased from 298K to 310K	1.1	1.1	1.1

1. Release calculations performed using SuperChems™
2. Base case is a two-phase release from 1/2" hole, T=298K, discharge coefficient = 0.62. Base case peak release rate was 1.0 kg/s
3. Sensitivity Factor is the "ratio to the base case" for ratios 1.0, and the inverse of the "ratio to base case" for ratios .0.

concern. Active as well as passive mitigation systems may be considered. The following suggestions are provided for selecting an alternative scenario for toxic substances:

- Use worst-case release and apply active mitigation to limit the quantity released and the duration of the release;
- Use information from the PHA to select a scenario;
- Review accident history and choose an actual event as the basis for the scenario; and
- Review operations and identify possible events and failures.

While the EPA's guidance provides general direction, the final decision on *scenario selection* is left up to the facility. Exhibit 2 provides a summary of advantages and disadvantages of several methods for selecting the alternative release scenario (Thompson Publishing Group, 1996). Most of these methods can be applied after a list of potential release scenarios has been developed for the covered process. This paper demonstrates how the risk-based selection method (Exhibit 2, Option D) can be applied to develop defensible alternative scenarios using the approach shown in Exhibit 3.

Importance of Scenario Selection in Preparing Hazard Assessments for EPA's RMP Rule　423

EXHIBIT 2. Possible Approaches for Selecting Alternative Release Scenarios

	Approach	Advantages	Disadvantages
A	Select alternative scenario(s) based on the potential application for emergency response preparedness and planning.	• Useful to first responders and agencies in the event of an actual release. • Good way to convey hazards to the public.	• Hard to select the most appropriate scenarios without additional criteria. • Results are only applicable for one set of meteorological conditions.
B	Group scenarios by level of impact and model a range of scenarios.	• Very sound approach for examining a spectrum of possible outcomes. • Analysis generates results for a range of possible releases.	• Resource intensive—requires a listing of potential scenarios and consequence analysis for several groups.
C	Select scenarios with the highest likelihood of reaching offsite.	• Approach considers likelihood and severity.	• Other less likely scenarios could have much worse offsite impacts; highest likelihood could be very small release. • Not the best approach for emergency planning.
D	Select alternative scenarios based on risk (likelihood/severity).	• Beneficial for designing risk mitigation. • Excellent approach for risk management. • Sound basis for supporting emergency planning.	• Difficult concept to communicate with public.
E	Use worst-case release scenarios with active mitigation.	• Excellent way to demonstrate how engineered systems (existing or planned) are used to control a release. • Easy to relate alternative release scenario to worst-case scenario.	• Active mitigation that reduces the likelihood of a release will not show any benefit over worst-case. • Other more appropriate alternative scenarios may be overlooked.

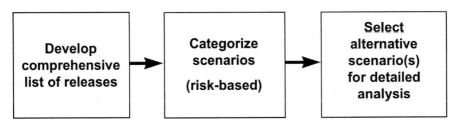

EXHIBIT 3. Key Steps in the Alternative Scenario Selection Process

Three case studies are used to demonstrate how the method can be applied. The case studies are based on studies previously conducted by Arthur D. Little. As such, the modeling parameters and the endpoints used were not in exact agreement with the RMP rule requirements. However, the available data are useful for demonstrating the risk-based selection method.

Developing a Comprehensive List of Release Scenarios

Developing a comprehensive list of potential release scenarios is a good way to start the selection process. Having this list available provides invaluable information for making decisions about the potential hazards and the candidate alternative scenarios. Also, documenting this as a starting point is also a good way to demonstrate to the Implementing Agency that a wide range of potential scenarios was considered in making the selection. The main disadvantage is that it may be difficult to compile a comprehensive list, and the quality of the information generated will depend on the assumptions made in developing the scenarios.

Several approaches can be used to compile the list of scenarios. For the case studies used in this paper, the approach varied depending on the information available to us. For example, in two cases, the process hazards analyses (PHAs) were available to us, so we were very familiar with the processes and the potential hazards. In the third case, the information was developed from a review of the process safety information, mainly the Process Flow Diagrams (PFDs) and the Piping and Instrumentation Diagrams (P&IDs).

Sources of information that provide useful input are presented in Exhibit 4. However, it should be kept in mind that certain sources may not provide the wide range of scenarios needed in the initial list. For example, a review of the

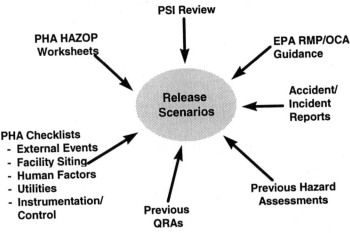

EXHIBIT 4. Resources for Identifying Potential Release Scenarios

past accidents and incidents may generate a representative range of smaller releases that have occurred in recent years, but it may not include more significant credible releases. Also, the range of scenarios included in a PHA might also be limited, given that many PHAs do not document potential hazards if the safeguards and controls are considered to be adequate.

From experience, we have found that a review of the process safety information (PSI) (and the PHAs, if available) by a risk analyst or a knowledgeable engineer can usually result in a representative range of scenarios in a relatively short period. The results are more reliable if they are reviewed by a PHA team, or if the scenarios are identified by the PHA team. A final check can be made to ensure that the various considerations highlighted by EPA in the RMP regulation and the OCA Guidance are addressed in the final list of scenarios.

Risk Ranking

Risk ranking is a highly defensible approach to scenario selection. Using risk as the differentiating criterion gets away from the argument of whether to select a "more likely" scenario or one with "more severe consequences." The main problems with the risk-based approach are: (1) it can be time-consuming to generate reliable data on release probabilities; and (2) it may be difficult to quantify the consequences adequately, depending on the assumptions for the scenarios that are postulated.

Despite the above limitations, it is usually possible to develop qualitative estimates of risk in a way that accounts for the relative differences in the release scenarios. Quantitative estimates are less important, given that the objective is to rank the scenarios on a relative basis. A qualitative review of potential hazards is often conducted by a PHA team as a way of ranking the recommendations and action items that are generated by the PHA. With some experience of the process and risk concepts, a qualitative assessment can be generated.

For the case studies presented in this paper, the following criteria were used to categorize the scenarios:

Frequency
    L Low likelihood of occurrence (unlikely to happen during the life of the process)
    M Moderate likelihood of occurrence (may happen during the life of the process)
    H High likelihood of occurrence (may happen on a regular basis)

Severity
    L Endpoint is reached within fenceline distance
    M Endpoint reaches offsite, but does not reach sensitive populations
    H Endpoint reaches sensitive populations offsite

Risk

A   Scenarios of highest concern
B   Scenarios of considerable concern
C   Scenarios of moderate concern
D   Situations of lowest concern

The risk categories were adapted from the Risk Management and Prevention Program Guidance document issued by the California Office of Emergency Services (OES, 1989). Again, the main objective of developing these criteria is to have a tool for differentiating between the scenarios, so the specific criteria are less important than the relative differences. Exhibit 5 shows how the frequency and severity data combine to form risk categories.

Scenario Selection

Once the release scenarios have been categorized on the basis of risk, a further round of screening may be needed to select the alternative scenario(s) for detailed analysis. For example, there may be several scenarios in the high risk category. In this case, a logical approach would be to select the scenario(s) with the highest severity rating. Again, if there are more scenarios than needed, then the ones with the highest frequencies could be selected. Finally, if there are still too many scenarios, then it may be desirable to use a consequence model to further screen the scenarios. This final filtering process is illustrated in the case studies described below.

EXHIBIT 5   Risk Matrix Used for Ranking Scenarios

FREQUENCY			
	D	A	A
	D	B	A
	D	D	C

SEVERITY

Frequency/Severity
H - High
M - Medium
L - Low

Risk
A - Serious concern
B - Considerable concern
C - May require planning
D - Minor concern

Case Studies

Toxic Gas (Hydrogen Sulfide)

A refinery sulfur unit is used to remove sulfur compounds contained in process streams. The unit consists of two amine reactivation units, two waste water strippers, two 100 ton per day sulfur plants (Claus Plants) and two tail gas recovery units (Beavon Plants). Usually, one sulfur plant and one tail gas plant are operated at a time, but the spare plants can be operated in parallel to process up to 200 tons per day.

The amine reactivation facilities are used to steam strip absorbed hydrogen sulfide (H_2S) from rich MEA streams. The lean MEA is recycled back to the H_2S absorbers. The MEA-acid gas produced is processed in the Claus and Beavon sulfur recovery plants. The waste water stripping facilities remove H_2S from refinery waste water. In the waste water stripper, the overhead is condensed and separated into oil, water and vapor.

The Claus process converts H_2S and ammonia removed from process streams into liquid elemental sulfur and nitrogen. The ammonia-acid gas from the waste water stripping facilities and one-third of the MEA-acid gas from the amine reactivation facilities are burned in an external combustion chamber to form sulfur dioxide and nitrogen. The sulfur dioxide is then burned with the H_2S in the remaining two-thirds of the MEA-acid gas in the reaction furnace to form elemental sulfur. The tail gas from the Claus Plant is fed to the Beavon Plant.

The Beavon Plant reduces the sulfur content of the tail gas to less than 1 ppm. A reducing gas is combined with the tail gas and is contacted with a catalyst bed where hydrogenation reactions convert the sulfur compounds to H_2S. The reacted gas is processed to the Stretford absorber where the active ingredient is reduced by oxidizing H_2S to free sulfur. The desulfurized tail gas is burned and vented to the atmosphere and the Stretford solution is regenerated by oxidation. Liquid sulfur is separated from the Stretford solution in a continuous autoclave.

Potential release scenarios for this process are listed in Exhibit 6. Of the 19 scenarios, seven fell in the highest risk category (A), five of which had a high severity rating (R4, R6, R8, R10, and R14). For this case, the best way to select the final alternative scenario would be to model the five high severity scenarios. For pure substances, release calculations would be adequate to make the final selection. However, in this case, with different H_2S concentrations in the stream compositions, consequence modeling would also be needed. Modeling the bulk flow gives the best physical representation of the release. With this approach, the toxic endpoint for hydrogen sulfide is adjusted appropriately based on the concentration of hydrogen sulfide in the host gas.

With consequence modeling, it was demonstrated that the range of hazard distances for the 19 scenarios ranged from 61 meters out to 1,441 meters (from R14). Given that R14 was one of the high risk scenarios, this would be the mod-

EXHIBIT 6. Hydrogen Sulfide

Ref #	Scenario	Frequency	Severity	Risk	Comment
R1	Flange leak and vapor release from amine reactivator #1	H	L	D	30 percent H_2S
R2	Line failure and vapor release from amine reactivator #1	M	M	B	30 percent H_2S
R3	Flange leak and vapor release from amine reactivator #2	H	L	D	32 percent H_2S
R4	Line failure and vapor release from amine reactivator #2	M	H	A	32 percent H_2S
R5	Flange leak and vapor release from reactivator OH condenser #1	H	L	D	30 percent H_2S
R6	Line failure and vapor release from reactivator OH condenser #1	M	H	A	30 percent H_2S
R7	Flange leak and vapor release from reactivator OH condenser #2	H	L	D	32 percent H_2S
R8	Line failure and vapor release from reactivator OH condenser #2	M	H	A	32 percent H_2S
R9	Flange leak and vapor release from reflux accumulator #1	H	M	A	83 percent H_2S
R10	Line failure and vapor release from reflux accumulator #1	M	H	A	83 percent H_2S
R11	Flange leak and vapor release from Rich MEA surge drum	H	L	D	3 percent H_2S
R12	Line failure and vapor release from Rich MEA surge drum	L	M	D	3 percent H_2S
R13	Flange leak and vapor release from reflux accumulator #2	H	M	A	83 percent H_2S
R14	Line failure and vapor release from reflux accumulator #2	M	H	A	83 percent H_2S
R15	Hydrogen sulfide release to flare from vent source	H	L	D	Assume H_2S combusted to form SO_2
R16	Rupture of water stripper vessel	L	H	C	46 percent H_2S
R17	Gas blow from water stripper to plant sewer	H	L	D	46 percent H_2S
R18	Seal failure on reflux pump	H	L	D	6 percent H_2S
R19	Release through absorber stack	M	L	D	3 percent H_2S

eled distance for the alternative scenario. For this case study, the distances were estimated for an endpoint of 300 ppm (H_2S).

Aqueous Solution (Aqueous Ammonia

Aqueous ammonia (30 percent solution) is used for selective catalytic reduction (SCR) at a power generating plant. The storage tank is a 20,000 gallon UST and the process consists of an ammonia feed pump to a vaporizer and injector. The UST is a horizontal tank with spherical heads buried to a depth of 12 feet. The equipment located on the surface immediately above the UST is retained by a curbed area with six inch high walls.

The UST meets the requirements of applicable UST regulations. It is equipped with instrumentation to monitor its pressure, temperature, and liquid level. Storage tank instrumentation also includes a fixed level indicator to avoid overfilling. High liquid level alarms and high pressure alarms are also included. There is one pressure relief valve on the UST set at 50 psig, provided to prevent the tank from exceeding its design pressure. In addition, ammonia vapor detectors are located throughout the storage area. If ammonia is detected, local and remote alarms are activated.

The storage tank is equipped with three submersible tank transfer pumps to transfer the ammonia from the UST to the ammonia vaporizers. Vaporization of the ammonia solution is done prior to injection of ammonia into the boiler flue gas. The ammonia solution is atomized by air in a two-phase-nozzle and is vaporized with hot flue gas in two vaporizers. Static mixers downstream of the vaporizers ensure proper distribution of ammonia in the flue gas. The mixture of ammonia and flue gas flows from each vaporizer to a header for distribution of the gas to the injection grid which has nozzles at numerous locations.

A delivery truck pumps aqueous ammonia into the storage tank through a liquid fill line while extracting ammonia vapor from the tank through a vapor return line. Gauges indicating level and pressure are present on the tank to prevent overfilling. The fill operations are attended by a station operator.

Potential release scenarios for this process are listed in Exhibit 7. There were eight potential release scenarios, most of them resulting in a spill of ammonia to the diked area. Only one scenario (R8) rated as a Category B, and one was Category C; all others fell in Category D. If all scenarios had ended up in Category D, the severity rating would provide a secondary basis for selecting the alternative scenario. (In this case there would be three candidates with medium consequences.)

For aqueous solutions, the selection process is unlikely to differ greatly from that used for toxic liquids, given that the approach is based on relative differences rather than absolute data. However, the evaporation and dispersion characteristics are different. The approach to modeling the aqueous solution needs to consider the different vapor pressures of the solution components, as this will influence the emission rate profile over time.

EXHIBIT 7. Aqueous Ammonia

Ref #	Scenario	Frequency	Severity	Risk	Comments
R1	Vapor release at truck unloading	M	L	D	Vapor return line
R2	Liquid leak from hose at truck unloading	M	L	D	May be outside containment
R3	Rupture of 2-inch hose at truck unloading	L	H	C	May be outside containment
R4	Vapor release from 2-inch line (unloading area to storage tank)	L	L	D	Vapor return line
R5	Liquid release from 2-inch line (unloading area to storage tank)	L	M	D	Spill to containment
R6	Liquid release at PSV from overfilling of storage tank	L	L	D	Spill to containment
R7	Liquid release to surrounding soil from storage tank failure	L	L	D	No release to atmosphere
R8	Flange leak with liquid release from 1-inch supply line (storage tank to vaporizers)	M	M	B	Spill to containment

Using a multi-component evaporation model (SuperChems™), the emission rates for all scenarios in this case study resulted in no offsite impacts. Only with "worst-case" meteorological conditions were the hazard distances for R3, R5 and R8 close to the fenceline.

Flammable Gas (Butane)

The butane handling and storage facilities at a refinery consist of the following:
- Ten storage bullets
- Two cryogenic storage tanks
- Two mixed butanes storage tanks
- Butane storage flare system

The horizontal storage bullets function as storage for butane production and as butane supply for the blending unit. Transfers are made from these tanks to the cryogenic tanks or vice versa as the storage situation warrants. Normal transfer rates are 3,000 barrels per day. Transfers into the cryogenic tanks pass through a water cooled exchanger to cool the butane to minimize vaporization.

Conversely, transfers from the cryogenic tank, require warming the butane by utilizing a steam heated exchanger. Butane blending pumps are used to transfer the butane to the cryogenic tanks.

The cryogenic storage tanks serve as excess storage for the storage bullets. Each tank is double-walled and double-roofed with an inner tank capacity of approximately 125,000 barrels. The space between the walls and roofs contains an insulating material plus a nitrogen blanket. The annulus nitrogen pressure is be held in an operating range of 1 to 2 inches of water pressure. Heating elements are used to control the ground temperature at 40°F to prevent freezing of ground water which could cause heaving and dislocation of the tank bottom.

The inner tank pressure is held in the range of 3 to 6 inches of water. If the pressure should rise to 10 inches of water, venting to the flare header occurs. If the tank pressure drops below 3 inches of water, the refrigeration compressor shuts down, and if the tank reaches a vacuum, the atmospheric vacuum breaker will open to admit air. Pressure is maintained by circulating butane vapor through a water-cooled exchanger.

When a transfer is made into a tank, increased pressure in the tank results. The additional compressor load is handled by a large filling compressor, which also incorporates an exchanger and accumulator. The filling compressor can also be unloaded by a low tank pressure signal; and, in addition, the compressor discharge gas can be diverted directly to the tank return line, bypassing the cooling exchanger. When a transfer is made out of a tank, tank pressure is reduced causing the holding compressor to run fully unloaded. Make-up butane vapor may be required to maintain tank pressure in the operating range. This will be provided by a butane vaporizer.

The two spherical mixed butane storage tanks are served with pumps and 4 inch lines which can be used to fill or pump out the tanks. Piping is arranged so that pumps, tanks, and line service can be interchanged, as it is likely that at times both tanks may be used to store butane-butene or isobutane. The pumps are equipped with low flow shutdown switches, flow rate instruments, and pressure control valves.

A smokeless flare stack is used to handle any hydrocarbon release. The flare requires a constant flare of hydrocarbon through the stack to prevent air from being drawn into it. This hydrocarbon flow is provided by the butane vaporizer on both pressure and flow control. Any additional hydrocarbon release enters the flare stack through a separate piping system provided with a knockout drum. The knockout drum collects any liquid hydrocarbon and vaporizes by means of a steam coil to prevent liquid carryover to the flare.

Potential release scenarios for this process are listed in Exhibit 8. Of the twelve scenarios, three fell in the highest risk category (A) (R6, R8, and R12), with just one with a high severity rating (R6). This is clearly a good candidate for the alternative release scenario.

EXHIBIT 8. Butane

Ref #	Scenario	Frequency	Severity	Risk	Comments
R1	Overpressure of butane storage bullets (8 vessels)	L	H	C	BLEVE or VCE
R2	Overpressure of vent vessel	L	H	C	BLEVE or VCE
R3	Overpressure of vaporizer	L	M	D	VCE
R4	Overpressure of a compressor accumulator	L	M	D	VCE
R5	Overpressure of butane storage spheres (2 vessels)	L	H	C	BLEVE or VCE
R6	Overpressure of cryogenic tanks (2 vessels)	M	H	A	BLEVE or VCE
R7	Release of liquid butane from cryogenic tank bottom	L	H	C	VCE
R8	PSV fails open; liquid butane to atmosphere from cryogenic tanks	H	M	A	VCE
R9	Butane release from pump seal leak	H	L	D	Flame jet
R10	Overpressure of butane storage cooler	L	M	D	VCE
R11	Release of butane from a refrigeration compressor	H	L	D	VCE
R12	Butane vapor cloud from flare stack	H	M	A	VCE

For this case study, the screening analysis is somewhat more complicated because of the range of "outcomes" that could occur for each release scenario (e.g., vapor cloud explosion, BLEVE, flash fire, etc.). For a complex facility, the selection process could be even more complicated given that only one alternative scenario is required (site-wide) for all flammable substances. While the risk-ranking approach could still be applied, it is somewhat harder to compare the relative severity of hazards from different processes. In this case, a screening model could be used to provide better estimates of the severity ratings.

With consequence modeling, it was demonstrated that the range of hazard distances for the twelve scenarios ranged from "no offsite impacts" out to 208 meters. For the selected scenario, the hazard distance was 80 meters. Both scenarios were based on the distance to an overpressure of 1 psi.

Summary

This paper summarizes a defensible approach for implementing the RMP requirements and the OCA Guidance regarding alternative scenario selection. The approach ensures that a wide range of scenarios is considered in the selection process. A risk-based approach provides a sound basis for selecting a representative scenario for detailed analysis.

References

California Office of Emergency Services (OES). 1989 Guidance for the preparation of a Risk Management and Prevention Program. November 1989.

EPA. 1996 Risk Management Program Offsite Consequence Analysis Guidance. May 1996.

John, J.I. and Henry Ozog. 1997. "Tools and techniques for conducting hazard assessments". Presented at the AIChE Practitioners Forum, Houston, Texas, March 11-12, 1997.

SuperChems™. 1996. Comprehensive scenario modeling package developed by Arthur D. Little, Inc. Available from Arthur D. Little, Inc. Safety and Risk Management Unit, Cambridge, Massachusetts.

Thompson Publishing Group. 1996. Risk Management Program Handbook, Accidental Release Prevention Under the 1990 Clean Air Act.

Joint EPA/OSHA Chemical Accident Investigation

Armando Santiago and Breeda Reilly
U.S. Environmental Protection Agency, Chemical Emergency Preparedness and Prevention Office (5104), Chemical Accident Investigation Team, 401 M Street, SW, Washington, DC 20460

ABSTRACT
An important initiative taking shape between the Environmental Protection Agency (EPA) and the Occupational Safety and Health Administration (OSHA) is a joint chemical accident investigation program. In 1996, an interagency Memorandum of Understanding (MOU) was signed to establish policy and general procedures for cooperation and coordination between EPA and OSHA. The MOU defines the criteria for defining a major chemical accident or release and for initiating a joint agency investigation. In general, major chemical accidents or releases are those that result in fatalities, serious injuries, significant property damage, environmental damage or other serious off-site impacts such as evacuation. Other less severe accidents or releases with the potential for grave losses may also be investigated. The purpose of these joint investigations is to uncover the root causes, make recommendations to prevent their recurrences, and to make the lessons learned available to other industrial facilities and the public.

Introduction

Accident investigation has long been recognized as an important element of chemical process safety. Serious or potentially serious accidents need to be investigated in order to learn about the causes and make improvements to prevent their recurrence. Accidents often occur because of problems in the safety management systems. Lessons learned from such investigations can then be considered in development of sound risk management strategies to prevent future accidents.

In major chemical accidents with potentially broad impacts (to industry, workers, communities and the environment), EPA and OSHA have a significant role in terms of investigating the circumstances, contributing factors, and root causes of the accident. The broader perspective of the joint investigation may lead to recommendations on what industry, communities, the government and other stakeholders can do to prevent similar events from occurring in the future. Lessons learned can also reach a wider audience through the joint report dissemination to all stakeholders.

Through a recently signed Memorandum of Understanding, the two agencies have agreed to work together in promoting accident prevention by identifying the root causes of accidents, evaluating ways to prevent such accidents from recurring, sharing lessons learned and increasing awareness of safety needs for the chemical processing industry. Both agencies have the authorities and the resources to efficiently investigate chemical accidents, capitalizing on the already established infrastructures throughout the country.

This paper describes the status of the two agencies efforts to develop a joint program to investigate chemical accidents for root causes. Because this is a recent initiative, many elements of the program (and related issues) are still in the developmental stage, so we can present the alternatives being considered. Other elements which are much further developed will be discussed: the Memorandum of Understanding between the two agencies; status of current investigations; and highlights of the first completed report that has been released to the public.

Background

The joint program was initiated in January, 1995, when the Administration asked OSHA and EPA under their own existing authorities to jointly investigate major chemical accidents and releases at fixed facilities. The two agencies authorities complement each other, since OSHA has the responsibility to regulate and enforce safety in the workplace, while EPA is responsible for protecting the general public and the environment.

The idea of a federal-level effort to investigate accidents had been proposed in the Clean Air Act Amendments of 1990. In these amendments, Congress established a national policy to prevent accidental releases of hazardous chemicals. To support this goal, Congress created a Chemical Safety and Hazard Investigation Board (CSHIB) to:

- investigate serious or potentially serious chemical accidents,
- identify probable causes of such accidents,
- make recommendations for corrective actions to prevent such accidents, and
- issue reports to the public.

These reports were expected to describe the facts, conditions, circumstances and the cause (or probable cause) of any accidental release. The CSHIB was also directed by Congress to coordinate its activities with other federal agencies. Over time, Congress did not appropriate funds for establishing the Board. Under the goals of reinventing government, the administration asked EPA and OSHA to jointly investigate chemical accidents using their existing authorities.

The EPA program office primarily responsible for implementing this joint program is the Chemical Emergency Preparedness and Prevention Office

(CEPPO). The mission of this office is three-fold: (1) to prevent and prepare for chemical emergencies; (2) to respond to environmental crises; and (3) to inform the public about chemical hazards in their community. CEPPO has undertaken various activities under its authorities to promote chemical accident prevention. For example, CEPPO collects information on accidental releases of hazardous chemicals through the Accidental Release Information Program (ARIP). This information is added to the ARIP database which is made available to the public with the goal of disseminating the information to help prevent future accidents. Also, CEPPO has conducted Chemical Safety Audits (CSA's) of facilities to identify successful practices and technologies for preventing and mitigating chemical accidents. The focus of this voluntary program has been on refineries, chemical manufacturers, ammonia refrigeration facilities, and water and wastewater treatment plants.

OSHA has authority under the Occupational Safety and Health (OSH) Act to promulgate and enforce mandatory safety and healthful working conditions for every worker in the nation and to investigate workplace accidents. In 1992, OSHA promulgated a the Process Safety Management (PSM) Rule to establish procedures for process safety management to protect employees by preventing or minimizing the consequences of chemical accidents involving highly hazardous chemicals.

EPA has authorities under several statutes which complement OSHA's mission. Under the Clean Air Act (CAA), EPA is directed to develop programs to prevent and minimize the consequences of accidental releases of extremely hazardous substances. Under Section 103 of the CAA, EPA is responsible for research and development, including conducting investigations into the causes, prevention, and control of air pollution. Section 103 also directs EPA to make the results available and to cooperate with other federal agencies. Under Section 112(r) of CAA, EPA is responsible for developing programs to prevent the accidental release and minimize consequences of releases of extremely hazardous substances. The CAA also provides EPA with the rights to enter a facility, to review records, to inspect any monitoring equipment, and to sample emissions when an accidental release of an extremely hazardous chemical occurs at a fixed facility. For any investigation, EPA also has authority to any issue subpoenas for attendance and testimony of witnesses or production of relevant papers, books, and documents. In addition to the CAA authorities, EPA has authorities under the Comprehensive Emergency, Remediation, and Cleanup Act (CERCLA) to respond to and investigate accidental spills.

Joint Accident Investigation Program

The first major milestone in creating the joint program was to negotiate an MOU between EPA and OSHA. The MOU was signed in December 1996 and establishes general policy for procedures and coordination between the two agencies, to ensure the most effective possible investigation and to limit duplication of efforts.

A key element of the MOU is the agreement between the two agencies as to the criteria for jointly investigating an accident. The MOU uses the term "major" chemical accident or release, and identifies criteria to define "major". In the MOU, "major" is defined as meeting one or more of the following criteria:

- Fatalities (one or more)
- Serious injuries (three or more)
- Significant property damage (on- or off- site)
- Serious threat to health or environment
- Serious off-site consequences such as evacuation
- Public concern

Other less severe accidents may also be investigated.

The MOU also provides the definition of "root cause" as agreed to by the two agencies.

The EPA/OSHA definition of "root cause" is adapted from the Center for Chemical Process Safety (CCPS) and appears in the MOU as "the underlying prime reasons, such as failure of particular management systems, that allow faulty design, inadequate training, or deficiencies in maintenance, which in turn lead to an unsafe act or condition and result in an accident." This definition of the term "root cause" was important to the agencies in focusing the scope of the joint investigations. With this definition, accident investigations expand from the traditional compliance and enforcement activities. The agencies believe that to prevent future accidents, the root causes must be uncovered and examined.

A major challenge to each of the agencies will be balancing responsibility to enforce regulations with the need to determine the root cause and effectively communicate recommendations to the public. According to consensus from Roundtable discussions (1995,1996) by leaders of industry, labor, government, insurance and public interest groups convened by the Wharton Risk Management and Decision ProcessCenter (University of Pennsylvania): "the accident investigation methodology most likely to yield information on the prevention of the future accidents, is also more likely to yield information that will lead to punishment of the firm under the present legal [and regulatory] system." Although the primary focus of each investigation will be determining root cause, information collected will be shared with enforcement and compliance personnel. EPA and OSHA may take enforcement actions if violations of their respective regulations have occurred.

Each agency plans to maintain a team of investigators with appropriate technical backgrounds and experience in engineering, science and project management. The team will be responsible for developing the infrastructure to support the program; and to participate and/or lead the accident investigations. The team consists of headquarters, regional and area office personnel as appropriate. Resources will be available to them, as needed, including government experts and outside consultants and government contractors to participate in the investigations.

The MOU also outlines some key features to ensure smooth coordination through the notification process. The initial notification of an accident is expected to reach Joint accident investigation team through existing channels, such as National Response Center and the OSHA hotline. The notification procedures include gathering information from regional and other personnel closer to scene and sharing the information between the agencies. Some critical decision points are also delineated in the MOU. First, within 24 hours of the initial notification, the agencies have to decide whether to send their investigators to the site as part of a joint team for fact-finding. Then, within 96 hrs, the agencies decide whether a full investigation is warranted. If a joint investigation is not warranted, individual agencies can elect to do so independently in the quest of possible "lessons learned" about imminent hazards.

Through the MOU, the two agencies have agreed to develop a joint training curriculum for their investigators. Elements of the training curriculum include: required personal health and safety training; accident investigation techniques, such as root cause analysis; chemical process safety; and training in the procedures and policies developed as part of the joint agency protocol.

Thus the MOU provides the basic framework for the joint program. The MOU defers to a joint agency protocol for additional details. The two agencies started working on the protocol shortly after the MOU was signed and it is anticipated to be completed by September, 1997. The intent of the protocol is to provide investigators with a structured framework for managing all aspects of the investigation in the field. Specific areas covered are: jurisdiction and coordination; notification and mobilization procedures; preparations for going on-site; personal safety considerations; assembling and reviewing evidence; documentation and follow up; and training curriculum.

In addition to the federal MOU, EPA also is pursuing supplemental agreements with states having OSHA-approved State Plans to administer their own health and safety programs. In those states with State Plans, EPA is negotiating the establishment of joint investigation procedures appropriate for the individual State agencies participating.

Accident Investigations

As of this date, the two agencies are conducting several accident investigations. These investigations are at different stages of completion. Some of the accident details and types of investigations are summarized in Table 1.

TABLE I
Summary Table of EPA/OSHA Chemical Accident Investigations

Event/Location	Date	Substance(s) Released	Consequences (fatalities, injuries, etc.)	Status
Terra Industries, Inc., Port Neal, Iowa	December 13, 1994	About 5,700 tons of anhydrous ammonia and 25,000 gallons of nitric acid released.	4 workers killed; 18 persons hospitalized; ammonia plumes offsite resulted in evacuations up to 15 miles away. Nitric acid was released to the ground resulting in groundwater contamination.	Major1 EPA investigation report completed in January, 1996. Expert review of report completed in September, 1996. Final report issued in January, 1997. Alert for Ammonia Nitrate—pending.
Powell Duffryn Terminals, Inc., Savannah, Georgia	April 10, 1995	Crude sulfate turpentine involved in fire. Hydrogen sulfide released.	2,000 residents evacuated [evacuation lasted 30 days for some residents]; an elementary school was temporarily closed; water in an adjacent marsh was heavily contaminated.	Major EPA investigation report- pending. Alert for carbon adsorption systems issued May, 1997.
Napp Technologies, Inc., Lodi, New Jersey	April 21, 1995	Mixture of sodium hydrosulfite, aluminum powder, potassium carbonate and benzaldehyde led to explosion.	Five employees were killed and numerous injured. One of the injured died a week later as result of injuries.	Major joint investigation report pending.
Pennzoil Products Company Refinery, Rouseville, Pennsylvania	October 16, 1995	Flammable hydrocarbons involved in explosion and fire.	3 workers were killed in the fire, and 3 others were injured. 2 of the injured died later as a result of their injuries. Workers at plant and nearby offices and the entire town of Rouseville (750 residents) were evacuated.	Major EPA investigation report pending. Alert for atmospheric storage tanks issued May, 1997.

Event/Location	Date	Substance(s) Released	Consequences (fatalities, injuries, etc.)	Status
Wyman-Gordon Forging, Inc. Houston, Texas	December 22, 1996	5,000 psi N_2	Eight workers killed; 3 hospitalized due to massive pressure release from large volume N_2 pressure vessel.	OSHA Investigation complete2. Hazard Alert planned.
Tosco Refinery Co., Martinez, California	January 21, 1997	Hydrocarbons involved in explosion and fire.	1 worker killed, 44 workers injured, and residents sheltered-in-place	Major EPA investigation in coordination with CAL-OSHA and local authorities. Field work ongoing.
Chief Supply Corporation, Haskell, Oklahoma	March 26, 1997	Flammable liquids, such as paints, oils, thinners, inks, and cleaning solvents involved in fire and explosion.	3 workers injured, 1 of injured later died as result of injuries. Nearby highway closed and residents within 1.5 miles of the facility were evacuated.	Field-level joint investigation. Fieldwork ongoing.
Surpass Chemica, Albany, New York	April 8, 1997	Hydrochloric (31%) acid spill	Eight workers treated for inhalation. Other employees (about 15) were evacuated. Reported evacuation of 39 residents.	Field-level joint investigation. Fieldwork ongoing.
National Vinegar Co., Inc., Houston, Texas	April 30, 1997	Flammable liquids released in explosion and fire.	One worker killed; another severely injured.	OSHA Field-level investigation in progress.
BPS Inc., West Helena, Arkansas	May 8, 1997	Fire and explosion involving several pesticides including azinphos-methyl.	3 firefighters killed as result of collapsing roof; 20 injured; area within a 3 mile radius initially evacuated.	Major joint accident investigation ongoing.

The earliest major investigation started in December, 1994 when an explosion occurred in the ammonium nitrate plant at the Terra International, Inc., Port Neal Complex (Iowa). The explosion resulted in the release of approximately 5,700 tons of anhydrous ammonia to the air and secondary containment, approximately 25,000 gallons of nitric acid to the ground and lined chemical ditches and sumps, and liquid ammonium nitrate solution into secondary containment. Off site ammonia releases continued for approximately six days following the explosion. Chemicals released as a result of the explosion contaminated the groundwater under the facility. After completion of the investigation, the team released its report in February, 1996. The investigation team concluded that the explosion resulted from a lack of written operating procedures for all chemical operations. Such procedures were not clearly established and resulted in unsafe conditions at the plant that allowed the explosion to occur. The full report (including expert review) and details of the of the accident investigation findings and recommendations can be found in the CEPPO Internet Website located at http://www.epa.gov/swercepp.

Conclusion

The main goal of the joint chemical accident investigation program is the prevention of recurring accidents in all chemical industries. It is essential to have a mechanism for sharing lessons learned form chemical accidents across all boundaries: individual process, individual companies and specific industry sectors. We aim to provide such prevention information to all stakeholders by investigating major accidents and providing recommendations conducive to continuous in-house examination and improvement of chemical operations. We have started this outreach effort to get reports and safety alerts out to interested persons. We encourage the chemical industry to use this information to anticipate possible accident scenarios and take actions to mitigate the risks. In addition, we extend an invitation to provide input to this program.

The views expressed in this paper are strictly those of the authors and do not reflect official agency policies.

OSHA Enforcement Actions for and the Defense of Facility Siting Citations

Mark S. Dreux
McDermott, Will & Emery, 1850 K Street, N.W., Washington, D.C. 20006-2296

Industry is struggling with an issue: how does it, in an effective and feasible way, fulfill the facility siting requirements in OSHA'S Process Safety Management Standard, 29 C.F.R.1910.119. This issue is very complex, involving various industry guidance documents, good engineering practices, and the exercise of professional judgment. Many facets of the issue are unresolved and subject to continuous debate within the Industry. For example, should qualitative or quantitative risk analysis be used; should credible or worst case scenarios be used; should occupancy criteria be 200 or 400 man hours per week; or should control rooms be moved or reinforced? Since the PSM Standard is a performance standard, individual employers and their Industry have a great deal of discretion in resolving these issues.

To date OSHA has provided very little guidance to Industry to assist it in finding reasonable solutions to these issues. The PSM Standard does not define "facility siting." Appendix C to the Standard, which is entitled "Compliance Guidelines and Recommendations," does not even mention it, and OSHA has not issued any interpretation letters explaining it. The sparse guidance from OSHA is limited to citations for failure to conduct proper facility siting studies and a brief paragraph in OSHA's PSM Audit Guidelines.

Between May 1992 and August 1996, OSHA issued approximately 30 citations for improper facility siting, which were based on two legal authorities: 29 C.F.R.1910.119(e)(3)(v) and Section 5(a)(1) of the OSH Act, the General Duty Clause. Twenty-one of the citations were characterized as "serious" and nine as "willful." The proposed penalties ranged from $875 to $70,000 per item. The citations alleged four types of violations:

1. No documentation that facility siting had been considered.
2. The facility siting study had simply not been done.
3. The facility siting study was inadequate.
4. The layout and spacing of buildings were improper.

In the citations, OSHA expressly relied upon four industry guidance documents. They were the Dow Fire and Explosion Index; IRI's General Recommendations For Spacing; Factory Mutual's Loss Prevention Data Sheet 7-44; and NFPA 496, Purged and Pressurized Enclosures for Electrical Equipment. None of these citations involved any alleged failure to adhere to API Recommended Practice 752, Management of Hazards Associated With Location of Process Plant Buildings.

Based on these 30 citations and PSM Audit Guidelines, a checklist can be developed of those factors that OSHA thinks should be properly included in a facility siting study. This checklist would include the following:

- location of control room
- construction of control room
- types and quantities of materials in the process
- types of processes/reactions
- sources of ignition
- location/spacing of other units
- layout of units/reactors
- location of reaction vessels
- location/adequacy of drains, dikes and sewers
- location of equipment such as compressors, cooling coils, valves, pipes, cable tray and pressure relief devices
- location of fresh air intake
- maintenance of positive pressure in control room
- installation of monitoring/warning devices for levels of contaminants, O_2, and positive pressure
- inspection of those monitoring/warning devices
- ensurance that the entry to control room is raised or diked
- effects of toxics upon adjacent sections of the facility
- fire protection /explosion response capabilities

While this checklist is, by no means, a definitive list of all facility siting factors or one that is necessarily proper for any specific facility, it does reflect the factors that OSHA has considered important enough to warrant a citation. Moreover, it can be readily included in any subsequent facility siting study or as part of a PSM audit. Lastly, it can provide some assistance in resolving the central issue to Industry: what factors should be considered in conducting a facility siting study in an effective and feasible way.

While there are many defenses to facility siting citations, two defenses frequently recur: the adherence to the relevant Industry standards and the significance of the risk of harm. As noted above, many facility siting citations are based upon an alleged failure to adhere to an Industry guidance document such as the Dow Fire and Explosion Index. For example, following an incident with significant damage to occupied buildings and serious injuries to employees, OSHA will usually allege that the employer failed to "comply with accepted industry guidelines for plant layout and spacing, including but not limited to the Dow ... Fire and Explosion Index...." The defense of such a citation focuses on the degree of compliance with the Dow Fire and Explosion Index and the proper utilization of generally acceptable safety and engineering principles for the proper layout and spacing of the process.

OSHA's authority to regulate hazards is limited to occupational hazards which poses a "significant risk of harm" to employees. The issuance of citations is one form of regulation. Given this, an employer will defend by proving that a particular hazard does not pose a "significant risk of harm" to employees, that OSHA lacks the authority to regulate it, and that the citation is invalid.

In the Benzene case, the Supreme Court held that "both the language and structure of the Act, as well as its legislative history, indicate that it was intended to require the elimination, as far as feasible, of significant risks of harm." This holding defines not only the limitations on OSHA's authority to regulate but also the level of acceptable risk for facility siting. Saying it another way, facility siting analyses must eliminate, to the extent feasible, those occupational hazards which pose a "significant risk of harm" to employees.

To defend on this basis, the central question becomes what is a "significant risk of harm"? There is no simple straight forward answer to this question. Each hazard must be evaluated on a case-by-case basis. The relevant OSHA caselaw suggests that qualitative or quantitative analyses may be utilized in making a "significant risk of harm" evaluation. For qualitative analyses, there is whole range of factors that should be included such as the likelihood of the harm, the severity of the harm, the engineering and administrative controls in place, the experience and training of the relevant employees, the inspection, auditing and maintaince practices, the operating history of the process and the Industry experience with that process, the nature of the process, the quantity and nature of the chemicals in the process, the magnitude and direction of a possible VCE, the predicted blast load, the strength of the occupied buildings, the occupancy load, etc. Considering all of these factors, any applicable Industry practices or guidance documents and any other relevant factor, the qualitative determination reduces to whether in the professional judgment of the engineer, the occupational hazard poses a "significant risk of harm" to employees. If the answer is yes, then the employer must "take appropriate steps to decrease or eliminate" the hazard. If the answer is no, then the employer has no OSHA driven duty to address the non-significant hazard and OSHA lacks the authority to regulate it through a citation.

In the Benzene case, the Supreme Court tacitly approved the use of quantitative analyses and suggested that a "significant risk of harm" is some where between one in thousand and one in a billion. Specifically, the Court stated that

> if . . . the odds are one in a billion that a person will die from cancer by taking a drink of chlorinated water, the risk clearly could not be considered significant. On the other hand, if the odds are one in a thousand that regular inhalation of gasoline vapors that are 2% benzene will be fatal, a reasonable person might well consider the risk significant and take appropriate steps to decrease or eliminate it.

To date, OSHA has not issued any guidance document or interpretation letter accepting or rejecting any quantification of acceptable risk. A common industry guideline, though, for determining whether a "significant risk of harm" is present for facility siting purposes is evaluating whether there is at least a one in ten thousand (1 in 10,000) chance per man year that an employee in a plant building will be killed or seriously injured as a result of a catastrophic release. Just as with the qualitative evaluation, if a hazard falls within the accepted quantitative risk (less than one in ten thousand), then the employer must eliminate or to the extent feasible reduce that hazard. If the hazard falls outside the accepted quantitative risk, then OSHA lacks the authority to regulate it and any citation for a non-significant risk of harm would be invalid.

WORKSHOP B
Root Causes and Failure Data Bases

Chair   **Brian D. Berkey**
Hercules, Inc.

Guidelines for Developing Equipment System Taxonomies and Data Field Specifications

Bernard J. Weber *chairs the CCPS database committee*
Det Norske Veritas, Houston, Texas.

ABSTRACT

An essential part of any equipment failure database is the specification of data fields. If the appropriate information is not collected, the resulting analysis may be limited in application or even flawed. For the reliability analyst in particular, the key fields for data collection are the failure modes and failure causes.

In order to determine meaningful failure modes and causes, a systematic method needs to be employed, so that the relationships that exist are adequately defined. This paper outlines the procedure used by the CCPS Equipment Reliability

Introduction

The CCPS is currently engaged in a project to develop a high-quality failure database for the process industry—the Equipment Reliability Database Project. To accomplish its objective, the CCPS has formed a group of sponsor companies to support and direct this effort. The sponsor companies, through representatives on a Steering Committee, are actively participating in the design and implementation of the database.

The Equipment Reliability Database is intended to collect data for many diverse types of equipment. The diversity of equipment in the database demands that equipment-specific fields are created to collect the appropriate data. The development of these equipment-specific data fields for all possible equipment is an arduous task. In order to expedite this task, the Steering Committee is contributing resources to develop the taxonomies and data field specifications for several equipment types.

This paper provides an overview of the standardized approach used to develop specific taxonomies and data fields for new system types. It also provides an understanding of the role of the taxonomy and data field specification in the database development effort through the use of examples.

Taxonomy/Data Field Specification Procedure

The following sections provide guidance on how to create a taxonomy and data field specification for a new system. Examples using Compressor Systems and Relief Valves will be provided as a samples of how the taxonomy should be developed and presented.

In general, the following steps are suggested to establish the data field specifications for a system being added to the Equipment Reliability Database:

Step 1. Verify/revise pick list for higher level (above system level) taxonomy tree
Step 2. Draw the boundary diagram
Step 3. Identify the functions and failures of the new system
Step 4. Identify the failure modes and their respective definitions
Step 5. Identify the possible failure causes for each failure mode
Step 6. Revise the system boundary diagram
Step 7. Revise the pick lists for general inventory data
Step 8. Define the new data fields and pick lists for the specific system inventory data
Step 9. Define any new data fields and pick lists for the specific system failure data
Step 10. Define event data fields necessary to establish failure mode per definition

Step 1. Verify/Revise Pick Lists for Taxonomy Tree

The CCPS has a pre-defined higher level taxonomy for equipment items. Before beginning work on the actual data field specifications, the taxonomy developer should review the current top-level taxonomy tree and prepare the recommended additions for the system of interest.

The seven levels for the CCPS database taxonomy are as follows:

1. Industry
2. Site
3. Plant
4. Unit Operations
5. System
6. Component
7. Part

The taxonomy development process will determine the make up of the taxonomy for Levels 5–7. It is important, therefore to define the scope of the system to be addressed as it fits into taxonomy levels 1–4 as part of the first step in this Procedure.

The taxonomy should be documented by specifying the following:

1. equipment class
2. system name
3. equipment type (if necessary)

The **equipment class** refers to one of several classes of equipment in the CCPS taxonomy. Some of the equipment classes are heat exchangers, piping, rotating equipment, and vessels.

The **system name** is the specific name associated with the system of interest. In the case of the rotating equipment class, there are seven different systems, such as pump, turbine and compressor, etc.

Some equipment systems can be further classified by **equipment type**. An example of this is the process sensor system, which has types denoting the parameter being sensed, such as displacement, flow, or level.

Example: Using the compressor system as an example, the taxonomy for system type would be as shown below:
    Equipment Class: Rotating Equipment
    System Name: Compressor
    Equipment Type: none

Step 2. Draw the Boundary Diagram

The system boundary defines the components that are to be included as part of the system. Keep in mind that separate failure modes and causes will be recorded for each component appearing as part of a system contained within the boundary diagram. As a starting point, it is helpful to refer to boundary diagrams shown in the ISO proposed standard, OREDA, or a similar database.

A few criteria are presented to assist in drawing a suitable boundary diagram:

1. For a component to be included, it must be necessary for the equipment to produce the intended function
2. If a component is also another system, it is necessary to establish a parent/child relationship to allow proper analysis of either system.
3. It is not necessary for the component to be present in all applications in order for it to be included in the system boundary

Consider these criteria as applied to a compressor. The boundary diagram for the compressor is shown below.

Note that the boundary diagram for the system satisfies Criterion 1; that is, all components shown within the boundary line are necessary for the compressor to deliver a certain quantity of fluid at a specified temperature and pressure, and without undue leakage to the atmosphere.

Example: the boundary diagram for compressors is shown in the figure below.

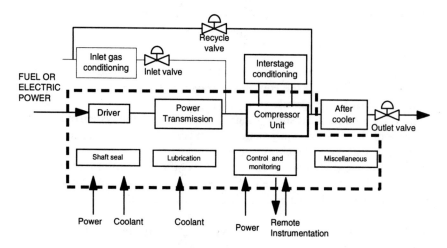

Note that interstage conditioning (cooling) is contained within the compressor boundary diagram and is also considered part of the equipment system for heat exchangers. In this case a parent child relationship exists, the parent being compressor and the child being heat exchanger.

Some of the equipment inside the boundary, for example the utility functions such as seal oil, may not be present for specific compressor applications, or will be present in vastly different levels of complexity. This does not merit their exclusion, as per Criterion 3.

Treatment of Subordinate Systems in the CCPS Database

Some systems, such as heat exchangers, motors, or control valves can function as an independent system in some applications; or as a support system in other applications. When supporting a larger system, the supporting systems function more like components of the larger system, rather than systems themselves. These supporting systems are referred to as *subordinate systems* when they function as components for a larger system.

To resolve this dilemma, the Equipment Reliability Database has included the option to record the subordination of systems, so that failure of a subordinate system accurately cascades to failure of the primary system. The systems are registered as independent systems, but they are tagged as being subordinate to a larger system. Using the example of a heat exchanger supporting a compressor, the user would enter the database in the compressor system and would then be automatically linked to the heat exchanger system. The heat exchanger would appear as a component when system-level events are recorded.

The taxonomy development task does not require any input regarding subordinate systems. The dependencies between systems is recorded as systems are registered with the database.

Step 3. Identify the Functions and Failures of the System, Components, and Parts

Before proceeding directly to failure modes, it is important that the analyst understands the nature of the functions and failures for the system. According to Rausand and Øien (Reference 3), "without evaluating failure causes and mechanisms it is not possible to improve the inherent reliability of the item itself."

This step is rather lengthy and therefore it is broken down into smaller sub-steps:

Step 3.1	Document Operating Modes
Step 3.2	Document the Function
Step 3.3	Document the Failure Modes
Step 3.4	Review and Update the Cause List
Step 3.5	Develop the List of Failure Mechanisms

STEP 3.1. REVIEW OPERATING MODES

In general, all operating modes for a system should be considered. The term *operating modes* refers to the state an item is in when it performs its intended functions. An item may have several operating modes. Since each mode has different functions (and functions lead us to the definition of failures), all modes should be reviewed when developing the data field specifications for an equipment item.

For instance, a valve may have several operating modes. A process shutdown valve may have four operating modes:

1. open
2. closing
3. closed
4. opening

Modes 1 and 3 are stable modes, whereas modes 2 and 4 are transitional. The essential function of this valve would be to close flow, that is, to operate successfully in modes 2 and 3. It is necessary to consider all operating modes to avoid overlooking some of the non-primary functions of an equipment item.

For illustrative purposes, the failure modes for Pressure Relief Valves have been developed. Pressure Relief Valves have the following operational modes:

Example: the operational modes for pressure relief valves are:

- closed
- relieving
- closing

STEP 3.2. DETERMINE THE FUNCTIONS

Functions are defined by using a classification scheme as an aid. The following scheme is recommended for development of the database field specifications:

Primary—the functions required to fulfill the intended purpose of the item. Simply stated, it is the reason the item appears in the process.

Auxiliary—functions that support the primary function. Containing fluid is a typical examples of an auxiliary function for a compressor. In many cases, the auxiliary functions are more critical to safety than the primary functions.

Protective—functions intended to protect people, equipment, or the environment

Information—functions comprising condition monitoring, alarms, etc.

Interface—functions that apply to the interface of the item in question with another item. The interface may be active or passive. For example, a passive interface is present if an item is used as the support base for another item.

Example: The functions derived for the relief valves are shown below.

Class of Function	Function for Relief Valves
Primary	Relieve pressure
	Relieve fluid at a desired rate
Auxiliary	Contain fluid within piping system
	Remain closed when no overpressure
	Relieve fluid without putting excess stress on piping
Protective	Maintain barrier between primary relief device and process fluid
Information	Monitor for leakage between protective barrier and process fluid
Interface	Connect process piping to relief effluent header

As noted in the referenced article, *primary functions* are "the reasons for installing the item." Hence, the primary functions listed above pertain to changing a fluid's pressure and flow rate. Auxiliary functions, such as containing fluid, are classified as such since these functions support the essential function of increasing flow rate or relieving pressure. They are not deemed to be less safety critical, however.

The process is repeated for all components and then all parts associated with the various components.

Step 4. Identify the Failure Modes and Their Respective Definitions

The next step is to develop failure modes, based on the functions declared in the previous step. The functions defined above now become the basis for the relief valve failure modes.

The CCPS database further defines classes of failure modes. These classes capture the information about the *timing* and *severity* of the failure mode. The following classes of failure modes have been defined:

Intermittent—failures that result in a lack of some function for a very short period of time. The item will revert to its full operational standard immediately after the failure

Complete and sudden—failures that cause a complete loss of a required function and cannot be forecast by testing or examination. These are typically referred to as *catastrophic* failures.

Complete and gradual—failures that caused a complete loss of a required function, but could be forecast by diligent testing or examination.

Partial and sudden—failure that does not result in a complete loss of function and cannot be forecast by testing or examination.

Partial and gradual—failure that does not result in a complete loss of function and could be forecast by testing or examination. These are commonly known as *degraded operation*. A gradual failure represents a "drifting out" of the specified range of performance values. The recognition of gradual failures requires comparison of actual device performance with a performance specification.

Example: for relief valves, the following failure modes are defined for the various failure mode classes:

Function	Failure Category	Failure Mode
Relieve pressure	Complete and sudden	Fails to open
	Partial and sudden	Opens above tolerance
	Intermittent	Chatter
Relieve fluid at desired rate	Partial and sudden	Fail to relieve at rated capacity
Remain closed when no overpressure	Complete and sudden	Spuriously open
Relieve fluid with no excess stress on piping	Complete and sudden	Pipework loss of containment
Maintain barrier	Complete and sudden	Spurious relief
	Complete and gradual	Leakage
Monitor barrier	Complete and sudden	100% output
		No output
	Partial and gradual	High output
		Low output
	Complete and gradual	Frozen output

To further enhance the quality of data, the CCPS database requires that all failure modes be specified in objective terms. This means that, for each failure mode, the taxonomy developer needs to provide a definition of the failure mode in concrete, measurable terms. This is intended to reduce interpretations required by field personnel as well as improve the overall consistency and quality of data.

In many cases, the failure mode need not be entered at all. Rather an actual value can be entered from which a failure mode can be determined automatically by the database software.

Example: the failure modes for relief valves are defined as follows:

Fails to open—failure to pop at 150% of setpoint
Opens above tolerance—pop pressure between 110% and 150% of setpoint
Chatter—rapid opening and closing damages valve
Fail to relieve at rated capacity relief—flow greater than rated capacity
Spuriously open—pop pressure less than 90% of setpoint
Pipework loss of containment—pipe supports fail to prevent loss of mechanical integrity
Spurious relief—total loss of mechanical integrity
leakage—flow greater than X%
100% output—100% output
No output—no output
High output—output more than 5% above actual
Low output—output more than 5% below actual
Frozen output—no change in output with change in actual process variable

Step 5. Identify Possible Failure Causes for Each Failure Mode

Causes are taken from a pre-defined list (Reference 3). A failure cause is defined in IEC 50(191) as " the circumstances during design, manufacture or use which have led to a failure." For the sake of data field specification, review and record failure causes for each cause category. Use causes from the pre-defined list wherever possible. If the list does not contain a necessary cause, then record the new cause. Upon sponsor approval, the new causes will be added to the CCPS database for programming.

The list below represents a first starting pass for failure causes.

Default Cause Category	Suggested Failure Causes
Design failure	Wrong material Wrong application Purchasing error Undersized
Weakness failure	Vibration Material defects Cavitation Cracking
Manufacturing failure	Defective manufacturing Installation error
Aging failure	Fouling Corrosion Erosion Cracking Fatigue
Misuse failure	Operated outside specification Improper environment
Mishandling failure	Lack of maintenance Delayed maintenance

Example: for one of the relief valve failure modes—fails to open—the following list of causes was derived:

Design failure	Wrong device for application Improper specification Purchasing error
Weakness failure	Material defect
Manufacturing failure	Defective manufacturing Wrong device shipped
Aging failure	Corrosion Fouling/plugging Mechanical wear Protective barrier leaks
Misuse failure	Operated outside specification Failure to fix identified barrier leakage
Mishandling failure	Wrong device installed Device installed improperly Block valve on inlet or outlet closed Blind left in place gag left in place

Step 6. Review and Redraw the System Boundary

At this point, the analyst will most likely have an improved state of knowledge about the scope of the system in question. The objective of this step, therefore is to re-examine the system boundary and challenge the components that are included or excluded.

Step 7. Revise the Pick Lists for General Inventory-Related Fields

A number of fields have set pick lists that may need revisions or new entries, based on the findings of the analyst. Check the fields listed in bold below to see if any changes need to made.

1. Review the current Pick List for **Unit Operation**. Determine if the new system type needs any new entries for this Pick List
2. Update the **System Type** Pick List to include the new equipment type

If necessary, add new items to the **Environment** Pick Lists (there are 3 P/L's for Environment)

For convenience, the CCPS Equipment Reliability Database uses a set of pre-defined tables for inventory-related information that is common to all types of systems. These are referred to generically as the "EQP" tables, since there code name begins with the three-letter designator EQP. Currently, EQP tables have been established for the following system parameters:

EQP1—Material Phase (liquid, gas, two-phase, etc)
EQP2—Corrosiveness
EQP3—Fouling Factor

If the taxonomy developer encounters the need to specify one of the above parameters, the developer must defer to one of the existing EQP tables, instead of creating a new table.

Step 8. Define Data Fields for System-Specific Inventory

This step is one of the most time-consuming in the data field development process. The purpose of the system-specific inventory fields is to be sure that all characteristics that could influence the reliability of a given type of equipment are recorded. When proposing a new inventory field, a key question to keep in mind is "will this field provide information that could realistically distinguish between failure rates for this equipment?"

For instance, if failure rates for compressors were known to have no dependence on the power consumption of the compressor driver, it would be a waste of time to collect that information.

When formulating the fields for this database, it is important to refer back to the results of Step 3. The functions, failure modes, and failure mechanisms should all be considered when specifying fields for the system inventory.

Refer to OREDA or the CCPS Gray Book for help in creating new fields for specific inventory.

After you have created the necessary data fields, create the Pick Lists for any fields that require standard inputs. If, for example, the developer chose to include an inventory field entitled "Shaft Seal Type" the Pick List of possible seal types needs to be developed.

Example: the system-specific inventory information is shown below for the compressor. *This table shows only a portion of the full table for compressors.*

Field Name	Description	Source
Subscriber ID	Unique identifiers	User
System ID	Unique identifiers	User
Service description	Type of application	Pick List, COM1
Type	Type of compressor	Pick List, COM2
Power Source	Source of power to the compressor driver	Pick List, COM3
Power	Power consumption to the compressor driver	User, in kW or HP
Utilization of capacity	Normal operating/design capacity.	User supplied, %
Fluid handled	Main fluid only	User supplied.
Molecular weight	Molecular weight of gas handled	User supplied, in g/mole.
Gas corrosiveness/ erosiveness	Qualitative assessment of the severity of the service.	Pick List, COM4
Shaft sealing	Type of seal used to minimize leakage of process fluid to the atmosphere along the drive shaft	Pick List, COM5
Discharge pressure (design)	Design pressure at the outlet of the compressor.	User, in psig
Suction pressure (design)	Design pressure at the inlet of the compressor	User, in psig
Number of casings	Number of casings in the compressor body	User, number
Number of stages	Number of compression stages	User, number
Bearing span	Equipment size, as measured by bearing span	User, feet
Radial bearing type	Type of radial bearing used	Pick List, COM6
Thrust bearing type	Type of thrust bearing used	Pick List, COM7

Step 9. Define Any New Data Fields and Pick Lists for the Failure Data Tables

There are several Pick Lists that need to be generated for each new system. The Pick Lists are used to collect the following kinds of information regarding the *failure of a system, component, or part:*

- Failure Mode
- Failure Cause

The three kinds of information, applied at the three levels of indenture produces the nine tables. For the sake of brevity, the Pick Lists for the system level are described below:

- Create the Pick List for the Failure Mode field. This field contains a list of failure modes for the system. The entries in this list come directly from the work performed in Step 1.
- Create a Pick List for Failure Mode Definition field. This field describes the possible failure modes for the equipment.
- Review and update the Pick List for Failure Cause.

Step 10. Define Event Data Fields for Failure Mode Definitions

Using the information generated in Step 4 above, the analyst should develop the necessary data fields to collect data that can be used to calculate the failure modes. This step should be fairly straightforward, as shown in the example for relief valves below.

Example: the failure modes for relief valves are defined as follows:
Fails to open—create a field to record actual lift pressure
Opens above tolerance—see "fails to open" failure mode
Chatter—field for—chatter: Y/N?
Fail to relieve at rated capacity relief—create field for relief flow
Spuriously open—see "fails to open" failure mode
Pipework loss of containment—create field for "pipe supports fail causing loss of mechanical integrity?"
Spurious relief—create field for total loss of mechanical integrity
Leakage—create field for leakage flow rate

References

1. IEC 50(191), *International Electrotechnical Vocabulary (IEV)*, Chapter 191—*Dependability and quality of service*, International Electrotechnical Commission, Geneva, 1990.
2. OREDA-1992, *Offshore Reliability Data*, DNV Technica, Høvik, Norway, 1992
3. Rausand M. & Øien, K., The basic concepts of failure analysis, In *Reliability Engineering and System Safety* 53 (1996), pp. 73–83.

An Ethylene Decomposition Event at Lyondell Polymers' Victoria Texas High Density Polyethylene Facility

Darrel E. Black
Lyondell Polymers, Victoria, Texas

ABSTRACT

In September 1996, an ethylene decomposition occurred in a reactor heater at Lyondell Polymers' Victoria High Density Polyethylene (HDPE) facility. The plant uses a solution process for HDPE manufacture, and, prior to this incident, believed that the reaction area was unlikely to experience an ethylene decomposition due to the large amount of solvent diluting the ethylene used in the process. The results of the investigation showed how this unrecognized possibility did actually occur, and gives insight into how a history of good process experience and changes within normal operating parameters can mask subtle process changes and result in a process safety incident.

The Lyondell Polymers Facility

Lyondell Polymers' facility at Victoria, Texas, about 125 miles southwest of Houston, is located on a 33 acre site within a DuPont Chemical complex, and has approximately 130 employees. It was started up in 1976 and has had capacity increases in several incremental steps from its original DuPont design for 150MM pounds per year to a new rate of 575MM pounds per year after completion of an expansion now in progress. The plant has an outstanding safety record, is an OSHA Merit Site, and has ISO 9002 certification. This year, the site completed 20 years with no lost-workday injuries, and has more than two years with no OSHA recordable injuries.

The Manufacturing Process

The Plant produces high density polyethylene (HDPE) using a solution process employing coordination catalysts. Ethylene monomer and octene copolymer are the basic ingredients for the polymerization, which is promoted using a Ziegler-Natta catalyst system.

Hexane is used as the solvent in the process. It also serves as the carrier to transport reactants and products through the process, and as a heat sink to carry heat evolved from the reaction away from the reaction area. After reaction is complete, the catalyst residue is removed by adsorption on alumina, and the solvent is separated from the polymer in a series of steps where the pressure of the hot polymer solution is dropped below the vapor pressure of the solution to flash evaporate a portion of the solvent.

Solvent and unreacted monomer and comonomer are recovered and recycled using a distillation step for separation, followed by purification in molecular sieve beds. The HDPE product is extruded to form pellets, stripped of residual solvent, blended and packaged in railroad hopper cars for shipment to customers.

Maximum pressures and process temperatures occur in feeds to the reactor area and in process flows from the reactor area. Pressures are up to around 3300 psig and temperatures vary up to around 200 degrees centigrade. Hydrogen is added to the reaction step to control polymer properties.

The Incident

At around 2:00 p.m. on Friday, September 27, a 6" piping elbow at the inlet to a heat exchanger discharging to one of the Victoria Plant's polymer reactors ruptured and released process fluids to the atmosphere. A loud noise was heard throughout the area, and witnesses described a cloud of black and sooty vapor and debris, about 100 feet high, originating from the top deck of the reactor area at the moment of the rupture. Flames were not immediately present, but the deluge system and fire protection interlocks were manually tripped from the control room within seconds. Ignition of the vapor cloud occurred quickly but with no blast effects. The initial flame was about the size of the vapor cloud, but seconds later was described as 20 to 30 feet high, and quickly became only about 6 feet high and three or four feet across. The fire was out within 5 minutes and minimal fire damage occurred.

The Approach to Investigating

As in most investigations for incidents of this nature, a team was assembled that provided expertise in many relevant areas. It consisted of plant operating and maintenance technicians, a plant safety department representative, and technical expertise from plant engineering professionals. It was supplemented by Lyondell's corporate engineering and development experts, DuPont site engineers with experience in the plant's HDPE process, and an outside consultant with considerable experience in ethylene decomposition studies.

One item to note in the study process was that the committee was serving two purposes: it was chartered to perform a normal incident investigatory function, but it was also requested to make immediate design recommendations to allow a quick and safe plant startup. Usually the investigation recommendations and the design elements to accomplish them are addressed separately and by different groups. Time and economics did not allow this to be the case here, so the investigative committee found itself in a dual role.

The committee proceeded to review "the facts" as they were known at the time. Care was taken to not draw conclusions from the facts until they were all brought forward. This was an attempt to avoid overemphasis of one suspected area to the possible exclusion of other important areas. Once all the facts related to the incident were listed, possible event scenarios were brain-stormed and listed. This resulted in two classifications of possibilities; mechanical causes and process causes. Six or seven possibilities were proposed under each classification. Each of these was discussed and its likelihood was debated.

The last HAZOP of the reaction area was reviewed. Although the HAZOP review committee had certainly been aware of the hazards of hot ethylene, they concluded that sufficient safeguards were in place, and they had proposed no recommendations that would bear on the current incident.

As pertinent details were brought to light, many of the proposed scenarios were dismissed as being unlikely. The committee came to agree on the stated causes and then proceeded to propose and critique recommendations. These were prioritized for consideration to implement before startup, and some after startup.

By Tuesday following the Friday incident, there was agreement within the committee to the most likely cause of the incident, as well as concurrence on the recommendations and their priority. A substantial amount of time was spent revisiting alternate scenarios or trying to propose additional ones, all with the idea of not overlooking other possibilities.

On Wednesday, a memo was written to the plant manager stating the committee's preliminary conclusions, and stating recommendations to be implemented before plant startup.

Summary of Supporting Facts and Observations

(Comments are in italics.)
- The plant was producing a broad molecular weight distribution resin at the time of the incident. This was a common product produced often throughout the year with no history of problems beyond quality control issues considered to be normal. However, this product was relatively recent in the design history of the plant, and required operating conditions that were greater in ranges than many of the plant's original products.

- Conditions around the heater just before the incident:
  —Controlled recipe conditions were allowing a small amount of process flow, hexane solvent with ethylene and hydrogen, to enter the heater.

- G.C. analysis of H_2 in the feed to the heater increased about 30 minutes before the event with no change being made by the operator. Flow meter readings did not show such a change.
  G. C. data support the conclusions that a reaction that consumed hydrogen was taking place many hours before the pipe rupture. This was probably an exothermic hydrogenation of ethylene. Minutes before the rupture, chromatographic readings suggest that the reaction mechanism became one that produces hydrogen, i.e. an ethylene decomposition.

- Prior to the increase in H_2 concentration from heater, the G.C. H_2 value had been running less than the calculated value based on flow measurements.
  See the previous comment above.

- Melt Index increased (based on observations of lower reactor system ΔP, extruder nose pressure, etc.) about 30 minutes prior to the event.
  This is consistent with an increase in hydrogen to the reactors.

- Heater control temperature decreased about 3°C just prior to the event.
  This was a significant event in the sequences leading to the pipe rupture. It is possible that the flow through heater became more restricted or totally plugged at this point. The temperature control could no longer compensate by opening the 2" valve to allow more flow through the heater. Stagnant flow of process fluids in the hot heater allowed reactions to accelerate or progress to decomposition stages.

- Metallurgical analysis of the failed elbow showed:
  —The elbow material conformed to its specified properties (ASTM A106, grade B).
  —No material or manufacturing defects were found in the ruptured elbow.
  —The rupture was determined to be an elevated temperature occurrence. This was evident from the diameter expansion and wall thickness reduction of the elbow in the rupture origin area.
  —The elbow may have been exposed to temperatures over 1100°F.
  This evidence supports a process problem, like a reaction that generates heat, rather than a mechanical failure, erosion, corrosion, etc.

- Eyewitnesses to the event reported a large (approx. 100 ft. high) cloud of black soot and debris originating from the top reactor deck. No flames were initially visible. Fire isolation and deluge systems were activated. The vapor cloud ignited, quickly diminished in size to 30 feet, and within about two minutes it was 6 feet high and smaller. Some flames were visible between the 2nd and 3rd decks. Reactors were depressured through the

Medium Pressure Separator (MPS) to the flare. The fire was quickly brought under control and fire damage was minimal.
The presence of the soot is indicative of a decomposition of ethylene.

- Process historian data showed no change in reactor pressures and temperatures during the event based on 10 second data.
  The reactions taking place in the heater piping were not tightly contained, so no excessive pressures would be produced. Also, molar volumes do not increase in some of the hydrogenation reactions, but a considerable amount of heat is produced.

- The six inch elbow in the piping leading to the heater (located about 1½ ft. from the heater head) failed apparently from excessive pressure and/or temperatures. Failure occurred as a 6" × 11" hole in the elbow at its outer radius. The initial pipe wall thickness of 0.864 (XX wall pipe) was reduced to about 0.45" by stretching at the point of failure.
  Subsequent metallurgical examination indicate that temperatures at or over 1100°F allowed the pipe to fail even though internal pressures did not exceed normal system pressure in the 2500 psig range.

- The pipe downstream of the failure point was bulged, showing signs of overpressure or overtemperature. Up-stream piping was clean and showed no apparent deformities.
  This supports a localized reaction and flame front, and perhaps supports a two-phase system that would trap the reaction area in the lighter phase. The lighter phase would be trapped in the head of the exchanger and the horizontal piping runs to/from it.

- Piping in the heater outlet was clean and showed no signs of deformation.
  Additional support for a very localized decomp reaction.

- Pressure relief valves immediately upstream of the heater were inspected and found to be clean. They did not lift.
  This is more evidence that normal system pressures were not exceeded.

- Ultrasonic thickness testing as part of the mechanical integrity program had not yet been performed on the elbow that failed. It was part of the original plant construction (approx. 20 yrs. old), but showed no signs of wear on its inner or outer surfaces.
  Mechanical failures due to wear, damage, or material defects are not supported by any evidence, either before or after the metallurgical testing.

Conclusions of the Investigating Committee

- A nearly stagnant reactor feed heater and operating conditions required by the polyethylene product recipe ultimately resulted in an ethylene decomposition and subsequent temperature-induced pipe failure.

- High ethylene concentration and high hydrogen concentration to heater was required by the product recipe.
- The reaction feed temperature required by the product recipe caused very low flow through the heater while the rest of the flow bypassed the heater.
- This resulted in some of the process flow contacting hot (~250°C) heater surfaces for an extended period of time.
- The low flow and piping configuration to the heater may have resulted in poor mixing of the reaction components, resulting in higher concentrations of H_2 and C_2H_4 in the heater and entrance piping.
- An exothermic reaction between ethylene and hydrogen started in the region of the heater and was indicated in analyzer data. Later analyzer data shows the reaction mechanism progressed to one that produced hydrogen. Furthermore, physical evidence of carbon buildup in the heater and the observations noting an initial release of a "cloud of soot" are consistent with the products of a decomposition reaction. Either or both of these reactions at the heater inlet produce enough heat to raise the inlet piping to or beyond 1100°F, and cause piping failure at normal process pressure (around 2500 psig).

Resulting Recommendations

The committee provided recommendations for two phases: those that had to be accomplished to assure safe operation before the plant was started back up, and those that could be deferred for later study. All of the pre-startup recommendations were implemented. Other recommendations have also been implemented, and additional work and study in this area is still under way at the time of this writing. Most noteworthy among the ongoing studies is an effort to define any points in the process and the required conditions that could allow a decomposition reaction to occur.

Recommendations to accomplish Prior to Startup

1. Consider revising the temperature control to maximize process flow to the heater at all times and minimize the temperatures achieved in the heater.
2. Consider relocating the hydrogen injection point downstream of the heater.
3. Consider installing skin temperature T/Cs on the inlet/outlet piping to the heater.
4. Consider changing the heater feed piping to flow only solvent (no ethylene or hydrogen) to the heater at low flow conditions.
5. Consider installing a pressure transmitter on the heater head.
6. Consider installing a minimum stop on the low flow solvent to heater control.
7. Consider drilling the heater baffle to provide a small hole that will vent the inlet space to the outlet space in the head, and install a valve on the existing head vent nozzle.

8. Consider pickling (for rust removal) the spare heater before use in the process and inspect the newly installed pipe for excessive rust.

Lessons Learned

This occurrence showed that in spite of thorough HAZOP reviews, and in spite of utilizing management of change procedures throughout the life of a plant with an exemplary safety record, a condition developed that resulted in a serious process incident.

The plant response to the event demonstrated that quick results and safe results can be achieved to serve the needs for process safety and to expedite plant repairs.

Results showed that a committee of experts, when convened to investigate an incident cause, can also efficiently provide design change elements for prevention of future similar incidents.

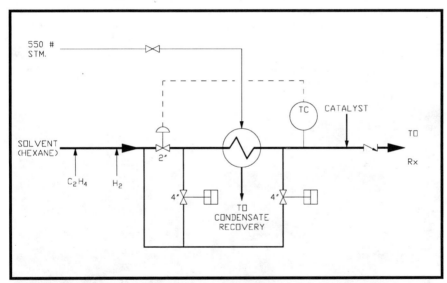

DIAGRAM 1. Schematic Representation of the Heater Before the Incident
Note the heat source is full pressure, 550 psig steam on the jacket of the heater. The ethylene and hydrogen are introduced to the solvent flow upstream of the heater. Depending on how the operators set the openings on the two 4" valves, some of the flow (most or all when feed low temperatures are called for by the recipe) bypasses the heater, and some flows into it. Often, all of the flow through the heater was only that passing through the 2" valve. At these low flow conditions, there is a long residence time and 249°C conditions for the hexane / ethylene /hydrogen process fluid inside the heater.

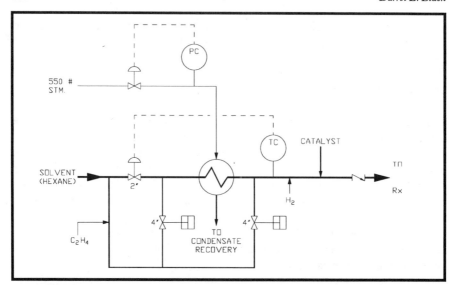

DIAGRAM 2. Schematic Representation of the Heater After Changes
This shows that the hydrogen stream has been relocated to the downstream side of the heater, the steam pressure is now controlled to minimize heater temperatures, the 2" valve now comes off the header upstream of the ethylene introduction so low flows through the heater will not contain ethylene, and a minimum flow stop has been put on the 2" valve to guarantee that stagnant flows will not result from closure of the 2" valve.

View of LOH #2 looking south, immediately after the fire.

Another view of the elbow entering the LOH. Note low amount of damage in the area

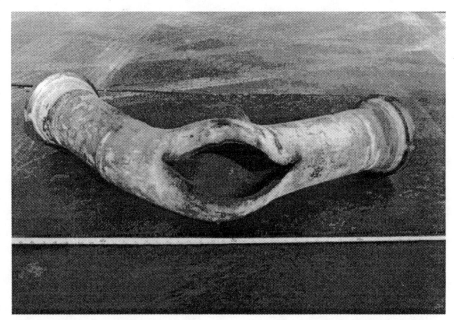

This is the pipe stripped of insulation, laying on the floor at the metallurgist's shop

Lessons Learned Databases—A Survey of Existing Sources and Current Efforts

G. Bradley Chadwell and Susan E. Rose
Battelle, 505 King Avenue, Columbus, Ohio 43201

ABSTRACT

Chemical accident prevention or mitigation is the major goal of process safety management and risk management programs. Accident histories and incident investigation reports go a long way toward realizing this goal. However, the true potential of such information is realized only when incident data is utilized beyond the confines of the facility or process where the incident took place. Lessons learned databases are effective vehicles for compiling and analyzing incident information. With a combined set of incident data, accident trends and common causes can be identified and addressed more readily. Efforts by industry, professional organizations, and government agencies are underway to develop lessons learned databases.

This paper presents a summary of available information pertaining to safety-related incidents in the Chemical Process Industries (CPI), including content, scope, and accessibility. In addition, the paper discusses the potential benefits of expanded availability and standardization of lessons learned databases.

Introduction

Process safety management and risk management programs both have the goal of preventing or mitigating the effects of accidents involving hazardous chemicals. To prevent or mitigate an accident requires anticipation—foreseeing potential hazards and predicting likely consequences. Past experience is a powerful tool for anticipating potential accidents. The usefulness of accident information is well established. Previous incident histories are addressed in process hazard analyses; incident investigations are an integral part of safety programs allowing companies to learn from past accidents and prevent similar occurrences in the future. However, the true potential of accident information is realized only when incident data is utilized beyond the confines of the facility or process where an incident occurs. Lessons learned databases are effective vehicles for compiling and analyzing incident information. With a combined set of incident data, accident trends and common causes can be identified and addressed more readily.

Several sources of accident information are available publicly in database form. The content, and hence the usefulness to the CPI, of any one of theses

database form. The content, and hence the usefulness to the CPI, of any one of theses databases depends on its intended purpose and scope. The next section details the content, scope, limitations, and availability of several existing databases. Background is provided to give insight into the intended purpose of each database and the reason for its structure and content. This is followed by a description of current efforts, along with a discussion of the potential benefits of expanded availability and standardization of lessons learned databases.

Existing Databases in USA

The largest publicly available incident databases in the United States are maintained by the U.S. government. The U.S. Occupational Safety and Health Administration (OSHA) and the Environmental Protection Agency (EPA) are governed by their own set of laws and regulations relating to incident reporting and each maintains separate incident databases. The laws and regulations administered under OSHA address recordable occupational injuries and illnesses, whereas EPA regulations address accidental releases of hazardous substances. An overview of these incident reporting requirements and associated data gathering systems are discussed in the following sections.

Occupational Safety and Health Administration (OSHA)

LAWS AND REGULATIONS

The Occupational Safety and Health (OSH) Act of 1970 directed the Secretary of Labor to issue regulations to require employers to maintain records on workplace injuries and illnesses. The Secretary of Labor was also directed to compile accurate statistics on occupational injuries and illnesses and to make periodic reports on such occurrences. The responsibility for collecting statistics on occupational injuries and illnesses was delegated to the Bureau of Labor Statistics (BLS).

The two main federal regulations promulgated under the Act that apply to the collection of incident-related data in the U.S. are 29 Code of Federal Regulations (CFR) 1904 and 29 CFR 1910.119 as described below. Each state must comply with these federal regulations by either directly adopting and adhering to the federal regulations as written, or by implementing state plans that are at least as stringent as the federal regulations.

29 CFR 1904—Recording and Reporting Occupational Injuries and Illnesses

Title 29 of the *Code of Federal Regulations* Part 1904 (29 CFR 1904) provides for recordkeeping and reporting by employers covered under the Occupational Safety and Health Act of 1970 as necessary or appropriate for enforcement of the Act, for developing information regarding the causes and prevention of occupational accidents and illnesses, and for maintaining a program of collection, compilation, and analysis of occupational safety and health statistics. Recordable occupational injuries and illnesses are any occupational injuries or illnesses which result in a fatality, lost workday cases (other than fatalities), or nonfatal cases without lost workdays.

Each year, the Bureau of Labor Statistics (BLS) conducts the Annual Survey of Occupational Injuries and Illnesses on a sample basis. Employers selected by BLS are required to complete the Occupational Injuries and Illnesses Survey Questionnaire. This survey form asks for summary data from the OSHA No. 200 form, along with information regarding the type of business activity and number of employees at the reporting unit.

29 CFR Part 1910.119—Process Safety Management of Highly Hazardous Chemicals

Title 29 of the *Code of Federal Regulations* Part 1910.119, referred to as the Process Safety Management (PSM) Standard, contains requirements for "preventing or minimizing the consequences of catastrophic releases of toxic, reactive, flammable, or explosive chemicals . . . where releases may result in toxic, fire or explosion hazards." The PSM Standard requires the employer to investigate each incident that resulted in, or could reasonably have resulted in, a catastrophic release of highly hazardous chemicals in the workplace. A *"catastrophic release"* is defined as a major uncontrolled emission, fire, or explosion involving one or more highly hazardous chemicals that presents serious danger to employees in the workplace. In general, the PSM Standard applies to processes that involve a chemical at or above the specified threshold quantities or to processes that involve a flammable liquid or gas onsite in one location, in a quantity of 10,000 pounds.

For each incident, an incident investigation report must be prepared at the conclusion of the investigation, reviewed with all affected personnel, and retained for five years. The regulation does not require an employer to release incident investigation results outside of the company. As such, this source of lessons learned data is available only within each individual company or where information is disclosed voluntarily.

DATABASES

The Bureau of Labor Statistics (BLS) maintains two separate databases, the Census of Fatal Occupational Injuries (CFOI) and the Survey of Occupational Injuries and Illnesses. Each database is described below.

Census of Fatal Occupational Injuries (CFOI)

Overview. The Bureau of Labor Statistics (BLS) Census of Fatal Occupational Injuries (CFOI) produces comprehensive, accurate, and timely counts of fatal work injuries. CFOI is a Federal–State cooperative program that has been implemented in all 50 States and the District of Columbia since 1992. To compile counts that are as complete as possible, the census uses multiple sources to identify, verify, and profile fatal worker injuries. Information about each workplace fatality—occupation and other worker characteristics, equipment involved, and circumstances of the event—is obtained by cross referencing source records, such as death certificates, workers' compensation reports, and federal and state agency administrative reports. To ensure that fatalities are work-related, cases are substantiated with two or more independent source documents, or a source document and a follow-up questionnaire.

Data compiled by the CFOI program are issued annually for the preceding calendar year. These data are used by safety and health policy analysts and researchers to help prevent fatal work injuries by

- Informing workers of life-threatening hazards associated with various jobs
- Promoting safer work practices through enhanced job safety training
- Assessing and improving workplace safety standards
- Identifying new areas of safety research

National data are published approximately eight months after the reference year in a news release, and more detailed data are published later in a bulletin. State-specific data on workplace fatalities may be requested from the state agencies participating with BLS in the census. Researchers may apply to BLS for access to the CFOI research file.

Data Sources. Data for the Census of Fatal Occupational Injuries are compiled from various federal, state, and local administrative sources—including death certificates, workers' compensation reports and claims, reports to various regulatory agencies, medical examiner reports, and police reports—as well as news reports. Multiple sources are used because studies have shown that no single source captures all job--related fatalities. Source documents are matched so that each fatality is counted only once. To confirm that a fatality occurred while the decedent was at work, information is verified from two or more independent source documents—such as death certificates, workers' compensation records, and reports to federal and state agencies—or from a source document and a follow--up questionnaire. The CFOI includes data for all fatal work injuries, both those that are covered by OSHA or other federal or state agencies or that are outside the scope of regulatory coverage.

Several federal and state agencies have jurisdiction over workplace safety and health. OSHA and affiliated agencies in states with approved safety programs cover the largest portion of America's workers. However, injuries and illnesses occurring in several other industries, such as coal, metal and nonmetal mining, and water, rail, and air transportation, are excluded from OSHA coverage because they are covered by other federal agencies, such as the Mine Safety and Health Administration, the U.S. Coast Guard, the Federal Railroad Administration, and the Federal Aviation Administration. Fatalities occurring in activities regulated by federal agencies other than OSHA accounted for about 20 percent of the fatal work injuries for 1995.

Fatalities occurring among several other groups of workers are generally not covered by any federal or state agencies. These groups include self-employed and unpaid family workers, which accounted for about 19 percent of the fatalities; laborers on small farms, accounting for about 2 percent of the fatalities; and state and local government employees in states without OSHA-approved safety programs, which account for about 4 percent. (Approximately one-half of the states have approved OSHA safety programs that cover state and local government employees.)

The BLS has formally acknowledged the efforts of the following agencies in assisting in incident data collection:

- Occupational Safety and Health Administration
- National Transportation Safety Board
- US Coast Guard
- Mine Safety and Health Administration
- Department of Defense
- Employment Standards Administration (Federal Employees' Compensation and Longshore and Harbor Workers' divisions)
- Department of Energy
- National Association of Chiefs of Police
- State vital statistics registrars, coroners, and medical examiners
- State departments of health, labor, and industries, and workers' compensation agencies
- State and local police departments
- State farm bureaus
- Federal Railroad Administration
- Federal Aviation Administration.

Contents. Approximately 30 data elements are collected, coded, and tabulated, including information about the worker, the fatal incident, and the machinery or equipment involved.

Limitations. A great deal of care is taken to ensure the accuracy of CFOI data through such measures as cross-referencing. However, the data are limited by the information available. The 1992 BLS Occupational Injury and Illness Classification Structures (OIICS) allow for ease of comparison and sorting, but naturally some clarity is lost in the coding. A narrative description of how an injury occurred is included in the file to retain this clarity. However, the extent of the information included in this field depends entirely upon the individual initially recording the incident.

Accessibility. The Census of Fatal Occupational Injuries (CFOI) Research File contains data collected from various data sources, in most cases under a pledge of confidentiality with the understanding that the information will be used for statistical and research purposes only. Although state codes and all personal identifiers have been excluded from the file, it may be possible to discover the identity of a decedent or business establishment. The CFOI Research File is available to those researchers who agree to protect the confidentiality of the data and who have safeguards in place to do so. Upon approval of this application, the Bureau of Labor Statistics (BLS) will prepare a Letter of Agreement that must be approved and signed by the Commissioner for Labor Statistics and by an official of the recipient's organization prior to release of the research file. By signing the Letter of Agreement, the researcher and the researcher's organization agree to adhere to the BLS confidentiality policy, as applicable to the CFOI Research File. In addition, all individuals who will have access to the CFOI data must sign a BLS Non-disclosure Affidavit (acknowledging their understanding of the BLS confidentiality policy) prior to accessing the CFOI data.

Initially, the recipient will be authorized to use the CFOI data for three years; however, the period may be extended to complete the project(s) if a request is made in writing to and approved by the BLS Project Coordinator.

The BLS Internet website (http://www.bls.gov/oshhome.htm) includes news releases and tables that summarize occupational injury and illness statistics. The data tables are searchable by keyword. Only summary statistics and industry averages broken down by various characteristics (e.g., event, type of injury, occupation, etc.) are available. Access to individual incident records is **not** available at the BLS website.

Survey of Occupational Injuries and Illnesses

Overview. The Bureau of Labor Statistics reports annually on the number of missed work days that work injuries and illnesses cause in private industry and the rate of such incidents since the early 1970s. The 1994 national survey marks the third year that BLS has collected additional detailed information on such cases in the form of worker and case characteristics data, including workdays lost, summarized in this release. (Counts and rates for cases without lost workdays and related measures also date back to the early 1970s. Because of limited resources, additional detail on less serious cases such as these has not been collected.)

In addition to the summary injury and illness data, the BLS survey provides details on the more seriously injured and ill workers (occupations, age, gender, race, and length of service) and on the circumstances of their injuries and illnesses (nature of the injury/illness, part of body affected, event or exposure, and primary and secondary sources of the injury/illness). "More seriously" is defined in the survey as involving days away from work.

Data Sources. The Survey of Occupational Injuries and Illnesses is a federal/state program in which employer reports are collected from about 250,000 private industry establishments and processed by state agencies cooperating with the Bureau of Labor Statistics. Occupational injury and illness data for coal, metal, and nonmetal mining are provided by the Department of Labor's Mine Safety and Health Administration and for railroad activities by the Department of Transportation's Federal Railroad Administration. The survey measures nonfatal injuries and illnesses only. The survey excludes the self-employed, farmers with fewer than 11 employees, private households, and employees in federal, state, and local government agencies.

Data Collected. OSHA works with state agencies to compile occupational injury and illness data from employers within specific industry and employment size specifications. The survey is based on a sampling of industry data; it is not a comprehensive census. Limitations of the sampling approach are discussed in a later section. Each year, survey forms are sent to employers. The Survey of Occupational Injuries and Illnesses (see Appendix E) is completed by copying information recorded on an employer's *Log and Summary of Occupational Injuries and Illnesses*, OSHA No. 200, and information about employment and hours worked from the employer's payroll records. Information taken from the OSHA No. 200 form includes the total number of injuries of various severity, the total types of illnesses, and the total number of illnesses of different types.

In addition to summary data on injury and illness counts, survey respondents also are asked to provide additional information for a subset of the most serious nonfatal cases logged, namely, those that involved at least one day away from work beyond the day of injury or onset of illness. Each individual case is recorded on the Survey's one-page form, *Case with Days Away from Work* (see Appendix F). Employers answer several questions about these cases, including the demographics of the disabled worker, the nature of the disabling condition, and the event and source that produced the condition.

Most employers use information from supplementary recordkeeping forms and state workers' compensation claims to fill out the Survey's "case form." Some, however, attach those forms when their narratives answer questions on the case form, an option the Bureau offers to help reduce respondent burden. Also, to minimize the burden of many larger employers, sampled establishments projected to have large numbers of cases involving days away from work receive instructions on how to sample those cases.

The Bureau developed a new Occupational Injury and Illness Classification System to permit standardized and uniform coding of the injury or illness involving days away from work and the way it occurred. The major code structures of that system are

- Nature of injury or illness, or the principal physical characteristic of the worker's injury or illness
- Part of the body directly affected by the injury or illness
- Source of injury or illness, that is, the object, substance, bodily motion, or work environment which directly produced or inflicted the injury/illness

- Event or exposure, which describes the manner in which the injury or illness was inflicted or produced
- Secondary source, which identifies the object, substance, or person that generated the source of injury or illness or that contributed to the event/exposure.

Data Limitations. The survey measures the number of new work-related illness cases that are recognized, diagnosed, and reported during the year. Some conditions, e.g., long-term latent illnesses caused by exposure to carcinogens, often are difficult to relate to the workplace and are not adequately recognized and reported. These long-term latent illnesses are believed to be understated in the survey's illness measures. In contrast, the overwhelming majority of the reported new illnesses are those that are easier to directly relate to workplace activity (e.g., contact dermatitis or carpal tunnel syndrome).

The survey estimates of the characteristics of cases with days away from work are based on a scientifically selected probability sample, rather than a census of the entire population. Two levels of sampling are used. First, establishments are selected to represent themselves and, in many instances, other establishments of like industry and workforce size that are not selected in a given survey year. Then, sampled establishments projected to have large numbers of days away from work cases are instructed before the survey begins on how to sample those cases to mini mize the burden of their response. An establishment expected to have 20 or fewer cases, however, is instructed to report on each case, regardless of the actual number it has for the year.

The number and frequency of these cases are based on logs and other records kept by private industry employers throughout the year. These records reflect not only the year's injury and illness experience but also the employer's understanding of which cases are work-related under current recordkeeping guidelines of the U.S. Department of Labor. The number of injuries and illnesses reported in a given year also can be influenced by changes in the level of economic activity, working conditions and work practices, worker experience and training, and the number of hours worked.

The data are also subject to nonsampling error. The inability to obtain detailed information about all cases in the sample, mistakes in recording or coding the data, and definitional difficulties are general examples of nonsampling error in the survey. Although not measured, nonsampling errors will always occur when statistics are gathered. However, BLS has implemented quality assurance procedures to reduce nonsampling errors in the survey, including a rigorous training program for state coders and a continuing effort to encourage survey participants to respond fully and accurately to all survey elements.

Data Accessibility. Data collected or maintained by the Bureau of Labor Statistics under a pledge of confidentiality is treated in a manner that will assure that individually identifiable data will be accessible only to authorized persons

(BLS employees) and will be used only for statistical purposes. As such, no direct access to the injuries and illnesses data is available to anyone outside of the BLS.

However, the BLS does issue annual reports on the injury and illness data collected. Summary data on nonfatal counts and rates for all recordable injuries and illnesses (separately for those with and without lost workdays) are issued in December of each year. The following April, summary data are issued on the characteristics of workers sustaining days-away-from-work injuries and illnesses and how those incidents occurred. Other regular publications presenting annual survey information include a summary of counts and rates by detailed industry and a comprehensive bulletin on counts, rates, and characteristics.

The BLS generates numerous tables that expand on information contained in the basic summary tables included in its annual news releases. The supplementary tables, for example, provide injury and illness counts for several hundred industries and occupations, as well as detailing the categories within the nature of injury or illness, part of body affected, source of injury or illness, and event or exposure. The characteristic featured in a table often is cross tabulated with selected categories of another characteristic, for example, detailed occupation by event or exposure. The latter table is useful in profiling the major ways that workers are hurt in high-risk jobs.

The summary tables are available on the Internet at the following locations:

- ftp://stats.bls.gov/pub/special.requests/ocwc/osh/c_d_data/
- http://stats.bls.gov/blshome.html

Environmental Protection Agency (EPA)

LAWS AND REGULATIONS

The laws and regulations administered under the U.S. Environmental Protection Agency that relate to the collection of incident data focus on accidental releases of hazardous substances. The primary federal statutes that cover accidental release reporting are the Comprehensive Environmental Response, Compensation, and Liability Act (CERCLA) and the Emergency Planning and Community Right-to-Know Act (EPCRA).

CERCLA, commonly referred to as Superfund, authorizes the federal government to respond directly to releases, or threatened releases, of hazardous substances that may endanger public health, welfare, or the environment. CERCLA also enables EPA to take legal action to force parties responsible for causing the contamination to clean up those sites or reimburse the costs of cleanup.

In 1986, CERCLA was updated and improved under the Superfund Amendments and Reauthorization Act (SARA). The reauthorized law made several important changes and additions to the program, such as increased criminal penalties for failure to report releases of hazardous substances. In addition to these improvements, SARA included a section called Title III that focused on strengthening the rights of citizens and communities concerning potential hazardous substance emergencies. SARA Title III, the Emergency Planning and Community Right-to-Know Act of 1986 (EPCRA), was designed to help communities prepare to respond in the event of a chemical emergency, and to increase the public's knowledge of the presence and threat of hazardous chemicals. Under SARA Title III, facilities must compile information about extremely hazardous substances they have onsite and the threat posed by those substances. In addition, those facilities must report any accidental releases of extremely hazardous substances. This information must be provided to state and local authorities, and more specific data must be made available upon request from those authorities or from the general public.

The laws and regulations governing accidental release reporting are described below, followed by details of the databases maintained by the EPA.

40 CFR 302.6—Notification Requirements

This regulation requires the person in charge of the facility to immediately report any release of a CERCLA "hazardous substance" that meets or exceeds the reportable quantity (RQ) listed in 40 CFR 302.4 to the National Response Center (NRC). Releases of mixtures or solutions (including hazardous waste streams) of hazardous substances are also subject to these requirements.

40 CFR 355.40—Emergency Release Notification

This regulation requires facility owners and operators to report to state and local authorities any releases of reportable quantity or more of a CERCLA hazardous substance (1 pound or more if a reporting trigger is not established by regulation) or an EPCRA "extremely hazardous substance" that results in *exposure of people outside the facility boundary*.

This regulation applies to any facility at which a hazardous chemical is produced, used, or stored and at which there is release of a reportable quantity of any extremely hazardous substance or CERCLA hazardous substance. Substances subject to this requirement are those on the list of 360 extremely hazardous substances as well as the more than 700 hazardous substances subject to the emergency notification requirements under CERCLA.

As soon as practicable after a release, the owner or operator must provide a written follow-up emergency notice (or notices, as more information becomes available) setting forth and updating the original information, including additional information with respect to actions taken to respond to and contain the release, any known or anticipated acute or chronic health risks associated with the release, and, where appropriate, advice regarding medical attention necessary for exposed individuals.

40 CFR Part 68—Accidental Release Prevention Requirements

Title 40 of the Code of Federal Regulations Part 68 (40 CFR 68) was promulgated under the Clean Air Act (CAA) Section 112(r)(7). The intent of the requirement is to "prevent accidental releases to the air and mitigate the consequences of such releases by focusing prevention measures on chemicals that pose the greatest risk to the public and the environment."

This regulation, often referred to as EPA's Risk Management Plan (RMP) Rule, is similar to OSHA's PSM Standard (29 CFR 1910.119) but includes additional requirements, such as the need to conduct a Hazard Assessment based on a "worst-case" accident scenario and submit the entire RMP, including a five-year accident history, to Local Emergency Planning Committees and other agencies.

The five-year accident history covers all accidents involving regulated substances, but only from covered processes at the source that resulted in serious on-site or certain known offsite impacts in the five years prior to the submission of the RMP.

DATABASES

Two databases are maintained by the U.S. EPA that contain information on accidental releases of hazardous chemicals. A brief description of each database follows.

Emergency Response Notification System (ERNS)

Overview. The Emergency Response Notification System is a national computer database and retrieval system used to store information on notifications of oil discharges and hazardous substances releases. ERNS provides a mechanism for documenting and verifying incident notification information as initially reported to the National Response Center (NRC), EPA, and/or the U.S. Coast Guard. The ERNS program is a cooperative data sharing effort among EPA Headquarters, the Department of Transportation Research and Special Programs Administration's (RSPA) John A. Volpe National Transportation Systems Center, other DOT program offices, the ten EPA Regions, and the NRC. The database includes information on release incidents occurring at fixed facilities, marine/offshore facilities, pipelines and transportation vehicles, and covers a broad spectrum of toxic chemical incidents, including chemical fires, explosions, spills, illegal dumping and air releases. ERNS provides the most comprehensive data compiled on notifications of oil discharges and hazardous substance releases in the United States. Since its inception in 1986, more than 275,000 release notifications have been entered into ERNS.

Data Sources. Notifications are received by the National Response Center or one of the ten EPA Regions and are documented in a report. The data usually include information about the material and the quantity released, the discharger, and the location of the release. This report is transferred to an appropriate agency, which evaluates the need for a response, and electronically transfers, once daily, all records to the ERNS database. In addition, each of the EPA Regions can update ERNS records if additional information becomes available after the initial notification. However, if a caller makes an additional report to update previous data, a second record is created and transferred to the ERNS database.

Examples of responding agencies are as follows:

- Environmental Protection Agency (EPA)
- United States Coast Guard (USCG)
- Department of Transportation (DOT)
- State Emergency Response Commission (SERC)
- Federal Emergency Management Agency (FEMA)
- Local Emergency Planning Committee (LEPC).

Contents. The types of release reports that are available in ERNS fall into three major categories: substances designated as hazardous substances under the Comprehensive Environmental Response, Compensation, and Liability Act of 1980 (CERCLA), as amended; oil and petroleum products, as defined by the Clean Water Act of 1972 (CWA), as amended by the Oil Pollution Act of 1990; and all other types of materials.

Parameters. ERNS contains, in addition to other data, information about the material and the quantity released, where the release occurred, when the release occurred, what agencies have been notified, and any information about property damage, injuries, and deaths occurring due to the release. In addition, when analyzing ERNS data, it is always important to consider that the information is typically based on the initial notification reported to a number of government agencies. Therefore, especially with historic data, there may be inconsistencies in the data because of different methods of data entry.

Limitations. Because ERNS is a database of initial notifications and not incidents, there are several limitations to the data. ERNS primarily contains initial accounts of releases, made during or immediately after a release occurs when exact details are often unknown. The data are usually not updated unless an EPA Region is involved in the response action. There may be multiple reports for a single incident, (this occurs when the caller makes a second report to update original data or a private observer reports a release that has already been reported by the facility). Because reports are taken over the phone, transcription errors (e.g., misspellings of discharger or location information), occasionally limit the quality of some data. According to the EPA, "Only about a third of the 193 information data fields are completed for most of the release notifications."

Accessibility. ERNS data are available in various forms. The cost of obtaining ERNS data is determined based on the medium used and the time and effort expended to fill the request. Data can either be delivered via First Class Mail or e-mail through the Internet. General descriptions of the information formats are provided below:

- *Summary Release Information.* These reports in table format offer a broad overview of data and are useful for analyzing trends in chemical and oil releases, or comparing groups (e.g., total release reports involving crude oil by year).
- *Standard ERNS Reports.* These reports are either dBASE files or one page reports in various word processing formats containing information about specific release notifications. This format is best for providing information on a small subset of data, such as notifications from a particular geographical region, on a specific chemical, or about an individual site.
- *ERNS database.* This is recommended for requestors who have extensive needs for ERNS data. The database is available two ways: (1) all data by year, which is available on magnetic data tapes from the National Technical Information Service (NTIS), or (2) data for each EPA Region by year, which may be downloaded from EPA via the Internet, direct dial-in, or FedWorld. The magnetic data tapes may be purchased from NTIS in Springfield, VA at (703)487-4650. Information on downloading data may be obtained by calling the ERNS Information Line at (202)260-2342.

Information on the ERNS database may also be obtained by calling the ERNS Information Line at (202) 260-2342, by sending an e-mail request to erns.info@epamail.epa.gov, by contacting the Freedom of Information Act (FOIA) Officer in the specific EPA Region of interest, or by contacting the ERNS Manager at the following address:

U.S. EPA
Mail Code 5203G
401 M Street, SW
Washington, DC 20460

Accidental Release Information Program (ARIP)

Overview. The ARIP database is a collection of information on accidental releases of hazardous chemicals at fixed facilities. The purpose of ARIP is to collect information from facilities that have had significant releases of hazardous substances, develop a national accidental release database, analyze the collected information, and disseminate the results of the analysis to those involved in chemical accident prevention activities. ARIP also helps to focus industry's attention on the causes of accidental releases and the means to prevent them. The ARIP data are collected by the EPA regional offices through the ARIP questionnaire, and then forwarded to EPA headquarters for inclusion in the ARIP database.

The ARIP database is maintained by the U.S. EPA Chemical Emergency Preparedness and Prevention Office (CEPPO), Office of Solid Waste and Emergency Response.

Data Sources. U.S. facilities are required by law, as discussed above, to report nonroutine releases of certain substances when those releases exceed a reportable quantity. The EPA compiles the reports into the Emergency Response Notification System (ERNS) database. The EPA then uses ERNS data to select releases for the ARIP questionnaire. The ERNS database includes a wide range of releases from both fixed facilities and transportation. Since the Department of Transportation is responsible for transportation accidents and OSHA is responsi ble for accidents affecting workers, ARIP targets those accidental releases at fixed facilities that resulted in off-site consequence or environmental damage. To focus on significant accidents, an ARIP questionnaire is sent to all facilities with releases that resulted in death or injury. If the release also resulted in off-site consequence or environmental damage, then the facility must complete the questionnaire.

Contents. The ARIP questionnaire consists of 23 questions about the facility, the circumstances and causes of the incident, and the accidental release prevention practices and technologies in place prior to, and added or changed as a result of, the event. The questionnaire focuses on several areas of accident prevention including hazard assessments, training, emergency response, public notification procedures, mitigation techniques, and prevention equipment and controls. The ARIP database contains information on incidents that occurred at fixed facilities in the United States from 1986 to 1995.

Parameters. Information contained in the ARIP database, in addition to data obtained from ERNS, covers the following categories: facility profile, hazardous substance release profile, and prevention profile.

Accessibility. The ARIP database can either be delivered via First Class Mail or e-mail through the Internet. The database requires about 9 megabytes of hard-disk space. It can be viewed and searched using any database software, such as dBase IV, FoxPro, or Access. The distributor is the Chemical Emergency Preparedness and Prevention Office. The point of contact for further information is as follows:

Emergency Planning and Community Right-to-Know Hotline
   U.S. EPA (5101)
   401 M St., SW
   Washington, DC 20460
   1-800-424-9346
   1-800-535-0202

The analysis of ARIP information and the resulting insights into the nature of chemical accidents are also published in EPA reports that are shared with interested individuals and organizations. In June, 1989, the EPA published a chemical accident prevention bulletin entitled, "Why Accidents Occur: Insights from the Accidental Release Information Program," which summarized the results and lessons learned from initial analysis of the ARIP data. This publication was targeted to state emergency response commissions (SERCs) and local emergency planning committees (LEPCs) and was designed to enhance their understanding of accident causes and steps used to prevent accidents.

Current Efforts

Most of the work on lessons learned databases is in professional and trade organizations and in the government. The American Petroleum Institute (API) and the American Institute of Chemical Engineers (AIChE) are actively involved in lessons learned database development. The government agencies primarily involved in lessons learned activities are OSHA and EPA, along with their associated research agencies, the National Institute for Occupational Safety and Health (NIOSH) and the Office of Research and Development (ORD), respectively. These agencies conduct research and administer programs intended to promote safety in the workplace and to protect the environment from accidental releases of hazardous chemicals.

API and AIChE Center for Chemical Process Safety (CCPS)

The API is in the midst of a two-year developmental program with the American Institute for Chemical Engineers, in particular the Center for Chemical Process Safety, to offer its members an incident database that will track refinery and petrochemical accidents and incidents. Thus far, 20 firms have agreed to participate. Each firm will be responsible for submitting to API updated summaries of incidents.

Joint OSHA and EPA Efforts

Both OSHA and EPA have a statutory responsibility to investigate major chemical accidents. Much of the same information is required to satisfy the requirements of the two agencies. As such, the U.S. government called for joint investigations that minimize duplication of effort. In December 1996, a Memorandum of Understanding (MOU) between the EPA and OSHA on Chemical Accident Investigation was issued. Together, OSHA and EPA will undertake investigations to determine the root cause(s) of chemical accidents and to issue public reports. A major chemical accident is defined in this MOU as one that meets one or more of the following criteria:

- Results in one or more human fatalities
- Results in the hospitalization of three or more workers or members of the public
- Causes property damage (on- and/or off-site) initially estimated at $500,000 or more in total
- Presents a serious threat to worker health or safety, public health, property, or the environment
- Has significant off-site consequences, such as large-scale evacuations or protection-in-place actions, closing of major transportation routes, substantial environmental contamination or substantial effects (e.g., injury, death) on wildlife or domesticated animals
- Is an event of significant public concern.

OSHA and NIOSH

In 1994, OSHA initiated a prioritization methodology, the Priority Planning Process, to take into account current incident trends in the workplace. The process is aimed at identifying the top-priority workplace safety and health hazards in need of either regulatory or non-regulatory action. The resulting set of priorities is intended to round out the agency's existing programs in order to ensure that the leading causes of occupational injuries, illnesses, and deaths are being effectively addressed. Incident data are used directly in determining the areas in which research and regulatory action are most needed. OSHA is responsible for implementing regulations and standards, while NIOSH is the government agency that conducts the related research. NIOSH investigates occupational health and safety hazards, develops specific control methods, and recommends federal hazard-reduction standards.

NIOSH played a significant role in the OSHA Priority Planning Process by providing technical expertise and assisting in the selection of priorities. Because many of the OSHA priorities both affect and depend upon NIOSH research, OSHA invited NIOSH to take a leadership role in developing action plans for these priorities. In addition, the results of the Priority Planning Process helped NIOSH develop the National Occupational Research Agenda (NORA). NORA is an ongoing project designed to set priorities for workplace research over the next decade. The agenda is the product of input from approximately 500 organizations and individuals outside of NIOSH.

The Priority Planning Process identified priority research areas that were grouped into three categories: Disease and Injury, Work Environment and Workforce, and Research Tools and Approaches.

EPA and ORD

The Office of Research and Development (ORD) is the scientific and technological arm of the EPA. Comprising three headquarters offices, three national research laboratories, and two national centers, ORD is organized around a basic strategy of risk assessment and risk manage ment to remediate environmental and human health problems.

Improved risk assessment and risk management are major goals of ORD's research. Risk assessment is the basis for both policy and technical decisions to determine priorities for risk management and to guide the actual process of managing risks. Risk assessment is conducted on both human health risks and ecological risks. Risk management is used to address the identified risks in both areas.

The assessment of risk to human health follows a widely accepted paradigm produced by the National Academy of Sciences (NAS) in 1983 and reiterated by the NAS in 1994 in Science and Judgement in Risk Assessment. The risk assessment paradigm is a framework for organizing analyses of pertinent data in a way that emphasizes the interdisciplinary nature of the work. The four elements of the framework are:

1. Hazard identification
2. Dose response assessment
3. Exposure assessment
4. Risk characterization.

Proposed Efforts

Although there are a number of incident-related databases currently available, the quality and level of detail is often not adequate for compiling and analyzing accident trends and common causes; nor are the existing databases specifically geared towards the CPI. Efforts should be made to build upon current data collection programs and to increase the chemical industry's use of lessons learned databases.

To be effective, an incident database should be widely available and should contain complete and standardized information. The ideal lessons learned database would compile incident-related data for the entire chemical industry. All CPI companies would contribute data and have access to the entire dataset. Appropriate measures could be taken to ensure anonymity. The information contained in individual records must be complete with emphasis on lessons learned. The incident data must provide all information necessary to determine likely causes and to identify trends. Standardized keywords would facilitate this process.

The database could be compiled from direct input from chemical companies. For facilities covered by RMP, accident histories will already be in electronic form per regulation requirements, thus the need for data re-entry would be eliminated. When a company submits its Risk Management Plan, it could forward an electronic copy of the accident history section to a central lessons learned database repository. If a standard format is maintained, a simple extraction program could import the incident data directly into the database.

Another approach could be a systematic review and upgrade of existing publicly available databases, such as the Bureau of Labor Statistics' Survey of Occupational Injuries database and EPA's ERNS database. These databases are large, but entries often are incomplete or lack the information required by a true lessons learned database.

Justifications

Expanding the use of lessons learned databases in the CPI is not an easy task. It will require much time and effort in gathering, standardizing, and analyzing data. In addition, logistic barriers will have to be overcome to provide access to the information throughout the industry. However, there are several reasons that justify this undertaking.

First, lessons learned databases are useful in identifying accident trends. For example, NIOSH, as described earlier, uses incident data directly in its Priority Planning Process in determining the areas in which research and regulatory action are most needed to address accident trends. NIOSH does this at a nationwide level, looking at data for every industry. A lessons learned database specifically geared toward the CPI could be used in a similar, but more focused, manner as NIOSH uses the occupational injury database. Thus, the specific needs of the CPI would be identified and addressed more rapidly and efficiently.

Second, the recent EPA Risk Management Program regulation, covering much of the chemical industry, requires a five year accident history to be included in each company's Risk Management Plan, which is available to the public. If incident information is already in the public's eye, why not show that the information is being put to good use? Instead of being merely individual records of past accidents, a lessons learned database would demonstrate to the public that the CPI is actively engaged in activities to learn from the past and prevent future incidents from occurring.

One last reason, simply put, is that anything that will save lives and reduce injury is worth the effort. Lessons learned databases have a tremendous potential to do just that. The information is available, the CPI should make every effort to use the information to its fullest extent.

References

1. Occupational Safety and Health Act of 1970, Public Law 91 - 596 (December 29, 1970).
2. United States Department of Labor, Occupational Safety and Health Administration, "Process Safety Management of Highly Hazardous Chemicals," 29 CFR 1910.119 (February 24, 1992).

3. United States Department of Labor, Occupational Safety and Health Administration, "Recording and Reporting Occupational Injuries and Illnesses," 29 CFR 1904 (July 2, 1971).
4. United States Department of Labor, Bureau of Labor Statistics, "Brief Guide to Recordkeeping Requirements for Occupational Injuries and Illnesses" (June 1986).
5. United States Department of Labor, Occupational Safety and Health Administration, "Memorandum of Understanding Between the United States Environmental Protection Agency and the United States Department of Labor Occupational Safety and Health Administration on Chemical Accident Investigation" (December 1, 1996).
6. United States Department of Health and Human Services, "National Occupational Research Agenda (NORA) Priority Research Areas" (April 1996).
7. United States Environmental Protection Agency, Office of Research and Development, "1997 Update to ORD's Strategic Plan" (April 1997).
8. "Comprehensive Environmental Response, Compensation, and Liability Act (CERCLA)," 40 CFR Parts 300-310 (1980).
9. "Emergency Planning and Community Right-to-Know Act (EPCRA)," 40 CFR Parts 302, 355, 372 (1986).
10. United States Environmental Protection Agency, "Notification Requirements," 40 CFR 302.6.
11. United States Environmental Protection Agency, "Emergency Release Notification," 40 CFR 355.40.
12. United States Environmental Protection Agency, "Risk Management Program Rule," 40 CFR Part 68 (June 20, 1996).
13. *Accidental Release Information Program (ARIP) FactSheet*, Chemical Emergency preparedness and Prevention Office, U.S. EPA, September 1994.
14. *An Overview of ERNS, Emergency Response Notification System Fact Sheet*, U.S. EPA, March 1995.
15. *Federal Register, Environmental Protection Agency, Part II, 40 CFR Parts 300 and 355, Extremely Hazardous Substances List and Threshold Planning Quantities; Emergency Planning and Release Notification Requirements; Final Rule*, Wednesday April 22, 1987.
16. *Federal Register, Environmental Protection Agency, Part II, Parts 117, 302, and 355 Reportable Quantity Adjustments*; Final Rule, Monday, June 12, 1995 (EPA).
17. *Accidental Release Information Program (ARIP) FactSheet*, Chemical Emergency preparedness and Prevention Office, U.S. EPA, September 1994.
18. *An Overview of ERNS, Emergency Response Notification System Fact Sheet*, U.S. EPA, March 1995.

Incident Database and Macroanalysis to Help Set Safety Direction

John A. McIntosh, III and Sarah Rogers Taylor
The Procter & Gamble Company, William Hill Technical Center, 6110 Center Hill Avenue, Cincinnati, Ohio 45224-1789

ABSTRACT

This paper describes the framework of an incident database and how incident data can be used to help set safety direction. It includes history, database design, and data collection and utilization. Examples illustrate how macroanalysis of incidents revealed inherently weak systems. These systems suffered disproportionate losses. Analyzing data from incident investigations improved the understanding of the risks associated with the processes. In partnership with the product category, corporate process safety organizations initiated equipment design modifications and procedural changes. These changes significantly reduced both the likelihood and consequences of incidents.

When safety incidents occur, most organizations conduct investigations and prepare reports. These reports generate much information. What happens to the collected information? Is the data reviewed frequently? Are incidents in similar systems analyzed on a macro basis? If incidents occur in similar systems, do reviews reveal lessons learned? Answers to these questions may depend on how easily the data can be accessed.

Databases offer an effective option for managing large amounts of information. Used to study process safety trends and underlying causes of incidents, databases can be powerful and effective risk management tools. Macroanalysis of incident data can reveal process safety weaknesses and help risk managers determine where to focus effort and resources

Introduction

When safety incidents occur, most organizations conduct investigations and prepare reports in an effort to learn the cause and to determine how to prevent similar incidents from occurring again. What happens to the collected information? Is the data reviewed frequently? If incidents occur in similar systems, do reviews of the data reveal lessons learned? Answers to these questions may depend on how easily the data can be accessed. Databases offer an effective option for managing large amounts of information. Used to study process safety trends and underlying causes of incidents, databases can be powerful and effective risk management tools. Macroanalysis of incident data can reveal process safety weaknesses and help risk managers determine where to focus effort and resources. Three case studies illustrate how Corporate process safety personnel, in conjunction with product category personnel, use incident data to reveal process safety weaknesses; to initiate equipment design modifications and procedural changes; and to improve risk management programs.

Background

A main objective of the corporate process safety organization is to improve the understanding of risks associated with processes and to reduce the likelihood and consequences of incidents. One strategy is to study incidents and reapply lessons learned. Incident reports have been collected at P&G for almost 20 years to gather information and report findings. Originally, incident reports were used primarily for documentation, but not much was done with the data in the reports. Manually reviewing reports and sorting incidents was not an effective method for analysis of incident data. Consequently, little data analysis was done, and potential lessons learned may not have been revealed. To improve data analysis capabilities, a process safety incident database was designed and developed. The process safety incident database provides an effective tool for managing the large amounts of information found in incident reports.

Data Collection

What data is important? How will the data be analyzed? What do we want to learn from the data? Answers to these questions help define what data to collect and how to collect it. For process safety incident analysis, the first step is to define what scenarios should be considered process safety incidents. This streamlines the contents of the database, and allows the analyst to focus on scenarios of interest. Procter and Gamble defines a process safety incident as "anything in a process or utility system which caused or could have caused a fire; an explosion; a release of flammable, reactive or hazardous material; or an overpressure condition (positive or negative)."

To develop an accurate and complete picture of process safety history, all incidents need to be investigated and reported. Consistent data reporting from process to process and from site to site maximizes the usefulness of the data. To ensure consistency in data collection and reporting, Procter and Gamble uses a standard incident investigation report form. The incident investigation form contains predefined data fields, including process, equipment, materials, costs, and incident category. The predefined data fields provide consistency and allow for easy queries on specific pieces of data. The report also contains sections for more detailed description of the incident and the causes. These sections provide more detailed information which may reveal critical insight on the incident.

Data Integrity

Incident investigations and reporting are critical components of process safety management. To reduce the likelihood and consequences of incidents, lessons learned from process safety incidents should be reapplied to similar systems. To do this, we need a complete, company-wide picture of process safety incidents. How do we ensure all process safety incidents are reported? How do we ensure data on the incident reports are complete and accurate? At Procter and Gamble, all risk program leaders are trained in conducting incident investigations and reporting results, including training on the use of incident investigation report forms. Following an incident, site personnel and corporate risk managers review incidents and discuss report content to ensure corporate risk managers fully understand the incident and its causes. Any unclear information or questions are resolved before database entry. Incident reporting is also checked via risk management program audits. Incident reporting is a specific line item in the audit. Auditors compare the number of incidents reported during a specific period of time and compare this with site records.

Database Design and Future Enhancements

The original Procter and Gamble process safety incident database was developed using commercially available, PC-based database software. Each field in the incident database corresponded to a field on the incident report form. The database structure provided the analyst with flexibility to perform ad hoc queries in addition to producing predefined reports. Flexibility to perform queries on any of the data fields is crucial. This allows the analyst to probe more deeply into cause and effect relationships. The next generation incident database is being developed on commercially available, mainframe-based software. While maintaining the flexibility of the original database, the change in software will align the incident database with our existing industrial health and safety database. This alignment will allow sites to enter data directly into the incident database, eliminating the need for a "hard copy" to be filled out and re-entered by corporate process safety.

Database Uses

The following Case Studies illustrate how Corporate process safety personnel, in conjunction with product category personnel, use the incident data to reveal process safety weaknesses, improve safety programs, and focus resources effectively.

Case Study 1: Risk Reduction—Reactive Releases

As we have pointed out, Procter and Gamble defines an incident as "Anything in a process or utility system which caused or could have caused a fire, explosion, release of flammable, reactive or hazardous material, or an overpressure condition (positive or negative)." The first step in dealing with large numbers of incident reports is to assure, as much as possible, that all incidents are classified into one of these defined categories. Figure 1 is a Pareto chart showing the distribution of the numbers of incidents in each of the defined categories.

Fires and explosions are combined because the distinction between the two is sometimes hard to make. Though "Release of Reactive Materials" is the third bar on this Pareto, it presents an interesting and fruitful example of how we used incident data to drive a risk reduction project. What does the data represented by the third bar of the chart in Figure 1 tell us? If all of the reactive releases are analyzed, an interesting picture begins to emerge. We took the data from the third bar on the Pareto chart above and created a new Pareto chart from that data.

The new chart is called a "nested Pareto" (Figure 2). The nested Pareto shows how Reactive Releases were distributed across our diverse product sectors. This chart shows us that over 75% of our releases occurred in one product sector. This is, of course, the classic "Pareto Principal"—80% of our problems are in 20% of our product sectors. So what should we do with this knowledge? Is there more knowledge to be gained? We repeated the process of developing another nested Pareto from the data represented by Product Sector "A" in the chart above. The incidents from that sector were distributed by the "Product Categories" within the sector. The chart shows this further subdivision of the data (Figure 3).

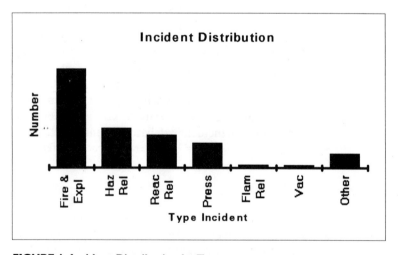

FIGURE 1. Incident Distribution by Type

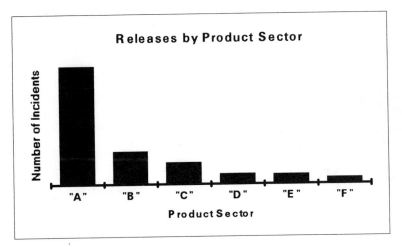

FIGURE 2. Releases by Product Sector

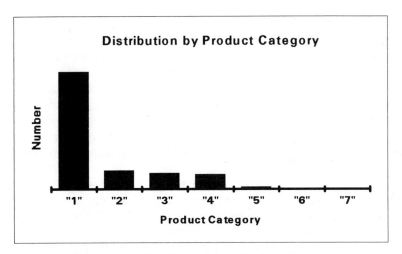

FIGURE 3. Distribution of Incidents Across Product Categories

This Pareto shows a clearer picture of where our problems were. Product Category "1" is responsible for most of our reactive releases. What did we do with this information? We know *where* we should act, but as of yet, we don't know *how* we should act. At this point, further classification of data helped us decide what we should do. The next step was to find out which chemicals were responsible for our releases. Again we used a nested Pareto (Figure 4) to discover that chemical "A" was the chemical of concern.

Every time we categorized the data and charted it we learned more. We now knew which product category and which chemical we should focus on. Further,

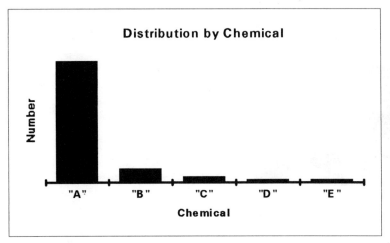

FIGURE 4. Distribution of Incidents by Chemical

from this information we knew which experts we needed to help us. We also had the information to show those resources why this effort was important to them. These pictures did, literally, say a thousand words. The charts were extremely powerful tools in convincing stakeholders, including upper management and engineering organizations, to dedicate resources to risk reduction efforts.

At this point we went to the leadership in Product Category "1" and presented this analysis, much the way we have presented it in this paper. We requested and were granted the formation of a task force made up of engineering, operations, maintenance and Process Safety personnel. The team had the specific goal of reducing releases of chemical "A."

The formation of the team was a milestone in our efforts to reduce reactive releases. Our analysis of incident data had given us specific direction. We knew which chemical and processes to concentrate on. It gave us an effective presentation tool to convince management to provide us resources. However, we still did not know exactly what we needed to do. This was the work of the task force.

The task force used the same method as we have used up to this point. Review the data, categorize the data, and chart the data. This is a simple concept, but this is not a trivial task. The task force reviewed all of the incident reports of concern and tried to decide how to categorize the different characteristics of the incidents. Several agreements were made on how to categorize data. However, when the data was charted on a Pareto, no clear 80/20 relationship would show up. When this occurred, we would go back and ask ourselves if there was another way to categorize this data. Ultimately we agreed on the categories shown on the chart below. Since all processes using this chemical were essentially the same design, categorizing the incident data by process components (Figure 5) was a successful strategy for us.

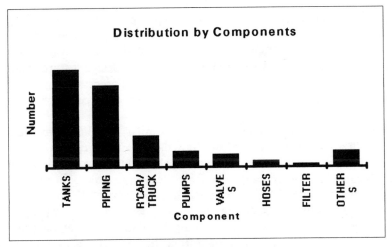

FIGURE 5. Distribution of Incidents by Component

We were beginning to zero in on the "what" we needed to do. The data revealed that we had a basic, systemic weakness in either the design or operation of our storage tanks and our piping systems. We further broke this data down and found that over 60% of our tank problems were simple tank over fillings (Figure 6).

Now we knew *what* we needed to do. We needed to stop trying to put two gallons of material in a one gallon bucket. The simple answer, of course, was to make a elementary design change and add a high level switch on the storage

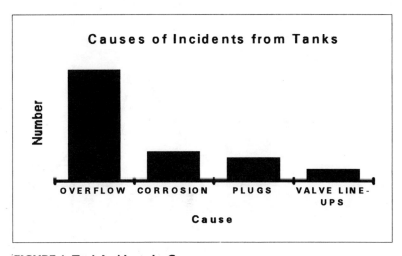

FIGURE 6. Tank Incidents by Cause

tanks. The switch was interlocked to the unloading pump and shut off the pump when the high level switch was made. When the tank is being filled, an input to the unloading pump logic is "High Level Switch Not Made". In retrospect, this seems as if a switch and interlock should have been a basic design feature of our unloading and storage system. But the fact was, a number of processes worldwide did not have this feature. Those without the feature were the source our overflow releases.

Analysis of the piping system failures yielded similar, fruitful information. We found that 70% of our failures were leaks from flanges and the remaining failures were due to corrosion. Further investigation showed that two thirds of our flange leaks occurred in the piping systems from the storage tank to the process. This piping conveys chemical "A" at relatively high pressure (>250 psig) compared to the unloading piping (<20 psig). So we would expect a greater propensity for flange leaks in the higher pressure piping. Similarly, all of our corrosion failures occurred in the low pressure piping from the unloading station to the storage tanks. Since chemical "A" is a corrosive, the procedure of connecting and disconnecting to railcars or trucks provided the perfect opportunity for moisture laden air to contaminate the piping. Again, the data had shown us what we might expect.

For what ever reason, years of Process Safety programs, with qualified engineers at each site and audits conducted biannually, had not revealed these facts to us. Only the analysis of data from our incident database showed us these critical pieces of information. The thrust of our risk reduction effort was focused on tank overflows and increasing the integrity of our piping systems. We began system improvements in 1992.

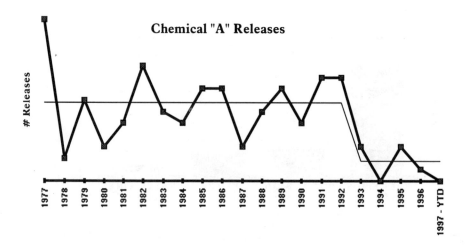

FIGURE 7. Releases of Chemical "A"; 1977–1997 YTD

The results our efforts are shown in the run chart in Figure 7. Prior to our risk reduction effort, we were suffering frequent releases of chemical "A." The range of our release numbers were as large as twice the mean. Since implementation of the task force improvements, our mean number of incidents has dropped by a factor of four. We are suffering only 25% of the releases we experienced prior to our risk reduction effort. Using the incident database and Pareto analysis, we have implemented changes which have reduced our incident frequency by 75%.

Though it will take several more years of data gathering to assure that the reductions shown on the chart above are not attributable to random variation, we are confident our initial results will continue.

Case Study 2: Risk Reduction—Process Heaters

Now we will look at a different case study dealing with Fires and Explosions. This case is interesting because it led us to new discoveries about one of our processes and completely changed our strategy for safe design. Several years ago, we suffered a significant incident in one of our processes involving a process heater used to prepare an agricultural commodity for packaging. Fortunately, no one was injured in the incident, but the equipment suffered significant damage due to an explosion. The results of the incident threatened our production capacity and reduced our flexibility. The impact on the process lasted for weeks as repairs and investigation proceeded. In order to fully understand what had happened, we went back to our incident database and began to evaluate all incidents that had occurred in this or similar processes.

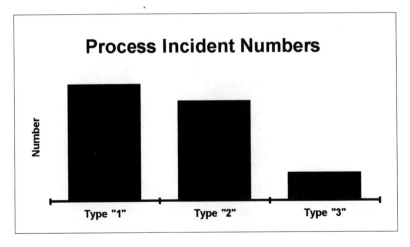

FIGURE 8. Number of Incidents per Process Type

We first looked at the numbers of incidents and categorized the incidents according to the process design in which they occurred. Twenty years of data are reflected in the distribution of incidents across three process designs (Figure 8).

The chart in Figure 8 taught us that while there was a much higher likelihood of an incident in two of the three types of processes, no clear 80/20 relationship existed with this categorization of incidents. As with the previous case study, we looked for another way to categorize the data. Our second cut at the data looked at the cost of incidents in each of the three types of processes instead of the number of incidents. Incident cost information began to provide us with more revealing information (Figure 9).

We now knew our problem was in Type "2" processes. As we had done before, we went to management, obtained resources, and established a risk reduction team. We combed the incident data looking for information which would help us understand the problem more clearly. The key piece of information we found was that though most of the incidents occurred during operation, a number occurred shortly after the process was shut down (i.e., when the gas to the hot air furnace burner was shut off). We again went through the distribution of incident numbers and found that incidents during normal operations outnumbered incidents during shutdown (Figure 10).

The chart in Figure 10 shows a fairly promising "Pareto" relationship. However, the decision to examine cost distribution (Figure 11) provided the real breakthrough for our team. This data was so overwhelming, we really had to understand why the data was so skewed.

The data shown in Figure 11 initiated a lot of activity from our team, the engineering organization, and the product development organizations. Various theories emerged and were discussed, but no clear hypothesis seemed to explain

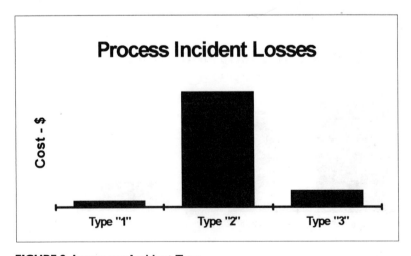

FIGURE 9. Losses per Incident Type

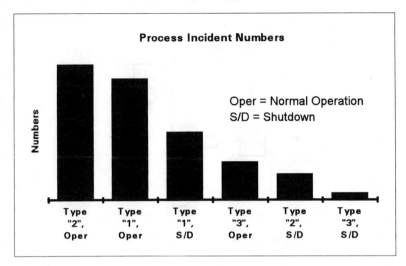

FIGURE 10. Incidents per Process Type at Various Stages of Operation

the data. Finally, process gas analysis revealed that while operating, these processes contained a flammable gas generated by the heating of the agricultural commodity being processed. Further, it was found that during operation the oxygen content in the process gas mixture was only about 12%. A simple schematic of the process is shown in Figure 12.

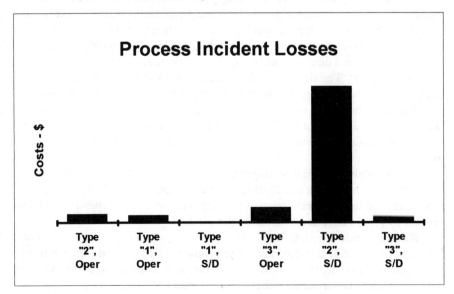

FIGURE 11. Total Incident Costs per Process Type at Various Stages of Operation

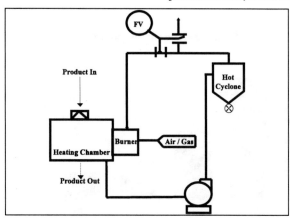

FIGURE 12. Process Schematic

We now began to investigate our operating procedures and practices. We found that the shutdown sequence simply shut off the gas supply and left the burner combustion air fan running. This allowed fresh air to enter the process gas stream and slowly raise the oxygen level. The process runs at temperatures in the range of 500 to 700°F and there are adequate ignition sources to ignite the flammable gas, if the explosive range is entered. This was our great discovery — this was a *gas* explosion we were suffering. This was counter to the conventional wisdom for these processes which held that the explosions were *dust* explosions. Across the board, venders supplying the processing equipment had always designed for a dust explosion and provided explosion venting only at the points in the process where dust accumulations were expected (e.g. the cyclones and the heating chamber). The significant damages that we suffered in our incident were due to inadequate venting, and the design bases for the venting of these processes were based on the wrong assumption. From this point on, our strategy was clear. We would redesign our safety features in these processes to minimize the likelihood of an explosion and minimize the consequences of an explosion if one were to occur.

Figure 13 shows a simple diagram of the operating environment to which these processes are subjected. During normal operation, with a depleted oxygen level, the machines operate relatively safely outside of the explosive envelope created by gases generated by heating the product.

When the unit was shut down, the combustion air fan continued to operate. Introduction of fresh air raised the oxygen content and provided the opportunity for the process to enter the explosive range. Our incident data says this will occur with significant consequences about once in every 40 operating years per unit. Figure 14 illustrates this phenomena.

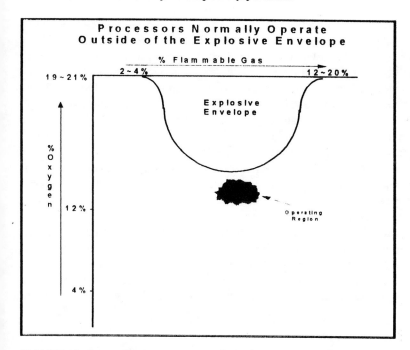

FIGURE 13. Operating Environment during Normal Operation

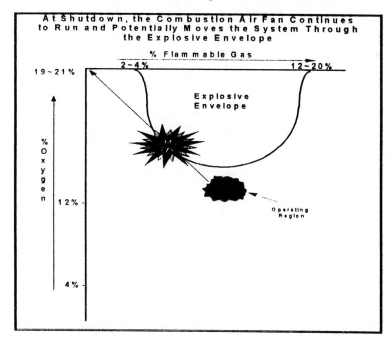

FIGURE 14. Operating Environment at Shutdown — Entering Explosive Range

Our first prevention action was to change our shutdown sequence. At shutdown, the gas to the burner is turned off, and the combustion control logics drive the combustion air fan inlet dampers to close. This minimizes the entry of fresh air to the machine. This strategy initially keeps us out of the explosive range. However, we still have an explosive envelope within the process. Our challenge was to figure out how to move from this condition to a safe, complete shutdown. Through research into Bureau of Mines publications on flammable gases, we learned that if we applied a cooling water mist to the process, we could narrow the explosive envelope by lowering temperature, and shrink the envelope by inerting the gas mixture. Cooling water mist gave us the ability to do both. Cooling water rapidly reduced the process temperature, and the resulting steam provided an inerting effect on the process. Figure 15 shows the effect of cooling water mist on the explosive envelope.

We established a desired cool down temperature set point that, when met, would stop water flow to the cooling mist. This cool down set point was well below the temperature where an ignition source could exist within the machine.

Our next step, with the explosive envelope significantly reduced, was to modify the process shutdown logics. The new logics called for the opening of

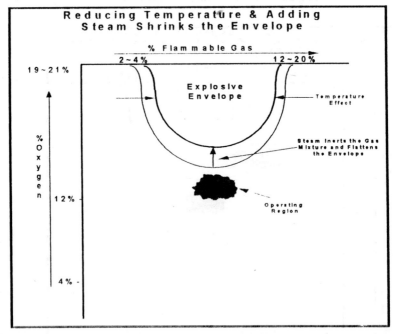

FIGURE 15. Effect of Cooling Mist on Process Temperature and Explosive Range

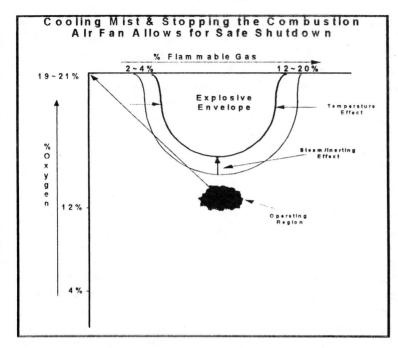

FIGURE 16. Achieving Safe Shutdown Conditions

both inlet and outlet dampers to provide rapid purge of our system. Figure 16 shows how we could move to a safe shutdown condition with normal oxygen levels and bypass the explosive envelope. Obviously, this drawing does not show that as we are introducing fresh air into the system, we are simultaneously removing the explosive gases.

These changes in operating procedures and shut down sequences greatly reduced our likelihood of suffering an explosion. However, we are still concerned that we may not understand our process completely and could not guarantee we would never experience another incident. Our next step was to reduce the severity of an explosion if one were to occur. For those who have dealt with explosion venting, it is often difficult to adequately vent an older processes in an existing building and still meet all of the requirements of NFPA 68 "Venting of Deflagrations." What is especially challenging is venting to a safe location (e.g., outside) with the geometry of the processes and their locations within the building. This is exactly the problem we faced. As mentioned, the explosion was assumed to be a dust explosion with venting located only at possible dust accumulation points. What we had discovered is the explosion was a gas explosion and required venting throughout the entire process. It was literally impossible to vent the machine at all of the required locations and direct those vents to the outside. We began to work with Rembe® GmbH, a German firm which manufactures explosion vents and a unique device called a Q-Rohr® Explosion Suppression device. The device is essentially a large flame arrestor attached to a rupture disc. It absorbs the energy of the explosion and quenches the flames. This looked like a promising solution to our venting dilemma.

The Rembe[a] device was designed for dust explosion and had never been tested for use with explosive gases. Procter and Gamble entered into a joint agreement with Rembe[a] to test the Q-Rohr®device using a flammable gas which replicated the gases generated during our processing. We completed those tests in early 1996 and proved the devices effectively performed in a flammable gas environment.

With the completion of these test, we had everything we needed to make a step change in the safety design of these processes. We had identified the problem, developed an effective prevention strategy, and determined the Rembe® devices could be utilized to safely vent these machines where we could not direct the venting to the outside.

This was an exciting project to work through. We learned much about our process, and we have made a real difference in assuring safe operation. It is important to note that the discoveries we made were a result of evaluating data. The existence of a database allowed us to review over 200 incidents in these processes and to categorize the data many different ways. The compelling information of the losses associated with a particular design shortly after process shutdown was the driving force of our investigations. We have left many of the details out, but this was a three year effort and we were assisted by some of the outstanding safety firms in the world. What drove us to go to the lengths we did was the data. We could not have tapped that data without an incident database.

Case Study 3: Strategic Direction—Hazardous Chemicals

The third case study will illustrate how incident data was used in two different ways to help set strategic direction for hazardous chemical management at Procter and Gamble.

Part One—Hazardous Chemicals Management Systems

In the early 1980s, Procter and Gamble developed a hazardous chemicals management system to improve safety and reduce risks associated with handling hazardous chemicals. In 1992 the OSHA Process Safety Management rule (PSM) and EPA Risk Management Program (RMP) rule, both of which established regulations for managing hazardous chemicals, were in the final stages of development and nearing implementation. We believed PSM and RMP represented the best practices for handling hazardous chemicals. The P&G hazardous chemicals management system closely paralleled OSHA PSM and RMP Prevention Programs, and applied to all P&G listed hazardous chemicals. The list of P&G hazardous chemicals included over 100 chemicals. Some of the P&G listed hazardous chemicals were covered by PSM and/or RMP (e.g., fuming acids). Others were not specifically covered by either regulation (e.g., caustics). To prepare our manufacturing sites for implementation of the new OSHA and EPA programs, we wanted to convince the sites that in addition to the legal requirement for implementation, the new management programs would actually improve process safety. Again utilizing incident data, corporate process safety demonstrated that, historically, P&G hazardous systems caused the greatest total losses. The chart in Figure 17 illustrates that 60% of process safety losses occurred in P&G hazardous systems.

This data helped convince our sites that these chemical management systems had merit and would help improve safety. Implementation of these management systems across all processes handling P&G listed hazardous chemical was the next step. However, closer examination of the data resulted in an unexpected learning.

Further subdivision of the process safety losses by category—OSHA listed chemicals, EPA listed chemicals, and "other" P&G listed hazardous chemicals—revealed that 50% of process safety losses were attributable to the first two categories. The third category, other P&G listed hazardous chemicals (over 80), contributed to less than 10% of total process safety losses. Clearly, not all P&G hazardous systems posed the same level of risk. Did all P&G hazardous systems require identical risk management programs? How should this group of "other" chemicals be managed? Relaxing process safety management requirements for these "other" chemicals was fundamentally a different idea for the risk manage-

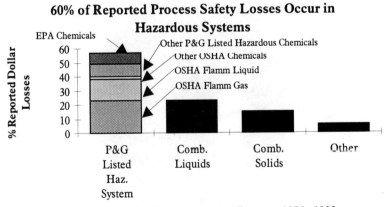

FIGURE 17. Percentage of Losses for Various Classes of Chemicals

ment organizations. This rarely, if ever, happened. As risk managers, we were good at asking for more. What we did not do so well is determine how to eliminate non value-added work. However, we were learning that managing all systems with the same set of requirements diluted process safety efforts and resources. We realized our hazardous chemicals management systems needed to be commensurate with the level of risk. The solution was the creation of a "tiered" hazardous chemicals management system, based on the level of risk posed by the particular process or chemical. This was a move away from "one-size-fits-all" risk management. Procter and Gamble now categorizes systems according to the level of risk, based on the chemical properties and quantities present in the system. The number of chemicals managed with the most stringent requirements—now known as P&G Class 1 chemicals—dropped from more than 100 to less than 20. This focuses process safety resources on the higher risk systems.

Part Two—Flammable Liquids Handling Practices

Another key learning evolved from this same incident data. Systems handling flammable liquids and gases were responsible for most process safety losses, as shown in Figure 18.

Why were these systems suffering disproportionate process safety losses? Were company process safety practices effective? Procter and Gamble process safety practices define company requirements for system design and operation. The practices are based on recognized industry standards, such as NFPA, API, and ASME codes, and company experience. A review of the practices for flamma-

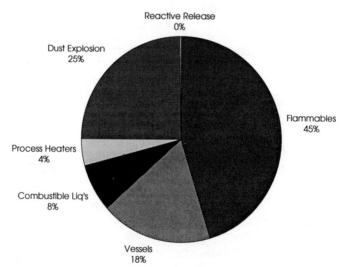

FIGURE 18. Process Safety Losses by Material Type; 1980 - 1994

ble liquids revealed some weaknesses. From a technical standpoint, the flammable liquids practices were written correctly and aligned with codes and current industry best practices, but they needed to be updated to reflect learnings from recent process safety incidents. The real weakness was how the information was communicated. Comments from engineers and plant personnel using the practices indicated there were too many options, and it was difficult to understand exactly what needed to be done. This decreased the overall effectiveness of the practices and resulted in the practices not being fully implemented at every site. Clear delineation of the requirements would improve system design and decrease the frequency of incidents in systems handling flammable liquids. This was clearly an opportunity for a strategic risk reduction effort—update the practices for flammable liquids to reflect learnings from recent incidents and make them more understandable and "user friendly." A major improvement to the practices was the addition of design checklists. These checklists can be used for both design bases and for assessing compliance of existing facilities and systems. The revised practices were issued in late 1996. Will the number of incidents involving flammable liquids handling systems decrease? As of today, we do not have enough post-revision data to make any assessment of the impact of these changes. However, we have chosen flammable liquids as a focus area for risk reduction efforts, and we will use incident data to track results and evaluate the effectiveness of the practices. We believe properly designed systems reduce overall risk. Use of the design checklists should lead to more consistent application of the design requirements, and hence, properly designed systems.

Conclusion

These Case Histories illustrate how much can be learned from process safety incidents and how powerful this information can be. As stated before, many of Procter and Gamble practices and operating procedures are based on company experience. The incident database provided a tool sorting and analyzing information from over 20 years of incident history. Without a database, analysis of this number of incidents would have been much more difficult, if not impossible. The incident data revealed process safety trends and pointed to opportunities for improvement. This allowed corporate process safety to eliminate non value-added work and focus on risk reduction efforts which would have the greatest impact.

WORKSHOP C
Risk Acceptance Criteria (Including Cost–Benefit) versus Corporate Liabilities

Chair **David Moore**
AcuTech Consulting

Implemented at Chevron

- general

A Risk-Based Approach to Addressing Recommendations from Process Hazard Analysis Studies

David A. Moore
AcuTech Consulting, Inc., 100 Pine Street, Suite 2240, San Francisco, CA 94111

Gregory L. Hamm
Applied Decision Analysis, Inc., 2710 Sandhill Road, Menlo Park, CA 94015

ABSTRACT

THE OSHA regulation, "Process Safety Management (PSM) of Highly Hazardous Chemicals" (29 CFR §1910.119) and EPA's rule, "Risk Management Programs (RMP) for Chemical Accidental Release Prevention," (40 CFR Part 68) require qualitative Process Hazards Analysis studies to be conducted. Neither regulation defines a model for making risk management decisions. And yet, management is held to an implied standard of care in making acceptable risk decisions. Taking a risk-based approach is sensible, but most PHA studies were not conducted with this type of decision process clearly in mind. Most studies were focused on the identification of hazards and did not benefit from using a formal risk management framework, or used only simple approaches.

As these studies are revalidated, or when new studies are conducted, companies would benefit from improved planning and a formal system for making these decisions. Three areas of improvement are recommended: better guidelines for conducting the studies that include rules for hazard analysis, a risk ranking system for ranking the hazard scenarios, and a decision process for addressing recommendations.

This paper discusses an approach aimed at streamlining the decision-making process and minimizing ambiguity in decisions. The characteristics and benefits of improved prioritization processes will be reviewed. Possible barriers to the implementation of improved processes and the future of risk-based prioritization of risk reduction actions will be discussed.

Introduction

Process Hazards Analysis (PHA) studies can be thorough, formalized, well documented studies. But often the good efforts of the PHA team are not focused and the discussions are not well captured. This, combined with a lack of a rigorous system for evaluating the PHA results and making decisions on PHAs, limits the effectiveness of PHA programs. The most critical process, that of managing the risks and/or responses to risks identified by PHA's, is too often done quickly and use overly simplistic tools. In this paper, the characteristics and benefits of improved risk management processes are reviewed. (Risk management process is defined as the method used to decide what portfolio of responses to risks will be budgeted and conducted.) Barriers to the implementation of improved processes, and the future of risk-based management of responses are also discussed.

What are some of the problems we see with current processes for deciding on responses to PHA's? The problems divide into two classes: organizational and technical problems. The most severe organizational problem is the lack of management involvement. It is critical that management sets clear objectives and measures by which risks will be evaluated. This is a management task, not a technical task. It is critical that management sets objectives clearly so that the PHA team members know what they are trying to achieve and the relative worth of those objectives to the organization. Another organizational problem is that decisions on which responses will be undertaken are often made using ad hoc procedures, and are not well aligned with corporate goals.

Technical problems with poor risk management decision processes include:

- Risks are improperly evaluated, which potentially leads to unnecessary risks being accepted or expenditures being made unnecessarily
- Informal or poorly designed processes are inefficient. Whatever the corporate objectives, the projects undertaken do not maximize the value per dollar spent.
- Another problem, linked to the first, is inconsistency. Poor processes lead the organization to spend inconsistently. For example, the organization may spend $100,000 in one case to prevent an injury, but not spend $70,000 in another case to prevent a similar injury with a similar likelihood of occurrence. (Theoretically an organization could be perfectly consistent in making bad choices, but this is rare.)
- Poor processes often work well for a period, but are subject to unexpected breakdowns when the procedure produces an obviously bad recommendation or when someone uses a lack of logic or formality as an excuse to attack specific decisions.
- Poor processes can waste time directly, or can waste time as the process breaks down and decisions must be revisited numerous times.
- Finally, and very important in an era when decisions are open to wide scrutiny, poor processes are neither auditable, nor defensible. When stakeholders cannot audit a decision, they will conclude that it is irrational and not in the public interest.

In the next section of the paper, we will discuss characteristics of good risk management decision processes. This discussion will cover characteristics of the general process, PHA's, risk evaluation, and decision making. The following section will outline the benefits from a good process, and the final section will talk about future directions for risk management decision processes.

Characteristics of Good Risk Management Decision Processes

Process Characteristics

The process must have management involvement and support to work. Management must "buy-in" and accept their role in setting objectives which will guide decisions.

The process must balance the effort expended on the system to the value achieved. We don't want to spend $100 making a $10 decision. To balance effort and value, we need a multilevel approach that makes the easy decisions quickly, but devotes greater resources to more complex and larger decisions. Many people fear that formal decision-making processes are too costly to implement. Actually, these processes often have payoffs over 10 to 1.

Another important characteristic of a good process is a focus on evaluating and prioritizing hazard reductions or responses to risks. How much risk reduction or benefit can be achieved for what cost? This is a key question, and this is not the same question as, "Where are my highest risks?"

Other good process characteristics are:

- Processes should have enough formalism and structure that evaluations are similar across risks. The process must be general enough that it is transportable from risk to risk, and people must be provided tools and training such that they can apply the process similarly to a wide scope of risks.
- In large organizations with broad implementations, data and evaluations will be developed by many people. Processes must be capable of producing consistent evaluations despite involvement from a large number of individuals.
- The process should be auditable. There is a great deal of scrutiny placed on organizations, and the best processes leave a paper trail that illustrates the firm has made a sincere attempt at spending money efficiently to reduce risks to workers and the public.
- Lastly, one of the general characteristics of a good process is that it provides feedback. Regular reviews are built into the process to determine if actions produced the anticipated results and to otherwise learn from the process.

PHA Characteristics

The primary purpose of the PHA is to identify and analyze the hazards of the process. The PHA team normally focuses on these objectives, and makes recommendations when the team collectively judges the team is that the risks are unacceptable. A key decision is how much authority to give the team. Some companies authorize the team to make final decisions on the hazards of the process and to make definitive recommendations for change. Other companies have the team identify the hazards of the process, leaving exact solutions and oversight of the recommendations to other groups or individuals. Experience shows that those companies that have a rigorous decision-making process, and that give their teams more authority, can save considerable time in the overall effort of hazard identification and resolution.

Usually the group is trusted with this evaluation without the benefit of rules on how the PHA should be conducted or how risks should be evaluated. Even in companies using explicit risk ranking systems, the underlying rules of how to identify and evaluate the risks are not normally well defined. Without rigor to the methodology, the possibility for inconsistent and inefficient risk decisions is troubling.

What are the attributes of a good PHA study? The PHA should be:

- **Complete.** It should identify all significant risks.
- **Well defined.** It should identify all risks of similar severity across the plant or organization under study. It should be consistent in following rules or guidelines on the level of risks of concern.
- **Fully informative.** It should cover all risks of concern and collect all the appropriate information to support the evaluation step of the process. It is undesirable making one review of the facility to identify risks to workers, another to identify risks to the public, another to identify actions that can reduce risks, another to estimate the costs of actions to reduce risks, and so on.
- **Documented.** It should be fully documented, yet clear and brief.
- **Risk-based.** To support clear communication, it should be based on well defined scales for measuring likelihood and severity of each type of cost or benefit.

To assure a good PHA study, rules and guidelines must be established for the PHAs. The purpose of these rules is to more concretely define each risk scenario so that improved decisions can be made. Without clear rules and guidelines, typical failures are:

- Completeness failure, where the study addresses only first order failures (simple failure of equipment and corresponding consequence with all safeguards functioning) versus second, third or ultimate level of failures (one or more safeguards fail to control the hazard). If this occurs, the con-

clusion of the team could range from "a trivial operational problem may occur" to "a catastrophic event could occur." The rules of cascading failures needs to be established or teams will make one of two mistakes—overstating or understating the risk.
- Completeness failure, where the study focuses on smaller hazards versus larger ones, or on operability versus safety concerns.
- Definition failure, where the study may judge the need for a recommendation inconsistently. Unless rules are established for ranking hazards and making recommendations where risks are determined to be above a certain level, the teams may not make recommendations. It is important to set rules where, for risks that fall within a certain range, a recommendation is required or is not required.
- Information failure, where the study could document insufficient information for the analysis. The team should be trained to document specific information, such as causes, consequences, and safeguards. Unless this is done, ambiguous information may be produced and the results could be unclear, or worse, misleading. Rules about how complete the PHA is expected to be are recommended. If a risk is high, for example, the details necessary to evaluate the recommendation and the underlying assumptions in determining the risk are needed.
- Information failure, where the study may not present a full range of responses. The teams should be encouraged to present alternative responses, especially when responses are high cost. This provides management with the most flexibility in creating a high value portfolio of responses. It is usually not difficult for a team to provide alternative risk reduction approaches. The team should consider responses that vary in cost, reliability, and impact on the organization (training and other costs, for example)

To simplify the analysis, the defined responses (recommendations) should be independent and similar in size (same magnitude).

- Independence means that neither the worth nor cost of one response depends on another. For example, two responses to a fire risk might be: install a new monitoring system and train employees to use the monitoring system. These responses will be much more valuable if done together. Their merits are dependent. Responses also need to be cost independent. For example, projects to redesign the interior of a building and to install new wiring in the building might be less expensive if done jointly. These project are not independent. The most frequent technique for dealing with dependence is to combine dependent projects.
- It is very difficult to evaluate projects that vary widely in size. To overcome this problem, we usually create groups of smaller projects with

common goals. For example, we may create groups of all fire prevention projects or training projects.

Risk Evaluation Process Characteristics

Most of the PHA processes now used recognize the severity and likelihood of the risk. This information is critical for proper evaluation. For some companies, this is a formal system; for others, it is simply the thought process that the teams employ.

If no action is taken with respect to a risk, certain outcomes are anticipated. If a response is taken, a different set of outcomes is anticipated. The evaluation process should rank the proposed responses to the risk, not only the risks themselves. To rank the responses, the process evaluates the change in the outcomes or the risk reduction. In most cases, attacking the biggest risks first will not have the greatest payoff in public, worker or environmental protection. The set of responses that offer the greatest improvement in outcomes per dollar spent must be found.

For most firms, the responses to risks identified by PHA's will affect several corporate objectives. Thus, decisions about responses are multi-attribute. A good process will clearly identify individual objectives, and measure the worth of outcomes with respect to each objective. The process will include an explicit weighting of the relative worth to the organization of changes in these objectives. A mathematical function (often a simple linear combination) will be used to combine the different measures and weights into a single value of the total worth (or cost) of outcomes.

Key components of a risk evaluation process are: a list of organizational objectives, scales for describing performance on each objective, a determination of the worth within each objective, and a determination of relative worth across objectives. Characteristics of each of these components are:

- **List of organizational objectives.** Organizational objectives that are almost always of concern in risk decision making are cost, worker safety, public safety, and environmental protection. Regulatory compliance is often a separate corporate objective. It is not unheard of that an action that is less protective of workers, the public, and environment may be seen as more compliant. Corporate reputation is often an important objective that we may take actions to protect. Worker, community, and customer relations, because they depend on perceptions and factors separate from health and safety, constitute separate objectives. Responses that improve these relations may appropriately be judged to have greater worth. Identifying organizational objectives is properly a management task.
- **Scales for describing performance.** After the core objectives of the organization are determined, a method of measurement is needed. Meas-

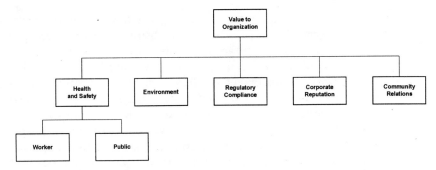

FIGURE 1. Example of a Value Hierarchy

urement requires clear descriptions of different outcomes and the relative worth of those outcomes. Outcomes can sometimes be described very briefly with numbers, 10 injuries or $10 MM. In other cases, longer, more detailed descriptions are needed to assure that the outcome can be clearly communicated. Defining scales that accurately describe potential outcomes is a technical task. Table 1 provides an example of a scale for measuring the severity of injury. In many cases, more detailed descriptions will be needed to assure that the scale will be absolutely clear.

- **Worth within each objective.** After we have a well understood method of measuring individual objectives, the worth of different levels of each objective and weights must be determined that describe the relative worth of different objectives. These questions of worth are issues of core organizational values, and thus should be determined by management, not technical teams. The simplest, though sometimes controversial, way to measure worth is in dollars. For comparing different outcomes, we set an appropriate dollar value for a specified state of public safety or corporate reputation. These dollars values are tools to aid the efficient allocation of funds, not measures of the intrinsic worth of these outcomes. If measuring

TABLE 1
Example of Defined Scale for Injuries

Level	Level Description
A	Permanent disability or death
B	Hospitalization, but not permanent injury
C	Temporary injury or discomfort
D	No significant health impact

value in dollars is unacceptable to the organization, the relative worth of different objectives can be measured in abstract units. Economists and decision scientists call these abstract units utilities (utils.) Figure 2 shows a graph of a utility function for the severity of injuries.
- **Relative worth across objectives.** After a method of measuring individual objectives is established, weights that describe the relative worth of different objectives to the organization are needed. Weights answer questions such as how important is an improvement in worker versus public safety. Weights allow us to make tradeoffs when actions improve one value but reduce another. For example, assume we are dealing with a hazardous process. One response places workers close to the process so they can act quickly to protect the public. Another response places workers further from the process to increase worker safety. Table 2 illustrates weights for different objectives and how weights and scores are combined to evaluate a response, Action A. Identification of objectives and placing relative worth on outcomes are management, not technical, functions.

One additional issue that must be dealt with in the evaluation step is timing. The importance of timing is most clear when a problem is rapidly getting worse. For example, a corrosion problem may currently create a minor risk. However, the corrosion problem can now be treated with a coat of paint, but in one year equipment replacement will be needed. Urgency needs to be identified during information gathering and urgent problems need to be given a higher priority rank during decision making.

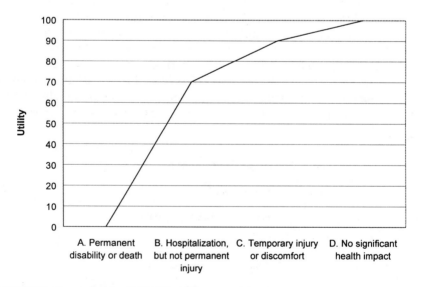

FIGURE 2. Example of a Utility Function

TABLE 2
Example of a Benefit Cost Calculation

Objective	Weight
Public Health and Safety	100
Worker Health and Safety	100
Environment	80
Regulatory Compliance	50
Corporate Reputation	50
Community Relations	10

Decision Making Characteristics

Decision making needs to be multi-level. That is there needs to be simple screening criteria and more complex decision criteria for actions remaining after screening. The screening criteria should identify risks that are really not of concern and risks that are so severe and immediate that they must be dealt with right away. Based on our experience organizations should be cautious about what is identified as an urgent, must-resolve risk. It is common for very large numbers of risks to be placed in this category. Usually the claim will not be that a response is essential due to health and safety, but that it is essential from a regulatory stand point.

We will pick responses to risks not screened into the not-of-concern and act-now categories. We then need decision rules to choose which responses will be taken. The two rules that seem to make the most sense are the net benefit rule and the benefit-to-cost ratio rule. The net benefit is the dollar value of all the benefits from a response minus the cost of the response. The highest positive net benefit response is taken. The benefit-to-cost ratio rule recognizes that budgets are often limited and seeks to determine the set of responses that provides the greatest total benefit for a fixed budget. An advantage of the benefit-to-cost ratio rule is that benefits do not need to be placed on a dollar basis.

After the basic rule for decision making is determined, there are lots of details that need to be considered. One of these details is the approach to discounting. The issue is how benefits, such as prevention of injuries, that occur in different time periods, should be compared. For example, what is it worth to prevent an injury ten years from now versus preventing an injury today? This is a value judgment that should be made by management.

Time horizon is another issue in decision making. How far into the future should costs and benefits be estimated? This issue has both technical and value components. The appropriate time horizon will usually depend on the useful life of technical solutions, understanding of future conditions, and the rate of change in the organization.

We try to define responses that can be evaluated independently; however, there is almost always some remaining dependence. The decision-making process can ignore this or can explicitly deal with it. Sophisticated mathematical techniques are available to identify optimal portfolios of responses when dependence is considered. For high cost portfolios, use of these techniques will usually have very high payoffs.

Treatment of uncertainty is another issue in decision making. Most risk evaluation processes deal with the uncertainty of a problem occurring, but many processes do not deal with uncertainty in the effects of the event, the effectiveness of the solution, or the cost of the solution. For complex problems and expensive solutions, it will usually be worthwhile to deal explicitly with these uncertainties.

Finally, a good evaluation process with appropriate information on risks and responses can have the capability of indicating when more information should be gathered prior to action. Such a capability requires statistical data or expert judgment about the ability of additional data to resolve uncertainty. With this information, relatively simple decision science tools, value-of-information calculations, can provide guidance on collecting additional information.

Outline of a Process

We have discussed the characteristics of a good process, but what is the series of steps that we go through to uncover risks, evaluate responses, and decide on a set of responses to undertake?

First, a certain amount of investment in building the system is required. Building the system includes:

- Establishing buy-in and support by management. Management should understand that they will need to set the objectives and be clear about the organizational values. Management should be committed to developing a good process and believe in the pay-back that the process can provide.
- Defining the objectives, risks and benefits; to be considered; the relative worth of different objectives, and how outcomes will be measured. Again, the first two of these will need management involvement.
- Putting the tools in place that assure consistency, accuracy, and completeness. This will be a set of forms, checklists, ranking protocols, procedures, and often computer programs.

- Establishing PHA teams and a system of assumption of responsibility and follow-up for the action items. The PHA teams are not expected to solve all problems, but it may be advisable to use the PHA team as a jury on the subsequent actions taken to address the findings they felt needed attention.
- Training the people to carry out the process.

Next, we move onto the steps that are part of the actual implementation. These would include:

- Identify and evaluate the risks via HAZOP's, siting studies, and risk assessments, and gather information needed in the evaluation process. (PHA team)
- Define possible responses (recommendations) to address risks. (PHA team)
- Screen on level of risk. Identify non-problems and must-act-now situations. (Evaluation team with oversight by PHA team)
- Evaluate responses to determine benefits and costs. For large sets of responses, the evaluations will be performed by multiple teams of experts. In this situation, the use of a review panel is recommended to assure consistency of evaluations. For high expense responses, it is likely that a probabilistic evaluation will also be worthwhile. The probabilistic evaluation can provide a more accurate estimate of costs and benefits and a better picture of risks. (Evaluation team)
- Do the calculations. Simple benefit-to-cost ratio calculations or net benefit calculations are needed at a minimum. Sensitivity analyses covering key assumptions and weights on objectives build insights and confidence in the process. The process works much more smoothly if people have tools that allow them to manipulate and test the results quickly and easily. For example, users may want to include or exclude particular projects from the portfolio, try certain fairness rules, etc. The simplest technique for analyzing portfolios of responses is to list the responses in order of benefit-to-cost ratio and implement the highest ratio projects that fall within the budget. However, for large, expensive portfolios of responses, more complex algorithms for constructing the portfolio will have high returns. This is particularly true if the projects stretch over several years, are very large, or if its success depends on common factors. (Evaluation team)
- Review the results and learn. The last step in the process is to review the results at regular intervals, and, as far as possible, test the performance of the process. Issues such as the accuracy of cost estimates, the effectiveness of responses, and the history of problems not addressed should all be reviewed. (Audit team)

Benefits of a Good Risk Management Decision Process

The chief benefits of a good risk management decision process are better alignment of decisions with corporate objectives and much higher benefits for the same or reduced levels of expenditure. Closely related benefits are: greater control over the decision making process, less politicization of decision making, greater consistency in decision making, and the ability to learn and improve. After the process is set up, most organizations will find that it is not more expensive than prior approaches, and is less apt to fail and require revisions of priorities.

A good risk management process also provides a good framework within which to interact with stakeholders. A good process, by providing a clear statement of objectives and beliefs about the efficacy of responses, is a great aid to communication. Often, consensus or negotiation can be blocked by value or information disagreements that in fact do not affect the decision. Sensitivity analyses can identify the relevant areas of conflict and focus discussions on the most critical elements.

Finally in an era of much considerable scrutiny, a good process is extremely auditable, highly logical, and, therefore, more defensible.

Future

In our practice and in the literature, we see more companies are adopting a benefit-to-cost framework to address risk management decisions. Also, we see more companies that use formal multiattribute decision criteria—realizing that benefits and costs come in many forms, that understanding the need to formally deal with uncertainty in costs and impacts.

Every organization that we have worked with to develop formal prioritization processes believes that the formal identification of criteria, the careful weighting of criteria, the careful ranking of activities versus criteria, and calculation of net benefit or benefit/cost ratios is extremely worthwhile, and has provided the organization with a combination of increased value and decreased costs.

However, we have also seen that logical systems are difficult to maintain in spite of the accolades they get when first implemented. The temptation to declare pet projects indispensable, and therefore outside the process, is enormous; the temptation to fall back on horse trading among decision makers to determine priorities is also very strong. The fact that these processes are oriented toward organizational goals, and are not constructed to bend to the politics of the organization, often causes their downfall.

A second problem is that, while we have described auditability and explicit decision rules as strengths of a good process, these are sometimes viewed as problems with these processes. They leave behind a clear track record of why some actions were taken and others were not. Some organizations are afraid that this leads to greater not lesser liability.

Finally, though it is very slow and enormously hampered by the distrust among the public, regulators, and industry, certainly there is a great deal of interest among regulators about benefit-to-cost approaches as a way of achieving for society a higher level of protection of human health and the environment at a lower cost.

Risk-Based Judgments in Process Hazard Analyses

Donald K. Lorenzo
JBF Associates, Inc., 1000 Technology Drive, Knoxville, TN 37932-3353

ABSTRACT

The primary product of a hazard analysis is a list of recommendations that management can implement to reduce risk. Unfortunately, the process by which analysis teams conclude that the risk of an accident is unacceptable is often highly subjective and inconsistent, as is the process by which management decides whether to implement the recommendations. This paper describes a simple scoring methodology by which risk-based judgments can be made more consistently, objectively, and defensibly. The methodology can be easily adapted to aid specific risk-based decisions, such as determining the required safety integrity level of an instrument system or the need for additional independent protection layers (decisions faced by all hazard analysis teams).

Introduction

Process hazard analyses (PHAs) use qualitative methods such as hazard and operability (HAZOP) analysis, failure modes and effects analysis (FMEA), or what-if/checklist analysis to identify hazards, to identify existing engineering and administrative controls (safeguards), and to identify the consequences of failure of those controls. The analysis team then judges whether the existing safeguards are adequate, and if not, recommends that management reduce the risk. Unfortunately, the common PHA methods do not lend themselves to accurate risk assessment, and few PHA teams have the skills and experience necessary to compensate for the flaws inherent in risk evaluations based on the simple PHA methods. This paper describes three variations of an analysis process that will improve teams' risk-based judgments and provides a practical methodology that both analysis teams and management can use to evaluate the merits of PHA recommendations.

Risk Concepts

A basic understanding of risk is fundamental to the process of making risk-based judgments. The risk (R) of an event is a function of the likelihood (L) that an event will occur and the consequences (C) of its occurrence as shown in the following equation:

$$R = f(L,C)$$

where L is expressed as either a frequency (events per unit time) or a probability, and C is expressed as dollars, lives lost, pounds of pollutants released, etc., per event. Figure 1 graphs risk as a function of L and C on a log-log scale. The solid line shows the most common form of the function, $\mathbf{R} = L \times C$. Arguably this is the most rational definition of risk because you can choose the threshold risk you are willing to tolerate and apply it in all circumstances. For example, if a loss of $100 is tolerable once per year, on average, then a loss of $100,000,000 should be equally tolerable if its frequency is once in a million years or less. In the real world, of course, the acceptability of risk decreases as the severity of the consequences increases. Individuals (and organizations) may tolerate the risk of a $100 per year *average* loss in minor damage to their property, yet these same decision makers will purchase insurance to protect themselves against the risk of $100 per year *average* loss from rare, catastrophic events such as large fires, tornadoes, or earthquakes. This aversion to risks with higher consequences can be modeled as a continuously decreasing function of higher consequences (illustrated in Figure 1 as a dashed curve) or as a piecewise-discontinuous jump(s) to a lower risk level(s) as consequences increase (illustrated in Figure 1 as a dotted line).

This aversion is expressed by PHA teams in recommending more protection against catastrophic, but rare, events while accepting less protection against less severe, but much more likely, events. When management fails to provide the

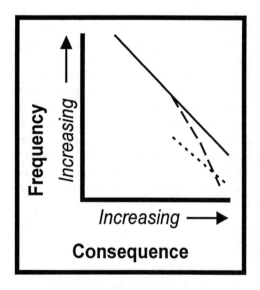

FIGURE 1. Risk Tolerance

PHA team with any guidance about the organization's risk tolerance, the team inevitably struggles to define its own criteria. Some define acceptable risk as conformance with laws, regulations, codes, standards, industry practices, company practices, etc. Others simply rely on the opinions and experience of the team members. The result is often a laundry list of recommendations to management encompassing everything the team thought might reduce risk, regardless of need or benefit. Management is then burdened with reviewing the dozens (hundreds?) of recommendations and justifying the resolution of each. Here again, when a clear definition of tolerable risk is absent, the resolution of recommendations is fraught with inconsistency because each manager's risk tolerance (and budget) is different.

Risk Matrices

Some organizations have developed a risk matrix to help standardize judgments about risk. A risk matrix is simply a chart of event frequency ranges versus consequence ranges that PHA teams and managers can use to determine the tolerability of risk in a given category. An example risk matrix is shown in Figure 2, and example definitions of likelihood and consequence categories are shown in Tables 1 and 2, respectively. The example matrix shows the same risk tolerance for all consequence types as defined in Table 2; some organizations prefer a separate risk matrix for each consequence type. Notice that if the frequency and consequence categories are defined as "order of magnitude" intervals, the upper left to lower right diagonals of the matrix correspond to the line of constant risk as shown in Figure 1; the aversion to more serious consequences is indicated by the reduction in risk tolerance as one moves from left to right on a specific diagonal.

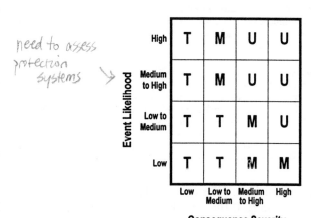

FIGURE 2. Risk Matrix

TABLE 1
Example Categories for Event Likelihood

Likelihood	Criteria
Low	The accident scenario is credible but highly unlikely. The probability is very low
Low to Medium	The accident scenario is considered unlikely, but similar things have happened once or twice in this industry. It would be surprising if it happened here
Medium to High	The accident scenario might occur once in the life of the facility. It would not be too surprising if it happened here
High	The accident scenario has occurred at this facility in the past and/or is expected to occur in the future

TABLE 2
Example Categories for Consequence Severity

Category	Personnel	Community	Environment	Facility
Low	Minor or no injury, no lost time	No injury, hazard, or annoyance to public	Recordable event with no agency notification or permit violation	Minimal equipment damage at an estimated cost of less than $100,000, and with no loss of production
Low to Medium	Single injury, not severe, possible lost time	Odor or noise complaint from the public	Release which results in agency notification or permit violation	Some equipment damage at an estimated cost greater than $100,000, or minimal loss of production
Medium to High	One (or more) severe injury	One (or more) minor injury	Significant release with serious offsite impact	Major damage to process area(s) at an estimated cost greater than $1,000,000, or some loss of production
High	Fatality or permanently disabling injury	One (or more) severe injury	Significant release with serious offsite impact, and more likely than not to cause immediate or long-term health effects	Major or total destruction to process area(s) estimated at a cost greater than $10,000,000, or significant loss of production

The risk matrix is very useful for a PHA team leader because it helps the leader manage the team's discussions. If the team has identified a hazard with high consequences and a high likelihood, the leader knows the risk is unacceptable and the team must generate a recommendation. Conversely, if the team has identified a hazard with low consequences, regardless of its frequency, the risk is tolerable and no recommendation is expected; therefore, the leader should push the team forward to a new topic. If the risk is marginal, the leader should press the team for any suggested improvements and move forward if it appears that the existing safeguards conform to best practices. Managers can also use the risk matrix to help prioritize their resolution of team recommendations. Recommendations addressing unacceptable risks should be given higher priority than those addressing marginal risks. Recommendations addressing risks deemed tolerable may be rejected because existing (or other newly installed) engineering and/or administrative controls provide a sufficient level of protection.

Unfortunately, in practice, teams seldom use a risk matrix correctly. The most common error occurs when the team identifies a hazard with potentially severe consequences, such as burning a facility to the ground. The team categorizes the consequences as high and then attempts to judge the likelihood of a fire. Perhaps the facility has experienced some small fires (flange fires, trash fires, fires from cutting or welding spatter, etc.) every year, so the team judges the likelihood of a fire to be high. The team then concludes that the risk of a fire is unacceptable (high likelihood and high consequences are an unacceptable risk per Figure 2), so they recommend additional protection against fires. But this judgment is wrong! They have combined the frequency of one event (a small fire) with the consequences of a different event (a large fire). Even though there have been frequent fires, there have NOT been frequent fires that burned down the facility. The existing safeguards have effectively prevented fires of that magnitude. So while the *potential* consequences of a fire may be high, the frequency of such large-consequence fires is low when the safeguard effectiveness is correctly considered. As shown in Figure 2, the high-consequence, low-likelihood cell in the matrix is classified as a marginal risk. Perhaps no further action is required.

The second common problem with using risk matrices is that the team is trying to judge the risk of a range of possible accident outcomes based on only one (usually the one with the worst-case consequences) accident scenario. PHAs performed in accordance with the OSHA process safety management regulation (29 CFR 1910.119) or the EPA risk management program regulation (40 CFR 68) lend themselves to this approach because they require teams to identify the consequences of failure of the engineering and administrative controls. Unfortunately, teams mistakenly assume that the accident scenario with the worst-case *consequences* is the scenario with the highest *risk*. In the previous example, the facility had experienced many small flange and trash fires that had been promptly extinguished (high frequency, low consequences). These fires are considered a tolerable risk per Figure 2. But one fire in the past 15 years burned up one of the batch reactors (medium-to-high frequency, medium-to-high consequences). In that case, a runaway reaction had discharged material through a rupture disk, and the release subsequently ignited. Assuming no new safeguards were implemented after that fire, the risk of another such fire is considered unac-

[handwritten note at top: protection systems usually only change likelihood not consequence]

ceptable per Figure 2. Thus, the accident scenario with the highest risk in this case is not the one with the worst-case consequences.

There is a better way to make risk-based judgments. The remainder of this paper describes three variations of a method for making such judgments with varying degrees of uncertainty in the results. First, standard event tree methodology is presented to show how to "correctly" solve the problem. However, this requires the greatest analytic effort and should be reserved for only the most important risk-based decisions. Then, qualitative and score-based versions of the event tree methodology are described. These methods are much simpler and faster to implement than the quantitative approach, yet can result in significant improvements in the risk-based judgments of PHA teams. The qualitative version is very similar to current practice, and should generally be used by PHA teams employing risk matrices. The score-based approach is somewhat more rigorous (resulting in less uncertainty) than the qualitative approach, but it requires more effort than the qualitative approach. It can be used by PHA teams to resolve disputes and by managers who want more justification for implementing or rejecting team recommendations.

Quantitative Risk Assessment Using Event Tree Logic

Event tree analysis[1] offers a structured way to evaluate the risk of a range of possible accident scenarios. As shown in Figure 3, an event tree starts with an initiating event and models branch points for the success/failure of each protective layer (safeguard) until the event is mitigated or some undesirable outcome occurs. If some phenomenological condition affects the progression of an accident (e.g., whether a flammable release ignites immediately or after a vapor cloud has developed; whether a fire impinges on adjacent equipment; whether the wind is from the north or the south), it can also be modeled as a branch point in the event tree. Each pathway from left to right through the tree is an accident scenario, each with its own frequency of occurrence, its own consequences, and, therefore, its own risk. The overall risk is the summation of the risk contributed by all the accident scenarios, not just the scenario with the worst consequences. Thus, to correctly judge the risk of a PHA deviation (or failure mode or question), the team needs to consider all of the accident scenarios and categorize the likelihood of scenarios in each consequence category of the risk matrix.

In addition to providing a more defensible basis for judging whether some action is necessary to reduce the risk, the event tree structure provides a basis for prioritizing any proposed improvements. Should a deluge system be installed to protect the compressor from a seal fire, should a high-high pressure alarm be installed in the reactor, or should the hot work procedure be revised? Given finite resources, the perfectly rational manager would choose to first implement the recommendation that offered the greatest risk reduction per dollar spent,

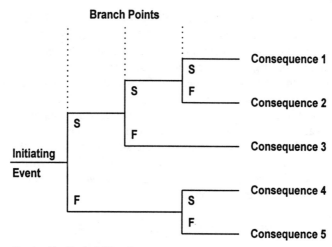

FIGURE 3. Event Tree Structure

then the one with the next highest risk reduction per dollar spent, and so on until the budget was exhausted. Mathematically, this can be expressed as ΔRisk/Cost = Priority. The highest priority recommendations are those that offer the highest risk reduction for the lowest cost; conversely, expensive recommendations that offer little risk reduction have low priority. The net present cost of a proposed change can be estimated using standard economic analysis techniques to account for capital outlays and ongoing operating and maintenance costs. However, estimating the net present value of the risk reduction is not so straightforward, particularly when safety and environmental risks are involved. Unless management is willing to assign a dollar cost to an environmental incident or a fatality, the analyst will have to prioritize safety, environmental, and economic improvement recommendations on separate scales.

The net risk reduction offered by any recommendation is the risk associated with the existing situation minus the residual risk after the change is made, *minus the risk associated with making the change.* This third term is often neglected, but it can be significant. For example, a bypass around a flow controller may reduce the risk of an unplanned shutdown, but the immediate risk associated with hot-tapping the pipe to install the bypass may outweigh any long-term benefit. To minimize this risk, the proposed change could be delayed until the unit is shut down for other reasons and de-inventoried. The event tree methodology discussed previously can be used to estimate the current risk, and the event tree can be reevaluated assuming the change has been made to determine the residual risk. However, the best methodology for estimating the risk associated with making the change depends upon the nature of the change.

To use the event tree methodology for calculating the benefit (risk reduction) of a recommendation, one must first identify all the accident scenarios affected by the proposed change. Not all scenario changes will be positive; for example, an inert gas fire suppression system may reduce the risk of fire while increasing the risk of asphyxiation. In addition to introducing new hazards, changes may alter the effectiveness of existing safeguards, alter the progression of accident sequences, and/or alter initiating event frequencies. Once all the affected accident scenarios are identified, the sum of the risk of all those accident scenarios gives the current risk, and the sum of the risk of the same scenarios after the change is made gives the residual risk. The **maximum** risk reduction afforded by that recommendation is the difference between the two sums. In algebraic form, the expression is:

$$\Delta \text{Risk}_{\max} = \sum_{n=1}^{N} F_{n,\text{before}} \times C_{n,\text{before}} - \sum_{n=1}^{N} F_{n,\text{after}} \times C_{n,\text{after}}$$

where N is the number of accident scenarios affected by the recommendation, F_n is the frequency of accident scenario n, and C_n is the consequence of accident scenario n.

This is the maximum risk reduction because (1) it does not account for the risk associated with making the change and (2) other higher priority recommendations, if implemented, may have already reduced the risk associated with the existing situation. (Hypothetically, other recommendations may have increased the risk of the accident scenarios addressed by the lower priority recommendation, but such situations are rare. If recommendations are closely intertwined, they should be evaluated together.) To be strictly correct, the prioritization of recommendations is an iterative process. Once the highest priority recommendation is implemented, the remaining recommendations should be reevaluated to determine the next highest priority, given the risk of some scenarios is now lower than it was during the initial prioritization. As a practical matter, of course, such reevaluations are seldom done unless the need for an expensive, low-priority recommendation is questioned by management.

The following example illustrates how to apply this methodology to prioritize recommendations. Assume a PHA has been performed on a reactive chemical process. There are 30 recommendations addressing a variety of hazards, including potential fires. The reactor vessel is protected from overpressure by a rupture disk that discharges potentially flammable material to the atmosphere. Among the recommendations is the suggestion that the reactor be equipped with a high pressure alarm to improve operator response to an upset.

Step 1. Select a proposed improvement. For this example, prioritize the proposed high pressure alarm.

Step 2. Identify each deviation, failure mode, or class of accidents affected positively or negatively by the change. Any accident scenario that could cause a runaway reaction (excess feed, contaminants in the feed, loss of reactor cooling, or excess catalyst) is affected by the change. No other classes of accidents are affected.

Step 3. Develop event trees describing possible accidents with and without the change. In the current design, the operator monitoring the reactor must detect any sign of a runaway reaction (such as an increasing temperature or a high feed flow rate) and respond to it. The operator has several response alternatives depending upon his/her assessment of the situation, ranging from increasing the cooling water flow to dumping shortstop (quench) solution into the reactor. If the operator fails to respond, or if the response is ineffective, the rupture disk should protect the reactor from bursting. Figure 4 shows an event tree for the current situation. Note that if the operator action is successful, the pressure rise is limited and the rupture disk is not challenged. Thus, there is no branch point for the rupture disk safeguard in the accident scenario (pathway) when the operator succeeds. The addition of a high pressure alarm would increase the probability of a quick operator response. Figure 5 shows how the event tree would change if an alarm were present. Notice that the accident scenarios labeled 4, 5, and 6 in Figure 5 are the logical equivalent of scenarios A, B, and C in Figure 4. The event tree logic given failure of the alarm is just like the event tree for no alarm.

Step 4. Estimate the initiating event frequency(s) and branch point probability(s) with and without the change. One cause of runaway reactions is a high feed rate, which could be caused by the feed flow control valve opening too much. From data books, mechanical failures of the valve causing it to be open too much can be expected three times in 1,000 years of operation (0.003 per year). Human errors in setting the flow controller setpoint too high are expected on 1 of

Runaway Reaction	Operator Response	Rupture Disk	Consequence
F_{IE}	$1 - P_3$		A - Runaway reaction, operator responds
	P_3	$1 - P_4$	B - Runaway reaction, operator fails, disk bursts
		P_4	C - Runaway reaction, operator fails, disk fails

FIGURE 4. Example Calculation—Step 3a

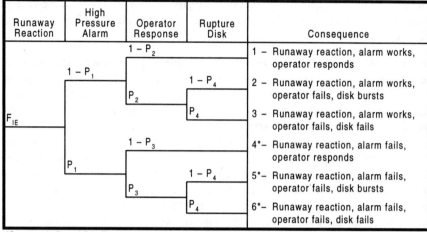

*Same as A, B, and C in Figure 4.

FIGURE 5. Example Calculation—Step 3b

every 10,000 attempts based on a human reliability analysis. Operators must adjust the setpoint about 70 times per year, so the expected frequency of operator-induced high flow rates is 0.007 per year. In this example, there is only one feed control valve, so common cause failure of several valves simultaneously is not an issue; hHowever, in redundant systems, common cause failure may be the dominant contributor to the initiating event frequency (or the safeguard failure probability). Table 3 summarizes this analysis, showing the total expected frequency of flow control valve failures is 0.01 per year. A similar analysis of other causes of runaway reactions added 0.09 failures per year, for a grand total initiating event frequency of 0.1 per year. The proposed change does not alter the initiating event frequency, so the same value will be used in both trees. However, the change will affect the branch point probabilities in the event trees. As shown

**TABLE 3
Initiating Event Frequency Data**

Initiating Event	Frequency (F_{IE})
Flow control valve opening	
— mechanical	.003/yr
— operator error	(.0001) (70/yr)
— common cause	N/A
Subtotal	.01/yr
Cause 2, 3, etc.	.09/yr
Grand Total	.1/yr

TABLE 4
Branch Point Probability Data

n	Branch Point	Probability of Success on Demand $(1 - P_n)$	Probability of Failure on Demand (P_n)
1	High pressure alarm	.99	.01
2	Operator response (with alarm)	.9	.1
3	Operator response (without alarm)	.8	.2
4	Rupture disk	.999	.001

in Table 4, the alarm itself (including all components from sensor to annunciator) is expected to alert the operator to 99% of all runaway reactions. Given an alarm, the operator should effectively respond to 90% of the demands; even without an alarm, the operator should still respond to 80% of the demands. (Sometimes the operator response given failure of an alarm is worse than for no alarm at all because the operator may not be as attentive to other indications if he/she is relying on an alarm; that was judged not to be the case here.) The rupture disk is unaffected by the change; it should protect the reactor against 99.9% of all runaway reactions.

Step 5. Estimate the consequences of each accident scenario. There are safety, environmental, and/or economic consequences associated with each accident scenario, and each of those consequences will cost something. That price may be paid immediately to replace capital equipment or inventory, or it may be paid eventually in business interruption losses, fines, legal expenses, increased insurance premiums, or lost market share. Table 5 summarizes the total cost assessed for each accident scenario in this example.

TABLE 5
Consequence Data

Scenario	Consequence			
	Safety	Environment (in thousands)	Economic (in thousands)	Total (in thousands)
1, 4	0	0	$10	$10
2, 5	0	$20 fine $30 cleanup	$50	$100
3, 6	Fatalities and/or severe injuries ($2 million settlement)	$50 fine $200 cleanup	$750	$3,000

Step 6. Calculate the risk for each accident scenario before and after the proposed change. If the event tree has been properly constructed, the frequency of each accident scenario is simply the product of the initiating event frequency and the probability of success or failure of each safeguard in that accident scenario. The accident scenario frequency can then be multiplied by the scenario consequences to determine the risk associated with that scenario. Figures 6 and 7 show these calculations before and after the proposed change.

Step 7. Estimate the risk associated with making the change. In this case, there is a negligible risk associated with adding the high pressure alarm.

Step 8. Calculate the net risk reduction. The sum of the accident scenario risks in Figure 6 yields the risk of the current situation ($2,900 per year). The sum of the accident scenario risks in Figure 7 yields the residual risk after the change is made ($1,900 per year). Subtracting both this value and the risk associated with making the change (~$0) from the current risk gives a net risk reduction of $1,000 per year.

Step 9. Calculate the priority score. Assume the unit has an expected remaining life of 20 years. The risk reduction benefit will accrue each year for a total benefit of $20,000. The alarm will cost $1,000 to install, and it will incur a $50 per year maintenance cost, for a total cost of $2,000. The resulting priority score is 10 ($20,000/$2,000). [Note: A more sophisticated economic analysis considering discount and inflation rates could be performed, but it would add nothing to the example. In practice, given the uncertainties inherent in any risk assessment, the more sophisticated analysis is seldom necessary to obtain a reasonable priority score.]

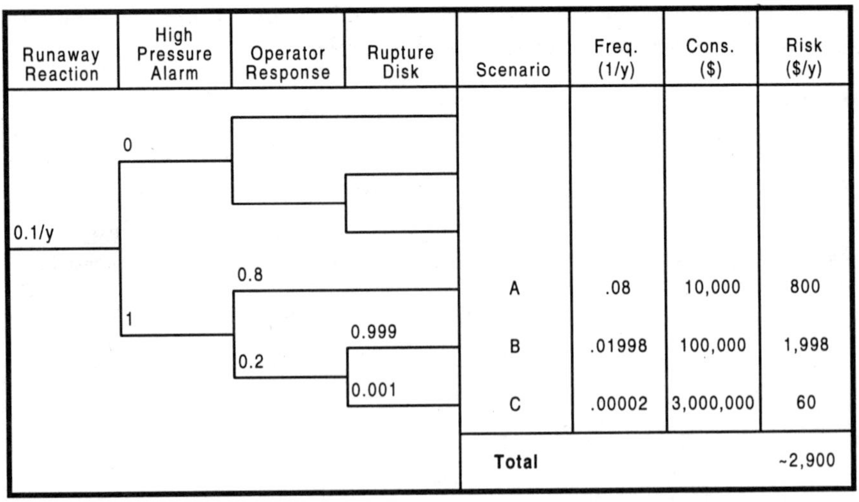

Runaway Reaction	High Pressure Alarm	Operator Response	Rupture Disk	Scenario	Freq. (1/y)	Cons. ($)	Risk ($/y)
0.1/y	0 / 1	0.8 / 0.2	0.999 / 0.001	A	.08	10,000	800
				B	.01998	100,000	1,998
				C	.00002	3,000,000	60
				Total			~2,900

FIGURE 6. Risk Before the Proposed Change

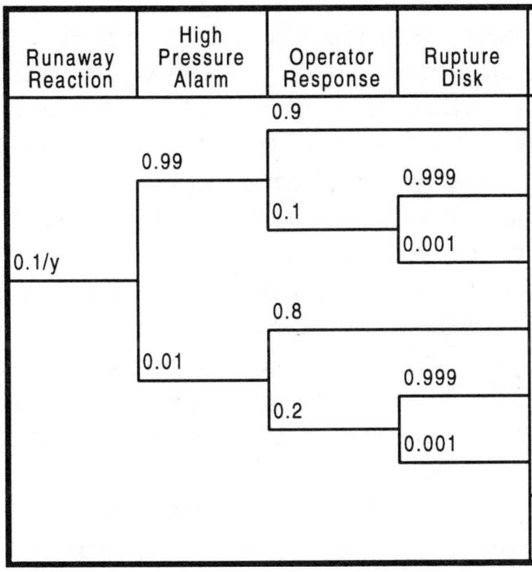

Risk-Based Judgments in Process Hazard Analyses

Runaway Reaction	High Pressure Alarm	Operator Response	Rupture Disk	Scenario	Freq. (1/y)	Cons. ($)	Risk ($/y)
0.1/y	0.99	0.9	0.999	1	.0891	10,000	891
		0.1	0.001	2	.00989	100,000	989
				3	.00001	3,000,000	30
	0.01	0.8	0.999	4	.0008	10,000	8
		0.2	0.001	5	.0002	100,000	20
				6	.0000002	3,000,000	<1
				Total			~1,900

FIGURE 7. Risk After the Proposed Change

If such an analysis were performed on every recommendation, a defensible priority could be assigned to each recommendation. Unfortunately, the quantitative risk analysis necessary for a "correct" prioritization may require significant effort, particularly as the number of affected accident scenarios increases. Unless the company has accumulated a significant body of plant-specific data, fault tree analyses and/or human reliability analyses may be required to estimate initiating event frequencies and branch point probabilities. Dispersion analyses, fire/explosion analyses, and/or effects models may be necessary to estimate the consequences for each accident scenario, and management may be reluctant to combine safety, environmental, and economic consequences on a dollar scale. And it would obviously be wasteful for management to spend more money justifying and prioritizing an inexpensive recommendation than it would spend to simply implement the change. Thus, for a variety of reasons, a complete quantitative risk analysis is impractical for most risk-based decisions. *However, the logic underlying such risk assessments is valid and must be maintained by any defensible analysis method.*

Qualitative Risk Assessment Using Event Tree Logic

A simplified approach to risk-based decision making can be structured around a risk matrix as initially discussed, but the team must judge the *total* likelihood of accident scenarios in *each* consequence category, not just the worst-case consequence category. Using the example risk matrix in Figure 2, the team would need to evaluate accident scenarios with high, medium-to-high, and low-to-medium consequences. (No evaluation of accident scenarios with low consequences is necessary because the risk is tolerable regardless of the likelihood.) In this relatively subjective approach, the team selects a deviation (or failure mode, or question) and uses its best effort to imagine the range of accident scenarios that could result in the highest consequence category. It then judges the likelihood of each scenario, considering the frequency of the initiating event(s) and the probability that the various safeguards will fail when needed. The likelihood of all such scenarios is then mentally summed to arrive at the total likelihood of consequences in the highest category. Just as before, the matrix guides the team decision as to whether the risk is tolerable or whether further risk reduction is required. If the risk is unacceptable, no further analysis is required because the team should make recommendations for improvements. However, if the risk is tolerable or marginal, the analysis must be repeated for each of the lower consequence categories until an unacceptable risk is discovered (at which point the analysis can stop), or until it is shown that the risk is tolerable in all consequence categories. By using the matrix in this manner, the team avoids the problems of (1) combining the frequency of one event with the consequences of another or (2) considering only the risk of the worst-case consequences. However, this approach only slightly reduces the general problem of risk matrices undervaluing a recommendation that reduces the risk of a large number of lower risk accident scenarios.

Consider the previous example of a chemical reactor with the potential for runaway reactions. The likelihood of accident scenarios resulting in a vessel rupture (high consequences) is very low because that would require both failure of the operator to respond and failure of the rupture disk to relieve the pressure. The high-consequence, low-likelihood cell in the matrix is categorized as marginal risk, so the team should consider further risk reduction, but it is not required. Moving to the next lower category, it is also possible that a runaway reaction could produce medium-to-high consequences if the operator fails to respond, but the rupture disk does work. The operator response does provide some protection, but the team judges that the accident scenario frequency is still medium-to-high. Because this risk is unacceptable, risk reduction is required. No analysis of accident scenarios for lower consequence categories is required once an unacceptable risk is discovered.

Score-Based Risk Assessment Using Event Tree Logic

A more rigorous approach to risk-based decision making can also be structured around a risk matrix, but the matrix must be slightly modified as shown in Figure 8. The purely qualitative likelihood category descriptions must be supplemented with quantitative values that define the boundaries between categories so quantitative judgments can be made. It is helpful, but not necessary, to

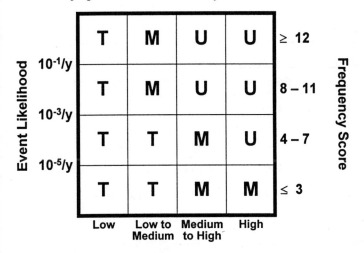

FIGURE 8. Modified Risk Matrix

define boundary values for the consequence categories as well. The matrix cells themselves can remain labeled as before or can be assigned a numeric score.

To use the revised risk matrix, the team (or analyst) must mentally step through the same logic one would use for a quantitative risk assessment. First, the team must select a deviation, failure mode, or question. However, rather than developing detailed event trees, the team picks the accident scenario(s) it believes to be the dominant contributor(s) to **each** consequence category (usually only one or two accident scenarios per category). This score-based methodology assumes the other accident scenarios in a given consequence category would not bump the total frequency into the next higher category (typically a factor of 10 or higher). If there are a large number of similar scenarios, there is a correction term that should be applied later in the analysis process.

The team must now assess the likelihood of each accident scenario by combining the frequency of the initiating event(s) and the probability that the safeguards will fail on demand (PFOD). If frequency and PFOD data are available, they should be used directly; otherwise, a generic score can be used. Table 6 lists an example scoring scheme for initiating events, and Table 7 lists an example scoring scheme for safeguard failures. Alternatively, charts such as Figure 9 can be developed to provide teams with better guidance on picking appropriate initiating event and PFOD scores or values.

To calculate the initiating event score, simply pick the cause with the highest score from Table 6. For example, an accident scenario that could be caused by a piping rupture (PEF, score of 2), an external fire (IEE, score of 3), or a human

TABLE 6
Example Initiating Event Scores

Score	Event Type	Examples
15	Any event (operating conditions, environmental conditions, human actions, equipment actions, etc.) expected to occur regularly (EE)	Switching pumps, unloading trucks
11	Human error on a frequent task (HEF)	Opening wrong valve
11	Active equipment failure (AEF)	Pump tripping off, valve failing closed
9	Human error on an occasional task (HEO)	Miscalibrating controller
3	Infrequent external event (IEE)	Lightning strike, external fire
2	Passive equipment failure (PEF)	Pipe rupture, vessel rupture

TABLE 7
Example Safeguard PFOD Scores

Score	Safeguard Type	Examples
1	Human response to indications	Noticing high level in a sight glass
2	Human response to alarms	Hearing a high level alarm
3	Safety Integrity Level 1 Interlock	High level switch closing an isolation valve
5	Safety Integrity Level 2 Interlock	Dual high level switches closing an isolation valve
7	Safety Integrity Level 3 Interlock	Separate high level switches closing two independent isolation valves in the same line
4	Active equipment (independent of the initiating event)	Level control valve opening in response to increased level
6	Standby equipment	Sump pump to transfer spilled liquid
8	Passive equipment	Dike to contain spilled liquid, rupture disk

error on an occasional task (HEO, score of 9) would score a 9 (conceptually corresponding to the most likely initiating event). [Note: Summing the scores would grossly overstate the likelihood because the scores are based on logarithms. In this case, a score of 9.012 (9+.009+.003) would be more accurate because the scoring system assumes the human error is 1,000 times more likely than the external event and 3,200 times more likely than the passive equipment failure.]

To take credit for the safeguards, the team can then select the PFOD score from Table 7 for every independent safeguard that would have to fail to cause the

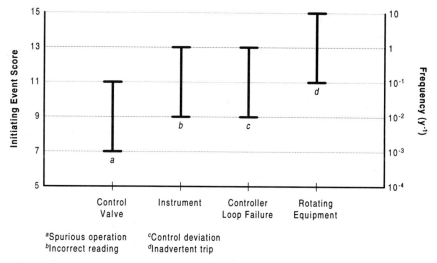

FIGURE 9. Common Events Involving Equipment Failures of Active Components

specific accident scenario. Subtracting the safeguard PFOD score(s) from the initiating event frequency score gives the mitigated accident scenario frequency score. [Note: To perform the same calculation with frequency and PFOD data instead of scores, add the initiating event frequencies and then multiply the total by the independent safeguard PFODs to get the accident scenario frequency value.] This process should be repeated for each accident scenario contributing to the specific consequence category, and the highest score should be selected (or the accident scenario frequencies should be summed). The team then maps the frequency score (or summed value) onto the matrix to determine whether the risk is tolerable. If there are a large number (>10 in the example matrix) of similar scenarios, the team should use the next higher likelihood category in the matrix.

Again, consider the example of the runaway chemical reactor. As previously discussed, the worst-case (high severity category) consequences result if both the operator and rupture disk fail to respond appropriately. The initiating events for this scenario included a flow control valve that could be open too far (AEF, score of 11; HEF, score of 11), contaminants in the feed stream (HEO, score of 9), a loss of reactor cooling (AEF, score of 11), and excess catalyst (HEF, score of 11). The highest score is 11, so that will be used as the initiating event frequency score. The PFOD score for the operator response is 1, and the PFOD score for the rupture disk failing is 8. Subtracting these from the initiating event score of 11 gives an accident scenario frequency score of 2 (11-1-8). For a high consequence event, this is a marginal risk according to Figure 8. Repeating this analysis for the medium-to-high consequence category (caused by the operator failing to respond, but the rupture disk working) gives an accident scenario frequency score of 10 (11-1), which is an unacceptable risk.

If the team recommended a high pressure alarm, management could use this same analysis technique to evaluate the benefit of that recommendation and prioritize it by repeating the calculation assuming the alarm was installed. For the high consequence category, the accident scenario score would drop to 0 (11-3-8), and for the medium-to-high category, the score would drop to 8 (11-3). [Note: A zero or negative score does *not* mean the accident cannot occur; it only means that further risk reduction for that scenario is probably not justifiable.] The alarm does drop the frequency one category, so it is probably worth doing if it is not too expensive. However, the risk is still unacceptable according to the risk matrix in Figure 8, so something more should be done.

Conclusion

Using a blend of the simplified and more rigorous approaches outlined in this paper will significantly improve the risk-based judgments of PHA teams and of managers evaluating their recommendations. During PHA team meetings, the qualitative approach should almost always be used, and it involves very little additional effort than is currently expended by teams using a risk matrix. Occasionally, if the team members cannot agree on the accident scenario risk, the semiquantitative approach may be used to arrive at a more defensible basis for their judgment of the need for risk reduction. For managers evaluating recommendations, the pure qualitative approach offers little insight beyond what is obvious with no further analysis. After resolving obvious, simple, and/or inexpensive recommendations such as installing properly sized relief valves, revising procedures, or improving labeling (which usually accounts for at least half the action items), management may want to use the semiquantitative approach to evaluate the benefit of the remaining recommendations. For those few expensive recommendations whose value is questioned (or whose budget must be justified to upper management for funding), a more detailed quantitative risk assessment may be warranted. In all cases, the analysis methodology selected should be the one "barely adequate" for making decisions, so money is not wasted on unnecessary analysis. However, the stakes are too high to use overly simplified methods based only on a worst-case scenario that may lead to the wrong conclusion about risk.

Reference

1. Center for Chemical Process Safety. *Guidelines for Hazard Evaluation Procedures, Second Edition with Worked Examples*, American Institute of Chemical Engineers, New York, 1992.

handwritten annotations: - equivalent to IRA, however this company / it qualitative / anonymous / 20 plants international / specialty company

Risk Acceptance Criteria and Risk Judgment Tools Applied Worldwide within a Chemical Company

William G. Bridges and Tom R. Williams
JBF Associates, Inc., 1000 Technology Drive, Knoxville, TN 37932-3353

ABSTRACT

From 1994 through early 1996, a multinational chemical company developed a standard for evaluating risk of potential accident scenarios. This standard was developed to help users (i.e., engineers, chemists, managers, and other technical staff) determine (1) when sufficient safeguards were in place for an identified scenario and (2) which of these safeguards were critical to achieving (or maintaining) the tolerable risk level. Plant management was held accountable for upholding this standard, and they were also held accountable for maintaining (to an extremely high level of availability) the critical safety features that were identified. In applying this standard, the users found they needed more guidance on selecting the appropriate methodology for judging risk; some used methodologies that were deemed too rigorous for the questions being answered and others in the company used purely qualitative judgment tools. The users in the company agreed to a set of three methods for judging risk and developed a decision tree, followed by training, to help the users (1) choose the proper methodology and (2) apply the methodology chosen consistently. The new guidelines for risk acceptance and risk judgment were taught to technical staff (those who lead hazard reviews and design new processes) worldwide in early 1996. This paper presents the evolution of the risk tolerance and risk judgment approach used by the company.

Background

This paper is written on behalf of a major chemical company headquartered in the USA. The company wishes to remain anonymous because of the litigious environment in the USA. This environment ultimately penalizes any company that recognizes the necessity of accepting or tolerating any risk level above "zero" risk. However, the only way to reach zero risk is to go out of business altogether. All chemical processing operations contain risk factors that must be managed to reasonably reduce the risk to people and the environment to tolerable levels, but the risk factors cannot be entirely eliminated. This chemical company has made significant strides in recent years in risk management; particularly, the company has implemented effective risk judgment and risk acceptance (tolerance) criteria. Because JBF Associates, Inc. (JBFA) has worked with the company in the training steps related to these criteria, the company has agreed to allow JBFA to share a synopsis of the company's approach in the hope that others can benefit from the lessons learned to date.

To understand the risk management systems described in this paper, a brief portrait of the chemical company is essential. The company conducts operations principally in North America, Asia, and Europe. The operations include more than 20 petrochemical, specialty chemicals, and polymer processing plants, along with several related terminals and blending facilities. The processes involve flammable, toxic, and highly reactive chemicals. The company is subject to OSHA process safety management (PSM) and EPA risk management program (RMP) regulations in the USA, and they have a corporate process safety standard that applies worldwide. Each plant has technical staff who implement the process safety standards and related standards and guidelines. The company has been successful in worldwide implementation of strategies described in this paper.

One key to this success is holding each plant manager accountable for implementation of the risk management policies and standards; any deviation from a standard or criteria based on a standard, must be preapproved by the responsible vice president of operation.

In our experience, many companies claim to hold plant managers accountable, but in the final analysis production goals usually take precedence over safety requirements; this company has shown equal vigilance in enforcement of safety- (risk-) related standards.

Chronology of Risk Judgment Implementation

Figure 1 and the following paragraphs present a synopsis of this company's efforts to implement a risk-based judgment system, which is now producing significant return for the company. Although other companies may follow a different path to achieve the same goals, there are valuable lessons to be learned from this company's particular experiences.

STEP 1: RECOGNIZE THE NEED FOR RISK-BASED JUDGMENT
The technical personnel who were responsible for judging risk of accident scenarios for the company recognized the need for adequately understanding and evaluating risk many years ago. However, most decisions about plant operations were made subjectively without comparing relative risk of the accident scenarios. Not until a couple of major accidents occurred did key line managers, including operations vice presidents, become convinced of the value of risk judgment and the need to include risk analysis in the decision-making process.

STEP 2: STANDARDIZE AN IMPROVED APPROACH TO HAZARD EVALUATION
The company realized that the best chance for managing risk was to maximize the opportunity for identifying key accident scenarios. Therefore, the first enhancement was to improve the specifications for process hazard analyses

Risk Acceptance Criteria and Risk Judgment Tools Applied Worldwide within a Chemical Company 547

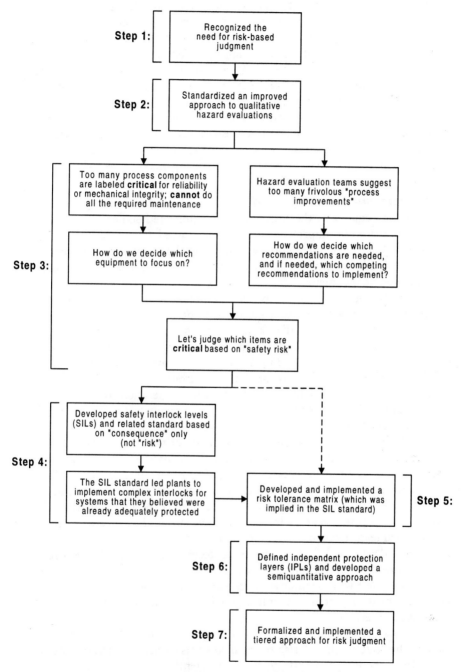

FIGURE 1 The Evolution of a Risk Judgment Approach

(PHAs) and provide training to PHA leaders to meet these specifications. A standard and a related guideline were developed prior to training. The standard became one of the process safety standards that plant management was not allowed to circumvent without prior approval. The guideline provided corporate's interpretation of the standard, and although all plants were strongly advised to follow the guideline, plant managers were allowed flexibility to develop their own plant-specific guidelines. The major enhancements to the PHA specification were (1) to require a step-by-step analysis of critical operating procedures (because deviations from these procedures lead to most accidents), (2) improve consideration of human factors, and (3) improve consideration of facility siting issues. The company also began using quantitative risk assessment (QRA) to evaluate complex scenarios.

STEP 3: DETERMINE IF PURELY QUALITATIVE RISK-BASED JUDGMENT IS SUFFICIENT

These improvements to the hazard identification methodologies led to many recommendations for improvements. Managers were left with the daunting task of resolving each recommendation, which included deciding between competing alternatives and deciding which recommendations to reject. Their only tool was pure qualitative judgment.

Simultaneously, the company began to intensify its efforts in mechanical integrity. Without any definitive guidance on how to determine critical safety features, the company identified a large portion of the engineered features as "critical" to safe operation. The company recognized that many of the equipment/instrument features listed in the mechanical integrity system did little to minimize risk to the employees, public, or environment. They also recognized that it would be wasting valuable maintenance and operations resources to consider all of these features to be critical. So, the company had to decide which of the engineered features (protection layers) were most critical.

With all of the impending effort to maintain critical design features and to implement or decide between competing recommendations, the company began a search for a risk-based decision methodology. They decided to focus on "safety risk" as the key parameter, rather than "economic" or "quality" risk. The company had a few individuals who were well trained and experienced in using QRA, but this tool was too resource intensive for evaluating the risk associated with each critical feature recommendation, even when the focus of the decision was narrowed to "safety risk." So the managers (decision makers) in charge of resolving the hazard review recommendations and deciding which components were critical, were left with qualitative judgment only; this proved too inconsistent and led many managers to wonder if they were performing a re-analysis to decide between alternatives.

Corporate management realized that they needed to make a baseline decision on the "safety-related" risk the company was willing to tolerate. They also needed a methodology to estimate more consistently if they were within the tolerable risk range.

STEP 4: PREVENT HIGH CONSEQUENCE ACCIDENT SCENARIOS

Many companies would not have this as the next chronological step, but about this time, the company recognized that they also needed a corporate standard for safety interlocks to control design, use, and maintenance of key safety features throughout their global operations. So, the company developed definitions for safety interlock levels (SILs) and developed standards for the maintenance of interlocks within each SIL. Then the company developed a guideline that required the implementation of specified SILs based solely on safety consequence levels (instead of risk levels). If a process had the potential for an overpressure event resulting in a catastrophic release of a toxic material or a fire or explosion (defined as a Category V consequence as listed in Table 1) due to a runaway chemical reaction, then a Class A interlock (triple redundant sensors and double redundant actuator) was required by the company for preventing the condition that could lead to the runaway.

However, basing this decision solely on the safety consequence levels, did not give any credit for existing safeguards or alternate approaches to reducing the risk of the overpressure scenario. As a result, this SIL standard skewed accident prevention toward installing and maintaining complex (albeit highly reliable) interlocks. The technical personnel in the plants very loudly voiced their concern about this extreme "belts and suspenders" approach.

STEP 5: MANAGE RISK OF ALL SAFETY-IMPACT SCENARIOS

Before the company's self-imposed deadline for compliance with the corporate SIL standard, the company agreed with the plants that alternate risk-reduction measures should be given proper credit. To make this feasible, the company had to begin to evaluate the overall risk of a scenario, not just the consequences. They decided to develop a corporate standard and guidelines for estimating the mitigated risk of accident scenarios. (This development had actually begun at the end of Step 3, but the momentum in this direction slowed when emphasis for risk control shifted temporarily to safety interlocks.)

First, a risk matrix was developed with five consequence categories (as were used for the SILs described earlier), and seven frequency categories (ranging from 1/year to 1/10 million years). Next, the company delineated the risk matrix into three major areas:

- **Tolerable Risk**—Implementation of further risk reduction measures was not required; in fact, it was strongly discouraged so that focus would not be taken off of maintaining existing or implementing new critical layers of protection
- **Intolerable Risk**—Action was required to reduce the risk further
- **Optional**—An intermediate zone was defined, which allowed plant management the option to implement further risk reduction measures, as they deemed necessary

Figure 2 shows the company's risk matrix.

Consequence Category / Frequency of Consequence (per Year)*	Category I	Category II	Category III	Category IV	Category V
10^{-0}	Optional (evaluate alternatives)	Optional (evaluate alternatives)	Action at next opportunity (notify corporate management)	Immediate action (notify corporate management)	Immediate action (notify corporate management)
10^{-1}	Optional (evaluate alternatives)	Optional (evaluate alternatives)	Optional (evaluate alternatives)	Action at next opportunity (notify corporate management)	Immediate action (notify corporate management)
10^{-2}	No further action	Optional (evaluate alternatives)	Optional (evaluate alternatives)	Action at next opportunity (notify corporate management)	Action at next opportunity (notify corporate management)
10^{-3}	No further action	No further action	Optional (evaluate alternatives)	Optional (evaluate alternatives)	Action at next opportunity (notify corporate management)
10^{-4}	No further action	No further action	No further action	Optional (evaluate alternatives)	Optional (evaluate alternatives)
10^{-5}	No further action	No further action	No further action	No further action	Optional (evaluate alternatives)
10^{-6}	No further action	No further action	No further action	No further action	No further action
10^{-7}	No further action	No further action	No further action	No further action	No further action

FIGURE 2. Risk Matrix

*For example, 10^{-2} is equivalent to 1/100 years.

TABLE I
Consequence Categorization

CATEGORIES I/II	
Personnel:	Minor or no injury, no lost time
Community:	No injury, hazard, or annoyance to public
Environmental:	Recordable event with no agency notification or permit violation
Facility:	Minimal equipment damage at an estimated cost of less than $100,000 and with no loss of production
CATEGORY III	
Personnel:	Single injury, not severe, possible lost time
Community:	Odor or noise annoyance complaint from the public
Environmental:	Release which results in agency notification or permit violation
Facility:	Some equipment damage at an estimated cost greater than $100,000, or minimal loss of production
CATEGORY IV	
Personnel:	One or more severe injuries
Community:	One or more severe injuries
Environmental:	Significant release with serious offsite impact
Facility:	Major damage to process area(s) at an estimated cost greater than $1,000,000 or some loss of production
CATEGORY V	
Personnel:	Fatality or permanently disabling injury
Community:	One or more severe injuries
Environmental:	Significant release with serious offsite impact and more likely than not to cause immediate or long-term health effects
Facility:	Major or total destruction to process area(s) estimated at a cost greater than $10,000,000, or a significant loss of production

Note: Later versions of the Standard defined Consequence Categories in terms of "quantity released" or "dollars of damage," rather than number of injuries or fatalities.

Some companies would have called this a semiquantitative approach, but in this company, the PHA teams used this matrix to "qualitatively" judge risk. Teams would vote on which consequence and frequency categories an accident scenario belonged (considering the qualitative merits of each existing safeguard), and they would generate recommendations for scenarios not in the Tolerable Risk area. This approach worked well for most scenarios, but the company soon found considerable inconsistencies in the application of the risk matrix in qualitative risk judgments. Also, the company observed that too many accident scenarios were requiring resource-intensive QRAs. It was clear that an intermediate approach for judging the risk of moderately complex scenarios was needed. And, the company still needed to eliminate the conflict between the risk matrix and the SIL standard.

STEP 6: DEVELOP A SEMIQUANTITATIVE APPROACH (THE BEGINNINGS OF A TIERED APPROACH) FOR RISK JUDGMENT

This was a very significant step for the company to take; the effort began in early 1995 and was implemented in early 1996. Along with the inconsistencies in applying risk judgment tools, there was still confusion among plant personnel about when and how they should use the SIL standard and the risk matrix. Both were useful tools that the company had spent considerable resources to develop and implement. The new guidelines would need to somehow integrate the SILs and the risk matrix categories to form a single standard for making decisions. And the plants also needed a tool (or multiple tools), besides the extremes of pure qualitative judgment and a QRA, to decide on the best alternative for controlling the risk of an identified scenario. The technical personnel from the corporate offices and from the plants worked together to develop a semiquantitative tool and to define the needed guidelines.

One effort toward a semiquantitative tool involved defining a new term called an independent protection layer (IPL), which would represent a single layer of safety for an accident scenario. Defining this new term required developing examples of IPLs to which the plant personnel would be able to relate. For example, a spring-loaded relief valve is independent from a high pressure alarm; thus a system protected by both of these devices has two IPLs. On the other hand, a system protected by a high pressure alarm and a shutdown interlock using the same transmitter has only one IPL. Class A, B, and C safety interlocks (which were defined previously in the SIL standard) were also included as example IPLs.

To ensure consistent application of IPLs (i.e., to account for the relative reliability/availability of various types of IPLs), it was necessary to identify how much "credit" plant personnel could claim for a particular type of IPL. For example, a Class A safety interlock would deserve more credit than a Class B interlock, and a relief valve would be given more credit than a process alarm. This need was addressed by assigning a "maximum credit number" for each example IPL (see Table 2). The credit is essentially the order of magnitude of the risk reduction anticipated by claiming the safeguard as an IPL for the accident scenario. The company believed that when PHA teams or designers used the IPL definitions and related credit numbers, the consistency between risk analyses at the numerous plants would improve.

Another (parallel) effort involved assigning frequency categories to typical "initiating events" for accident scenarios (see Table 3); these initiating events

TABLE 2
Credits for Independent Protection Layers (IPLs)

Example IPL	Maximum Credit Number for IPL
Basic Process Control Systems	
Automatic control loop (If failure is not a significant initiating event contributor and is independent of the Class A, B, or C interlock [if applicable] and final element is tested at least once per 4 years)	1
Human Intervention	
Manual response in field with more than 10 minutes available for response (If sensor/alarm are independent of the Class A, B, or C interlock [if applicable] and operator training includes required response)	1
Manual response in field with more than 40 minutes available for response (If sensor/alarm are independent of the Class A, B, or C interlock [if applicable] and operator training includes required response)	2
Passive Devices*	
Secondary containment such as a dike (If good administrative control over drain valves exists)	2
Spring-loaded relief valve in clean service	3
Safety Interlocks	
Class A interlock (Provided independent of other interlocks)	3
Class B interlock (Provided independent of other interlocks)	2
Class C interlock (Provided independent of other interlocks)	1

* Claiming passive devices, such as a relief valve, in conjunction with the interlock in question, should be the exception.

were intended to represent the types of events that could occur at any of the various plants. The frequency categories were derived from QRA experience within the company and provided a consistent starting point for semiquantitative analysis.

Finally, a semiquantitative approach for estimating risk was developed, incorporating the frequency of initiating events and the IPL credits described previously. Although this approach used standard equations and calculation sheets not described here, the basic approach required teams to:

- Identify the ultimate consequence of the accident scenario and document the scenario as clearly as possible, stating the initiating event and any assumptions

TABLE 3
Initiating Event Frequencies

Event	Estimated Frequency
Loss of cooling (Standard simplex system)	1/year
Loss of power (Standard simplex system)	1/year
Human error (Routine, once per day opportunity)	1/year
Human error (Routine, once per month opportunity)	1/10 years
Basic process control loop failure	1/10 years
Large fire	1/100 years* 1/1,000 years

*Fire frequency for an individual process system of 1/100 years is conservative.

- Estimate the frequency of the initiating event (using a frequency from Table 3, if possible)
- Estimate the risk of the unmitigated event and determine from the risk matrix if the risk is tolerable as is:
  —If the risk is not tolerable, take credit for existing IPLs until the risk reaches a tolerable level in the risk matrix; use best judgment in defining IPLs and deciding which ones to take credit for first
  —If the risk is still not tolerable, develop a recommendation(s) that will lower the risk to a tolerable level
- Record the specific safety features (IPLs) that were used to reach a tolerable risk level

The company demanded "zero" tolerance for deviating from inspection, testing, or calibration of the documented hardware IPLs and enforcement of administrative IPLs. (Any deviation without prior approval was considered a serious deficiency on internal audits.) Other features not credited as IPLs could be kept if they served a quality, productivity, or environmental protection purpose; otherwise, these items could be "run to failure" or removed because doing so would have no effect on the risk level.

This semiquantitative approach explicitly met a need expressed in Step 3: determining which of the engineered features were critical to managing risk. PHA teams began applying this approach to validate their qualitative risk judgments. However, the company still needed to (1) formalize guidelines for when to use qualitative, semiquantitative, and quantitative risk judgment tools and (2) standardize the use each tool.

STEP 7: FORMALIZE AND IMPLEMENT THE TIERED APPROACH

The company decided that the best way to standardize risk judgment in all of the plants was to (1) revise the risk tolerance standard, (2) revise the SIL standard, (3) formalize a guideline for deciding when and how to use each risk judgment tool, and (4) provide training to all potential users of the standards and guidelines (including engineers at the plants and corporate offices, PHA leaders, maintenance and production superintendents, and plant managers). The formal guideline and training would be based on a decision tree dictating the complexity of analysis required to adequately judge risk. The company's first attempt at a decision tree is shown in Figure 3.

After the training needs were assessed for each type of user, the company produced training materials and exercises (including the decision tree) to meet those needs. The training took approximately 1 day for managers and superintendents (because their needs were essentially to understand and ensure adherence to the standards) and approximately 4 days for process engineers, design engineers, production engineers, PHA leaders, and QRA leaders. The training was initiated in early 1996, and early returns have shown strong acceptance of this approach, particularly in Europe, where the experience in the use of quantitative methods is much broader. The most significant early benefits have been:

- A reduced number of safety features (IPLs) labeled as "critical"
- Less frivolous recommendations from PHA teams, which now have a better understanding of risk and risk tolerance
- Better decisions on when to use a QRA (because there is now an intermediate alternative)

Path Forward

The next steps are to continually evaluate the current approach and modify it as necessary to meet the changing needs of the corporation and the plant personnel. For instance, the decision criteria for when to use the semiquantitative or the QRA method may change; the credit given to IPLs may need to change. More training is probably necessary on selected topics; for example, the personnel in the United States need additional training on the use of the semiquantitative approach and on how to mesh risk-based judgments with OSHA PSM and EPA RMP compliance efforts (there is an excellent opportunity for synergy here). A computer program may be developed to simplify some of the decisions, calculations, tracking, and reporting.

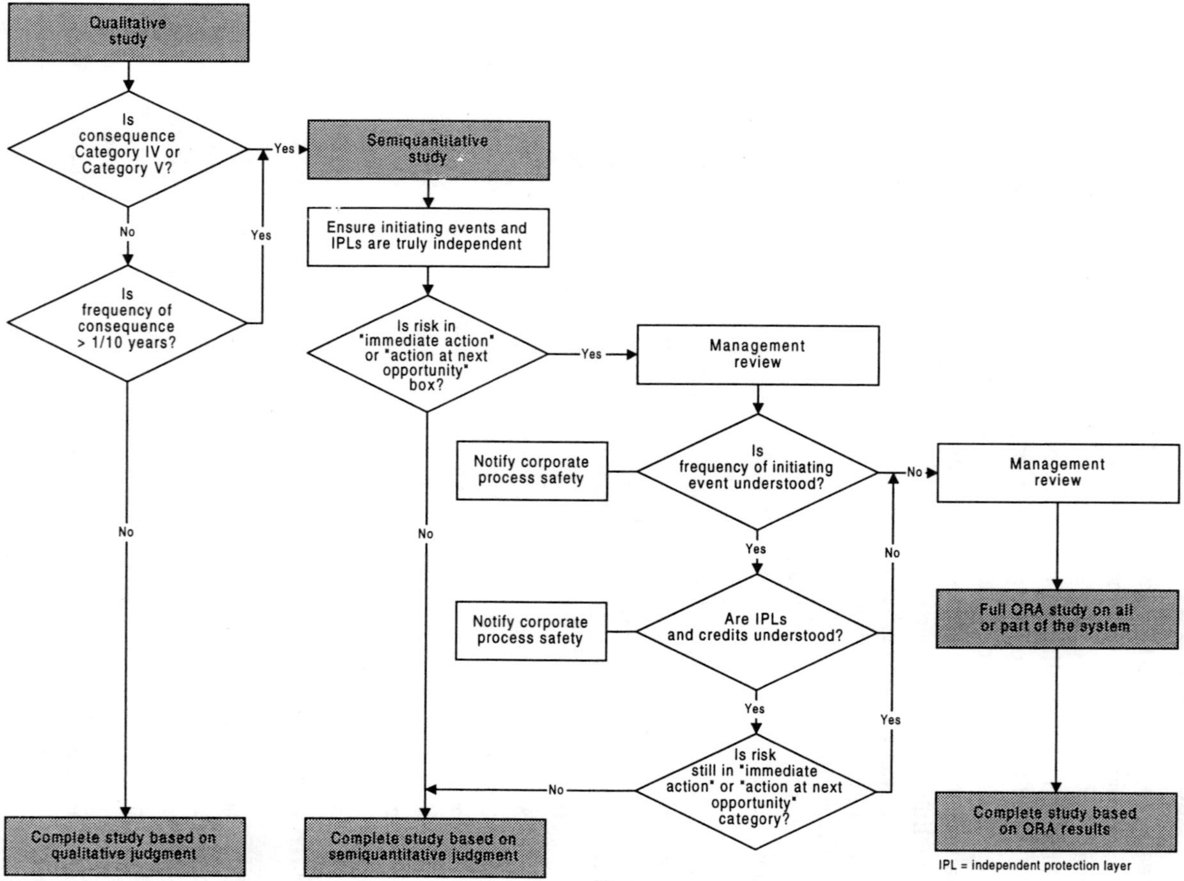

FIGURE 3 Decision Tree Dictating Which Risk Judgment Tool to Use

IPL = independent protection layer

Conclusions

The company believes they have experienced major reductions in risk throughout the stepwise implementation of this approach. The approach helps the company manage their risk control resources wisely and helps to more defensibly justify decisions with regulatory and legal implications. The key to the success of this program lies beyond the mechanics of the risk-judgment approach; it lies with the care company personnel have taken to understand and manage risk on a day-to-day basis. Company management has developed clear, comprehensive standards, guidelines, and training to ensure the plants manage risk appropriately. This is reinforced by company management taking an aggressive stance on enforcing adherence by the plants to company standards. The risk judgment standards and guidelines appear to be working to effectively reduce risk while minimizing the cost of maintaining "critical" safeguards. This company's success serves as only one example that risk management throughout a multinational chemical company is possible, practical, and necessary.

Bibliography

Advanced Process Hazard Analysis Leader Training, Process Safety Institute, Knoxville, TN, 1993.

Guidelines for Chemical Process Quantitative Risk Analysis, Center for Chemical Process Safety, American Institute of Chemical Engineers, New York, NY, 1989.

Guidelines for Hazard Evaluation Procedures, 2nd Edition with Worked Examples, Center for Chemical Process Safety, American Institute of Chemical Engineers, New York, NY, 1992.

F. P. Lees, *Loss Prevention in the Process Industries*, Vols. 1 and 2, Butterworth's, London, 1980.

D. F. Montague, "Process Risk Evaluation—What Method to Use," *Reliability Engineering and System Safety*, Vol. 29, Elsevier Science Publishers Ltd., England, 1990.

Startup Challenges for New Risk Management Programs

Chuck Fryman
FMC Corporation

Introduction

Many companies want to use risk-based and cost–benefit approaches in assisting them to make decisions of "how safe is safe enough." This requires the establishment of an overall framework or model for managing risks, including the establishment of risk acceptability criteria. Furthermore, the use of quantified risk assessment allows a company to perform cost–benefit analysis of risk reduction options, thus identifying the options providing the optimal risk reduction for the amount of money spent.

The development and implementation of risk management programs presents significant challenges to any company, and the objective of this paper is to firstly outline the main components of a comprehensive process safety risk management program and then the challenges faced by companies in developing and starting up these programs.

Components of a Comprehensive Risk Management Program

The overall framework or model for a comprehensive process safety risk management program is shown in Figure 1. The model begins with the identification of potential risks. These potential risks include risks to people (including workers, contractors and members of the general public), offsite impacts (including environmental cleanup and an estimate of any outrage costs), property damage, business interruption, and liabilities (including punitive and civil damages).

An awareness of risks is maintained through a careful analysis of the incidents experienced throughout the company. One should also be interested in learning from incidents that have happened to others in our industry, and thus have in place a system to collect information on incidents occurring outside of their company and a system to share that information throughout the company. Finally, one should also review reports of studies and reviews, including those completed by regulators, activists, special interest groups, and competitors. All

Process Safety Risk Management Model

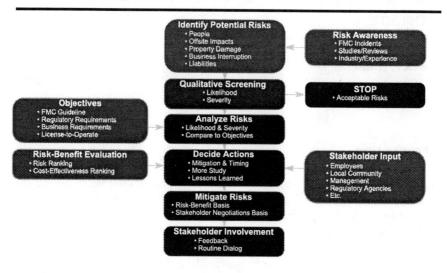

FIGURE 1

of this information is used to help maintain what is hopefully an evergreen view of the risks facing the company's operations.

The next step in the risk management process is a qualitative screening of the potential risks. Risk is the product of consequence and frequency—it answers the questions of "How big?" and "How often?" This screening identifies the best estimate of the likelihood of the event occurring anywhere in the company's operations and the severity of the event, allowing one to decide whether further analysis is required.

If further analysis of the risk is deemed necessary, then one should conduct a more detailed analysis of the severity of the losses and likelihood of the event occurring. This analysis might be done via a detailed quantified risk assessment (QRA) study or some other risk assessment methodology. If a company has significant experience with QRA studies, they can often extend existing QRA studies from similar process units to the current process unit under study. Other methodologies often include HazOp studies, "what-if reviews," and project safety reviews.

The objectives of the risk analysis will drive the scope of the risk assessment and often the methodology used. Objectives for these studies include compliance with company guidelines, regulatory requirements, business requirements, and license-to-operate issues. A company's risk acceptability guidelines usually set maximum individual risks for our workers and contractors, and for the offsite public. Regulatory requirements, like the OSHA PSM and the EPA RMP regulations, require an identification of the consequences due to accidental releases; although neither of these regulations require it, one could perform a risk evaluation to determine where and when to spend our money. Business requirements include a company's concerns for operating safely and without damaging the environment. Finally, license-to-operate issues are usually driven by the specific concerns of the local community or workforce.

After the risks have been assessed, a decision needs to be taken on the actions needed to address the issues. A decision must be made on which risk mitigation measure to implement as well as the implementation timing. Sometimes additional risk assessment study or other input is needed in order to decide the actions. One should capture and apply lessons learned coming from the decision making process, for example, incorporate in a company's engineering and operating practices, and in a lessons learned database.

There are two inputs for deciding the actions—the risk–benefit evaluation *and* stakeholder input. The risk–benefit evaluation considers the risks (severity and likelihood) from the initiating event, ranks those risks with those from other initiating events, considers the risk reduction associated with the mitigation options, and the cost-effectiveness ranking of the mitigation measures. This allows the decision to consider the relative magnitude of the risk and the amount of risk reduction for the monetary spend.

The other important input for deciding actions is from the stakeholders. The stakeholders include the employees, contractors, local community, company management, regulatory agencies, special interest groups, and anyone else who has an interest in the outcome of the decision. Stakeholder input is really a continuous process and is one of the most important elements of the entire risk management process. The final decision must also consider normal industry practices and our concerns about liability.

Mitigation of the risks will usually be done based on a combination of outcomes from the risk–benefit evaluations and stakeholder negotiations. One should use the risk–benefit evaluation as the starting point for stakeholder negotiations. The risk–benefit evaluation presents the company's view of what is safe and appropriate—recognizing that not everything can be done at once because of limited resources, and that everything that could be done is not cost-effective or appropriate in every situation—so something must be done to prioritize spends. However, at the end of the day, the outcome of stakeholder negotiations usually determines which risk mitigations get implemented for those few issues holding strong interest to the stakeholders. For those issues, it is of course important to maintain a continuing dialog with the stakeholders to see that their interests are addressed.

Development and Startup Challenges

Setting the stage for introducing risk-based techniques within a company sometimes requires a significant amount of work. Some of the basic points that must be covered during this formative stage include the following:

- *Using common terminology to discuss process safety risk.* Often when people talk about risk they are really referring to consequences. As a starting point, it is important that everyone understand the differences between likelihood, consequences and risk.
- *Understanding the resources available for risk assessment studies.* Although risk assessment is not a new technique, companies new to risk assessment often do not understand the resources available and widely used by other companies.
- *Demonstrating how risk assessment can help make decisions.* Case studies conducted for a company are often used to demonstrate the usefulness of the risk assessment technique to making decisions.
- *Move the discussion from a philosophical basis to dealing with real world issues.* Discussions of philosophy often seem to go nowhere quickly, but having a real world issue to work tends to focus the mind to allow an evaluation of the risk assessment technique.
- *Understanding the role of quantified risk assessment (QRA).* QRA techniques allow decisions to be made against quantified measures, as opposed to the more subjective qualitative measures used in non-QRA risk assessment studies. QRA studies are both expensive and time consuming, and they should not be even attempted for evaluating every risk. However, QRA studies do provide important information to help manage a company's major risks. They also allow cost–benefit decision making on risk reduction alternatives. Risk acceptability criteria must be set by the company before a QRA study is undertaken, otherwise the results of that study will not be easy to interpret.
- *Speak in financial terms whenever possible.* Since the main concern of business is money, risks should be expressed in monetary terms and decisions made using standard financial arguments. Losses (people, equipment, business interruption, liability, etc.) can be expressed as a negative annual cash flow, and risk mitigation measures can be handled as capital investments depreciated over time and subject to current hurdle rates used by the company; operating expenses for risk mitigation measures can be expressed as annual expense costs.
- *Legal liabilities may not as high as some will think.* The discussion of legal liabilities often is very contentious, many time derailing QRA efforts. Few will argue that there is some liability inherent in any type of study or audit that identifies where actions might be needed. But few will also argue against the value of taking pro-active steps to manage risks, so the judg-

Startup Challenges for New Risk Management Programs 563

ment usually involves the balancing of liability for doing the study and losses and potential negligence liability associated with not doing the study. In my experience, I have never seen or heard of a case where a QRA study was used in litigation against a company.

In successfully addressing the aforementioned challenges, a company will develop a firm foundation for a comprehensive risk management program. From that point forward, my experience has been that the risk management program will grow and mature driven by line management's desire to use more objective risk evaluation information in its decision making processes.

[Handwritten notes:]
USCS: $10m for 1 fatality
$1m for 1 injury public
$30K per 1 injury NWA

WORKSHOP D
Methodology for Comparing Risk Assessment

Chair **Patrick J. McNulty**
The Wharton School of the University of Pennsylvania

An Internet Thesaurus/Dictionary for Analyzing Risk Assessment Processes, Laws, and Regulations

A.J. Ignatowski
HazCom Consulting, 765 Wooded Road, Jenkintown, PA 19046

I. Rosenthal
Risk/Decision Processes Center, Wharton School of the University of Pennsylvania, 1325 Steinberg-Dietrich Halls, Philadelphia, PA 19104-6366.

L. D. Helsing
Environmental Protection Agency, Chemical Emergency Preparedness and Prevention Office, 401 M Street, SW, Washington, DC 20460

ABSTRACT

A computer-based thesaurus/dictionary was developed in response to a need expressed at an OECD workshop on risk assessment in the context of accident prevention, preparedness and response. The workshop concluded that there is a lack of consistency in the use of terms in the field of risk assessment. In response to this need, the System described in this presentation consists of an analytical engine that parses the risk assessment process into its essential components, reducing these into elementary units called operational descriptors. The thesaurus is an online facility accessible on the Internet. This paper describes how the analytical structure of the System breaks down into their elementary components the regulations, laws, guidelines, and definitions that relate to accident risk assessment. The reports that can be generated from the System allow for a comparison of laws, regulations and the like using a common denominator of uniform descriptor terms.

Introduction

The familiar technical term "risk assessment" means different things to different people. Thus, when a risk assessment process is applied to accidental releases at fixed installations by different people, a wide disparity among risk assessment conclusions can be expected because of the large number of technical considerations that may or may not be included in their assessments.

When a multinational body such as the Organization for Economic Cooperation and Development (OECD) attempts to promote its charter in the siting of chemical facilities in member nations, it is evident that an even wider range of divergence in "risk assessment" terminology and practice is possible. Laws and regulations that deal with such situations in various countries are naturally influenced by local cultural factors, regulatory attitudes, regional experiences, and language usage, which together can lead to different conclusions even though, on the face of it, the assessment process is said to be the same. Terms of art and even the fundamental understanding of what properly constitutes the risk assessment process varies from country to country, and even from regulatory agency to regulatory agency within a country. In the face of these difficulties, it is not surprising that the OECD called for the development of a method that will transcend these idiosyncratic difficulties, and would promote better understanding, communication, and comparison of the various laws, regulations, codes of practice, and definitions around risk assessment processes that exist among various jurisdictions.

The Risk Assessment Thesaurus/Dictionary for analyzing risk assessment processes is a proposed solution to the OECD call for an unbiased and uniform means of promoting understanding of the intended meaning of regulatory agencies, national bodies, or individuals seeking to communicate "risk assessment" processes. This paper describes how the Thesaurus/Dictionary performs the analysis of risk assessment in regard to accidental releases of chemicals at fixed installations. Briefly stated, the Thesaurus/Dictionary consists of an analytical engine that parses a particular risk assessment process into its fundamental constituents called descriptors for which near universal meaning is understood. That is, state-of-the-art terms are often laden with hidden or un-stated assumptions that may vary from user to user. By reducing these terms to neutral or operational language with unambiguous meaning, the usage of such terms can be compared among different systems. The absence of certain factors is also of great interest. This paper will describe the analytical scheme that is used to analyze the risk assessment process into its fundamental constituents.

Analysis

One often sees risk as being defined as a function of "hazard" and "exposure," or possibly as a function of "incident frequency" and "consequence." These simple expressions mask the fact that innumerable factors may or may not be embodied in any particular implementation of risk assessment. It is the intent of the Thesaurus/Dictionary analytical engine to tease out those factors that different practitioners include in a given risk assessment, and by implication, reveal those factors that are absent.

The Thesaurus/Dictionary is divided into four major sections and two supplemental sections. The major sections are referred to as *"generic elements."* The four generic elements correspond roughly to the four major considerations that the literature generally includes in the scheme for doing a "risk assessment" of a release of a chemical from a fixed facility. These are:

1. Consideration of the nature of the material being released; who or what might be affected by the release; and what is considered to be a detrimental impact to the subjects affected by the release;
2. Consideration of the ways that the harmful release could occur and thereby lead to "exposure" with subjects of concern;
3. Consideration of the assumptions related to the type and magnitude of interaction of the harmful release with the subjects affected by the exposure; and
4. Details of how the "risk" is estimated and characterized.

Note: Of necessity, up to this point, we have been using the terms "hazard," "exposure," "risk," and "risk assessment" in the most generalized senses. The point we have been trying to make is that these terms often mean different things to different people. This is the reason these terms have been enclosed within quotation marks. To undercover the intended meaning of these terms in a particular context is the purpose of the Thesaurus/Dictionary.

Generic Element I

Generic Element I is perhaps the easiest element to understand as it generally encompasses those concepts that are usually encompassed by the term "hazard identification." Generic Element I specifies the nature of the hazardous release, the specification of the harmful outcome that might ensue from some type of contact with the hazardous release, and the identification of the potential targets of the release. The terminology used in the Thesaurus/Dictionary is designed to be as neutral as possible. The text is:

Element I: *Identification of the sources with the potential to cause undesired outcomes to the subjects of concern.*

Generic Element II

A consideration of the concept of "exposure" illustrates that this term consists of several individual components. This term encompasses concepts commonly related to consequence or exposure analyses. Exposure includes an understanding of the distribution of the hazardous release itself and some understanding of how the exposed entities are distributed. In practical terms, the assumption of hazard distribution implies some model of a possible release scenario, and some assumed boundary value that delimits the extent of the distribution analysis. Generic Element II specifies how the release may distribute itself in relationship to the distribution of people or other systems that may be effected by the release. It includes a consideration of feasible scenarios and is delimited by boundaries of consideration.

Element II: *Identification of possible sequences of events leading to the loss of containment of the potential to cause undesired outcome to a subject of concern resulting in its entry into a domain of the ecosystem. Estimation of possible distributions of both the released potential and the subjects of concern over time periods within a domain delimited by specified boundaries or end-points.*

Generic Element III

Risk is often spoken of as being a function of hazard and exposure. The exact functional relationship between hazard, exposure and the resulting outcome is rarely stated. In any particular implementation of risk assessment, there must be some assumption or understanding of the relationship between the undesired outcome and the mode of interaction with the subjects of concern because of the exposure. This relationship is defined explicitly in Generic Element III which focuses on the manner in which the exposed subject of concern is related to the undesired outcome:

Element III: *Identification and description of how the specified undesired outcome is related to the intensity, time and mode of contact of a specified potential to cause the undesired outcome to the subject of concern.*

Generic Element IV

Finally, Generic Element IV speaks to the nature of the left hand side of the equation relating hazard and exposure: the expression of risk. Estimates or conclusions regarding risk can be expressed any number of ways, including expressions of the chance of occurrence of an event to an individual or to certain groupings of a population. Because of the uncertainties and the many assumptions made in performing the risk assessment, the uncertainty of the estimate is often as important as the final estimate of risk itself. Underlying any risk assessment is some notion of a comparison to a standard of acceptability. Indeed, inclusion of the notion of comparison is used by some to differentiate risk assessment for risk analysis (1). In many situations it is also important to know what might be the effect of alternative assumptions. Generic Element IV deals with the manner in which the risk is expressed, including consideration of the uncertainty of the risk assessment and reference to standards or guidelines. Specifically:

Element IV: *Identification of the basis for estimating and expressing the likelihood that a specified undesired outcome of a specified magnitude for a specified subject of concern will occur and description of the quality/uncertainty of such estimates; comparison of the estimates with relevant standards and guidelines; and evaluation of the impact of specified alternative assumptions on the estimates.*

Supplemental Elements

There are also two supplemental elements that are provided in the Thesaurus/Dictionary that are intended to capture any other aspect of the risk assessment process that may not be adequately included in the previous four elements. Element 0 allows for the capture of those aspects of the risk assessment process that are not captured by Generic Elements I through IV, and are judged to precede them. This element may be used to include statements of the scope or purpose of the risk assessment process that others may feel are not properly an integral part of the risk assessment process *per se*.

Element V allows for the identification of aspects of the risk assessment process that have not been captured in previous elements, and are judged to follow them. Items of this type may be descriptions of the risk assessment/risk management interface or the risk assessment/risk characterization interface.

Further Analysis

The four generic elements are at the highest level of analysis and therefore are not particularly revealing of the details of risk assessment process. Hence, each generic element is further divided into varying numbers of sub-elements. For example, the three concepts contained within Generic Element I are separately identified and analyzed in three sub-elements:

- Sub-element i is: Identification of sources with the potential to cause undesired outcomes;
- Sub-element ii is: Identification of subjects of concern; and
- Sub-element iii is: Identification of undesired outcomes.

Finally, in order to reduce each risk assessment concept to its lowest, most concrete and operational basis, each sub-element is further characterized by very specific *"categories"* and *"descriptors."* *The categories and descriptors listed in the Thesaurus/Dictionary are not intended to be definitive but are merely proposed as possible examples of the items that may apply in a particular situation.* They are supplied to suggest possibilities of the type of item that may apply, but the user is free to reject any or all possibilities and to enter other specific descriptors that more closely convey the users' intended meaning in a particular situation.

For example, the "identification of *sources* with the potential to cause undesired outcomes" of Sub-element i of Generic Element I may be chosen from the list of following categories and descriptors that are very specific, operational, and almost universally understood equivalently:

Substances
Explosive
Flammable
Reactive
Toxic to humans
Toxic to ecosystems
Teratogenic substances
Other substances
Undefined substances

Energy
Pressure
Thermal flux
Radiation
Dynamic energy
Other form of energy
Undefined form of energy

Physical Situations
Systems containing flammable substances
Systems containing reactive chemicals
Systems containing regulated chemicals
Areas prone to natural disasters
Other physical situations
Undefined physical situations

Legally defined
Listed substances
Other legally defined sources
Undefined legal sources

Other Sources
Other categories of sources
Other undefined categories of sources

By selecting as many of these specific descriptors as is appropriate, one is quite concretely conveying to others exactly what is meant by "sources with the potential to cause undesired outcomes" in a particular case.

Operation of the Thesaurus/Dictionary

The Thesaurus/Dictionary is accessible on the Internet. In this interactive environment, the individual who is entering a description of a particular risk assessment process (the "client") is presented with a series of computer screens that leads the client through the analysis. In its most basic operation, the client selects all appropriate descriptors that apply to the client's understanding of the risk assessment process under consideration. This is done one sub-element at a time as illustrated above. Following this selection, the client is presented with a computer screen that allows the client to characterize the choice as being either *"explicitly"* or *"implicitly"* included in the meaning of the law or regulation being entered. In addition, the client is presented with the possibility of stating whether the law or regulation also provides *"criteria"* that are to be used in deciding whether the descriptor is appropriate in a given case. The client is also provided with the opportunity to indicate that *"tools"* may be used to determine whether the particular source is to be included in the descriptor class. Examples of criteria may be a flash point value that would determine whether a chemical is to be regarded as "flammable," or a LD50 range that determines whether a chemical is to be considered toxic to humans. Examples of tools may be the specification of an OECD test method for acute lethal dose determination, a formula for classifying chemical mixtures, or the mention of an industry consensus standard for flash point determination.

For every descriptor chosen, the client is finally presented with a screen that allows for providing a literature citation that references the basis for the descriptor selection. In most cases this will be a certain paragraph or sub-paragraph in a particular section of the specific regulation or law under consideration. If desired, explanatory comments or the actual language of the law/regulation may be entered in a text box provided to further document the selection.

The intent of the list of descriptors is to provide examples of the type of description that may characterize a particular element/sub-element/category. In every case, the client is offered the possibility of selecting "other" when none of the listed descriptors apply. The *"other"* choice consists of a special text field in which the client is permitted to write in any descriptor that the client believes more properly characterizes the situation on that point. In other cases, the law or regulation may not be specific at all, but the client understands that the category is appropriate for the current situation. In this case, the client may select *"undefined."*

Following completion of the first sub-element, the client is offered the opportunity to make selections in the following sub-elements in turn. Finally, completing the first Generic Element, the client is presented with the ability to make selections from subsequent Generic Elements in like fashion as appropriate. There is no obligation to make selections from any Generic Element, Sub-element, or category.

All Generic Elements follow a similar scheme. Some Elements are more complicated, and some sub-elements are further divided into multiple *"terms"* that are ultimately characterized by the categories and descriptors. Because of the complexity of the overall Thesaurus/Dictionary, a detailed discussion of each Generic Element is not feasible here. For reference purposes, all elements, their sub-elements and categories are provided in the Appendix.

All entries, that is, the particular selection of descriptors, sub-elements, elements, and all reference information associated with the risk assessment process being described, are associated with the individual who has performed the mental analysis in making the entry. The total of all this information results in a valuable database regarding risk assessment that can be the subject of future analysis and reporting.

Other Types of Input

Four types of input are possible. The foregoing has described the use of the Thesaurus/Dictionary to input risk assessment processes based on law or regulation. It is also possible to enter the risk assessment practices used in *"specific risk assess-*

ment studies." These may be academic or industry developed methodologies or applications to assess the risk at a particular chemical site or a specific process. The scheme of inputting this information is identical to that described for the law/regulatory scenario. In this case, the inputter selects only those elements that were utilized in the particular risk assessment process. A third type of input is the description of more generic risk assessment *"guidelines, policies, or codes"* and the like that are not specific to any law or regulation, and do not apply to any particular chemical or case. The procedure for this type of input follows the same scheme as is used for inputting laws and regulations.

It is also possible to input specific *"definitions"* that apply in the context of the risk assessment process. In this case, the client is provided with a special screen that identifies the term being defined, a free-text field in which the definition is stated, and then the same sequence of screens as applies to the law/regulatory input situation from which the client may select all appropriate descriptors that characterize the definition. Again, as with the entry of risk assessment processes, the client is asked to characterize the descriptor as being "explicitly" or "implicitly" included in the intended meaning of the definition. However, for definitions, screens are not presented for criteria, tools, or reference citations.

Output

Up to this point, only the inputting of information into the Thesaurus/Dictionary has been described. Although some individuals who have used the Thesaurus/Dictionary to input regulatory information have commented on the personal insights gained in very act of analyzing the requirements of the law for purposes of data entry, the ultimate value of the Thesaurus/Dictionary is to be measured from the information and insights that can be gained from an examination of output reports. The Thesaurus/Dictionary is equipped with several pre-designed output reports that can be used to compare different risk assessment processes. Comparison can be conducted at various levels of detail, that is, a reporting and comparing of elements, sub-elements, terms, categories of descriptors, and descriptors. Two types of comparison can be made: comparison between or among different regulations, say, between a state regulation and the federal regulation nominally addressing the same concern, or a comparison of the profile of entries of the same regulation made by two or more individuals. The first kind of report can compare the risk assessment regulations in two jurisdictions, or it can compare the law and its implementing regulations within a jurisdiction. Differences in the selection of the characterizing descriptors reveal regulatory differences at the most concrete level of application.

The second type of comparison report can be even more interesting. When the entry has been performed by two or more clients who are nominally equally versed in the law/regulation, differences reveal areas of ambiguity in the law/regulation that is subject to varying interpretation. It is common experience that there are areas of the law that become the focal point of disagreement, often between the regulators and the regulated community. When the dispute arises from ambiguity in the law, eventually the issue is settled by the development of case law in the matter. A tool such as the Thesaurus/Dictionary might be used to "pre-test" the law (regulation) for areas of obvious ambiguity before final promulgation. Other papers in this series will demonstrate the utility of the Thesaurus/Dictionary in various contexts.

Appendix

Generic Element I

Identification of the sources with the potential to cause undesired outcomes to the subjects of concern.

Sources with the potential to cause undesired outcomes

- Substances
- Energy
- Physical Situations
- Legally defined sources
- Other Sources

Subjects of concern

- People
- Ecosystems/environment
- Cultural assets
- Property and physical systems
- Facilities
- Other subjects of concern

Undesired outcomes

- Undesired outcomes for people
- Undesired outcomes for ecosystems/environment
- Undesired outcomes for cultural assets
- Undesired outcomes for society
- Undesired outcomes for facilities having the release
- Undesired outcomes for property
- Undesired outcomes for other classes of subjects

Generic Element II

Identification of possible sequences of events leading to the loss of containment of the potential to cause undesired outcome to a subject of concern resulting in its entry into a domain of the ecosystem. Estimation of possible distributions of both the released potential and the subjects of concern over time periods within a domain delimited by specified boundaries or end-points.

Basis for generating possible sequence(s) of events

- Sequence of events based on past events and experience
- Sequence of events based on technical analysis
- Sequence of events legally determined
- Other sequence of events categories

Basis for estimating distribution of the released potential

- Distributions based on real-time monitoring
- Distributions based on technical analysis
- Distributions legally determined
- Other approaches to estimation of distributions

Basis for estimating distribution of subjects of concern

- Real-time monitoring
- Historical data
- Distributions legally determined
- Other approaches to estimation of distribution of subjects

Basis for boundaries and endpoints of distribution estimates

- Boundaries or end-points based on toxicological analysis
- Boundaries or end-points legally determined
- Other approaches to setting boundaries or end-points

Generic Element III

Identification and description of how the specified undesired outcome is related to the intensity, time and mode of contact of a specified potential to cause the undesired outcome to the subject of concern.

Mode of contact between the potential to cause the undesired outcome and the subjects of concern

- Mode of contact between the potential to cause undesired outcomes and people

- Mode of contact between the potential to cause undesired outcomes and ecosystems/environment
- Mode of contact between the potential to cause undesired outcomes and cultural assets
- Mode of contact between the potential to cause undesired outcomes and property and public infrastructure systems
- Mode of contact between the potential to cause undesired outcomes and other classes of subjects of concern

Basis for the relationship used to predict how the specified undesired outcome is related to contacts with the potential to cause the undesired outcome

- Relationship to humans
- Relationship to ecosystems/environment
- Relationship to cultural assets
- Relationship to property and public infrastructure systems
- Relationships to other subjects of concern

Dimensions/measurement units of the potential used in predicting undesired outcomes

- Concentration of the substance with the potential to cause the specified undesired outcome that interacts with the subject over a specified time
- Amount of the substance with the potential to cause the specified undesired outcome that interacts with the subject over a specified time
- Over-pressure delivered to a specified area of the subject over a specified time
- Energy flux delivered to the subject over a specified time and area
- Other measures/dimensions of input potential

Dimensions/measurement units used to express the predicted undesired outcome

- Magnitude of the undesired outcome response experienced by the specified individual subject of concern
- Number of the undesired outcome events experienced by the specified population of subjects
- Number of the undesired outcome events experienced by the specified population of subjects expressed as a frequency
- Likelihood of the undesired outcome given the specified interactions with the potential to cause the undesired outcomes by specified subjects
- Other categories of outcomes

Generic Element IV

Identification of the basis for estimating and expressing the likelihood that a specified undesired outcome of a specified magnitude for a specified subject of concern will occur and description of the quality/uncertainty of such estimates; comparison of the estimates with relevant standards and guidelines; and evaluation of the impact of specified alternative assumptions on the estimates.

Basis for estimating that the likelihood that specified undesired effects will occur

- Quantitative assessment
- Semi-quantitative assessment
- Qualitative assessment
- Assessments based on historical data
- Judgments based on degree of conformance with criteria
- Other estimation processes for likelihood

Method of expressing the likelihood that an undesired effect(s) will occur

- Quantitative expressions
- Semi-quantitative expressions
- Qualitative expressions
- Expressions of likelihood defined in law
- Other expressions of likelihood

Description of the quality/uncertainty of estimates of likelihood

- Characterization of estimate

Undesired outcome of a specified magnitude for a specified subject of concern

- Specified undesired outcome for one specified member of the concerned population
- Specified undesired outcomes simultaneously experienced by a group of N or more of the subjects of the concerned population
- The presence at specified location(s) of the specified potential at concentrations/intensities over time that are sufficient to cause a specified undesired outcome for specified subjects of concern if they were present at the location
- The presence of a specified potential at a specified location at concentrations/intensities over time that exceed a specified limit
- Other resulting undesired effects

Nature of standard or guideline to which estimates are compared

- Type of standard or guideline

Type and form of information needed for comparisons of estimated likelihood

- Comparison Metrics

Identification of specified alternative assumptions whose impacts on likelihood are to be evaluated

- Alternative assumptions

Evaluation of the impact of alternative assumptions on estimates of likelihood

- Evaluation metrics

References

1. D.A. Jones, ed., *Nomenclature for Hazard and Risk Assessment in the Process Industries*, 2nd ed. Institution of Chemical Engineers, Rugby, UK, 1992, p. 27.

http://grace.wharton.upenn.edu/oecd

Europeans are entering data via Internet

Evaluation of a Proposed Thesaurus/Dictionary for Risk Assessment Using an Industrial Quantitative Risk Analysis

Dennis C. Hendershot
Rohm and Haas Company, Bristol, PA 19007

Stanley J. Schechter
Wharton Risk and Decision Processes Center, University of Pennsylvania, Philadelphia, PA 19104

ABSTRACT
The OECD has developed a thesaurus/dictionary for describing risk assessment processes, laws, and regulations. The thesaurus/dictionary provides a generic structure which can be applied to a specific quantitative risk analysis to understand how it relates to other analyses and also to risk management laws or regulations. The proposed OECD risk structure will illustrated and evaluated using an example industrial risk analysis of a chlorine unloading facility.

Introduction

The Organisation for Economic Co-operation and Development (OECD) has developed a prototype Risk Assessment Thesaurus/Dictionary with the help of a group organized by the Wharton Risk and Decision Processes Center of the University of Pennsylvania and supported by the United States Environmental Protection Agency (EPA) (Rosenthal, et al., 1996; Ross and Ignatowski, 1996, 1997). The Thesaurus/Dictionary is intended for use in describing and comparing various risk management laws, regulations, policies, and risk analysis methodologies, with particular attention to catastrophic industrial risks. The OECD Thesaurus/Dictionary was used to characterize an industrial quantitative risk analysis (QRA) as a test of its utility. In this discussion, we will briefly describe the QRA and summarize its description using the OECD Thesaurus/Dictionary. The QRA was done a number of years ago, and the study is characterized in the context of the legal requirements, regulations, and corporate policies which were in place at the time the QRA was actually done.

581

The Quantitative Risk Analysis

A QRA of a chlorine unloading facility was used to evaluate the OECD Risk Assessment Thesaurus/Dictionary. The QRA was done in the mid 1980s by Rohm and Haas Company. A summary of the risk analysis and its use as a risk management and decision making tool was published by Hendershot (1991). This QRA was done as part of an expansion project for a plant, which would result in a doubling of chlorine usage, hence increasing the potential risk. The QRA was done to assist the design engineers in the selection of an appropriate design for the new and expanded chlorine unloading facility. The plant is located in a mixed industrial, commercial, and residential area. A general map of the plant and its surrounding area, including the location of the chlorine facility on the site, is shown in Figure 1. Liquid chlorine is unloaded from tank trucks using nitrogen pressure, and transferred to a manufacturing building several hundred feet away from the truck unloading facility. The original chlorine facility, which had been in use for a number of years, is shown in Figure 2, and a proposed improved and expanded design is shown in Figure 3.

The QRA was not required by any local or national regulations at the time, and a QRA would not be required for this facility today. Rohm and Haas Company chose to use QRA as a tool for selecting the best design for the new facility, and also for understanding and managing the risk associated with the design and operation of the expanded chlorine unloading facility. The QRA estimated both individual and societal risk using several risk measures described by CCPS

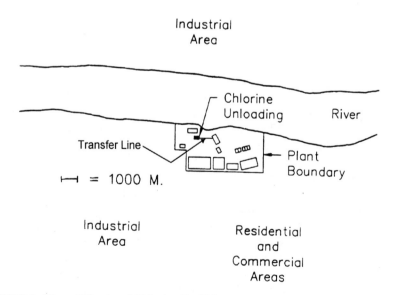

FIGURE 1. Map of Plant and Chlorine Facilities

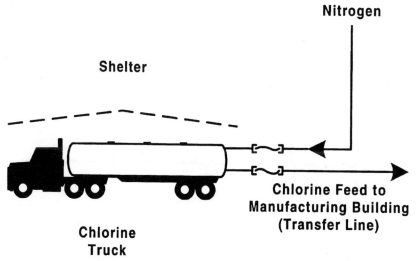

FIGURE 2. Original Chlorine Handling Facility

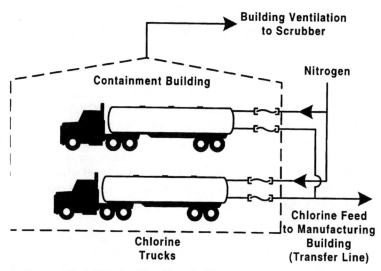

FIGURE 3. Modified Chlorine Handling Facility

(1989). The QRA scope included the tank truck unloading facility and the transfer lines to the manufacturing building. The movement of the chlorine trucks to the facility and the manufacturing process were considered in separate risk management studies. Individual risk estimates were expressed as individual risk contours, illustrated in the sample results shown in Figure 4. Societal risk estimates

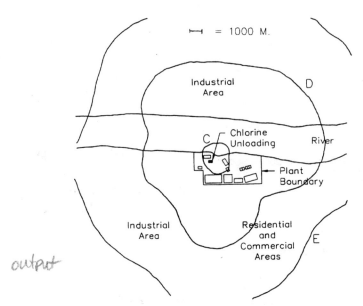

FIGURE 4. Example Individual Risk Contours for the QRA (C > D > E)

were in the form of F-N Curves. An example of the societal risk estimates for several design options is shown in Figure 5. The results of the QRA showed that the proposed design reduced risk associated with the tank truck unloading system, because of the addition of the enclosure and scrubber. However, modifications to the transfer pipelines to support the increased chlorine usage resulted in an increased estimated risk from potential incidents associated with transfer pipe leaks. This understanding of the most important contributors to risk focused the efforts of the design team on redesign of the pipeline, and a number of design options were identified and evaluated.

This was one of the early QRAs done by Rohm and Haas, and the company did not have formal guidelines on how to use the results of a QRA at that time. Since then, Rohm and Haas has developed and published guidelines on the use of QRA as a tool for risk management, as a part of its Major Accident Prevention Program (Renshaw, 1990; Hendershot, 1996). In fact, this QRA was one of the benchmark studies used to aid in establishing the Company's formal risk management programs. The Rohm and Haas Company internal risk guidelines would, and do, apply to the chlorine unloading facility today, but they had not been adopted at the time the QRA was actually done.

In the mid 1980s, when the QRA was done, it was used in a comparative manner for risk decision making. Different design alternatives were compared to

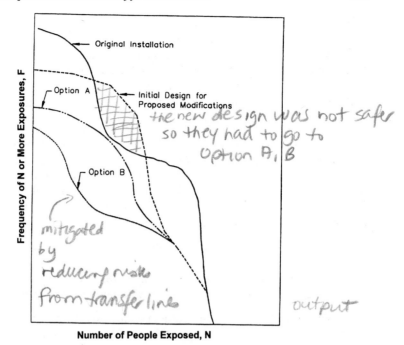

FIGURE 5. Example of F-N Curves from the QRA

each other, and to the original installation. The key risk management guideline was that the new, expanded facility would have an estimated risk no higher than the original installation. Both societal and individual risk measures were to be considered. Design options which met this requirement were then evaluated, and a final design selected, based on consideration of a number of safety, environmental, and economic factors.

Application of the Thesaurus/Dictionary to the Example QRA

Although the OECD Thesaurus/Dictionary was intended as a tool to aid in understanding risk management policies, risk regulations, and risk assessment methodologies, it can be applied to a specific QRA study. This can help in understanding the methodology of the study, and might be useful in understanding how well a particular study meets the requirements of a risk management policy or regulation.

The Thesaurus/Dictionary postulates that all risk assessments include four basic elements:

Element I: Sources with the potential to cause undesired outcomes to subjects of concern
Element II: Basis for generating possible sequences of events
Element III: Mode of contact between the potential to cause the undesired outcome and the subjects of concern
Element IV: Basis for estimating the likelihood that specified undesired effects will occur

The details of the structure of the Thesaurus/Dictionary are reported in another paper in this workshop (Ignatowski et al., 1997).

Tables 1 through 4 summarize the characterization of the example QRA, including the tools used to fulfill the various elements and subelements used to describe risk analysis processes in the Thesaurus/Dictionary. The descriptions can be summarized as follows:

- **Element I:** Sources with the potential to cause undesired outcomes to subjects of concern:
  The QRA considered the acute toxicity of chlorine to people, including on-site population, off-site population, and several sensitive population locations

- **Element II:** Basis for generating possible sequences of events
  Potential accident scenarios were identified based on past experience in the company and industry, and using analytical tools such as Hazard and Operability studies (HAZOP), What If, fault trees, and event trees. Vapor cloud dispersion models were used to estimate the chlorine concentration resulting from identified accident scenarios. Population distribution was based on census data and direct contact with local businesses.

- **Element III:** Mode of contact between the potential to cause the undesired outcome and the subjects of concern
  The primary mode of contact considered was inhalation of chlorine vapor. The toxic effects of chlorine were estimated using a probit model developed by Withers and Lees (1985) to estimate the dose-response relationship. The undesired outcome was measured in terms of the likelihood of fatality at a particular location, and as the estimated number of fatalities.

- **Element IV:** Basis for estimating the likelihood that specified undesired effects will occur
  Quantitative fault tree analysis, quantitative event tree analysis, and historical data were used to estimate the likelihood of occurrence of the undesired outcome. For this QRA, best estimate data were used, and the uncertainty of the estimates was not evaluated. The analysis was done for the total population, as well as for several sub-groups of interest (for example, on-site employees, nearby residents). Several different individual and societal risk measures were used (CCPS, 1989), and a number of design options were considered to help the process designers select the best overall system.

TABLE I
Subelements and Descriptors for Element I for the Example QRA

Element I: Sources with the potential to cause undesired outcomes to subjects of concern			
Subelement		Descriptors	
i	Sources	Substances	Toxic to humans	The human toxicity of chlorine is well known (Withers and Lees, 1985).
			Legally defined	Although the chlorine inventory in the facility is below the threshold quantity for coverage by the relevant national and local regulations, the fact that chlorine was a listed substance in those regulations identified this facility to Rohm and Haas management as one requiring special attention.
		Energy	Pressure	A Rohm and Haas internal prioritization, since published in a modified form (Renshaw, 1990), highlighted the contribution of elevated pressure to the system hazard.
ii	Subjects of concern	People	Nearby residents, employees of neighboring factories and businesses, and on-site workers were considered.	
iii	Undesired outcomes considered	People	The QRA was done using immediate fatality as the measured outcome. Other potential outcomes (for example, injury, delayed fatality, legal sanctions, economic loss, anxiety) were not considered explicitly in the QRA, although management was certainly aware of them in the decision making process.	
		Facilities	Sensitive population locations (schools, hospitals, a pedestrian tunnel) were considered in the QRA	

TABLE 2
Subelements and Descriptors for Element II for the Example QRA

Element II: Basis for generating possible sequences of events			
Subelement		Descriptor	
i	Basis for generating accident scenarios	Past experience	Insurance and industry records, company records, professional judgment
		Technical analysis	Fault tree analysis, event tree analysis, HAZOP, What If? analyses as defined in the original edition of the CCPS *Guidelines for Hazard Evaluation Procedures* (CCPS, 1985)
ii	Basis for estimating distribution of potential for undesired outcome over space and time	Technical analysis	Event tree analysis PHAST® consequence analysis program and SAFETI® risk analysis program (from Technica, Inc., now DNV-Technica)
iii	Basis for estimating distribution of subjects of concern	Historical data	National census data, plant population data, contact with neighboring plants and businesses to obtain population data
iv	Basis for boundaries or endpoints	Other toxicologically derived endpoints	Probit relationship for inhalation toxicity of chlorine (Withers and Lees, 1985)

TABLE 3
Subelements and Descriptors for Element III for the Example QRA

Element III: Mode of contact between the potential to cause the undesired outcome and the subjects of concern			
Subelement		Descriptor	
i	Type of contact	People	Inhalation
ii	Basis for relationship used to predict undesired outcome	People	Human epidemiological data, animal data, weighted combination of data (Withers and Lees, 1985)
iii	Measurement units of the potential specified	Concentration of substance over specified time	Dose-response relationship over 0-60 minute time period based on Withers and Lees (1985) probit
iv	Measurement units of undesired outcome	Total number of outcome events in the specified population	Estimated number of deaths; likelihood of fatality at a particular location

TABLE 4
Subelements and Descriptors for Element IV for the Example QRA

Element IV: Basis for estimating the likelihood that specified undesired effects will occur			
	Subelement	Descriptor	
i	Basis for estimating the likelihood that the undesired events will occur	Quantitative assessment	Event tree analysis, fault tree analysis, SAFETI® risk analysis program, consultant and in-house vapor cloud dispersion models
		Historical data	Equipment failure rate data and human error rates from various chemical industry databases for use in fault tree and event tree analyses
	Expressions of likelihood of undesired event	Quantitative	Frequency of releases; probability of various sequences of events resulting in various specific outcomes; probability of specific undesired outcomes at a particular location in a specified time period
	Quality/uncertainty	Characterization of estimate	Best estimate
	Undesired effects	Specified for a single member of the population	Average member Member closest to potential Members at specified locations (individual risk contour map)
		Specified for a group of N or more people in the specified population	F-N Curve for entire population and specified sub-groups of the population (on-site workers, nearby residents)
ii	Comparison with relevant standards and guidelines	Type of standard or guideline	Other standards or guidelines - internal company guideline that the risk of the new facility would not exceed the risk of the existing plant.
		Comparison metrics	Individual risk at specific locations F-N Curves Average individual risk Rate of death Individual risk contours Consideration of all practical measures to reduce magnitude and likelihood of undesired events (case studies for various design options)
iii	Impact of alternative assumptions	Alternative assumptions	Impact of sheltering in place was considered

Conclusions

The OECD Dictionary/Thesaurus provides a logical description of the example QRA, and documents the assumptions used to do the analysis. The description of the QRA clearly illustrates a sound and logical risk management process, and clearly describes the basis for decision making.

The Thesaurus/Dictionary is an initial prototype, and work is in progress for improvement. Data entry is overly repetitive, and a way of entering a reference to a document or policy once, and cross referencing it when needed would be a significant improvement. Additional thought needs to be given to the reports which can be generated by the Thesaurus/Dictionary. The current reports are too long, and it is difficult to locate specific pieces of information. Tables 1 through 4 represent a substantial condensation of the actual Thesaurus/Dictionary reports, but they also allow a quick understanding of the basis and assumptions used for the example QRA.

The Thesaurus/Dictionary has potential usefulness in the following areas:

- For documenting the basis, tools, and assumptions used for a particular QRA
- For understanding the requirements of risk management laws, regulations, and policies
- For comparing a particular QRA to the requirements of laws, regulations, or policies
- For comparing the methodologies of QRAs done by various organizations or agencies

References

Center for Chemical Process Safety (CCPS) (1985). *Guidelines for Hazard Evaluation Procedures*. New York: American Institute of Chemical Engineers.

Center for Chemical Process Safety (CCPS) (1989). *Guidelines for Chemical Process Quantitative Risk Analysis*. New York: American Institute of Chemical Engineers.

Hendershot, D. C. (1991). "The Use of Quantitative Risk Assessment in the Continuing Risk Management of a Chlorine Handling Facility." *The Analysis, Communication, and Perception of Risk*, ed. B. J. Garrick, and W. C. Gekler, 555-565. New York: Plenum Press.

Hendershot, D. C. (1996). "Risk Guidelines As a Risk Management Tool." *Process Safety Progress* 15, 4 (Winter), 213-218.

Ignatowski, A. J., I. Rosenthal, and L. D. Helsing (1997). "An Internet Thesaurus/Dictionary for Analyzing Risk Assessment Processes, Laws, and Regulations." *International Conference and Workshop on Risk Analysis in Process Safety*. October 21-24, 1997, Atlanta, GA. Workshop D: Methodology for Comparing Risk Assessment. New York: American Institute of Chemical Engineers.

Renshaw, F. M. (1990). "A Major Accident Prevention Program." *Plant/Operations Progress* 9, 3 (July), 194-197.

Rosenthal, I., B. J. M. Ale, and L. Helsing (1996). "An Outline of the Approach Being Used in Developing the OECD Thesaurus-Dictionary of 'Risk Assessment' Terminology." *Conference on TransBoundary Risks*, October, 1996, Warsaw, Poland.

Ross, W. C., and A. J. Ignatowski (1996). "OECD Risk Assessment Thesaurus/Dictionary System Instructions and Workbook." Version 1.20. Wharton Risk and Decision Processes Center, University of Pennsylvania, Philadelphia, PA. (December 30).

Ross, W. C., and A. J. Ignatowski (1997). "OECD Risk Assessment Thesaurus/Dictionary System Instructions and Workbook Supplemental Notes." Version 1.20. Wharton Risk and Decision Processes Center, University of Pennsylvania, Philadelphia, PA. (January 30).

Withers, R. M. J., and F. P. Lees (1985). "The Assessment of Major Hazards: The Lethal Toxicity of Chlorine." Parts 1 and 2. *Journal of Hazardous Materials* 12, 3 (December), 231-282 and 283-302.

Use of the OECD Dictionary/Thesaurus to Encode Delaware's Law for Process Safety

Patrick J. McNulty
The Risk Management and Decision Processes Center, The Wharton School of the University of Pennsylvania, 1325 Steinberg Hall-Dietrich Hall, Philadelphia, PA 19104-6366.

Robert A. Barrish and Richard C. Antoff
State of Delaware, Department of Natural Resources and Environmental Control, Division of Air and Waste Management, 715 Grantham Lane, New Castle, DE 19720-4801.

ABSTRACT

The OECD and the EPA have undertaken to develop a Dictionary/Thesaurus to promote understanding and communication of the risk assessment processes used by various practitioners and government agencies in the determination of risks associated with accidental chemical releases. This paper describes the use of the OECD Dictionary/Thesaurus to encode regulations under the State of Delaware's "Extremely Hazardous Substances Risk Management Act." The Dictionary/Thesaurus proved to be a valuable tool in understanding Delaware's risk assessment process and confirmed that the regulation addressed all requirements of the law. The Dictionary/Thesaurus used in this study was a prototype that needs additional work to improve the ease of data entry, clarify ambiguities and simplify report output. The final form of the Dictionary/Thesaurus should be a useful analytical tool.

Introduction

The process of analyzing, assessing, managing and communicating risks has become exceedingly complex, probably because of the increased complexity of technology and its development within specialized groups. Risk analysis, historically the province of specialized guilds, increasingly is conducted by scientists and engineers within a variety of scientific disciplines, regulatory agencies and specialized businesses. This specialization imposes on practitioners the necessity to understand the thinking, nomenclature and tools of fellow practitioners in order to reach an understanding about the nature and magnitude of a given risk. The process becomes more difficult when a practitioner must communicate across disciplinary lines (Jones, 1992).

In addition to differences in terminology and methods, there are differences in the use of the risk assessment process to manage risk (Rosenthal, 1991). Specialists concerned for human health symbolically express risk using the equation,

$$\text{Risk} = f(\text{"Hazard"})(\text{"Exposure"}),$$

where "hazard"[1] is understood as a property of a material but response to dose is problematic. Practitioners concerned with chronic disease spend considerable energy devising ways to reduce exposure.

Specialists concerned for safety symbolically express risk using a similar equation,

$$\text{Risk} = f(\text{Incident frequency})(\text{Consequence}),$$

where consequence is understood in terms of the degree of injury to people, property, etc., but the incidence frequency, i.e., the probability of the release of the "hazard" in a given incident, is problematic. Practitioners concerned with safety are likely to devise ways to change consequences as a result of process hazard review and also to reduce probability as a result of prevention programs.

When differences in terminology, content and philosophy are compounded by differences in language, culture and legal jurisdiction, significant barriers to understanding exist. Attempts to utilize technology to generate wealth but to protect against unreasonable adverse effects results in a variety of regulatory approaches. To learn from the experiences of different countries and to facilitate cooperation in addressing common problems, the Expert Group on Chemical Accidents of the Organization for Economic Co-operation and Development (OECD) and the Environmental Protection Agency's (EPA) Chemical Emergency Preparedness and Prevention Office (CEPPO) have undertaken to develop a Dictionary/Thesaurus to facilitate an understanding of the processes used by different practitioners in determining the risks associated with major accidental chemical releases (Rosenthal et al., 1996; Ross et al., 1996, 1997; and Ignatowski et al., 1997).

This paper describes the use of the Dictionary/Thesaurus to examine the content of the State of Delaware's "Extremely Hazardous Substance Risk Management Act" and its "Regulation for the Management of Extremely Hazardous Substances."

The OECD Dictionary/Thesaurus

The OECD approach has been to develop an overall characterization of the risk assessment process used by various practitioners, without using terms of art. The OECD intent is not to standardize the risk assessment process but to enhance mutual understanding in the context of chemical accidents. The OECD hopes to accomplish this objective by mapping out the steps in the risk assessment

---

[1] Terms of art, such as hazard and exposure, are non-precise but are used as a convenience in the narrative of this paper; they are not used in the OECD Dictionary/Thesaurus.

process, the approaches and methodologies various practitioners use, and the factors that influence choice in particular approaches (Rosenthal et al., 1996). The result is a tree-like Dictionary/Thesaurus (Ignatowski et al., 1997) comprised of four risk elements.

ELEMENT I—"HAZARD" IDENTIFICATION
Element I identifies sources[2] with the potential to cause undesired outcomes ["hazards"], the subjects of concern [who or what gets injured] and the undesired outcomes [the consequences].

ELEMENT II—"HAZARD" RELEASE/"EXPOSURE" SCENARIO
Element II identifies the basis for arriving at scenarios that will predict the release of a source, the basis for predicting the distribution of the source once it is released, the basis for predicting the distribution of the subjects of concern in the space occupied by the source, and the basis for defining the extent of the distribution of the released source.

ELEMENT III—SOURCE AND SUBJECT INTERACTION
Element III identifies the relationship between the intensity, time and mode of contact of a source and the undesired outcome in subjects of concern. It identifies the dose/response relationship between the source and subject.

ELEMENT IV—BASIS FOR ESTIMATING LIKELIHOOD
Element IV identifies the basis for estimating the likelihood that a specified undesired outcome will occur to subjects of concern as a result of a specified sequence of events. It compares that likelihood to relevant standards, and evaluates the impact that alternative assumptions would have on the estimate.

Within each element is a hierarchy of sub-elements, terms, category of descriptors and descriptors, the sum of which characterize the risk assessment process. Figure 1 presents an overview of Element I. One of the **Terms** (*Source with the potential to cause undesired outcomes*) is expanded into five **Categories of Descriptors** (*substances, energy, physical situations*, etc.). One **Category of Descriptors** (*Substances*) is expanded into six **Descriptors** (*explosive, flammable, reactive*, etc.). By using successively all four elements it is possible to parse the risk analysis process used by a given practitioner or regulator.

Delaware Regulation to Prevent Accidental Chemical Releases

In 1988 the Delaware legislature passed the "Extremely Hazardous Substances Risk Management Act" and added it to the Delaware Code, Title 7, Chapter 77

---

[2] In some federal regulations the term "source" is used to indicate a regulated facility. In the OECD Dictionary/Thesaurus the term "source" refers to a physical situation which has the potential to cause an adverse outcome, such as, chemicals with hazardous properties, chemicals listed by authoritative agencies, storage tanks containing chemicals, etc.

> o Element I - "Hazard" Identification
> - Source with the potential to cause undesired outcomes
>   - *Substances*
>   Explosive
>   Flammable
>   Reactive
>   Toxic to humans
>   Toxic to ecosystems
>   Teratogenic to humans
>   - *Energy*
>   - *Physical Situations*
>   - *Legally Defined*
>   - *Other Sources*
> o Subjects of Concern
> o Undesired Outcomes
>
> o Element II - "Exposure" Scenario
> o Element III - Source and Subject Interactions
> o Element IV - Basis for estimating Likelihood

FIGURE 1. Structure of OECD Dictionary/Thesaurus

(*Delaware Code*, 1989). In 1989 "Regulation for the Management of Extremely Hazardous Substances" was created to carry out the law (Barrish, 1990). In this investigation Delaware's law was chosen as a vehicle to study the utility of the OECD Dictionary/Thesaurus because it is an established state law with a history of practical enforcement and because it preceded but is similar to important major federal laws for preventing accidental chemical releases. The application of the OECD Dictionary/Thesaurus to the Delaware regulation was intended to address two questions:

1. How well does the Dictionary/Thesaurus help one understand the risk assessment process.
2. How well does Delaware's "Regulation for the Management of Extremely Hazardous Substances" meet the intent of the Delaware legislature stated in the "Extremely Hazardous Substances Risk Management Act?" (This comparison is important because the law provides the legal outline for preventing major chemical releases but the regulation expands and enhances the law by providing specific details necessary to meet the intent of the law.)

Delaware's law was enacted before the Occupational Safety and Health Administration's (OSHA) Process Safety Standard (*Federal Register*, 1992). Schaller (Schaller et al., 1996) has suggested that Delaware's regulation is similar in many respects to the OSHA Process Safety Standard recognizing that OSHA's Standard is to protect people inside a facility and Delaware's law is to protect people outside a facility. Both use a performance-oriented approach requiring that management controls are in place to ensure appropriate accident prevention measures. Both require that facilities have risk management programs, process hazards analyses, incident investigations, maintenance programs, inspections, training and emergency preparedness programs. Both establish criteria for identifying sites subject to the standard or regulation, have provisions for inspections and oversight and require compliance audits. (In this investigation we did not apply the OECD Dictionary/Thesaurus to a comparison of Delaware's regulation with the OSHA PSM.)

Delaware's regulation also preceded the passage of the EPA's rule on chemical releases (*Federal Register*, 1996), section 112(r) of the Clean Air Act Amendments, by several years. Regulated facilities in Delaware have had to comply with chemical process regulations under the Delaware law for some time period, whereas, they will not have to comply with section 112(r) on the Clean Air Act Amendments until June 20, 1999. For this reason Delaware's regulation has been of interest to researchers studying the EPA's rule (Schaller et al., 1996; McNulty et al., 1996). (In this investigation we did not use the Dictionary/Thesaurus to compare Delaware's regulation to the EPA).

Analysis of Delaware's Regulation Using the OECD Dictionary/Thesaurus

The use of the OECD Dictionary/Thesaurus to analyze the Delaware law and regulation provided different levels of comparisons. Comparisons can made at the level of Risk Elements (simplest comparison), Terms, Categories of descriptors, or Descriptors (most complete comparison). A complete summary is given in Tables 1, 2, 3.

The OECD Dictionary/Thesaurus was designed to accommodate risk assessments of varying complexity. Because the Delaware's law and regulation were designed to protect people from chemical release and do not include provisions for protecting ecosystems, public buildings, etc. the analysis used only a portion of the Elements, Terms, Categories of Descriptors and Descriptors available.

	Available in the Dictionary/Thesaurus	*Used in the Delaware Law*	*Used in the Delaware Regulation*
Elements	4	3	3
Terms	19	4	9
Category of Descriptors	70	4	15
Descriptors	368	9	25

TABLE I
Element I—"Hazard" Identification

Term	Category of Descriptor	Descriptor Law	Descriptor Regulation
Source with the Potential to Cause Undesired Outcomes	Substance	Explosive Flammable Toxic to humans	Explosive Flammable Toxic to humans
	Energy	Pressure Thermal flux	Pressure Thermal flux
	Physical situation	—	Systems containing regulated chemicals
	Legally defined	—	Listed substances
Subjects of Concern	People	Undefined People	Undefined People
Undesired Outcomes	People	Death Acute injury Irreversible health effects —	Death — — Other undesired outcomes

TABLE 2
Element II—"Exposure" Scenario

Term	Category of Descriptor	Descriptor Law	Descriptor Regulation
Basis for Predicting Release of Source.	Past experiences	—	Technical Guidance
	Technical analysis	Undefined technical analysis	—
		—	Failure Mode
		—	Fault Tree
		—	Hazop
	Legally defined	—	What If
		—	Check List
		—	Worst Case
Basis for Estimating Distribution of Potential	Technical analysis	—	Other technical analysis
	Legally determined	—	Look Up Tables
Basis for Estimating Distribution of Subjects	Other estimates	—	Other

TABLE 3
Element III—Source and Subject Interaction

Term	Category of Descriptor	Descriptor Law	Descriptor Law
Mode of Contact Between Source and Subject.	People	— —	Pressure wave Radiant flux Undefined
Basis for Predicting the Undesired Outcome	Relationship to Human		Undefined
Units Used to Predict Undesired Outcomes	Number of Events	—	Undefined

In the simplest comparison of law and regulation, comparing **Risk Elements**, one finds the following:

Element I. *Both the law and regulation address sources with the potential to cause undesired outcomes and subjects of concern.*

Element II. *Both the law and regulation include generation of possible sequences of events. The regulation expands to estimate release potential and distribution of subjects of concern.*

Element III. *The law is silent on how subjects will react with the source but the regulation addresses specifics.*

Element IV. *Both the law and regulation are oriented towards practical prevention rather than theoretical calculations of probabilities. Consequently, Element IV is not part of the law or regulation.*

In the more complete comparison of law and regulation, comparing **Descriptors**, one finds the following:

Element I. *Both the law and regulation include identical sources, i.e., explosive substances, flammable substances, materials toxic to humans, pressure waves, and thermal exposures[3] as potentials to cause undesired outcomes; both reference undefined people as subjects of concern; and both list death to people as an undesired outcome. The law also*

---

[3] Even if the law did not include thermal exposure the regulation could include it because explosive substances would produce a thermal flux. This would have resulted in the descriptor "thermal flux" being implicit in the law but explicit in the regulation. Version 1.2 of the OECD Dictionary/Thesaurus does not distinguish between implicit and explicit requirements but Version 2.0 will have such capacity

lists acute injury and irreversible health effects as undesired outcomes but the regulation is more general and includes these under other undesired outcomes.

Element II. *The law lists undefined technical analysis as a basis for determining the release of a source but the regulation is more detailed and offers five specific methods of analysis. The law does not define worst case release but the regulation does. Having defined worst case release, the regulation specifies how to estimate the distribution of the source, using look up tables, and the distribution of subjects of concern, using other technical analysis.*

Element III. The law is silent on the basis for estimating the mode of contact between the source and the subjects but the regulation provides guidance.

At the descriptor level there are 21 differences. However, the differences are understandable and all indicate that the regulation expands and enhances the law.

Conclusions

The use of the OECD Dictionary/Thesaurus to compare Delaware's law and regulation for preventing accidental release of chemicals was a useful exercise. The Dictionary/Thesaurus demonstrated both the logic of the risk assessment process used in Delaware and the agreement that exists between the law and the regulation.

The Dictionary/Thesaurus used in this investigation was Version 1.2, a prototype. In addition to the limitations mentioned earlier, it was not easy to encode data into the database or to produce well-formatted reports. These limitations will be corrected in Version 2.0. Even in its present form, however, the Dictionary/Thesaurus is a powerful analytical tool for understanding the risk assessment process.

The comparison of Delaware's regulation with the EPA rule would be useful and would illustrate where the State of Delaware would have to modify its regulation to become as stringent as the federal rule and where state regulation are already more stringent.

The OECD Dictionary/Thesaurus should be useful whenever there is need to conduct a risk assessment because its application helps clarify the scope, nomenclature, tools and assumptions used in the risk assessment process. This clarification should help practitioners and regulators think more objectively about the risk assessment process and identify subjective judgments that reflect personal bias and which make communication and acceptance of risk more difficult. Its use should help the regulated community, especially private firms that have facilities located in various legal jurisdictions, understand regulatory differences and develop comprehensive compliance strategies and help the public understand risks imposed on it and evaluate the acceptability of risk assessment findings.

References

Barrish, R. A. (1990). "An Overview of the Delaware Regulations for the Management of Extremely Hazardous Substances." 1990 Spring National Meeting, March 18-22, 1990, Orlando, Florida. American Institute of Chemical Engineers, New York, New York.

Delaware Code, Title 7, Chapter 77 (September 25, 1989). "Regulation for the Management of Extremely Hazardous Substances".

Federal Register (June 20, 1996). Volume 61, Number 120, 31667-31733, Part III. EPA, Accidental Release Prevention Requirements: Risk Management Programs Under Clean Air Act Section 112(r)(7); Final Rule. (40 CFR Part 68).

Federal Register (February 24, 1992). Volume 57, Number 36, 6356-6417, Part II. Department of Labor, Process Safety Management of Highly Hazardous Chemicals; Explosives and Blasting Agents; Final Rule. (29 CFR Part 1910).

Jones, D.A. (1992). *"Nomenclature for Hazard and Risk Assessment in the Process Industries"*, D.A. Jones, editor, second edition, p. 1. Institution of Chemical Engineers, Rugby, Warwickshire, UK.

Hendershot, D.C. and S.J. Schecter (1997). "Evaluation of a Proposed Thesaurus/Dictionary for Risk Assessment Using an Industrial Quantitative Risk Analysis", *International Conference and Workshop on Risk Analysis in Process Safety*, October 21-24, 1997, Atlanta, GA, Workshop D: Methodology for Comparing Risk Assessment, American Institute of Chemical Engineers, New York, New York.

Ignatowski, A.J., I. Rosenthal and L.D. Helsing (1997). "An Internet Thesaurus/Dictionary for Analyzing Risk Assessment Processes, Laws and Regulations," *International Conference and Workshop on Risk Analysis in Process Safety*, October 21-24, 1997, Atlanta, GA, Workshop D: Methodology for Comparing Risk Assessment, American Institute of Chemical Engineers, New York, New York.

McNulty, P.J., L.C. Schaller and K.R. Chinander (1996). "Communicating Under Section 112(r) of the Clean Air Act Amendments", Working Paper, *Risk Management and Decision Processes Center*, The Wharton School of the University of Pennsylvania, Philadelphia, PA.

Rosenthal, I., B.J.M. Ale and L.D. Helsing (1996). "An Outline of the Approach Being Used in Developing the OECD Thesaurus-Dictionary of 'Risk Assessment' Terminology," *Conference on TransBoundary Risks*, October, 1996, Warsaw, Poland.

Rosenthal, I. (1991). "Management of Major Health and Safety Risks - Some Principles and One Firm's Practice," *Proceedings of the First Czecho-Slovak and American Workshop*, October 5-9, 1991, Bratislava, Czechoslovakia.

Ross, W.C. and A.J. Ignatowski (1996). "OECD Risk Assessment Thesaurus/Dictionary System Instruction and Workbook." Version 1.20. Wharton Risk Management and Decision Processes Center, The Wharton School of the University of Pennsylvania. (December 30, 1996.)

Ross, W.C. and A.J. Ignatowski (1997). "OECD Risk Assessment Thesaurus/Dictionary System Instruction and Workbook Supplemental Notes." Version 1.20. Wharton

Risk Management and Decision Processes Center, The Wharton School of the University of Pennsylvania. (January 30, 1996.)

Schaller, L.C., P. J. McNulty and K. R. Chinander (1996). "Impact of Hazardous Substances Regulations on Small Firms in Delaware and New Jersey", Working Paper, *Risk Management and Decision Processes Center*, The Wharton School of the University of Pennsylvania, Philadelphia, PA.

Risk Assessment for Toxic Catastrophe Prevention: New Jersey's Risk Assessment Method Culminates in an Appropriate Risk Reduction Plan

Reginald Baldini and Peter Costanza
New Jersey Department of Environmental Protection (NJDEP)

ABSTRACT

The Toxic Catastrophe Prevention Act (TCPA) rule, first effective in June 1988, places risk assessment as the preeminent program element of risk management. According to the 1986 TCPA statute (1) the single most effective effort toward prevention of toxic catastrophes to be made is anticipating the circumstances that could result in their occurrence and taking those precautionary and preemptive actions required. This paper presents an overview of risk assessment under TCPA in brief and in the language of the European Economic Community's Office of Economic Cooperation and Development (OECD) Risk Assessment Thesaurus. As part of a regulatory enforcement program TCPA risk assessment provides results that are consistent among the diverse industries that apply it through the use of quantitative techniques based on standardized inputs. These techniques, along with the requirements for documenting assumptions, unmask inputs of subjective judgements and give confidence to the appropriateness of the determined risk reduction measures.

Introduction

A risk management program as defined in the TCPA statute consists of eight elements (see Table 1). Two of the elements deal with anticipating the circumstances that could result in the occurrence of catastrophic releases—safety review and risk assessment. An overview of the relationship of safety review and risk assessment within the TCPA risk management program is shown in Figure 1. In a safety review
 the design and operating procedure documents of the facility are first affirmed to conform with applicable codes and standards. Next, the physical plant and the operating actions are checked to conform with those documents. In a risk assessment those same documents are examined again to determine the scenarios of potential release and whether such releases could have catastrophic consequences.

TABLE I
Eight Elements of a Risk Management Program under the TCPA Statute NJSA 13:1K-21

1. Safety review of design for new and existing equipment

2. Standard operating procedures

3. Preventive maintenance programs

4. Operator training

5. Accidents investigation

6. Risk assessment for specific pieces of equipment or operating alternatives

7. Emergency response planning

8. Internal or external audit procedures to ensure programs are being executed as planned

Background on Risk Assessment

The 1982 oil companies study group for conservation of clean air and water—Europe (CONCAWE) paper on Risk Assessment (2) provided the theoretical basis for TCPA risk assessment. Table 2 presents the CONCAWE definitions of hazard, risk, analysis and assessment and Figure 2 presents a flow diagram overview of the CONCAWE risk assessment overall procedure. TCPA risk assessment reflects these CONCAWE concepts, especially that hazard analysis is seen as being more technically specific than "risk assessment", and is part of it.

TCPA Risk Assessment Method

Exhibit 1, excerpted from a TCPA guidance document (3), employs a flow chart to give an overview of the TCPA risk assessment method as incorporated in the current TCPA rule since 1993. First, determine the quantity rate/duration of the potential release. Next, determine the cloud defining concentration for the substance that reflects the duration of the release. Next, obtain down wind distance of the cloud for the release using the criteria concentration and also the downwind distance of the cloud defined by five times that criteria concentration. If the five times cloud extends beyond the property line, perform a state of art risk reduction study on that release. If only the criteria concentration cloud extends

Risk Assessment for Toxic Catastrophe Prevention

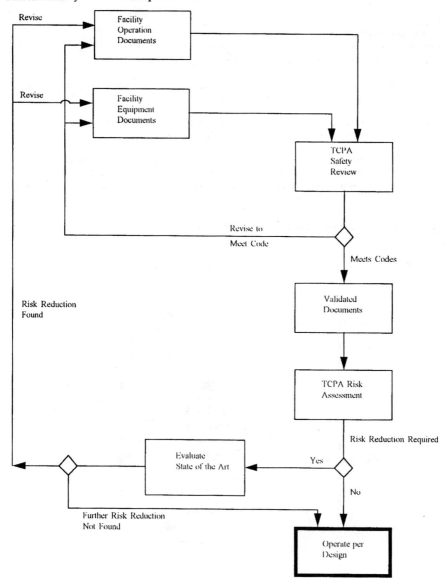

FIGURE 1 Safety Review and Risk Assessment

Table 2
Risk Assessment or Hazard Analysis

Analysis	A technical procedure following an established pattern
Assessment	Consideration of the results of an analysis in a wider context to determine the significance of the analytical findings
Hazard	An inherent property of a substance or situation which has a potential to cause harm (e. g., hydrogen fluoride or a falling stone)
Risk	Related to the consequences of a hazard potential being realised and causing harm. It is expressed with "likelihood, e.g., the chances that a hydrogen fluoride containment system will fail and cause hydrogen fluoride to escape and cause damage to persons."

Source: CONCAWE (2) with edited text in quotes by the authors.

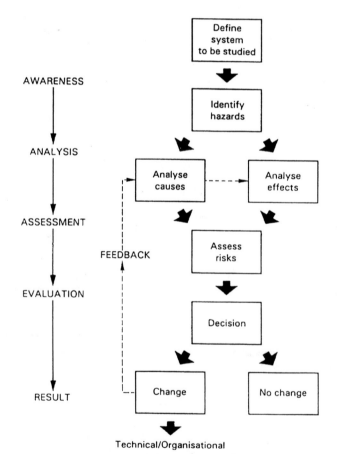

FIGURE 2 Overal Procedures — Hazard Analysis, Consequence Analysis, Risk Assessment

Risk Assessment for Toxic Catastrophe Prevention

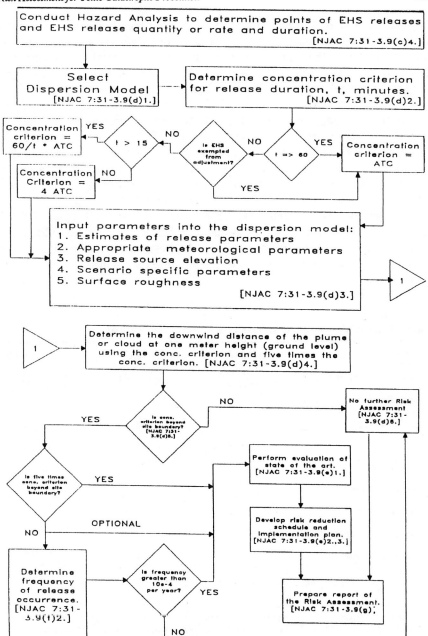

EXHIBIT I Risk Assessment Procedure

beyond the property line elect to proceed with a risk reduction study or to determine frequency of release occurrence. If the frequency is greater that 10E-4 per year, proceed with the study of risk reduction on the potential release. Exhibit 1 includes citations of the TCPA rule, N. J. A. C. 7:31 effective 1993, which prescribes the several steps of TCPA risk assessment.

Tables 3 and 4 give excerpts from TCPA program source and guidance documents that provide important basic data to perform TCPA risk assessment. Table 3 presents an excerpt of a larger table in a TCPA guidance document (4) that gives the acute toxicity concentrations for each of the 104 currently TCPA regulated substances upon which the cloud defining concentration values are based. Table 4 presents an excerpt of a larger table of likelihood of failure data for equipment components and operating errors in a TCPA guidance document (5) from which likelihood of release values may be estimated. Full copies of TCPA guidance guidance documents are available. A TCPA document providing guidance on dispersion modeling for TCPA risk assessment is also available (6). It provides guidance so that input to dispersion models is consistent with TCPA criteria.

TABLE 3
Acute Toxicity Concentrations (ATCs) of Selected Substances

Substance	Acetaldehyde	Acrolein	Boron trichloride
Chemical Abstracts Number	00075 07 0	00107 02 0	10294 34 5
Acute toxicity concentration ppmv	150	5	20
g/m^3, at 20°C	0.2749	0.0117	0.0975
ATC basis	LC50/10	IDLH	LCLo
Test animal	Rat	—	Rat
Test time (hours)	4	—	7
ATC data source	1,2	3	1
Vapor pressure (torr, at 20°C)	750	214	>760
Acute toxicity effects	C,D,E,I,K,L,T	B,C,E,H,I	C,D,E,I

Excerpted from NJDEP EHS Basic Data Document, July 19, 1993

Key for toxic effects: B=blood irregularities; C=burns (corrosion of tissue); D=central nervous system (CNS) depression; E=pulmonary edema; H=hemorrhage; I=irritation, sensitization, mutagenic, carcinogenic; K=kidney damage; L=liver damage; T=teratgogenic (fetal).

Key for ATC data source: 1=RTECS (Registry of toxic effects of chemical substances, USDHHS/NIOSH 1981-82 and supplement 1983–84); 2=HSDB (Hazardous substance databank, National Library of Medicine, Washington, D. C.); 3=NIOSH/OSHA pocket guide to chemical hazards by Mackinson, F.W., Stricoff, R.S. and Partridge, L.J., Jr., editors, GPO, Washington, D.C., 1978 and 1985.

Risk Assessment for Toxic Catastrophe Prevention

TABLE 4
Likelihood/Frequency Data, Selected

	Base Failure Data	
Item	Mode of failure	Failure rate
Pump	Fails to start	$4 \times 10E\text{-}5$ per hour
Unloading hose, heavily stressed	Rupture	$10E\text{-}3$ per demand
Pressure safety valve	Lifts/light leakage	0.06 per year

Excerpted from NJDEP Source Document for Risk Assessment, April 5, 1993.

Data included in the NJDEP Source Document is from Appendix IX of the Report to the Rijnmond (Netherlands) Public Authority by Cremer and Warner reprinted with permission from Kluwer Academic Publishers, B. V.

Acute Toxicity Concentration

The TCPA program developed the concept of "acute toxicity concentration" (ATC) originally to determine the registration quantity of substances that were added to the program's list of extraordinarily hazardous substances. The concentration represents that level that would cause death or permanent disability to a mammal during a test duration of approximately one hour. However, it should be noted that some values are from tests of up to eight hours. ATC values are used in TCPA risk assessment because exposure to a cloud with such a concentration could result in significant consequence to members of the public, namely, fatality or permanent disability. Table 3 presents examples of the three ATC bases used.

OECD Generic Risk Assessment Elements

In 1996, Rosenthal, Ale and Helsing (7) reported that the OECD determined in 1995 that the ". . . risk assessment process be mapped out in a generic way. In addition, the detailed approaches or methodologies for the different steps should be described so that stakeholders can see a range of possibilities open to them, characterize their risk assessments and better understand the work and results of others." To that goal the OECD prepared an "OECD Risk Assessment Thesaurus/Dictionary" in late 1996. Presented here in Appendix A from the Thesaurus version 1.20, 12/30/96, the OECD defining terms as underlined are shown to apply to TCPA risk assessment.

Application in New Jersey

Approximately 105 owners or operators handling extraordinarily hazardous substances (EHSs) apply TCPA risk assessment methods in New Jersey. They are at 119 sites in 75 municipalities where they operate 223 facilities. Chemicals manufacturing is the largest single category of such owners and operators with 18 of top 100 U. S. chemicals producers represented. In addition, there are four major U. S. and European pharmaceuticals producers represented. Other manufacturing owners and operators produce food products, paper products, plastics and rubber products, petroleum products and secondary metals. Non manufacturing among these owners and operators covers electric, gas and water utilities and wholesale trade of food and chemicals.

Risk assessment is used by these owners and operators most importantly to deal with the issues found at smallish sites in a densely populated state. Thirteen sites are in municipalities with population density ranging between 10,000 and 17,000 persons per square mile; 25 sites between 4,000 and 10,000; 33 sites between 2,000 and 4,000 and the rest between 70 and 2,000. Proximity of facilities to a site boundary is widely found in New Jersey. Fifty of the 223 facilities are 3 to 80 feet from their site's boundary; 45 facilities, 90 to 200 feet; 21 facilities 200 to 300 feet and 92 facilities 300 to 5,000 feet. An accidental release to the atmosphere even of modest quantity can be projected to extend beyond the property line at many of these sites,

TCPA risk assessment has led to mitigations that reduce the quantity or rate or duration or likelihood of accidental releases. Such mitigations, many "low tech," include placing the formerly outdoor operation indoors; providing a building air scrubber to reduce the release quantity to the atmosphere; providing remotely operated block valves to interrupt the release without "suiting up;" increasing the frequency of non destructive testing of equipment; providing barriers to impact by external force from site vehicle traffic; elevating the discharge point of the potential release, etc. Some of these mitigations are covered by codes and standards, but very often not the codes applicable in New Jersey, for many cases of extraordinarily hazardous substance. The state of art study portion of the TCPA risk assessment leads owners and operators to check the codes and standards of other parts of the nation and of nations around the world for the risk reduction concepts they include.

Conclusions

TCPA risk assessment mandates practices that employ quantitative data to analyze the hazard and its consequence and likelihood to assess the need and benefit of risk reduction. Consistency of analysis, its primary goal, is achieved by dealing with facility and substance specific variables. The practices also address existing-mitigation dependent parameters at a facility. A significantly sized potential release at one facility (say a thousand pounds of substance X) for which a risk reduction should be considered is treated differently from the same size release of another substance at another facility. Short time releases whose duration reflect existing safeguards are also analyzed differently than those of longer duration. Documenting the assumptions used to estimate the quantitative inputs permits scrutiny of inputs of subjective judgement so that confidence in the appropriateness of risk reduction measures is obtained.

References

(1) New Jersey Statutory Authority (N.J.S.A.) 13:1K-20.
(2) CONCAWE Report 10/82 (December 1982), Methodologies for hazard analysis and risk assessment in the petroleum refining and storage industry (the oil companies' international study group for *con*servation of *c*lean *a*ir and *w*ater—*E*urope), pages 2 and 3.
(3) Background Document for Toxic Catastrophe Prevention Act Program Risk Assessment, September 1994, NJDEP, Trenton, NJ 08625-0424.
(4) Extraordinarily Hazardous Substance Basic Data Document, June 19, 1993, NJDEP, Trenton, NJ 08625-0424.
(5) Source Document for Risk Assessment: Acute Toxicity Concentration Data and Likelihood/Frequency Data, April 5, 1993, NJDEP, Trenton, NJ 08625-0424.
(6) Dispersion Modeling Guidance on TCPA Risk Assessment, March 1995, NJDEP, Trenton, NJ 08625-0424.
(7) Rosenthal, I., Ale, B., and Helsing, L., Report to the OECD Steering Committee, February 1996, Wharton Risk Center, University of Pennsylvania.

APPENDIX: TCPA Risk Assessment in OECD Terms

The OECD Risk Assessment Thesaurus divides the risk assessment process into four generic elements, I, II, III, and IV plus an optional element V. Each element is further divided into subelements that analyze the various concepts contained in that element. The terms of the subelement are finally defined by descriptors that characterize the term with operational definitions. TCPA risk assessment steps according to the OECD Risk Assessment Thesaurus are:

Generic element I is the identification of *sources with potential to cause undesired outcomes to subjects of concern.*

Sources with the potential to cause undesired outcomes. The TCPA rule requires owners or operators handling substances toxic to humans that are "extraordinarily hazardous" to implement a consistent risk management program which includes risk assessment. That risk assessment shall cover systems containing regulated chemicals as included in a legally defined list [N. J. A. C. 7:31-2.3(a)].

Subjects of concern. The category of people addressed by a TCPA risk assessment is the *public outside the boundary of the owner's site* where equipment is located and the operating alternatives are conducted. Public outside the site boundary includes such categories as residents, patrons or employees of commercial, service or manufacturing establishments, people in transit on foot or in vehicles. Subjects not addressed by TCPA are ecosystems, cultural assets, property or public infrastructure (except water facilities).

Undesired outcomes. The undesired outcomes for people addressed by a TCPA risk assessment are *death or permanent disability (irreversible health effects).* The TCPA statute defines a toxic catastrophe to be the "...release into the environment (of sufficient quantity of substance) that would produce a significant likelihood that persons exposed will suffer acute health effects resulting in death or permanent disability..." [N.J.S.A. 13:1k-21e]. Undesired outcomes not addressed by TCPA are those to ecosystems, cultural assets, property or public infrastructure.

Generic element II is the *identification of possible sequence of events leading to loss of containment* resulting on the entry of the potential to *cause undesired outcomes* into one or more compartments of the ecosystem. OECD also includes the estimation of the possible distributions of both the released potential and the subjects of concern over time periods within a compartment delimited by specified boundaries or end-points.

Sequence of events. TCPA risk assessment uses *release scenario* to specify sequence of events. A scenario is based on information from several sources. First are the owner records of its equipment failure data, equipment reliability data, or toxic substance accident reports, as well as industry wide and insurance records of sequence of events [NJAC 7:31-3.9(c)4ii]. Second are the results generated by technical analysis which may be performed using hazard analysis methods of Hazop, Failure mode and effects analysis, qualitative Fault tree analysis, What if/Checklist or any other systematic examination of both human and equipment failure that may result in an accidental release from a review of updated process flow diagrams, piping and instrument diagrams, electrical one-line diagrams, standard operating procedures, maintenance procedures, accident investigations and equipment reliability studies [NJAC 7:31-3.9(c) 1].

The legally specified sequence of events is potential basic (initiating) or intermediate event sequences and includes estimates of the released quantity or rate/duration [NJAC 7:31-3.9(c)3]. "Worst Case," "Most Credible Worst Cases" or other predefined cases are not addressed by a TCPA risk assessment.

The basis for estimating distributions of the release potential for undesired outcomes in a TCPA risk assessment is by any dispersion model that meets specifications of the TCPA rule [NJAC 7:31-3.9(d)1]. TCPA risk assessment estimates are not based on real time monitoring. The use of "look up" tables is accepted when the models on which those tables are based meet the rule requirement.

Distribution of subjects of concern. The basis for estimating distribution of subjects of concern within compartments of interest in TCPA risk assessment is historical data, namely, United States Geographic Survey topographical maps and the owner site plan drawing. [NJAC 7:31-3.9(g)5]. Distributions determined legally include site centers of residential areas, major highways and areas of sensitive populations such as hospitals, schools and nursing homes.

Identification of basis of boundaries or end-points to delimit estimates of distributions. TCPA risk assessment is based on toxicologically derived end-points, namely, the ATC of each listed substance. The definition of ATC is included in the TCPA rule [NJAC 7:31-1.5]. Inhalation of air containing the ATC or greater would cause acute health effects resulting in death or permanent disability.

Generic Element III consists of the sub elements of intensity, time and mode of contact to cause the undesired outcome to the subject of concern and how the specified undesired outcome is related to the subject of concern.

Mode of contact. The TCPA risk assessment addresses released clouds of vapor or gas of the listed substances dispersed in the atmosphere near ground level. The acute health effect that would be experienced by the subject people would be experienced from their *inhalation* of the cloud. The value of the ATC of a listed substance was determined by experiment with such contact by test animals. Skin absorption and ingestion that occurs with an inhalation exposure may also be significant in the case of some substances.

Relation of the specified outcome to the subjects of concern. For TCPA risk assessment the ATC value for a particular substance is the lowest value of either LC50 divided by 10, or LClo determined for mammalian species or IDLH (immediately dangerous for life and health) determined by NIOSH for humans.

Dimensional/measurement units of the potential to cause the undesired outcome that are used to predict those outcomes. LC50, LClo and IDLH values are reported in the literature as parts per million by volume (ppmv) to produce the undesired outcome in 60 minutes or less. These values are translated to grams per cubic meter using the ideal gas law for 20 degrees Celsius.

Description of the dimension/measurement unit of response to express the undesired outcome. TCPA risk assessment employs death or permanent disability of one person as the measure of undesired outcome. Measures not used are percentage of skin burn or percentage reduction of respiratory capacity.

Number of undesired outcomes experienced by the specified population of subject. TCPA risk assessment employs one or more deaths in a population of 20 persons or more.

Likelihood of the undesired outcome given the specified interactions with the potential to cause undesired outcomes by specified subjects. A criteria likelihood of accidental release is specified for TCPA risk assessment [NJAC 7:31-3.9(f)2].

Generic element IV consists of identification of the basis for estimating and expressing the likelihood that the specified undesired effects will occur and description of the quality/uncertainty of such estimates. it also includes comparison of the estimates with relevant standards and guidelines and evaluation of the impact of using specified alternative assumptions on the estimates.

Basis for estimating that specified undesired effects will occur. TCPA risk assessment employs a quantitative criteria of the likelihood of the release scenario. Releases with the criteria likelihood or greater where the cloud defined by the ATC extends beyond the site property boundary are required to be addressed for risk reduction [NJAC 7:31-3.9(f)2]. However the TCPA risk assessment does not involve translating the estimate of likelihood of release to likelihood of undesired effect, although owner may voluntarily elect to do so for his own purposes. Semi quantitative or qualitative assessments of likelihood are not employed in TCPA risk assessment. The likelihood of release and the probability of other events that yield the likelihood of undesired effect are from the analysis of published or owner recorded data. Technical criteria are set forth by the TCPA rule at NJAC 7:31-3.9(f). The are no defined means of characterizing quality estimates, except that the calculations shall be documented and references be cited for failure rate and probability data.

WORKSHOP E
Risk Perceptions and Communications (Regulatory, Industry, Public)

Chair **Vin Boyen**
Mallinckrodt Chemicals

Community Participation in Risk Acceptance Criteria

Richard G. Runyon
SCIENTECH, Inc.

ABSTRACT

Acceptable risk can not be determined by the analysts alone. The level of acceptable risk will differ from community to community and from individual to individual. The input of the local community, particularly the local regulatory community, prior to performing the analysis may be critical in the ultimate acceptance of the level of risk. By discussing the process with the local community at the beginning and throughout the analysis you can describe what it is you are trying to accomplish in the study and what types of results can be expected. You can also gain an understanding of the concerns of the local community which may assist you in focusing your analysis. Although it is unlikely that a quantifiable level of acceptable risk will be established in the discussions, an understanding of the concerns of the community by the analysts can be critical in ultimate acceptance of the results by the local community. During discussions with the local community it is critical that good risk communication techniques be utilized to insure that the analyst perspective is appropriately conveyed to the community.

Introduction

With more requirements on industry to present the results of the analysis of their facility risk to the regulatory agencies and the local public, the issue of acceptable risk will become of greater importance. Much of the discussion to date has been on the technical aspect of how the analysis was to be accomplished, what scenarios would be acceptable, the level of passive mitigation allowed or what consequence models would be allowed. Little, if any, discussion has occurred on what level of risk might be acceptable to the agencies, or more importantly, the communities surrounding the facilities. The direction that is evolving appears to be that acceptability criteria would be established between the community and the facility. It is important to remember that the ultimate acceptance of the risk of your facility may not be the EPA but the local community.

Community Risk

Ideally the community's level of acceptable risk should be established early within a land use process. This establishes, both for the residents and industry, an agreed upon level of acceptable risk for the community before industry has established its operation. However, this process rarely occurs and the process will require a great deal of education, patience, and time. Further, this process will only assist new industries or modifications of existing facilities. For the established industries, discussion of the risk of current operations with the community will be much different.

The acceptability of a level of risk is very much a local, and even more so, a personnel issue. Any acceptance of risk is based on issues of perception, familiarization with the operation, real or perceived personal benefit from the operation, level of personal control of the operation and many other real issues. Further the public is much more likely to accept a level of risk for which they feel they have some understanding and involvement in the process of developing the risk acceptance criteria. Therefore, the time to discuss acceptable levels of risk is not after the analysis but as early in the process as possible.

It is important to remember that the perception of the risk by the community is likely to be much different than the results of an objective analysis. This, however, does not make the communities concern any less real. As risk acceptability is a very personal issue, what may be an acceptable level of risk for a facility manager is likely to be different for a regulator and different again from the mother of three children who gains no direct benefit from the facility.

For existing facilities the earlier the communities are brought into the process the better. Those industries with an active CAER program are in a much better position to discuss the issues of community risk and may have already resolved many of the issues. For those industries with significant risks who have not begun the process of involving the community with their operation, they need to begin as soon as possible. With the EPA Risk Management requirement to analyze and communicate a "worst case" scenario to the local community the earlier you bring the community and the local regulators (Fire, Police, Health) into the process the more likely you will reduce problems associated with the "shock" of your analysis.

The community and the local regulators should be consulted early on as to what can be expected from a hazard analysis and a consequence analysis. By receiving community and local regulatory input on the type of analysis and the data presentation format, the facility can reduce many of the problems of explaining the results at the time of analysis presentation. If the community knows what to expect and the data is presented in a format and form useful to them, they will be less likely to be confused or mistrustful of the results. It is unlikely that you will obtain a specific acceptable risk level by the community, however the discussions of the concerns of the community will provide you an opportunity to tailor your analysis to both the concerns of the community and to the local and federal regulatory agencies.

A discussion of the Level Of Concern (LOC) or toxicity target level of the material used in the modeling should occur. The community and the local regulators should understand the facility's EPA mandate for the use of LOC 2 and the criteria under which this level was developed. If other levels are used (i.e. no LOC developed) the facility should be prepared to show the bases for this level and how it compares to the criteria for the development of the LOC.

You may find that some of the more sophisticated local regulators will ask some probing questions on why you have chosen a particular hazard analysis ("What-If" as opposed to HazOp, for example) or why you chose a particular model or meteorological conditions for the consequence analyses. You should be prepared to answer them during the preliminary discussion. They may request that the results be displayed in a particular format to assist them in understanding the risk or to better use the data in preparing for an emergency. You may receive requests for analysis of scenarios that you feel have a low or no risk to the community. However, serious consideration to the analysis of these community concerns should be undertaken. This will help to provide credibility and assurances to the community that you take their concerns seriously.

The community/regulators should be kept informed during the process, particularly if it is expected to take some time to complete. In this way the community is kept part of the process and changes or updates can be easily communicated.

Once the analysis is completed, the results should be provided in such a manner to be understood and utilized by the community and the regulators. It may be necessary to produce maps with plume overlays showing projected dispersion patterns under prevailing and perhaps abnormal meteorological conditions. An estimate of the likelihood of these scenarios should be provided in understandable terms. The use of technological jargon should be avoided. If the community has requested that specific hazard be analyzed, these analysis should also be presented.

In addition to the risk analysis information, additional specific information that would be of assistance to the local emergency responders should also be provided. This could include possible mitigation techniques, meteorological data, physical and chemical information on the chemicals of concern, any census data on sensitive population that may be of interest to emergency responders. This information should also include the preparations the facility has undertaken to prepare for an emergency and, to the local response agencies emergency contact names and phone numbers at the facility.

Conclusion

By having discussions with the community on their concerns you will obtain a better understanding of what types and levels of risk will be acceptable. Although it is not likely that a quantitative level of acceptable will be established for the local community initially, this goal may be worth attempting for the future. By understanding qualitatively the concerns and fears of the community, you can better develop your program to directly respond to these specific community concerns and assist them in understanding risk and the risks of your facility.

The East Harris County Manufacturers Association Program for EPA RMPlan Communications: Progress 1994–1997

S. E. Anderson and J. W. Coe
Rohm and Haas Texas Incorporated

Steve Arendt and David Hastings
JBF Associates, Inc.

ABSTRACT

The East Harris County Manufacturer's Association (EHCMA) will present a summary and status of activity in developing a system for dealing with the communications and technical issues related to promulgation of the Environmental Protection Agency's Risk Management Plan (RMP) Rule (40CFR68). The objective is an integrated plan involving over 100 individual plants and facilities representing about 40,000 direct employees in seven Outreach Areas, which correspond to the seven communities and Local Emergency Planning Committees (LEPCs) in the Houston Ship Channel Area. Consistent communications, educational efforts, release modeling, and RMP preparations are the keys to the effort. Efforts are directed toward leveraging and cooperating on these efforts to minimize costs. Formation, operations, and accomplishments of the Risk Management Communications and Technical Committee (RMCTC) will be discussed. Communications strategies and preparations will be emphasized.

Introduction

The East Harris County Manufacturer's Association (EHCMA) appreciates the opportunity to present an update of our activity in developing a system for dealing with the communications and technical issues that will be brought to the forefront by the promulgation of the EPA RMP Rule (40CFR68). Some information about EHCMA will help to understand our approach to the issues.

Description of EHCMA

EHCMA membership now includes over 100 individual plants and facilities representing about 40,000 direct employees. Facilities associated with EHCMA account for about 30% of the total chemical production in the United States. The organization is governed by a Board of Directors having 17 members, all of whom are Managers from the member facilities. There are seven Outreach Areas in the Association which correspond to the seven LEPCs in the Houston Ship Channel Area. Board members come from all Outreach Areas. The Board appoints various committees from time to time to address needs that may arise. Currently, four committees are active:

1. Environmental
2. Community Relations
3. LEPC
4. Risk Management Communications and Technical

Risk Management Communications and Technical Committee Formation

The Risk Management Communications and Technical Committee (RMCTC) was formed in 1994 by combining members from the Responsible Care and Community Relations committees (including the Chairs of each). The committee was initially made up of about 15 Industry professionals, some having technical expertise in Release Modeling, some experts in Process Safety Management, and some experts in Community Relations and Communications. This committee was given the task of developing ways to address the Risk Management Plan rule and communicate with the public about (among many other things) risk, the Chemical Industry, and the risk management systems each of the facilities has, and coordinating and managing the communications effort to maximize the benefits and minimize the liabilities. We believe that a proactive, coordinated response to the EPA Rulemaking is essential for several reasons:

1. There are about 600,000 people living within 10-mile of the Houston Ship Channel.
2. As mentioned above, in this area there are 7 cities and communities having 7 Local Emergency Planning Committees (LEPCs). These LEPCs are very active; much more than seems to be common in the rest of the country.
3. There are seven active Community Advisory Panels in the area (one for each LEPC).
4. With 100 facilities distributed among 7 communities and LEPCs, it is evident that LEPCs, CAPs, and communities will receive communications from and will be impacted by more than one member company.
5. It is important that we manage our communications so that the same message gets to various audiences from all of the various sources at the same time.

6. The news media will be involved in whatever we communicate about RMPs.
7. Several member companies will probably be developing RMPlans for the same compounds (*e.g.*, Chlorine and Ammonia).
8. It is a way to help us fulfill several important commitments under Responsible Care®.

The current vision and mission statement of the RMCTC is as follows:

VISION: The Manufacturing facilities of EHCMA will join with the surrounding communities in a spirit of cooperation and information sharing to foster understanding of risk management, and to promote and encourage risk reduction. As part of this vision we will provide guidance for the development and communication of risk management plans as defined by the EPA's Risk Management Program Rule.

MISSION: The industry and community members of the Risk Management Communications and Technical Committee will facilitate a consistent, coordinated, consensus-based approach to both regulatory compliance and communication with the community by:

- Ensuring that all technical work, is accurate, scientifically defensible, accessible, and is peer reviewed.
- Interpreting, explaining, and applying regulatory requirements as consistently as possible throughout East Harris County.
- Planning and carrying out a communications strategy that includes broad employee and community involvement.
- Communicating these plans to the people in our communities in a clear and understandable manner and seeking to provide resolutions to questions concerning risks and our plans to manage those risks.
- Seeking public participation by encouraging its support and meaningful involvement in the development of the risk management plans and the strategies for communicating these plans to the community.
- Effectively utilizing and managing all resources provided through EHCMA

Additional conceptual materials from the Committees formative stages will be found in Appendix A.

Committee Operations, 1994–1995

One of the first things we did was to survey the membership. We found that there were a large number of chemicals which were common to many members. For example, at that time, 38 of our members handled Chlorine, and 35 handled Ammonia. So the opportunity for leveraging our technical work was real.

The next item on the list was how to involve the public in these efforts. We decided on a Community Advisory Group with members from a variety of backgrounds. (Our objective is to have participation from all of the LEPC areas.) To the initial group of 15 we have added another 30 industry professionals and 27 citizens. We have housewives, college professors, business people, and Community Advisory Council members and Facilitators. The CAG became involved as early as we could arrange it and, in fact, they have become an integral part of the Committee's operations.

As we worked these issues, we became aware that the work was probably bigger than a group of part-time volunteers could manage effectively. Accordingly, we decided to enlist the services of a contractor to help in all three of the phases. JBF Associates, Inc. was chosen to work with us as prime contractor. The contractor team also includes Radian International and Erin Donovan Communications. The CAG participated in the choice of the contractor.

During the first year we established a good basis for reasonable and useful communications. We had good and productive meetings, and began to establish trust and credibility in both directions. However, near the end of 1995 it became evident that we needed a change in our committee structure and meeting format. We decided that we needed even more citizen participation, and the resulting committee would have been too large to manage effectively. We also saw the need for a lot of activity by each half of the Committee (Communications, Technical) that would have meant half of the Committee would have been idle half the meeting time. Accordingly, we have split into two subcommittees and are now meeting separately. We do come together to coordinate and share information and progress as needed. Minutes of all meetings are shared by both subcommittees.

Committee Operations, 1995–1996

In August of 1995 we held a one-day workshop for the Committee. At that workshop we defined the directions that the two subcommittees would take for the rest of 1995 and all of 1996. One of the most important decisions made was that the Committee would perform essentially all RMPlan work on three widely-used chemicals as a Pilot Project to better learn what the various aspects of the Rule actually meant and how they needed to be addressed by using them in preparation for a Pilot Communications effort. Chlorine, Ammonia, and Propylene were chosen for the Pilot Project because they represented characteristics that addressed a wide range of modeling issues (e.g., liquefied gases), they represented toxic and flammable classes of chemicals, and they were the most widely-used materials as determined by the initial 1994 survey.

The Technical Subcommittee provided a forum for sharing information about RMP Coverage interpretations, RMP tier screening, and RMP registration assistance. More important, however, was the formation of Work Groups to define protocols for model selection and use, the 5-year accidental release data collection, meteorological data collection, and prevention program summary development. All of this work has been done for the three Pilot Chemicals, the work has been checked and validated by our Contractors, and the information is available for the Communications Subcommittee for their package development and testing.

During late 1995 and 1996, the Communications Subcommittee was quite active, also. Several Work Groups were operating during 1996 to review communications products, develop presentations for the various audiences, and to develop Chemical Fact Sheets. The Education research work group is ensuring that all communications products are understandable. The News Media work group is developing media strategies, Plant Manager Talking Points, Information Kits, and a Photo file. The Editorial Review Board is ensuring consistency, clarity, and that the products are suitable for each audience. A brochure describing the total EHCMA RMP effort was issued, along with two articles suitable for inclusion in Plant Newsletters, and a quarterly RMP Project Bulletin. A communications package for employees is being prepared, which will include a videotape about the RMP Rule.

Throughout 1995 and 1996, a total of 15 presentations to various groups interested in the effort have been made. We have made presentations about what we are doing and planning, as well as what the EPA RMP Rule is to each LEPC and CAP. Presentations were also made to the Area Organization of Mayors and Councils, the South Texas Section of AIChE, the American Society of Safety Engineers, and others.

Committee Operations, 1997

In January 1997, we held a workshop to bring all participants up to speed and to establis Work Groups for the remaining chemicals in EHCMA. The intent is to provide a Work Group for every chemical which has as many as 5 users. A single work group will be provided for "all others" so that all EHCMA members will have the benefits of the common chemical work group experience. For the rest of 1997, we plan to continue in a number of areas and to add others. Now that the technical work on the Pilot Chemicals has been completed, we will begin to test the communications tools developed by the Communications subcommittee using focus groups to determine their effectiveness. A general list of tasks we intend to continue or initiate is given below:

- Develop RMPlan Software for member facility guidance—Coverage
  —Tier screening
  —5-year history criteria
- Prepare outreach area stakeholder groups for pilot RMP communications
  —Pilot Studies done
    »Develop infrastructure to facilitate common work patterns
    »Common Met data (1996)
    »Proof-test protocols
    »Learn each other
    »Learn the ramifications of the programs by using them
- Facilitate Member Hazard Assessment Work groups
  —Protocol
  —Data
  —Modeling assistance
  —Common chemicals
- Help with other required programs
  —Prevention Programs
  —Generic PSM for non-covered facility
  —Enhance emergency response and mitigation programs.
- Pre-RMPlan risk communication products
  —Communications products review group (1996)
- Detailed RMPlan communications for each outreach area
  —Order
  —Delivery mechanisms
  —Demographics (1996)
  —Analysis of needs (1996)
  —Presentation work group
  —Chemical Fact Sheet group
  —Education research work group (1996)
    »See if content is understandable
  —News Media work groups
    »Media strategy
    »Manager talking points
    »Information kits
    »Photo file
  —Editorial review board
  —Employee communications package (1996)
  —Test RMPlan messages with focus groups
  —Make brochure and quarterly bulletins (1996)

Results and Conclusions

The East Harris County approach to the EPA Risk Management Plan rule will help the industry present the information required in a unified and cost-effective manner. Results to date show that the concepts behind the formation of the Risk Communications and Technical Committee are sound, and many of the benefits of the common effort are beginning to be seen. Many areas of the country have begun to form similar groups, and the EPA supports the concepts. We expect to be able to fulfill our mission.

Appendix A: Early Concepts for Committee Communications

The vision we had for the kinds of communications we needed to make was as follows:

1. Develop communications package that is modular and can be presented in 15 to 30 minute sections.
2. Each module builds on the overall information base.
3. The program will contain a blend of video, oral, and written presentations.
4. Possible messages to be communicated:
    a. What is risk
    b. Industry **does** pose a risk
    c. Risk can be managed
    d. Industry has systems in place to manage risk
        (1) "Process Integrity" —a form of Process Safety Management
    e. Benefits of the Chemical Industry
    f. Emergency Preparedness
        (1) Plant emergency plans
        (2) Plant emergency drills
        (3) Channel Industries Mutual Aid
        (4) Integration with Community Emergency Plans

Some of the target audiences for the messages which were identified are listed here:

1. Employees
2. Retirees
3. LEPCs
4. CAPs
5. Fire Departments
6. Community/Civic Groups
7. Churches

8. Schools
   a. Administration
   b. Teachers
   c. Parent/Teacher Organizations
9. Immediate Neighborhood Groups
10. Elected Officials (Federal, State, County, City)
11. Environmental Groups
12. Senior Citizen Groups
13. Health Department
14. Parks and Wildlife Department

The Plant Managers will be the keys to the communications effort.

We expect this effort to be divided into three phases:

1. Phase I
   a. What is this all about?
   b. Risk and its management
   c. General background and education
2. Phase II
   a. Examples of two release consequence models
   b. Define "conservative" and be sure the audience understands how it applies in "Conservative Modeling"
   c. Environmental Impact
   d. Glossary of terms
   e. Worst Case vs "Most Credible"
3. Phase III
   a. Retell and Recap
   b. Meet regulation "with minimal outrage"

WORKSHOP F
Technical Considerations in Choosing Multiple Levels of Safeguards

Chair   **Jan Windhorst**
Nova Chemicals Ltd.

Over-pressure Protection by Means of a Designed System Rather Than Pressure Relief Devices

Jan C. A. Windhorst
Strategic Initiatives, NOVA Chemicals, Red Deer, Alberta T4N 6A1

ABSTRACT *Aug 96 approved*

Code case 2211 of ASME VIII allows the substitution of a pressure relief device by a system such as a safety-instrumented system. Relying solely on pressure relief devices does not necessarily result in a safer facility. The ASME ruling allows improvements in the overall safety and environmental performance of a facility by utilizing the most appropriate tools for pressure protection.

Companies that use risk based design methods are in a good position to take advantage of this ruling since they have processes in place to make necessary decisions.

The approach taken consists of working through three basic levels of analysis: a quick risk assessment, an assessment of the requirements and the application of specific standards. This approach follows closely the DIN, IEC, Namur and ISA standards and guidelines for achieving process and environmental safety in process plants through process control systems. The paper provides a detailed and overall defensible method for using systems for over-pressure protection in jurisdictions that follow ASME either directly or indirectly.

Introduction

In many jurisdictions, e.g., Canadian Provinces, many US States, etc. the authority having jurisdiction on pressure vessels often requires, if not mandates the need and use of mechanical pressure safety devices. This requirement sometimes forces the operators of process equipment to install token or impractical pressure relief devices (see [1]). Token relief devices contravene inherent safety, since installing more ancillary equipment will make systems more and sometimes unnecessarily complicated. In individual cases pressure relief valves are made functional through the addition of measurement and control (MC) equipment, e.g., for purging of blowdown lines, and/or management systems (car-sealed valves). This practice compromises the true independence of a mechanical relief system, which is normally considered an independent protection layer (or IPL). The overall reliability of such a system is most likely less than the reliabilities of

631

its individual components: the MC equipment and the pressure safety valve. Present and future environmental requirements to treat certain types of blowdown from pressure safety valves (PSVs) will diminish this IPL character further. When encountering valves that either don't work reliably or are part of a large complicated PSV-environmental system the question needs to be asked if it would not be preferable to prevent a blowdown rather than mitigate it. In other words improve the reliability of the basic process control systems (BPCSs) and the safety-related systems (SRSs) rather than to try to make ancillary PSV equipment work.

Code Case for Section VIII 1 and 2 Pressure Vessels with Over-pressure Protection by System Design [Case 2211 (Bc 95-019)]

Until recently ASME held the position that only mechanical relief devices would satisfy UG 125(a) of Section VIII; however, in 1995/1996 Code Case 2211 was approved. ASME adopted the opinion that a pressure vessel may now be provided with over-pressure protection by system design in lieu of a pressure relief device as required by UG-125 (a) of Section VIII Division and by AR-100 of Section VIII, Division 2 under the following conditions:

- The vessel is not exclusively in air, water, or steam service.
- The decision to provide a vessel with over-pressure protection by system design is the responsibility of the User. The Manufacturer is responsible only for verifying that the User has specified over-pressure protection by system design, and for listing this Code Case on the data report.
- The User shall ensure that the MAWP (see UG-98) of the vessel is greater than the highest pressure which can *reasonably* be expected to be achieved by the system. The User shall conduct a detailed analysis, which examines all credible scenarios which could result in an over-pressure condition.

  This analysis shall be conducted by engineer(s) experienced in the applicable analysis methodology. Identified over-pressure concerns shall be evaluated by an engineer(s) experienced in pressure vessel design and analysis. Results of the analyses and the SRS (including the identification of all truly independent redundancies and a reliability evaluation) shall be documented. The documentation shall be made available to the authorities having jurisdiction at the site where the vessel will be installed and the User of this Code Case is cautioned that prior Jurisdictional acceptance may be required.

The existing situation in many North American companies is best described as follows:

TABLE 1
Typical Imposed Requirements

Legislated and/or imposed by the governmental body having jurisdiction	Internal company standards
Often Prescriptive, e.g., ASME codes	Deterministic/Probabilistic/Possibilistic

This situation has several disadvantages:

- There are grey areas, e.g., environmental requirements versus safety requirements which are complicated by a rigorous interpretation of the existing ASME code.
- It does not give full credit to the technological developments which have occurred over the last decade, such as those shown in Table 2.
- It does not specify upset conditions under which the process can exceed the maximum allowable working pressure (MAWP, see UG-98) of the vessel by more than the maximum 21 percent allowed under fire conditions such as in the case of deflagrations (see [15] and [16]).
- It does not specify maximum relief conditions under upset conditions, e.g., an allowance to exceed the MAWP up to the points of deformation or rupture.
- It does not provide quantifiable criteria for phrases such as *unreasonable* for Code Case 2211.

TABLE 2
Technology Developments Over the Last Decades   *ie. NEW developments*

Area:	Specific Developments:
Process Safety Designs	Dynamic process simulations, DIERS technology that leads to a much better understanding of the physical phenomena involved in a process (see [2]).
MC-safety systems, e.g., Safety-instrumented or Safety-interlock systems (SIS) or a Electrical/Electronic/Programmable Electronic System (E/E/PES)	ISA S84.01, IEC 1508 (see [3], [4], [5], [6], [7], [8], [9], [10], [11], [22], [23], and [27])
Management Systems	OSHA 1910.119 (see [12])
Risk	CCPS QRA books (see [13]), DIERS' Risk Committee (see [14])
Mechanical Integrity	NFPA 68, NFPA 69 and VDI 3673 (see [15]. [16] and [17], respectively) *allows vessel deformation*
ASME Code(s)	Case 2211 but generally lagging

Authorities having jurisdiction as well as operating companies have to address these difficult technical as well as sociopolitical issues. A few simple rules; however, could be introduced, such as:

- mechanical devices do not need to be installed in services that render them impractical;
- the reliability of the system needs to be as good as or better than the reliability of the mechanical relief device it replaces; and
- operating companies are responsible for a safe and environmentally friendly operation.
- A functional approach

Based on the wealth of information that is now available it would seem reasonable to re-focus on why we have code requirements in the first place and what goal they are intended to serve. The objectives ought to be the following:

1. protection of human life,
2. protection of the environment,
3. protection of other generally valued material properties, and
4. protection of property/production, i.e., economics.

An operating company has to satisfy the first three requirements and the fourth one where it concerns noncompany assets. Internal financial risks; however, should be a company's responsibility. This would allow companies to eliminate "token" equipment and improve personnel and environmental safety. For example, a PSV intended to protect a certain type of polymerization reactor performs unsatisfactorily. When called upon it releases a three-phase reacting mixture that plugs the tail pipe. Such a system is better served with a more reliable reactor control system than by trying to come up with an innovative solution for making the relief valve work. The "DIERS (Ad-Hoc) Committee for reducing the number and the size of relief valves" looked at other jurisdictions for a suitable PSV system substitution process. Certain jurisdictions outside North America allow such "swaps", e.g., AD-Merkblätter, TRB 403 clause 3.7, etc. (see [18], [19], [20], [21] and Figure 1).

Namur NE 31 (see [22]), which is a PSV specific application, uses a qualitative risk-based process to derive at the requirements for a SRS to replace a PSV. ISA S84.01 and IEC 1508 describe a similar generic process that can be used. Namur NE 31, ISA S84.01, CCPS (see [17]) and IEC 1508 all capture the Code Case 2211 requirements for documentation, validation, verification and safety integrity.

German Standard

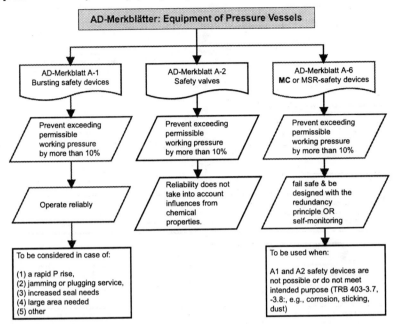

FIGURE 1 Summary and comparison of AD-Merkblätter

The Process

A flowchart showing the decision making process for a possible substitution of a PSV, rupture disk (RD) or other pressure safety device by system(s) is laid out in **Figure 2**. This high-level decision making process follows the initial stages of the lifecycle process described in [3] and [5] (see **Figure 3**). We distinguish:

- The first stage includes the hazard and risk analysis review, which has to identify the hazards of the equipment under control (EUC) in terms of cause, effect and a risk ranking. A FMECA (see [23], [24], [25], [26] and [27]) or similar type of review is recommended, this review should be conducted prior to *completion* of the process flow sheets. This is considered to be an optimum time in terms of being able to make cost effective changes, to enhance inherent and general safety, to define engineering specifications and to define an operating philosophy.
- The second stage involves the following activities:
  » Securing specified operations by allocating safety measures to one or more *independent* SRSs (see [27]). The allocation process aims at reducing the overall high pressure risk to or below a tolerable level by dividing the risk in different incremental risk portions for each identified

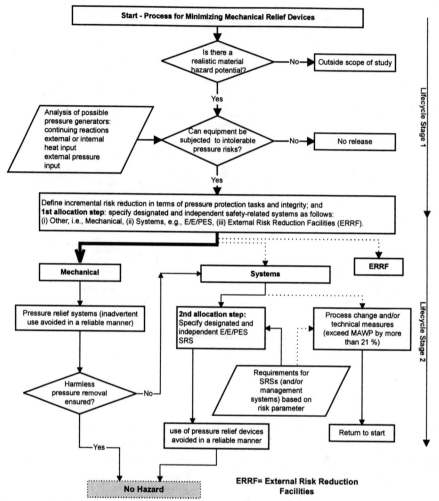

FIGURE 2 Decision chart for selecting relief devices or alternate systems (see [30])

pressure safety hazard. Incremental risks are defined in either a qualitative or quantitative manner (see [13], [14] and [28]) and are negated by one or more *independent* SRS(s). The process is two-fold, first a selection has to be made between different types of SRSs then secondly one or more *independent* SRSs need to be selected from that SRS group (see **Figure 4**). Reliability and economics drive the allocation process. SRSs can be pneumatic, E/E/PESs, other protection techniques, management/organizational systems or a combination of such systems. For example, TRB 403, clause 3.7 (see [21]) mentions that where a

 p. 638

nova method.

Before Hazop.

FMEA
↓
DIN 19250 → narrative ←
↓
VDI/VDE 2180
DEC 84 part 3 ←
↓
IEC 1508
type Allocation process
↓
process as
described in → Training, Testing, procedures
IEC 1508
↓
Process Hazard Review Figure 6

Over-pressure Protection by Means of a Designed System Rather Than Pressure Relief Devices 637

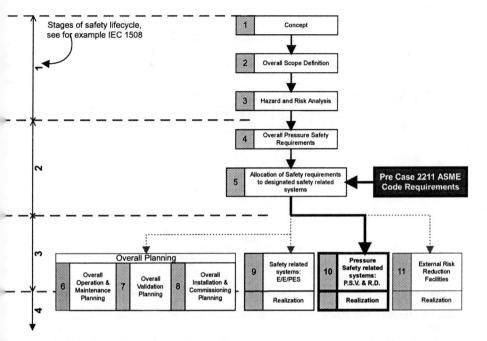

FIGURE 3 Lifecycle Approach

mechanical relief device is not practical it could be replaced by an alarm system provided the operational measures taken, upon reception of the alarm, will prevent the equipment from being over-pressured by more than 10 percent (*alarm management!*);

» The development of the pressure safety integrity requirements or safety integrity levels (or SILs, see **Figure 5**).

The allocation review process is **essential** for a safe and economic process; it needs to be performed by an interdisciplinary team of chemical, process/mechanical, safety, and instrument engineering operators, and where appropriate, other professionals (Namur NE 31).

Where a E/E/PES SRS has been selected to replace a PSV it needs to be defined in terms of:

- safety objective;
- availability;
- function, i.e., the necessary incremental risk reduction to meet the tolerable risk level;
- conceptual process design;
- nature and frequency of the scheduled functional testing; and
- associated management systems (e.g., scheduled maintenance).

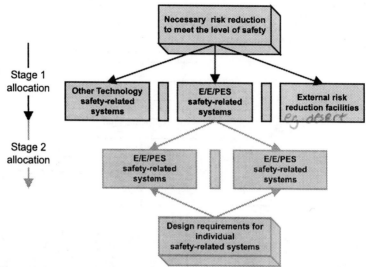

Notes:
Stage 1 allocation: Allocation of safety requirements to independent safety-related systems and external risk reduction facilities
Stage 2 allocation: Allocation of safety requirements to independent E/E/PES safety-related systems

☐ Indicates independence at Stage 1 allocation between safety-related systems and/or external risk reduction facilities.

☐ Indicates independence at Stage 2 allocation between E/E/PES safety-related systems.

FIGURE 4 Allocation process (see [5])

The SRSs should be distinguished from and have priority over (in case of emergency) basic control and monitoring instrumented systems.

Risk Requirement Category Determination

DIN V 19250 (see [29] and [9]) allows a qualitative/semi-quantitative risk assessment, which provides graduated requirement categories (see **Figure 5**). This method is only applied to the incremental risk negated by the MC SRS and not to whole systems.

Parameter C "extent of damage" is defined as follows:
- C1. Slight injury to one person or harmful environmental effect(s) which is not, for example, covered by legislation.
- C2. Serious, irreversible injury to one or more people or death of one person; temporary greater harmful environmental effects, e.g. according to accident regulations.

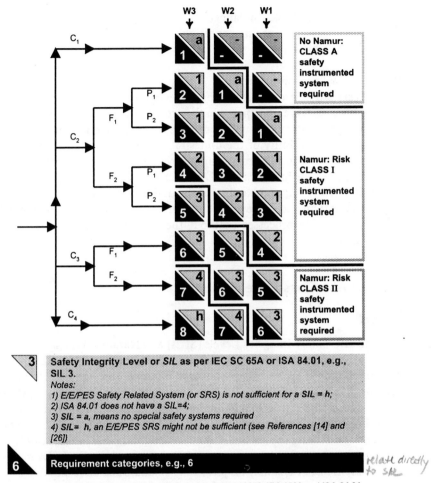

FIGURE 5 DIN V 19250 Qualitative risk chart (see [29])

C3. Death of several people; protracted greater harmful environmental effects, e.g. according to accident regulations.
C4. Catastrophic effect, many deaths.

Parameter F "duration" of presence:
F1. Rare to more frequent presence in the danger area.
F2. Frequent to continuous presence in the danger area.

Parameter P "avoidance" of danger:
P1. Possible under specific conditions.
P2. Hardly possible.

Parameter W "probability":
W1. Very slight probability of the undesirable event; only very few undesirable events are anticipated in the process under evaluation or in comparable processes (without the presence of MC protective equipment).
W2. Slight probability of the undesirable event; few undesirable events are anticipated in the process under evaluation or in comparable processes (without the presence of MC protective equipment).
W3. Relatively high probability of the undesirable event; undesirable events are anticipated more frequently in the process under evaluation or in comparable processes (without the presence of MC protective equipment).

Instrument Class Determination

General

ISA and IEC Safety Integrity Levels (SILs) and Namur NE 31 safety classes can be found once the "Requirement Classes" (according to DIN V 19250) have been determined through the linkage shown in **Figure 5**. ISA and IEC 1508 standards provide details for E/E/PES SRS based on the safety integrity level (SIL) determined. IPLs, SRSs with their SILs should be compiled in a document such as a hazardous event severity matrix (see [9], [27] and **Figure 6**), which also provides a means of verifying the results. *A more fundamental result (see [5]) can be obtained by following a (potentially elaborate) probabilistic approach via Fault Tree and/or Event Tree Analysis and determining the probability of failure and associated SILs.*

Namur NE31 Classifications

In this paper we follow the Namur NE 31 guideline, which as a PSV application specific standard is considered the best practice but do a severity matrix assessment at the end. Namur NE31 distinguishes two levels of SIS requirements, "low risk" and "high risk."

NAMUR CLASS I "LOW RISK APPLICATION"
(Risk is considered to be only slightly above the risk limit).
Requirements: A covert fault must be identified and eliminated within a time span during which a disturbance of the specified operation is unlikely to occur. However, the risk might already be below the limit risk if other (non-instrumented

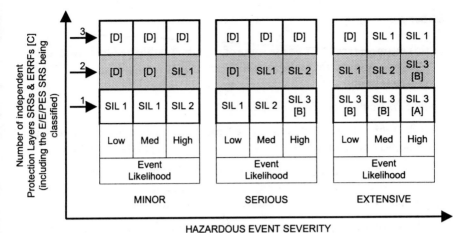

NOTES:
[A] One SIL 3 E/E/PES does not provide sufficient risk reduction at this risk level additional risk reduction measures required.
[B] One SIL 3 E/E/PES may not provide sufficient risk reduction at this risk level additional risk reduction measures required.
[C] Event likelihood: Likelihood that the hazardous event occurs without any SRSs or ERRFs.
[D] An E/E/PES SRS is probably not required.

FIGURE 6 Hazardous Event Severity Matrix

protective measures, e.g., management system) protective measures were implemented. Measures include a single channel SIS with:

- a short fault detection time (for instance, frequent functional testing, permanent plausibility check) or
- a low probability for covert faults within the SIS.

NAMUR CLASS II "HIGH RISK APPLICATIONS"
(DIN V 19250 Requirement Class 5 and higher)

Requirements: One covert fault must not impair the protective function of the SIS. Regardless the status of the process, the fault must be detected and eliminated within a time span during which a second independent covert fault is unlikely to occur. Measures include redundancy of safety devices. Generally, a 1 out of 2, or a 2 out of 3 (for simultaneous high production availability) suffices when combined with regular function testing. Diverse redundancy (e.g., for transmitters) does not automatically produce increased safety versus a homogeneously redundant configuration. The latter provides an alternate measure for the prevention of possible systematic errors. Fail safe systems or systems with self-diagnostics are considered equal to redundant systems.

Example

In this example (see [22]) we use some of the tables published by Jörg Loock (see [30]) as part of a reactor survey study done by TÜV.

Collect Physical, Chemical/Kinetic Data

This data should be collected for all feedstocks, intermediate products, end products and wastes involved in the process (see [27] and [30]) and include:

- maximum amount(s) of material present during the process;
- phase of the material (solid, liquid or gaseous);
- hazardous properties, including toxic and enviro-toxic data.

Based on the data conclusions need to be drawn concerning the presence of material hazard potential(s).

The example (see [22] and **Figure 7**) shows a semi-batch reactor charged with a material 2 to which a material 1 is added. The reaction, e.g., an alkylation reaction, is exothermic and a cooling jacket absorbs the released heat. A reaction that does not start, a high addition rate, agitator failure or failure of the cooling system can cause very high rates of pressure and temperature rise. A safety relief valve or rupture disk is *not* practical due to the toxicity of released materials. The protection should therefore be provided by means of instrumentation (E/E/PES). The following criteria were established:

- Interruption of the addition of material 1 will bring the reaction to a halt.
- A low reaction temperature will cause a deterioration of product's 3 quality, which can be avoided by the addition of hot water (to the cooling jacket). This will not lead to critical reactions since the hot water is tempered.
- The ratio of [(reaction rate)/(addition rate)] within the whole temperature range is high enough that an accumulation of material 1 can be excluded.

Determine the Presence of Pressure Generators

The safety objective is to prevent personnel injury and environmental damage by an eventual rupture of the stirred reactor. The safety issue is to prevent a rapid and uncontrollable rate of pressure increase that would result in a vessel rupture and the release of toxic material (see Table 3).

Qualitative Risk Analysis

CONSEQUENCE ANALYSIS

The next step is the listing of over-pressure generators and their consequences, as shown in Tables 4 and 5 (see [30]).

TABLE 3
Listing of Potential Over-pressure Generators

Excessive over-pressure through:	Incident possible ?	PSV sized for event?	Remarks	Probability (see Fig.5)
Continuing reaction Erroneous reactor conditions:				
• temperature	yes		Is captured under accumulation and other.	
• accumulation of reactant	yes	no	If the reaction temperature is too low, accumulation of material 2 will occur and the reaction will not occur. A stepwise *management* procedure with condition verification is needed.	W1
• accumulation of reactant	yes	no	If the addition rate is too high, accumulation of material 2 will occur and the reaction will run away. Material 1 determines the chemical reaction rate and the intolerable increase in pressure. Interruption of the flow of material 1 is therefore suitable to avoid operation outside the specified range.	W1
Continuing reaction Agitator failure	yes	no	Causes an accumulation of material 2 but the concentration will be less than in the case of accumulations given above.	W2
Conclusion:	Over-pressure generator present			

TABLE 4
Rating of Consequence(s) with Respect to Reference Criteria

| Incident | Materials released | Amount and duration of release | Maximum emission concentrations/Lower Flammable Limit | | Assessment Criteria; e.g., OSHA | DIN V 19250 Consequence parameter |
			On-site	Off-site		
Accumulation of reactant	X1, X2, X3	Total inventory upon rupture of vessel	2,011 kg of X3 with LFL @ 186 m			C3
Agitator failure	X1 and X2	1.5 kg of X1 for 20 minutes				C2

RISK ANALYSIS

TABLE 5
Determination of Requirement Class(es), Namur Classes and SIL Classes

Incident	Protection Measure	Cons.	Prob.	Duration	Avoid	Requirement/ Namur/ SIL Class
Accumulation	Operation outside the operation range (pressure and temperature too high) can be recognized by measuring values for T1 and P1. Both can be used for protective measures for the plant and are therefore safety relevant process values.	C3	W1	F2	—	5 / II / 3

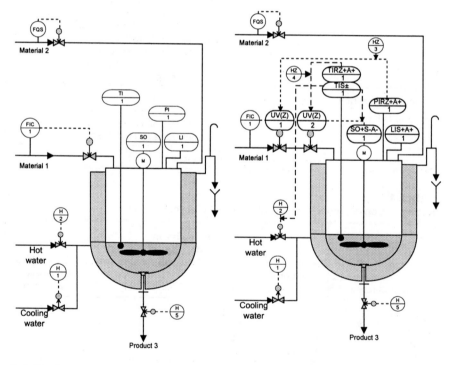

FIGURE 7. Example of the PSV substitution process (see [22])

DESIGN OF PROTECTIVE MEASURES BY MEANS OF INSTRUMENTATION

The safety relevant process values pressure P1 and temperature T1 are (in this example) for all intent and purposes equivalent. This allows the realization of a diverse and redundant safety instrument system. The final control elements are installed in series in the piping for material 1 since a 1oo2 (1 out of 2) system satisfies the requirements. If one of the two final control elements is also used for flow control then it needs to be continuously monitored by operations. Common fault failures of the protection systems (e.g., loss or decrease of energy supply) are monitored and will trip the protection system (if they occur). The actuators are of a fail-safe design, in case of a loss of auxiliary energy a spring action will move the valves into a safe position.

HAZARDOUS EVENT SEVERITY MATRIX ASSESSMENT

A matrix assessment shows for:
No. of IPLs = 1;
event severity = extensive (possibility of a fatality), and
event likelihood = low;
SIL = 3. This supports the analysis done where we found in a (more) rigorous manner that without PSVs the before-mentioned E/E/PES SRS needs to have a SIL = 3 or a Namur = II classification. It also provides a *caution* that in case of a single IPL additional risk reduction may be required, or that an additional PHA review is required (see [27]). *If* there had been a PSV the SIL would have been bumped down to a value of 1.

Organizational Measures or Management Systems

Despite following all these design criteria, failures of the protection system cannot always be completely excluded. Technical protective measures are therefore complemented with management systems (organizational measures) which are documented in an operating manual. Continuous monitoring, regular function testing and immediate repairs can only eliminate covert faults. In the example given, temperature and pressure have a certain relationship which is characteristic for the specified operation of the plant. Operators have to check this relationship in terms of plausibility. Pressure measurement and the tightness of the valve seals are checked during shutdowns by pressurizing the vessel. A temperature trip method is devised, e.g., by connecting a test resistor to the thermometer head. Actuators must close during tests. The test procedure needs to be performed at agreed to fixed intervals Test methods, their execution and the test results are to be documented in a test procedure manual.

Companies operating in North America should develop their management systems based on [27] and [12]. "Guidelines for safe automation of chemical processes" (see [27]) contains guidelines for Reliability Block Diagrams (RBDs), validation, verification, auditing, documentation, training, etc., which are geared towards a North American environment.

The Role of (mechanical) Pressure Safety Devices in Combination with MC Equipment

The use of MC safety equipment in combination with pressure safety devices is primarily aimed at reducing the probability of activating these mechanical devices and or treating blowdown material. PSVs and RDs will continue to play a role since they are simple and essentially independent mechanical devices that are not subject to the common faults we might encounter in SISs. The role of PSVs could therefore change from focusing on continuous production to a mitigation role by preventing vessel ruptures (see [30]) as part of a multi-layered safety strategy.

The safety levels include:
preventing high pressures by pressure safety valves, rupture disks and/or MC systems;
1 release prevention measures by MC systems; and
mitigation measures by pressure safety valves and MC systems. *Note:* the P_{set} of a PSV will in that case be determined by $P_{rupture}$ or $P_{deformation}$ and not the MAWP, it can therefore be set at a much higher pressure; potentially P_{set} could be as high as $0.75 \times P_{burst}$ *(for carbon steel).*

Conclusions/Recommendations:

Contact your authority having jurisdiction in matters pertaining to pressure safety devices before putting a great deal of effort into the substitution of PSVs by systems (they might not like it).

MC systems should only be considered when other inherently safer measures and/or independent systems, such as pressure safety valves, are either not applicable or insufficient or not economically viable for risk reduction.

- Adopt a structured and functional (risk-based) process for substituting PSVs and RDs with systems, including documentation.
- The decision to replace pressure safety devices with systems should be made by a multidisciplined team whose members have the *knowledge* and *experience* to appreciate the risk as well as technical and organizational ramifications.
- When adopting standards, especially from different legal jurisdictions, watch out for implicitly implied requirements, e.g., alarm management, and inconsistencies between different standards.
- Verify SIL levels obtained via [29] with the safety interlock integrity level qualitative SIL approach given in [27] and [9] as a matter of ensuring accuracy.

- Where functional, permissible (by the authorities having jurisdiction) and only economic factors need to be considered; pressure relief valves or rupture disks should be used to avoid vessel rupture (or deformation) and provide an extra IPL.

ACKNOWLEDGMENTS

The author likes to thank the members of the DIERS "Risk" Sub-committee and the DIERS Ad-Hoc Committee "to reduce the number and size of relief valves" who participated in many, long and stimulating discussions concerning these issues. Special thanks go out to Donald E. Meadows (Bayer Corporation), John Noronha (Eastman Kodak Company) and Ian Nimmo (Honeywell) for discussing these issues with me; and to Della Wong and Brian Jones (NOVA Chemicals) who reviewed the original draft. The views are the author's and do not necessarily represent those of the subcommittees mentioned or their members.

References

1. Letter of the American Petroleum Institute (API) to the ASME Boiler and Pressure Vessel Committee, May 10, 1995.
2. "Emergency Relief System Design Using DIERS Terminology", Fisher, H.C., Forrest, H.S., Grossel, S.S., Huff, J.E., Muller, A.R., Noronha, I. A., Shaw, D.A, Tilley, B.J., AIChE/DIERS Project Manual, TP155.5E63 (1992).
3. ISA-S84.01-1996, "Application of Safety Instrumented Systems for the Process Industries", Instrument Society of America S84.01 Standard, Research Triangle Park, NC, 27709.
4. ISA-dTR84.02, Electrical (E)/Electronics (E)/Programmable Electronic Systems (PES)—Safety Integrity levels (SIL) Evaluation techniques, Part1, Version 1, March 1997.
5. Draft IEC 1508-1: Functional safety—Safety-related systems—Part 1: "General requirements"/DIN IEC 65A/180/CDV (VDE 0801 Teil 1): 1995-12 Titel: "Funktionale Sicherheit".
6. Draft IEC 1508-2: Functional safety—Safety-related systems—Part 2: "Requirements for E/E/PES"/ DIN IEC 65A/180/CDV (VDE 0801 Teil 2): 1995-12 Titel: "Anforderungen an elektrische/elektronische/programmierbar elektronische Systeme (E/E/PES)".
7. Draft IEC 1508-3: Functional safety—Safety-related systems—Part 3: "Software requirements"/DIN IEC 65A/181/CDV (VDE 0801 Teil 3): 1996-01 Titel. "Anforderungen an Software".
8. Draft IEC 1508-4: Functional safety—Safety-related systems—Part 4: "Definitions"/DIN IEC 65A/182/CDV (VDE 0801 Teil 4): 1995-09 Titel: "Definitionen".

9. Draft IEC 1508-5: Functional safety—Safety-related systems—Part 5: "Guidance on the application of Part 1"/DIN IEC 65A/183/CDV (VDE 0801 Teil 5): 1996-05 Titel: "Anwendungsrichtlinie für Teil 1".
10. Draft IEC 1508-6: Functional safety—Safety-related systems—Part 6: "Guidance on the application of Part 2 and 3"/DIN IEC 65A/185/CDV (VDE 0801 Teil 6): 1996-07 Titel: "Anwendungsrichtlinie für die Teile 2 und 3"
11. Draft IEC 1508-7: Functional safety—Safety-related systems—Part 7: "Bibliography of techniques and measures"/DIN IEC 65A/185/CDV (VDE 0801 Teil 7): 1996-09 Titel: "Anwendungshinweise über Verfahren und Maßnahmen".
12. CFR 1910.119-1992 Process Safety Management of Highly Hazardous Chemicals, Explosives, and Blasting Agents, (Final Rule; February 24, 1992).
13. Guidelines for Chemical Process Quantitative Risk Analysis, 1989, published by the Center for Chemical Process Safety (CCPS) of the American Institute of Chemical Engineers (AIChE), 345 East 47 Street, New York, NY 10017.
14. "Risk Considerations for Runaway Reactions," dated April 4, 1994 by the Risk Committee of the Design Institute for Emergency Relief Systems Users Group.
15. NFPA 68 "Venting of Deflagrations".
16. NFPA 69 "Explosion Prevention Systems".
17. VDI-Richtlinie 3673, "Druckentlastung von Staubexplosionen—Pressure Release of Dust Explosions" (bi-lingual), Blatt 1: Beuth-Verlag GmbH, Burggrafenstrasse 6, D-1000 Berlin 30; 1992.
18. AD-Merkblatt A 1, Safety devices against excess pressure—Bursting safety devices, January 1995.
19. AD-Merkblatt A 2, Safety devices against excess pressure—Safety valves, November 1993.
20. AD-Merkblatt A 6, Safety devices against excess pressure—MCR-safety devices, June 1986.
21. Technische Regeln Druckbehälter (TRB) 403, "Ausrüstung der Druckbehälter Einrichtungen zum Erkennen und Begrenzen von Druck und Temperatur", February 1989.
22. NE 31—NAMUR Empfehlung: "Anlagensicherung mit Mitteln der Prozeßleittechnik", version 1.1.1993, published via Bayer A.G; and NE 31—NAMUR recommendation: "Safety of Process Plants Using Process Control Systems", NOVA Chemicals translation.
23. DIN V VDE 0801/01.90, "Grundsätze für Rechner mit Sicherheitsaufgaben-Änderung A1, 1994-10.
24. DIN V 19251 "MSR-Schutzeinrichtungen", 1995; and DIN V 19251 "Requirements and measures for protection functions", NOVA's English translation, 1997, NOVA Chemicals Strategic Initiatives.
25. DIN/VDE 2180, Sicherung von Anlagen der Verfahrenstechnik mit Mitteln der Meß-, Steuerungs- und Regelungstechnik, Part 1 (draft), Part 2, Part 3, Part 4 and Part 5.

26. "Guidelines for Hazard Evaluation Procedures", 1985, published by the Center for Chemical Process Safety (CCPS) of the American Institute of Chemical Engineers (AIChE), 345 East 47 Street, New York, NY 10017.
27. "Guidelines for safe automation of chemical processes", 1993, published by the Center for Chemical Process Safety (CCPS) of the American Institute of Chemical Engineers (AIChE), 345 East 47 Street, New York, NY 10017.
28. M.R. McPhail and J. Windhorst, "Integration of risk management into process safety standards and application of those standards", International Process Safety Management Conference and Workshop, Center for Chemical Process Safety of the American Institute of Chemical Engineers, San Francisco, September 22-24, 1993, pp. 101-115.
29. DIN V 19250, "Grundlegende Sicherheitsbetrachtungen für MSR-Schutzeinrichtungen"; and the English Manuscript, "Basic Safety Evaluation of Measuring and Control Protective Equipment".
30. "Anforderungen an technische Einrichtungen an Reaktionsbehältern zur Vermeidung von Emissionen gefährlicher Stoffe über Druckentlastungseinrichtungen" or (English: "Prevention of emissions of dangerous substances via pressure relief systems"), Dr. Jörg Loock of the Dept. of Safety and Environmental Protection (TÜV-Bavaria). This study which was initiated by the Bavarian government as a result of the Hoechst incident in Griesheim (December, 1994).
31. J. Windhorst, "Technical Requirements For Reactors to Prevent Hazardous Emissions via Pressure Relief Devices", DIERS Users' Group, September 1996 Calgary.

Handwritten notes:

This could be used for flare debottlenecking projects, where
a) you cannot exceed 1.5 or 2x MAWP
b) you will allow deformation
c) highly reliable safety instrumented system

Actual damage to vessels appear at 6x MAWP
Deflagration scenarios better handled via rupture disks
PSV in tough scenarios is better/probably only good for fire case (ie simple).

This methodology has been used in Germany, Texas.

An important related question is the additional hazard of asphyxiation by N_2 if blanketing is implemented. If this hazard exceeds the reduction in flammability risk then it is clearly better not to implement N_2 inerting. Operating experience at Eastman Chemical suggests that asphyxiation can be a potential problem in some enclosed operating areas, and a better solution may be a properly instrumented process without inerting, if there is data to support such a conclusion.

There are two ways to approach the relative hazard issues:

(i) Calculate the risk of explosion from a noninerted vessel and then consider the reduction in risk due to inerting. At the same time, the additional risk of asphyxiation to operators and mechanics must be added to give the total risk associated with the change. In this way the benefits of inerting can be compared with the noninerted case.

(ii) Look only at a noninerted system with various types of instrumentation and control, developing experimental data until an acceptable level is found. These controls would be aimed at reducing the explosion risk other than by inerting. The meaning of "acceptable" would also have to be defined objectively.

The aim of this paper is to present a logical review of flammability issues with regard to batch process vessels. This will be done by consideration of a specific process operated by Eastman Chemical and will include consideration of flammability principles, operating procedures and reactor controls.

The fatality risk from an explosion within a vessel will be quantified to give an "absolute" measure of the hazard. This level will be compared with "acceptability" criteria typical for the chemical industry in order to objectively decide on an acceptable process design.

Process Description

The process selected for evaluation is a typical esterification/hydrolysis type using cellulose (pulp) as the starting material, and involves three consecutive operating steps in three vessels:

- Mixer vessel
- Primary aging vessel
- Secondary aging vessel

Due to lack of space only the mixer vessel will be addressed. In the vessel, approximately 1000 pounds of cellulose and 500 gallons of acetic acid are charged into a horizontal vessel fitted with two mixers consisting of ribbon blades, which approach the walls of the vessel, and each other, to within a fraction of an inch. A single temperature probe is installed in the center of the vessel. To the pulp, acetic acid and other liquids including acetic anhydride and water are charged as the temperature is controlled in the range of 90 to 160°F (32 to 71°C) by alternate uses of hot and cold tempered water. The maximum available tempered water temperature is 180°F (82°C).

During a typical batch cycle, the mixer vessel undergoes the following concentration variations in the ullage:

- An initial charge of pulp is added to an empty reactor containing flammable vapors. The pulp is full of air gaps and nearly fills the reactor vessel, leaving only a 6-inch space at the top. This is the maximum extent of the flammable volume.
- As acetic acid and acetic anhydride are added, liquid displaces the air in the pulp; the flammable space is still limited to the volume above the pulp/liquid surface and is, therefore, quite small.
- The liquid concentration changes during the batch period (~1.5 hours) as does the liquid temperature. The most significant period is between the first and second catalyst additions, because the temperature is largely below the value needed to give a flammable mixture. Toward the end of the batch the temperature is around 140°F (~60°C). However, during this entire period the vapor space is limited to less than 6 inches above the pulp/liquid surface. The operator would normally take a sample of the product at this point in the batch cycle.
- At the end, the final product is discharged from the bottom of the vessel, an acetic acid/water wash is added to the vessel, and the temperature gradually raised to around 160°F (71°C). As the product is dumped out, its volume will be replaced by air being drawn in through the manhole. At the end of this operation, a fairly concentrated mixture of acetic acid/air (plus some water) may be expected in the entire vessel space. The period in which this will occur is about 30 minutes normally, but could be up to 2 hours if there is a process upset.

Hot tempered water used to control reactor temperature will stay in the vessel jacket at 160°F for quite some time, and it is not unreasonable to assume that the temperature in the reactor will remain significantly above the liquid flash point until the next batch is started.

Throughout a batch and thereafter, an induced draft fan ensures that fresh air is drawn into the vessel and out through a scrubber. This air movement prevents a build-up of a flammable concentration, particularly at the top of the vessel close to the manhole area.

One possible (and very significant) departure from normal operation is either forgetting to add the pulp, or choking the feed duct. Should the feed duct become choked, the operator would not know this until the end of the batch when a sample is normally taken. Acetic acid and other liquids however would continue to be added as normal, resulting in a flammable vapor space much larger than normal.

After approximately 1.5 hours the product is transferred by gravity to the primary aging vessel. A sample is taken and analyzed before the product transfer occurs.

Flammability Aspects

Flammability Data

The main flammability information available for the chemical mixtures involved in the selected process is summarized in Table 1, based on acetic acid as the predominant liquid.

In view of the large variations in temperature through a typical batch cycle, it is important to know where the concentration lies in relation to stoichiometric concentration on the one hand and the flammable limit on the other. An estimate based on acetic acid vapor pressure data is presented in Figure 1. This is a plot of liquid temperature against a dimensionless parameter that varies from 0 to 1 as the concentration in the vapor phase varies from stoichiometric to the lower flammable limits. The definitive assumptions made with regard to the operating periods that will lead to a flammability risk are summarized in Table 2.

Explosion Pressure

In terms of severity, the explosion pressure is related directly to the fuel-air concentration. The maximum occurs at approximately stoichiometric condition, and this falls gradually as the vapor space concentration moves outward toward the two flammable limits. Applying theoretical results based on typical data from hydrocarbons, the initial/maximum explosion pressure ratio is as high as 4 at the

TABLE 1
Summary of Relevant Flammability Data

Flammability Variable	
Lower Flammable Limit (Vol. %) in air	4 to 5.4%
Temperature at Lower Limit (100% acid)	97°F (36°)
Temperature at Upper Limit (100% acid) (estimate)	176°F (70°C)
Minimum Ignition Energy	2 to 3 mJ
Minimum Energy with 20% water (estimate ONLY)	10 mJ
Flash Point	103°F (39.4°C)
Temperature to form stoichiometric mixture in air (approximate)	140°F (60°C)

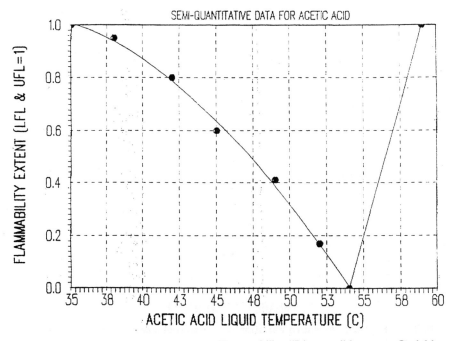

FIGURE 1 Liquid Temperature versus Flammability "Distance" between Stoichiometric Concentration and UEL/LEL.

TABLE 2
Flammable Concentration Assumptions used in Study

(a) **Near-liquid Full Vessels (Mixer and Secondary Aging Vessel)**
• Nonflammable mixture in small space above liquid while fan is in operation.
• Potentially flammable if scrubber system fails, but the consequences of an explosion are of little concern, due to the small explosion volume.
Hence, under normal operation, the mixer vessel presents little explosion risk.
(b) **Partially Full or Empty Vessels**
Significant explosion risk exists between batches (i.e. vessel empty), due to the large volume of gas. The temperature at the end of a batch is high enough to produce an explosive mixture in the vessel vapor space.

flammability limits, and increases to 8 to 9 near stoichiometric mixtures. The change in explosion pressure ratio as you move between the flammability limits is shown in Figure 2, and should be a conservative estimation of acetic acid/air and similar mixtures. The x-axis goes from 0 (stoichiometric) to 1 at the flammable limits; i.e., it is the fractional distance between the worst condition and the flammable limit. Also plotted in Figure 2 is the reduction in pressure ratio as the

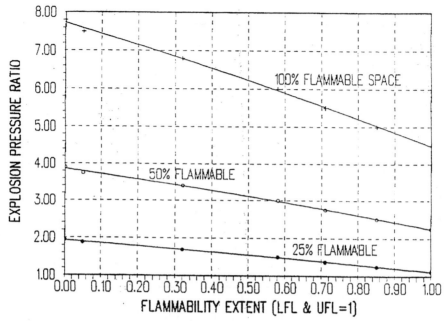

FIGURE 2. Explosion Pressure Variation with Flammability Extent and Percent of Flammable Volume

flammable volume fraction in a container is reduced. Thus, the line marked 50% indicates the result if half the vessel contains the flammable mixture and the remainder is nonflammable gas.

The severity of an explosion is expressed in terms of both the maximum pressure and the rate of pressure rise. The latter is directly proportional to the burning velocity of a fuel-air mixture, and this is a fundamental property for a given fuel and varies with concentration. The expected change in burning velocity between stoichiometric concentration and flammable limits is shown in Figure 3. This is based on published data for methane and propane in air.

One practical way of using the information in Figure 3 is to evaluate the explosion pressure rate of rise parameter KG used to size vents. (KG is equivalent to Kst parameter for dust, which is commonly used for vent sizing). The information in Figure 3 could be used to plot the change in KG for a fuel as a function of concentration. This could then be used to decide if the vent on a vessel (in this case the unrestrained manway) could cope with an explosion of any specified composition.

The key assumptions used in this study with regard to the explosion severity are summarized in Table 3.

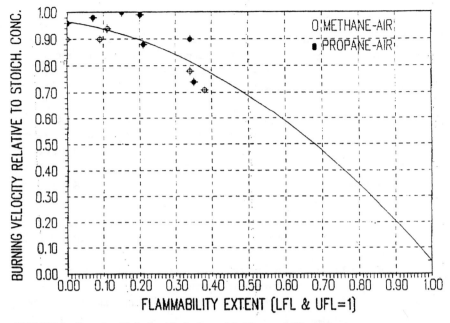

FIGURE 3. Burning Velocity Variation with Flammability Extent

TABLE 3
Assumptions in Relation to Explosion Severity

The following position is adopted with regard to the severity of an explosion, if the vapors are somehow ignited:

- *Near-full vessel*—no risk of vessel damage.

- *Part-full or empty vessel*—considerable risk of damage regardless of vapor composition within flammable limits. However, damage to vessel only occurs if the vent (24" manway cover and the vent line to the scrubber) is incapable of relieving the pressure. In situations where the vessel agitator is still running, there is assumed to be a 50% chance that the explosion will be successfully relieved. In situations where the vessel space is not agitated, there is only a 25% chance that venting is unsuccessful. (These percentage figures are not arbitrary, but relate directly to the flammability of the gas mixture that explodes. Thus, if the concentration range from LFL to UFL is 100%, then half these concentrations (typically 25% to 75%) will be explosive enough to exceed the venting capacity of the gas if stirred. If the vessel is not stirred, the dangerously explosive range is narrower. The reason for this assumption is that agitation increases the burning rate.)

Ignition Energy

The minimum energy required to ignite a flammable mixture varies sharply with concentration. The ratio of the minimum energy at worst conditions to that at other concentrations as you move toward the flammable limits is plotted in Figure 4 for a number of hydrocarbons. This shows that the increase in ignition energy correlates relatively well with the fractional distance between the "worst point" and the LFL or UFL. (The "worst" point in this case is the concentration that gives the lowest ignition energy or the MIE.) Typically, a tenfold increase in ignition energy can be expected as you move toward the flammability limits.

Although ignition can, in general, occur in many different ways, it is very limited in the context of the process being considered. The following sources are identified as requiring consideration:

- Static electricity discharges from operators approaching a vessel.
- Static electricity discharges while transferring liquids in or out of vessels.
- Mechanical heat at agitator seals.
- Sparks due to tramp metal objects caught up in agitator blades.
- Welding and similar maintenance operations.
- In the case of the mixer vessel, the specific potential risks of smoldering pulp being added.

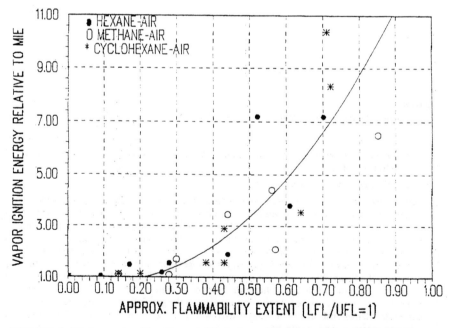

FIGURE 4. Minimum Ignition Energy Variation with Approximate Flammability Extent

Mechanical sources of ignition in the mixer vessel are relevant due to the design of the impeller blades and the fact that a packing gland seal is used.

The main static ignition source of relevance is charge on operators as they approach the reactors.

A summary of the key assumptions with regard to ignition risk is given in Table 4.

Risk Quantification

Methodology

The frequency of a major explosion with the mixer vessel will be affected by the likelihood of achieving the following conditions at the same time:

- Generation of a flammable mixture capable of rapid combustion.
- The reactor vessel is not full, so that the vapor space is sufficiently large to cause a major explosion.
- The presence of an ignition source of sufficient energy to ignite the flammable vapor.
- An operator being present when this combination of events takes place.

A fault tree analysis will be used to express the combination of conditions needed to satisfy the above requirements. A fault tree is a graphical presentation of the logical relations between a selected accident and the primary causes of the accident. In the present study, the accident (so-called TOP EVENT) is a damag-

TABLE 4
Assumptions Used in Relation to Ignition Likelihood

In terms of ignition, the following approach is adopted:
• Ignition by static discharge (from operator only) is limited to vapors with ignition energy less than 25–30 mJ; this has been used to justify the fact that vapor composition needs to be at least half-way between stoichiometric and the flammable limit.
• In vessels where the scrubber fan is in operation, but otherwise no mixing of the vapor space, the gas layer at the top of the vessel (close to the cover) will be outside, or close to the flammable limit and, therefore, not capable of ignition by static.
• If the scrubber fan is not in operation and the vessel temperature is high enough to give a composition significantly in the flammable range, ignition by static discharge will be possible after a delay of about 30 minutes to allow for the liquid and vapor spaces to equilibrate.
• If the vessel temperature is sufficiently high to generate a flammable vapor significantly inside the flammable range, then static discharge can cause ignition, provided there is some gas mixing (e.g., agitator running), even if the scrubber fan is in operation.
• If the vessel temperature is sufficient to give a mixture anywhere in the flammable range, then all mechanical ignition sources will be capable of initiating the event.

ing explosion in one of the vessels, and the primary causes are failures of equipment such as the scrubber fan, reactor temperature control, overheating of bearing seals, etc. The various causes (or events) are related by logical "OR" or "AND" gates; thus, if two events must occur simultaneously to satisfy a particular condition, then they will be connected by an "AND" gate, while if either can satisfy the conditions, then an "OR" gate will express the relationship.

Failure rate data is then obtained (or assumed) for each low-level event that eventually makes up the "TOP EVENT," and this data is combined mathematically to arrive at the frequency of the top event. This frequency can then be compared with some suitable benchmark number.

Development of Fault Tree

The combination of plant failures needed to generate a destructive explosion in the process vessels being considered are summarized in Tables 2 to 4. This breaks up the batch cycle according to the liquid charge in the vessel and then presents the set of conditions needed to cause ignition by static electricity, by mechanical friction, and due to certain maintenance operations. These conditions are based on the need to produce not just a flammable mixture, but also one that can be ignited by the appropriate ignition source. Thus, to ignite by means of static charge, the flammable concentration must be closer to stoichiometric, but a mechanical source will probably ignite any gas mixture. The general logic tree that applies is shown in Figure 5. The top event (violent explosion in mixer) can occur (in principle) when the vessel is almost full, when it is partly full, or when it is empty.

For the trivial case of almost full, the causes are developed simply in terms of:

- Flammable concentration present.
- Ignition source.
- Sufficiently large flammable volume to cause vessel damage.

These items are not extended any further, since it is assumed that the small vapor volume renders this part of the fault tree irrelevant.

Explosions while reactor are partially, or totally empty are divided into the causes of ignition: static, mechanical and "other". Static ignition can cause a problem (for completely empty or partially empty vessel) when the following occur simultaneously:

- Scrubber fan is not operating, or the vessel agitator is left running in an empty vessel.
- The temperature is in the range that gives concentrations close to stoichiometric.
- An operator (the source of static charge) opens up the manhole while significantly electrostatically charged.

— there are 1800 other vessels in 'similar' service

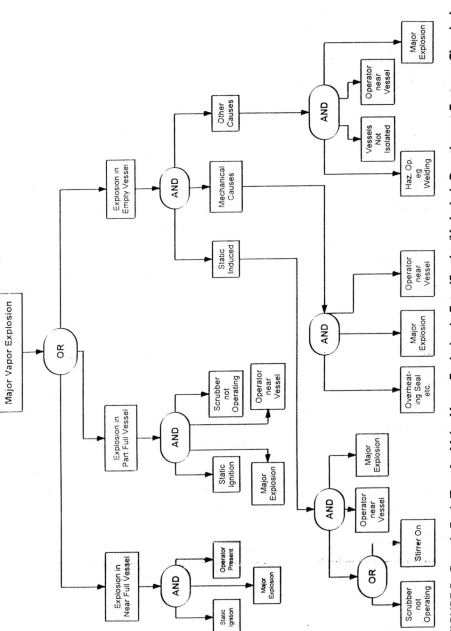

FIGURE 5. Generic Fault Tree for Major Vapor Explosion in Esterification/Hydrolysis Reactions at Eastman Chemical

Explosion due to mechanical ignition can occur only if the vessel is completely empty. In order to get this problem, the following has to occur:
- Agitator is left rotating while vessel is empty and,
- Either tramp metal from pulp or postmaintenance operations is left in the vessel, or overheating of the mechanical seal takes place.

For the sake of completeness, ignition by "other" sources is included to allow for things such as maintenance operations. These cases are only applied to empty vessels, as these operations would not be carried out when there is liquid in the vessels. Another known ignition contributor, which is identified separately, is cellulose, which is smoldering prior to addition to the mixer. This condition can be created in the cellulose shredding operation prior to transfer into the mixer vessel.

Results of Fault Tree Calculations

The frequency of a serious (fatal) incident due to an explosion, based on the fault tree described above, is 1.5×10^{-4} per year, using readily available generic data.

It is also interesting to review the contributors to risk from the events that lead to ignition, and this is done in Table 5.

In terms of the basic events behind the ignition causes, the following are the most significant:

- Agitator left running while reactor is empty.
- Reliability of the scrubber fan.
- Operator interaction with vessels, particularly when the vessels are empty.

Risk Reference Points

The calculated fatality frequency values can be put into context by comparing with the value typical for the chemical industry, and by reference to more common nonindustrial activities. The fatality risk to an operator working in the chemical industry in the USA is about 5×10^{-4} per year. This includes all forms of risk, not just from fire and explosion. The criteria normally used for assessment of plant is one tenth of this value, i.e., about 5×10^{-5} per year. Thus the fatality risk posed by an explosion in the mixer vessel is a factor of around 3 higher than

TABLE 5
Contributors to Risk in Terms of Ignition

Vessel	Ignition Contributors to Risk			
	Mechanical	Static	Other	Total
Mixer	60.8%	33.7%	5.5%	100%

TABLE 6
Relative Fatality Risk for Various Activities

Activity	Relative Risk ($\times 10^{-5}$)
Chemical Industry	5
Coal Mining	40
Home Life	3
Motorcycle Riding	660

$Mixer = 1.5 \times 10^{-4}$

this target. This risk is compared with other activities in Table 6, and shows that in absolute terms the chemical industry is relatively safe.

Improvement in Risk

Available Options

The preliminary risk evaluation points to the need for further work in two primary areas:

IMPROVE THE EXISTING EVALUATION
This will require:

- Evaluate the ignition energy change as a function of concentration (or temperature) to confirm the assumed relationship obtained from literature.
- Obtain explosion pressure-rise data (so-called KG values) for acetic acid vapors in air, at different concentrations, and check the adequacy of the existing vents on the reactors in coping with these explosions.
- Confirm the frequency of certain operator actions by a more thorough discussion and compare the result against the assumption in the present study.
- Check by calculation and/or on-line sampling, the variation in concentration and temperature in the process vessels, both as a function of time and vessel freeboard height. This should be done with, and without, the scrubber fan in operation.

IMPROVE CERTAIN OPERATIONS
The study points to the need for stricter control of the following:

- Agitator operation when the vessels are empty: if possible, the agitator should be automatically switched off through the control system.

- Close control of maintenance operations, with use of effective permits-to-work systems.
- Training and education of operators to make them more aware of the flammable hazards while vessels are empty, specifically with a view to eliminating the operator opening the man-hole covers and making them touch near-by metal rails, or other grounded objects, before they open the covers.

Recalculation Based on Improvements

The fatality risk was re-calculated assuming that the three sets of controls identified as weaknesses are improved. The following changes in risk are obtained:

(a) If the agitator is switched off as soon as the vessel is emptied, so that the chances of it being left running are reduced by a factor of 10, the annual fatality risk is reduced from 1.5×10^{-4} to 3.9×10^{-5}.

This is a very important control because it influences three key risk factors:
—uniform mixing of the flammable vapor space
—a potential mechanical ignition source
—increases the explosion severity.

(b) Improvement in permits-to-work procedures to reduce the possibility of incidents during maintenance operations that involve sparks. Reduction in the probability by a factor of 10 only reduces the total annual risk from 1.5×10^{-4} to 1.42×10^{-4}. Clearly this is not a very influential change because it affects only the ignition frequency when the vessel is empty.

(c) Reduction in operator approach of vessel while it is empty: if this frequency is reduced tenfold, the annual risk is reduced to 1.04×10^{-4}. This has a moderate impact because it affects ignitions when the vessel is both full and partially empty.

If all of the above changes are incorporated, the annual fatality risk is reduced to 1.24×10^{-5} which is below the risk acceptance threshold used for this study. However, other operator risk exposures must be considered before a final determination can be made.

Conclusions

The use of QRA for the assessment of explosion risk in a process vessel containing flammable liquids can produce very valuable results. The logical and combined use of flammability, process operating and process design information can lead to a fairly definitive statement about the explosion hazard. Additionally, it can show that with reasonable controls in place, the risk to an operator can be

Safe Handling of Flammable Liquids In Process Vessels: A QRA Approach 665

kept to an acceptable level without the need for a retrofitted nitrogen inerting system.

The flammability data used as the basis for the risk calculations needs to be carefully considered as this has an important bearing on the results. The methods used in the present study will be checked against new, experimentally generated data for specific mixtures of interest at the relevant conditions.

Multiple Safeguarding Selection Criteria or How Much Safety Is Enough?

George Mehlam [handwritten]

R. P. Stickles, S. Mohindra, and P. J. Bartholomew
Arthur D. Little, Inc.

When the consequences are catastrophic, designing for double and triple jeopardy is generally necessary. Certain codes and standards (e.g., ISA, API, etc.), which are essentially a codification of risk management criteria, provide guidance to a point. Ultimately, the application of failure analysis coupled with critical event tolerability criteria may be the best means of reaching a decision. This paper will explore existing standards for explicit and implicit examples of criteria for multilevel safeguards. It will conclude with a case study involving fault tree analysis, and the use of critical event criteria.

Much has been written about the intolerance of multiple fatality events (1,3,6), even though quite often the probability of such events (e.g., tunnel fires, bridge collapse, nuclear reactor containment failure) is very low. However, it has been demonstrated in many situations (Canvey Island, Bhopal, Pan Am 104, etc.) that the general public has a low tolerance to multiple-fatality accidents. This nonlinear shape of our "value-function" (2) is exemplified by the public's preference for ten single-fatality accidents over one accident with ten fatalities.

This effect is not new and in the past has influenced the development of certain engineering and design standards. The following are examples of codes/standards that have been developed to avoid catastrophes the can result in multiple injuries/fatalities.

TABLE I
Codes/Standards to Avoid Catastrophes — *standards addresses consequences only* [handwritten]

Code	Description	Avoided Incident
ASME Section VIII	Unfired Pressure Vessels	Pressure vessel mechanical failure
ANSI B31.3	Chemical/Refiner Piping	Piping mechanical failure
NFPA 68/69	Venting/ Containment of Deflagrations	Equipment failure due to internal explosion
API 650	Atmospheric Storage Tanks	Tank mechanic failure
API 752	Facility Siting	Plant building fatalities
API 2510	LPG Installations	Vapor clouds & BLEVEs

In this context, codes/standards can be viewed as the codification of mitigation measures (engineered controls) for avoidance of high consequence risks.

Until recently, most of the risk-avoidance codes/standards have been prescriptive in nature. That is, they prescribe in exacting detail how to design a system or piece of equipment. Now, performance-based design guidelines and standards are beginning to appear that include risk management decision-making concepts. Examples of performance-based design guidelines and practices include:

- ISA-S84.01-1996 Design of Process Safety Instrumentation
- CCPS Guidelines for Design Solutions for Process Equipment Failures
- API-RP752 Management of Hazards Associated with Location of Process Plant Buildings

These practices provide designers and managers some degree of flexibility to trade-off risk-reduction benefit, design complexity, and cost. The presumption being, that overall safety will not be compromised. To achieve this goal, the application of these practices requires a risk tolerability benchmark against which to the judge risk level achieved by a given design.

Demming said about manufacturing processes, "What you don't measure, you can't manage." The equivalent for the risk management process would be, "Without risk tolerance criteria, you can't make rational decisions." While it is not the intent of this discussion to address the setting of suitable risk tolerability criteria, it should be obvious, that organizations intending to adopt performance-based design practices need to tackle this issue. Without tolerability criteria, it will be impossible to obtain consistent decisions regarding safety design. Furthermore, once such criteria have been established, it allows the proper use of other quantitative tools such as fault tree analysis, reliability analysis and QRA to be applied in risk management, as will be demonstrated below.

Designers familiar with the traditional prescriptive codes know that they focus heavily on the risk of mechanical and electrical causes of initiating events, and less on process control and human error induced causes. However, in today's operating environment of highly automated processes, and fewer operators (more demands per individual), the risk of incidents caused by process control failure or human error must also be managed. The application of fault tree analysis can be effective in establishing the relative frequency of potential incidents associated with base-case and alternative design concepts. The technique has the versatility to handle equipment and control failures along with human errors. A good example of the application of fault tree and reliability analysis for evaluation of safety interlock systems has been reported by R. Freeman (4). Different integrity levels for an interlock can be established such as:

Class 1   Fully redundant
Class 2   Redundant final element
Class 3   No Redundancy

TABLE 2
Unreliability of Level Interlock Systems with Consideration of Common Cause Failures (Freeman, 1994)

Mission Time	Mission Time (Hours)	Unreliability Class 3	Unreliability Class 2	Unreliability Class 1
1 shift	8	0.010%	0.007%	0.005%
1 day	24	0.029%	0.020%	0.016%
1 week	168	0.200%	0.140%	0.110%
1 month	720	0.870%	0.610%	0.490%
1 quarter	2160	2.610%	1.840%	1.490%
6 months	4320	5.220%	3.690%	3.030%
1 year	8760	10.580%	7.540%	6.390%
18 months	12,960	15.660%	11.220%	9.780%
2 years	17,520	21.160%	15.270%	13.720%

The article demonstrates the level of analysis that can be applied, and as Table 2 (reproduced) illustrates, provides the decision-maker with a good picture of the reliability trade-offs, for a given mission requirement.

This methodology also provides a means of setting reliability tolerance criteria for different classes of interlock integrity level (e.g., Class 1—fully redundant). For example, Table 3 presents the interlock reliability (1—unavailability) for the three level interlock classes as a function of proof testing interval.

TABLE 3
Interlock Reliability

Integrity Level	Redundancy	Test Interval	Reliability, %
Class 1	Fully	Monthly	99.5
		Quarterly	98.5
		Annually	93.6
Class 2	Final Element	Monthly	99.3
		Quarterly	97.8
		Annually	90.9
Class 3	None	Monthly	99.1
		Quarterly	97.4
		Annually	89.4

The data accounts for common mode failures. As seen, there is a trade-off between testing frequency, and the advantage gained by selecting the next integrity Class. With monthly proof testing, the gain in reliability between Class 1 and Class 3 is only, 0.4%. Therefore, cost–benefit considerations would suggest quarterly or yearly proof testing. By performing similar analyses for flow, temperature and, pressure interlocks for a specified test interval, one could set generic reliability criteria for each interlock integrity level, and allow the designer latitude for achieving the required reliability.

Fault tree analysis can also be applied to situations involving a combination of human error and DCS failures is shown in Figure 1. The process involves batch chemistry in which one of the reactants reacts violently with water. The concern is feeding an aqueous phase into a reactor filled with chemical "X." One of the scenarios investigated was caused by a DCS failure. The fault tree depicts the failure sequences for the an uncontrolled process venting due to such a scenario. The frequency of the top event is less then 1×10^{-5}/yr., which complied with the established critical-event tolerability criterion.

The example presented involved the use of a critical-event tolerability criterion. For process facilities, our experience suggests that values for critical-event criteria fall in the range of 10^{-4} to 10^{-5}/yr. That is, the frequency of all fault tree top events with potential for fatality or off-site health impact, should be less than the specified criterion. Some companies have a maximum individual risk (individual at maximum risk) criterion of 10^{-5}/yr., and use the same value for critical-event. The rationale being that, if the event frequencies are 10^{-5}/yr., the individual risk criterion for all events should be achieved. One of the worked problems in API RP 752 uses the following individual risk criteria:

$>1.0 \times 10^{-3}$ Risk mitigation or further risk assessment is required.
1.0×10^{-3} to 1.0×10^{-5} Risk reduction should be considered.
$<1.0 \times 10^{-5}$ Further risk or assessment reduction need not be considered.

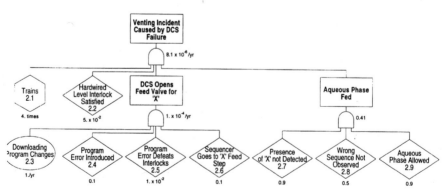

FIGURE I.

While these criteria are presented in API RP 752 as an example, the magnitude of the values in consistent with other references (6).

When catastrophic events at process facilities (1) have the potential to impact the general public, (2) require regulatory agency involvement, or (3) senior management needs reassurance regarding its risk exposure, a fully quantified risk assessment is appropriate. The QRA can be thought of as a tool for analyzing tertiary levels of safeguarding, since it should account for other primary and secondary safeguards design into the facility equipment and management systems. Again the issue of suitable tolerance criteria is presented. Since QRAs often address the risk to society in addition to individual risk, the criteria needs to be appropriately crafted (i.e., frequency of one or more impacts).

Two common methods of presenting the results of QRA studies include Individual Risk Contours and F-N (Frequency-Number) for societal risk assessment. A typical individual risk criterion might be that the 10^{-6}/yr. contour should not exceed the facility fenceline. For societal risk, criteria often take the form of line of demarcation or bands on an F-N plot, as illustrated in Figure 2. Calculated risk profiles that fall above the top line are considered intolerable and

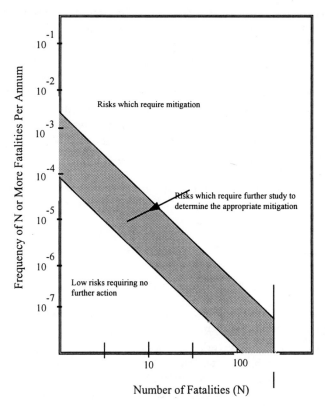

FIGURE 2. Criteria for Societal Risk. An F-N curve can be used to prioritize risks into three categories

TABLE 4
Safeguarding Selection Criteria

Level of Safeguarding	Risk Management Mechanism	Risk Criteria Type	Range
Primary	Prescriptive Codes/Standards	None	—
	Performance Codes/Standards	Critical Event, Individual	10^{-4}–10^{-5}/yr.
Secondary	Design Selection by FTA/RA	Critical Event, Reliability	10^{-4}/yr., 90–99%
Tertiary	Risk Evaluation by QRA	Societal, Individual	5×10^{-3}/yr. @ 1 or more, $<10^{-6}$/yr. offsite

must be corrected. Profiles that fall below the bottom line require no further mitigation. For profiles between the lines, additional mitigation should be investigated, with the goal of moving the profile closer to the tolerable region.

In summary, multiple levels of safeguarding can be employed to manage process risk, depending on the potential consequences. In general, the greater the perceived catastrophe, the more layers of protection that are employed. The trend is away from only relying on prescriptive codes and standards for protection from major incidents, toward use of performance-based standards for process risk management. In order to make rational risk decisions regarding the necessity for various levels of safeguarding, risk criteria of the type shown in Table 4 are required.

Such criteria are not only useful, but essential for making rational cost-benefit decisions regarding how much safeguarding is reasonable.

References

1. Royal Society, "Risk Assessment, Report of a Royal Society Study Group," London, (January, 1983).
2. Covello, V.T. et al., "Risk Appraisal—Can It Be Improved by Formal Decision Models?" In *Uncertainty in Risk Assessment, Risk Management, and Decision Making* (V.T. Covello, ed.). New York: Plenum Press (1987).
3. Griffiths, R., "Problems in the Uses of Risk Criteria." In *Dealing with Risk* (R.F. Griffiths, ed.). New York: John Wiley (1982).
4. Freeman, R.A., "Reliability of Interlocking Systems," *Process Safety Progress*, Vol. 13, No. 3 (July 1994).

5. American Petroleum Institute, "Management of Hazards Associated with Location of Process Plant Buildings," *Recommended Practice 752* (May 1995).
6. Bendixen, L. M., "Risk Acceptability Criteria," presented at ASCE Spring Convention & Energy Conference (1987).
7. Bendixen, L. M., "Risk Acceptability in the Chemical Process Industry—Working Toward Sound Risk Management," *Spectrum*, Arthur D. Little Decision Resources (March 1988).

POSTER SESSION

Methodology for Focusing a Transportation Risk Analysis

Paul E. McCluer
H&R Technical Associates, Inc., 151 Lafayette Drive, Suite 220, Oak Ridge, TN 37831

ABSTRACT

For the transportation risk analysis case study described in this paper, the analysts used an innovative system of screening thresholds and aggregating assumptions to reduce a large hazardous material shipment data base to a manageable number of bounding shipments in terms of potential consequences. The goal was to identify shipments that were significant from a consequence perspective for later quantitative analysis.

This study began with a data base of approximately 5,000 hazardous materials shipped by truck and air to and from a large industrial site during a period of 5 years. Six factors were judged important to analyzing the transportation risk: (1) hazardous material property (e.g., toxicity, volatility, or flammability), (2) population density along the route traveled, (3) type of packaging, (4) distance a shipment travels, (5) historic accident rates, and (6) inventory of the shipment.

The project timetable did not allow for a detailed examination of each of these factors for the 5,000 hazardous material shipments. In addition, the shipment records were not sufficiently detailed for such an examination. In fact, these records were not well characterized: for instance, only a few material shipments specifically identified the inventory. Many shipments were classified only on the basis of the U.S. Department of Transportation (DOT) hazardous material classes (listed in 49 CFR 173, "Shippers—General Requirements for Shipments and Packagings"). These DOT material classes apply to large categories of materials rather than specific substances.

By applying the methodology to the shipment data records, analysts identified five hazardous material shipments that bounded the consequence spectrum of possible truck transportation accidents and identified three hazardous material shipments that bounded the consequence spectrum of air transportation accidents. Although employed for transportation risk analysis in this case study, the methodology can easily be adapted to other types of risk analyses (e.g., for storage and processing of hazardous chemicals).

I. Introduction

A client asks your consulting firm to conduct a transportation risk analysis for one of its sites. The data available are a large data base that documents shipments of hazardous and radiological material shipments to and from the client's industrial site over a 5-year period. Of course, the client needs the analysis quickly. How can you qualitatively reduce a large shipment data base to the most important shipments that can be later analyzed from a bounding consequence perspective?

Using the methodology described in this paper, bounding shipments from a consequence perspective were identified from a shipment data base containing about 5,000 entries. The risk factors important to transportation are described in Section 2. The specific screening steps and assumptions described in Section 3 were used to reduce the data base to eight shipments judged to bound the spectrum of possible consequences that could result from a transportation accident. Finally, conclusions are presented in Section 4.

2. Transportation Risk Factors and Beginning Data

Risk is defined as a combination of the probability or frequency of an undesirable event occurring and the severity of the consequence or direct effect of the undesirable event. In transportation risk, the frequency of an undesirable event occurring is defined in terms of number of accidents occurring per mile traveled or number of accidents occurring per year. According to the Federal Highway Administration (1992), 13 elements have been identified as risk indicators when considering highway routes for hazardous material shipments:

- population density,
- type of highway,
- type and quantities of hazardous materials,
- emergency response capabilities,
- results of consultations with affected persons,
- exposure and other risk factors (such as distance to hospitals),
- terrain considerations,
- continuity of routes,
- alternate routes,
- effects on commerce,
- delays in transportation,
- climatic conditions, and
- congestion.

Of these risk factors, population density along the route traveled and the type and quantity of the hazardous material were judged most important. The other factors varied too much with different situations to be useful in a generic study. In addition to these two factors, three others were used in this analysis of the transportation risk: type of packaging used, distance a shipment travels, and accident rate for the transportation mode (truck or air).

This study began with a data base of approximately 5,000 hazardous material shipments by truck and air to and from a large industrial site during a period of 5 years. The data base contained the following information: (1) the origin and destination of a shipment, (2) the transportation mode and the specific carrier, (3) the date of shipment and the commodity being shipped or its DOT hazardous material classification, (4) the packaging type and number of packages in the shipment, (5) the weight of the shipment, (6) the number of curies and the specific radionuclide present, in the case of radiological shipments, and (7) the physical form of the material (e.g., solid, liquid, or gas). Some shipments identified the specific substance, but unfortunately, many shipments were only identified by the applicable DOT hazardous material classification listed in 49 CFR 173, "Shippers—General Requirements for Shipments and Packagings." These classifications apply to large categories of materials rather than specific substances.

3. Screening Steps

In this section, the screening steps and aggregating assumptions that comprise the focusing methodology and the way these steps and assumptions were specifically applied are described (Figure 1). The first screening step is to group similar shipments or data entries into categories. The second step involves comparing the material groups with any regulatory thresholds such as those presented in

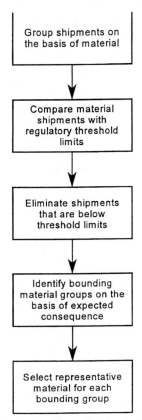

FIGURE 1. Flow Chart of Transportation Risk Analysis Focusing on Methodology

40 CFR 302, "Designation, Reportable Quantities, and Notification" and 40 CFR 355, "Emergency Planning and Notification." After eliminating any groups that are below the regulatory limits, the remaining groups are placed into broad categories on the basis of the types of consequences expected from a material release. For example, flammable materials are all grouped together and toxic materials are grouped together. Next, identify those materials for which a release presents the worst possible outcome among the groups present. Finally, identify the specific material in each material group that presents the worst possible outcome in the event of a release and quantitatively analyze possible accidents involving these specific materials.

The specific application of each of the screening steps to the subject transportation risk analysis case study are described in the following subsections.

3.1. Grouping Shipments by Material Type and Mode of Transport

The first step taken to reduce the 5,000 entries in the hazardous material data base to the bounding cases was to group these shipments on the basis of the commodity being shipped. Very little specific information was available on most shipments. The 5,000 entries were arranged into 81 different groups on the basis of the type of material or materials being shipped (see Table 1 for example groups). The first column shows the material group. The second column shows the mode of transportation. The third column lists the total number of shipments in a material group and the total weight of all the shipments in that group. The fourth column presents the maximum weight of a shipment. The fifth column lists the maximum number of curies for material groups that include radiological shipments. The seventh and eight columns list the average weight of a radiological shipment and the average number of curies in a radiological shipment, respectively.

3.2. Regulatory Threshold Comparison and Elimination of Shipments

In the second screening step, the maximum weights (for material groups containing hazardous, nonradiological shipments) and maximum activities (for material groups containing radiological shipments) were compared with regulatory thresholds. The regulatory thresholds chosen are as follows: for toxic materials, the reportable quantity (RQ) values listed in Table 302.4, 40 CFR 302, "List of Hazardous Substances and Reportable Quantities," or threshold quantity (TQ) values listed in 40 CFR 355, "Emergency Planning and Notification"; for radiological materials, the RQ values in Appendix B, 40 CFR 302.4; and for flammable materials, a general TQ value of 10,000 lb as listed in 29 CFR 1910.119, "Process Safety Management of Highly Hazardous Substances." Because few of the shipments identified their specific contents, only a few of the material groups were eliminated in this screening step.

TABLE I
Grouping of Hazardous Material Shipments by Commodity and Mode

Commodity	Shipment mode	No. of shipments (total pounds)	Max. weight	Max. curies	Average weight	Average curies
Explosives, NOS (Class 1.1)	Truck	90 (102,894)	13,801	0	1,143	0
	Air	7 (331)	195	0	47	0
	Government truck	12 (2,311)	1,040	0	193	0
Explosives, NOS (Class 1.2)	Truck	5 (9,078)	4,930	0	1,816	0
	Air	1 (240)	240	0	240	0
	Government truck	—	—	—	—	—
Explosives, NOS (Class 1.4)	Truck	14 (8,862)	2,674	0	633	0
	Air	92 (6,559)	563	0	71	0
	Government truck	6 (279)	63	0	47	0
Waste explosives (Class 1.1)	Truck	2 (32)	30	0	16	0
	Air	—	—	—	—	—
	Government truck	—	—	—	—	—
Non-flammable gas	Truck	544 (840,211)	22,000	0	1,545	0
	Air	46 (6,397)	2,195	0	139	0
	Government truck	9 (128,940)	33,000	0	14,327	0
Ammonia, anhydrous	Truck	7 (638)	255	0	91	0
	Air	2 (40)	35	0	20	0
	Government truck	—	—	—	—	—

NOS: Not otherwise specified.

An example of the results obtained from the regulatory threshold level comparison for the 81 material groups is documented in Table 2. In the first column of Table 2, a group number is assigned to a material group. In Column 2, a material classification from the data base is presented. In the third column, the maximum weight for shipments by truck (and maximum activity for radiological shipments by truck) is presented. In the fourth and fifth columns, the results of the regulatory threshold comparison screening step for the truck mode and the

TABLE 2
Comparison of Maximum Shipment Weights and Maximum Shipment Activities with Regulatory Limits

Group	Commodity	Truck mode max wt. (lb)	Screen for truck?	Explanation	Air mode max. wt. (lb)	Screen for air?	Explanation
1	Explosives, NOS (Class 1.1)	13,801	No	This class of explosive contains mass explosive materials. No regulatory threshold levels found.	195	No	This class of explosive contains mass explosive materials. No regulatory threshold levels found.
2	Explosives, NOS (Class 1.2)	4,930	No	This class of explosive includes explosion projection hazards. No regulatory threshold levels found.	1	No	This class of explosive includes explosion projection hazards. No regulatory threshold levels found.
3	Explosives, NOS (Class 1.4)	2,674	No	This class of explosive contains all explosives not classified as mass explosives or projection explosives. No regulatory threshold levels found.	533	No	This class of explosive contains all explosives not classified as mass explosives or projection explosives. No regulatory threshold levels found.
4	Waste, explosives (Class 1.1)	30	No	Assumed to be mass explosive material (Class 1.1). No regulatory levels found.	N/A	Yes	No shipments.
5	Nonflammable gas, NOS	33,000	No	No specific material specified and therefore, no regulatory threshold level found.	2,195	No	No specific material specified and therefore, no regulatory threshold level found.
6	Ammonia, anhydrous	255	No	RQ value is 100 lb.	35	Yes	RQ value is 100 lb.
7	Chlorine	2,000	No	RQ value is 10 lb. The observed quantities for chlorine are several thousand pounds.	7 (See explanation entry)	No	RQ value is 10 lb. The observed quantities for chlorine are several thousand pounds.

NOS: Not otherwise specified.   RQ: Reportable quantity.

rationale for truck mode screening decisions, respectively, are presented. In column 6, the maximum weight for shipments by aircraft (and maximum activity for radiological shipments by aircraft) is presented. In the seventh and eighth columns, the results of the regulatory threshold comparison screening step for the aircraft mode and the rationale for aircraft mode screening decisions, respectively, are presented.

After the comparison with regulatory threshold limits, 73 material groups remained. Of these material groups, 49 groups were commodities transported by truck, and 24 groups were commodities transported by aircraft.

3.3. Broad Material Categorization and Bounding Category Identification

Many of these material groups could not be eliminated because the broad DOT classifications did not specify an individual material; therefore, no regulatory threshold value comparison could be made. Therefore, some assumptions to further screen the material groups were necessary. First, the material groups were placed in general categories on the basis of type of consequence expected if a material release were to occur. Five general categories were assumed to bound the spectrum of consequences that could result from a worst-case transportation accident scenario. These categories were explosives, toxic materials, flammable materials, etiologic agents or infectious substances, and radiological materials. It should be noted that the selection of these bounding consequence categories is specific to this case study. Other applications of this methodology may yield different group selections.

The remaining groups included shipments of pure materials such as nonflammable gases, oxidizers, etc. The consequences of a material release from one of these material groups was assumed to be bounded by the consequences of a material release from one of the five selected categories. In addition, some of the remaining material groups included mixed shipments of commodities such as flammable materials and oxidizers. These mixed commodity shipments generally included waste materials. The specific percentages of the hazardous materials composing the mixed commodity shipments were not given in the data base. Therefore, the waste materials were assumed to be spent materials and as such much less potent than pure materials, or the amounts of the hazardous components in the mixed commodity shipments were assumed to be very low or dilute. Because of this dilution, the consequences of a material release for a mixed commodity shipment accident were assumed to be bounded by the consequences of a material release from one of the five selected categories.

For the truck transport mode, four groups were explosive materials, four groups were flammable materials, three groups were etiologic agents or infectious substances, twelve groups were toxic materials, and five groups were radiological materials. These groups are listed in Tables 3–7. For the air transport mode, seven groups were toxic materials, three groups were explosive materials,

TABLE 3
Truck-Mode Explosive Material Category

Group	Commodity	Maximum weight (lb)
1	Explosives, NOS (Class 1.1)	13,801
2	Explosives, NOS (Class 1.2)	4,930
3	Explosives, NOS (Class 1.4)	2,674
4	Waste, explosives (Class 1.1)	30

NOS: Not otherwise specified.

TABLE 4
Truck-Mode Flammable Material Category

Group	Commodity	Maximum weight (lb)
15	Flammable gas, NOS	106,106
16	Hydrogen gas	45,000
20	Waste, flammable gas	2,737
21	Flammable solid, NOS	6,014

NOS: Not otherwise specified.

TABLE 5
Truck-Mode Etiologic Agent Category

Group	Commodity	Maximum weight (lb)
48	Etiologic agent	730
49	Infectious substance	280
50	Waste, infectious substance	4,275

and five groups were radiological materials. These air-mode groups are listed in Tables 8, 9, and 10.

No etiologic agents or infectious substances were transported by air according to the shipment data base. No flammable air shipments exceeded the regulatory threshold value of 10,000 lb.

3.4. Selection of Representative Material from Each Bounding Category

In the last screening step, categories of materials for each transportation mode were selected that bounded the spectrum of consequences for a worst-case transportation accident scenario. A representative material from each of these catego-

TABLE 6
Truck-Mode Toxic Material Category

Group	Commodity	Maximum weight (lb)
6	Ammonia, anhydrous	255
7	Chlorine	2,000
13	Toxic gas—inhalation hazard	5,000
14	Waste, poisonous gas	4,570
24	Beryllium metal	75
33	Class A poison	2,450
34	Class B poison	840
35	Mercuric nitrate	32
36	Poison, solid NOS	170
37	Poison, liquid NOS	1,688
38	Waste, poisonous material	1,731
41	Nitric acid	2,206

NOS: Not otherwise specified.

TABLE 7
Truck-Mode Radiological Material Category

Group	Commodity	Maximum activity (Ci)
67	RAM, empty package	0.46 (^{238}U); 200 (^{3}H in gaseous form)
68	RAM, fissile NOS	5,000 (^{239}Pu)
74	RAM, LSA NOS	0.0014 (^{232}Th)
76	RAM NOS	700 (^{3}H in solid form)
78	RAM, NOS special	610.2 (^{252}Cf); 160 (^{60}Co); 2.9 × 10^4 (^{3}H in solid form)

RAM: Radioactive material   NOS: Not otherwise specified.   LSA: Low specific activity

ries for each transportation mode must be chosen. At this point, the criteria for choosing representative materials become analysis-specific; thus, the analyst should make selections on the basis of the current situation.

The criteria used to choose each representative material are as follows: (1) a list of chemicals used at the industrial site that are of concern to plant personnel, (2) conversations with plant experts and vendors, (3) the hazard presented by the material, (4) site knowledge (gained from previous experience and walkdowns at the industrial site), and (5) maximum group weight or activity. In addition, the

TABLE 8
Air-Mode Toxic Material Category

Group	Commodity	Maximum weight (lb)
7	Chlorine	7
24	Beryllium metal	85
33	Class A poison	1
34	Class B poison	49
36	Poison, solid NOS	30
39	Irritant, NOS	1
43	Hydrofluoric acid, spent	144

NOS: Not otherwise specified.

TABLE 9
Air-Mode Explosive Material Category

Group	Commodity	Maximum weight (lb)
1	Explosives, NOS (Class 1.1)	195
2	Explosives, NOS (Class 1.2)	1
3	Explosives, NOS (Class 1.4)	563

NOS: Not otherwise specified.

TABLE 10
Air-Mode Radiological Material Category

Group	Commodity	Maximum activity (Ci)
71	RAM, instruments and articles	1200 (^{3}H)
72	RAM, limited quantity, NOS	200 (^{3}H)
74	RAM, LSA, NOS	2.0 (mixed fission products)
76	RAM, NOS	9.7×10^5 (^{3}H)
78	RAM, NOS special	0.4 (^{241}Am)

RAM: Radioactive material   NOS: Not otherwise specified.   LSA: Low specific activity

physical form of the material as well as an estimated release fraction were considered. For solids, liquids, and gases, the release fractions were assumed to be 0.001, 0.01, and 1, respectively. On the basis of these criteria, the representative materials chosen for each category and transportation mode are presented in Table 11. The basis for the representative material choices are presented in the following paragraphs.

TABLE II
Representative Materials

Transportation Mode	Category	Material
Truck	Flammable material	Hydrogen
Truck	Toxic material	Chlorine
Truck	Radiological material	^{239}Pu
Truck	Explosive material	HMX
Truck	Etiologic agent	Ebola virus
Air	Toxic material	Selenium hexafluoride
Air	Explosive material	HMX
Air	Radiological material	^{3}H

The representative radiological materials were chosen on the basis of the maximum activity, dose conversion factor, and physical form for a data group. For the truck mode, Group 68 included a shipment of fissile material containing 5,000 Ci of ^{239}Pu, and Group 78 included a shipment containing 2.9×10^4 Ci of ^{3}H. Both are in solid form, but the former has a dose conversion factor orders of magnitude higher than the latter. Therefore, ^{239}Pu was chosen as a representative radiological material for the truck mode. For the air mode, Group 76 included a shipment containing 9.7×10^5 Ci of ^{3}H; therefore, ^{3}H was chosen as the representative radiological material.

The representative explosive materials were chosen on the basis of maximum amount of material present and type of explosive. Three classes of explosives (Class 1.1, Class 1.2, and Class 1.4) have been shipped to or from the site. Class 1.1 explosives are mass explosion hazard materials; Class 1.2 explosives are projection explosion hazard materials; and Class 1.4 explosives are explosives that are not considered mass explosion or projection explosion hazards and generally are considered minor hazards. Class 1.4 explosives are thus eliminated from further consideration as the bounding class because the other two classes of explosive are judged more hazardous. The Class 1.1 explosive shipments have higher maximum weights compared with Class 1.2 explosive shipments. For the air mode, the maximum weight for a Class 1.1 explosive (Group 1) is 195 lb. For the truck mode, the maximum weight for a Class 1.1 explosive (Group 1) is 13,801 lb. Of the Class 1.1 explosive materials shipped to the site, HMX was judged by explosive experts at the industrial site to be most sensitive to impact; therefore, HMX is the material chosen.

According to DOT regulations, etiologic agents and infectious substances are synonymous. The bacteriological and viral organisms that fit into this category are listed in 42 CFR 72.3, "Transportation of Materials Containing Certain

Etiologic Agents; Minimum Packaging Requirements." No virulency comparison for the listed organisms was found, and no data on the specific etiologic agent shipped was found. Therefore, the Ebola virus was arbitrarily chosen as the representative material. The largest shipment in this category was a waste, infectious substance shipment. Its maximum weight was 4,275 lb. This shipment was eliminated from consideration as the bounding shipment because the waste was assumed to be materials exposed to an etiologic agent; therefore, considering the entire shipment as the pure material was not prudent. The largest pure etiologic agent shipment was 730 lb (Group 48) and this shipment was used for the bounding case.

The flammable material was chosen on the basis of how a material is shipped and a material's explosive concentration range in air. The largest shipment in the flammable material category was 106,106 lb (Group 15). The commodity shipped was listed as flammable gas, not otherwise specified. A brief literature search indicated that the gases with the largest explosive concentration ranges in air are acetylene, hydrogen, and carbon monoxide. Acetylene has a lower explosive limit (LEL) of 2.5% and an upper explosive limit (UEL) of 82%. Carbon monoxide has an LEL of 12.5% and an UEL of 74.2%. Hydrogen has a LEL of 4.1% and UEL of 74.2%. All these explosive limits are volume concentrations in air, and the values may be found in the *Pocket Guide to Chemical Hazards* (National Institute for Occupational Safety and Health 1994). Although acetylene has a larger explosive concentration range, it was not shipped in bulk. Hydrogen was shipped in bulk as a liquefied gas. Therefore, hydrogen was chosen as the representative flammable material.

The representative toxic materials were chosen on the basis of maximum weight, estimated release fractions, and toxicity. For the truck mode, the maximum weight (5,000 lb) occurred in Group 13 and the shipment contained toxic gas. The most toxic gas used at the plant site was selenium hexafluoride. However, this material was not shipped in bulk. Site knowledge indicates that chlorine, another gaseous chemical of concern, was shipped in bulk. Therefore, chlorine was chosen as the representative toxic gas for the truck mode.

For the air mode, the toxic material group included shipments of Class A Poisons, which included toxic gases. The maximum weight for a Class A Poison shipment was 3 lb. Selenium hexafluoride was the most toxic gas of concern identified at the plant site; therefore, it was chosen as a representative toxic material for the air transport mode.

4. Conclusion

Using the screening criteria and aggregating assumptions described in this paper, the bounding materials from a worst-case consequence standpoint can be identified from a large number of diverse chemicals.

It should be noted that the weights used in this study included the weight of the shipping container. Using this extra weight adds a layer of conservatism to the study. In addition, the data for the larger shipments that are identified as representative material candidates should be verified against independent shipping records to ensure that no errors exist in the data base. For example, one radiological shipment was listed as having an inventory of approximately 1×10^{10} Ci of ^{241}Am. However, further investigation revealed that this entry was in error. Therefore, the analyst should always keep the big picture in mind and not accept data at face value.

References

Federal Highway Administration 1992. "Transportation of hazardous materials; highway routing," *Federal Register* 57 (169):395222–39533.

National Institute for Occupational Safety and Health 1994. *Pocket Guide to Chemical Hazards*, Publication 94-116, Cincinnati, Ohio.

Environmentally Sensitive Flares

John F. Straitz III
NAO Inc., 1284 E. Sedgley Ave., Philadelphia, PA 19134

ABSTRACT

Flares are essential safety systems. They must work right, each and every time, or the consequences may be catastrophic. The two most important considerations in flaring volatile, carcinogenic or other dangerous wastes must always be safety and reliability. Protection of the regional and global environment is another important factor. *With state-of-art enclosed thermal oxidizer flares, energy recovery from flared liquid/gaseous waste streams is now possible.*

Emergency Flares

For emergency disposal of waste gases in "open" elevated flare systems: (1) ambient air must be kept out of flare burners and stacks *to prevent burn-back, burning inside, and burn-through, which can cause potentially catastrophic explosions*; (2) liquids must be kept out of waste streams going to elevated flares *to prevent "burning rain"* from splattering personnel and equipment; and (3) reliable pilots with individual windshields and with flame retainers on elevated flare burners must assure dependable, stable flame patterns, regardless of wind or weather conditions, including hurricane-force winds, *to prevent deadly explosive clouds* from floating down into refineries or chemical/petrochemical plants — or drifting further afield to threaten residential communities.

The author will discuss the drawbacks and dangers of conventional molecular (labyrinth) seals, positioned below refractory-lined flare burners; single-cone seals that concentrate burn-through conditions inside flare burners of various shapes; patented multi-baffle kinetic seals that turn back *all* intruding air before it can penetrate open flare burners; and the importance of properly sized liquid disentrainment drums. He will also discuss other effective ways to improve safety and reliability of elevated emergency flares; to reduce maintenance on *open* flare systems; and to extend the working life of flare burners, pilots, and both electronic and flamefront ignition systems.

Enclosed Flares

State-of-art enclosed flares for onshore and offshore installations, including energy exploration/production platforms as well as floating production, storage and off-loading (FPSO) vessels will be discussed. With these state-of-art enclosed flares, also called thermal oxidizer flares, ground flares or populated area combustors, there is no need for costly steam, assist gas, power blowers or purge gas; and there are no steam-pulsation, noise or associated maintenance problems.

The author will emphasize the importance of natural draft, fully staged enclosed burners (designed, manufactured, installed and serviced to ISO-9001 quality standards), which are protected from searing heat, thermal expansion, frequent maintenance outages and costly replacement of redundant burners. (*Natural draft operation assures dependable, safe and complete destruction of waste streams in the event of a power outage.*) Mr. Straitz will also stress the need for ultra-reliable ignition of both fully enclosed flares and open emergency flares.

Today's full-feature enclosed thermal oxidizer flares conserve energy and reduce maintenance by eliminating all noisy, trouble-prone components utilized to reduce smoke from ineffective enclosed flares. Fully-enclosed, steam-free smokeless flares also combine improved safety and reliability with superior protection of the global environment by completely destroying *both* gaseous and liquid wastes — with no visible flame, no objectionable light, no dangerous thermal plume, no smoke, no odor and no thermal radiation/heat shield problems.

Paybacks for these steam-free, low-maintenance enclosed flares will be cited.

Properly designed, manufactured and installed to worldwide ISO-9001 quality standards, enclosed flares are extremely safe, ultra-reliable, easy to service and environmentally friendly. They combine dependable, complete destruction of waste gases/offgases (as well as waste liquids and watery, low energy slurries) with no visible flame, no smoke, no odor and very low noise.

Smokeless destruction of hard-to-burn wastes (i.e., olefins, diolefins, heavy paraffins, toluene, xylene) is accomplished in a state-of-art enclosed flare without any assist gas, purge gas, expensive steam, inefficient water spray, or troublesome air blowers. Hence, there are no steam pulsation/noise problems, no complex, maintenance-prone steam headers, no smoky emergency discharges, and no dangerous maintenance on elevated stacks—including cutting off and welding on replacement flare burners, high above a refinery, process plant or offshore platform, where volatile, carcinogenic gases may be present.

Other advantages of enclosed flares include: No possibility of dangerous flame lift-offs; no blow-offs of potentially catastrophic clouds of volatile, unburned deadly gases; no possibility of "burning rain" (liquid carryover) splattering personnel and equipment; no thermal radiation (heat shield) problems; and very convenient and safe service of all components at grade level.

Two Types of Enclosed Flares: In-Ground and Low Profile

There are two basic types of environmentally sensitive enclosed flares: (1) in-ground earthen enclosed flares, also called iso-fluidic oxidizers; and (2) low profile, enclosed "ground flares," commonly called populated area combustors or thermal oxidizer flares. Both basic types typically employ hundreds of small staged burners, instead of the single-, dual-, or limited array of burner configurations, utilized on elevated open flares.

TABLE I
Environmentally Sensitive Enclosed Flares Protect Local and Worldwide Environments

	Earthen Enclosed Flares	Enclosed Low Profile Flares	Elevated Open Flares
Visible Flame	no	no	yes
Smoke	no	no	yes[1]
Odor	no	no	yes[1]
Objectionable noise	no	no	yes
Flame lift-offs, blow-offs, blow-outs	no[2]	no[2]	yes[2]
Burning rain	no	no	yes[3]
Thermal plume	yes[4]	no	yes
Purge gas required	no	no	yes[5]
Steam, water spray or air blowers required	no	no	yes
Steam pulsation problems	no	no	yes

[1] Steam or other assist media can reduce, but not eliminate, smoke and odors from elevated flares.
[2] Flame lift-off, blow-off and blow-out are eliminated with state-of-art burners equipped with flame-retention rings, dependable pilots, and special design features which may be mandated by the waste stream.
[3] An adequately sized liquid knockout/disentrainment drum will preclude burning rain from an elevated or angled flare, onshore or offshore.
[4] Thermal plume must be allowed to rise up and disperse safely—with no elevated structures in immediate vicinity of an earthen enclosed flare.
[5] *Patented Fluidic Seals*™ (2700 installations, worldwide) substantially reduce or completely eliminate the need for costly purge gas by preventing air penetration at the open end of an elevated flare burner to prevent burn-back and burn-thru in burner tips and flare stacks.

Table 1 summarizes the basic advantages of in-ground earthen enclosed flares and low profile, fully enclosed flares, compared to elevated open emergency flares.

Fully enclosed flares are hybrids, developed from open flares and enclosed incinerators. Consequently, the design and materials of construction reflect a cost that is higher than a standard elevated emergency flare. Higher initial costs, however, are offset by substantial savings in steam, purge gas, maintenance and replacement expenses, which will be discussed in detail.

As an example, an enclosed ground flare designed to burn up to 700,000 lb/hr of 20 psig waste gas/offgas with a 40 to 50 molecular weight, where flaring is regulated to (a) a maximum noise level of 75 decibels at 3 ft from the flare, (b) no nuisance light, (c) no nuisance heat, and (d) complete compliance to all federal, regional and local emission restrictions would consist of:

- 402 vertically fired staged burners, operating at 5 psig
- a 64-ft outside diameter enclosure, 120-ft high, and
- a 114-ft diameter fence, 22-ft high

Such a flare system would cost about U.S. $3,840.000, installed. However, by eliminating the need for steam (approximately $1.5 million/yr), steam-related maintenance costs, purge gas, maintenance/replacement expenses for obsolete fire brick or refractory, and frequent replacement of traditional elevated flare burners and their pilots, the payback for a low maintenance, environmentally sensitive, enclosed flare can be very favorable.

Guideline information, including dimensions and estimated prices, for ISO-9001 certified enclosed flares appears in Tables 2 and 3. These tables provide data for four maximum flows (200,000 — 500,000 — 700,000 — and 1,000,000 lbs/hr) of the previously cited 20 psig waste gas/offgas with 40 to 50 molecular weight. For these examples: Maximum noise levels are 70, 75 or 85 dB(A) for the completely enclosed low-profile ground flares and 70 or 75 dB(A) for the in-ground earthen enclosed flares.

New Materials Are Required

To meet noise, nuisance and environmental regulations, new materials are required and several old materials must be rejected. Low density, 3" thick, ceramic blankets replace castable refractory linings; many small staged burners replace a single large burner; an acoustical fence and gravel floor are installed; and expensive, troublesome steam, water spray, and forced air blowers are eliminated.

Using the preceding example of an enclosed ground flare for up to 700,000 lbs/hr of gas, a 64-ft diameter enclosure is constructed of 28-ft wide, carbon steel flanged panels, which bolt together onsite. Support legs form an air gap at the bottom of the combustion chamber, thus allowing air into the chamber for natural draft operation.

Ceramic blanket anchor pins made of 310 stainless steel are welded to the inside of the enclosure and fitted with high temperature ceramic cuplocks to protect the pin tips, which otherwise would be exposed to the flare's flame. Three inches of insulation, rated at 2,400°F for intermittent use (and 2,250°F for continuous use), maintain the temperature of the carbon steel combustionchamber shell at a typical 300°F. A 304 stainless steel raincap on the top edge of the ceramic blanket is installed with overlapping adjacent components to prevent wind and rain from getting between the blanket and shell.

This state-of-art design eliminates the need for refractory and its frequent, costly repairs. There are no refractory spalling problems; no need for slow heat-

TABLE 2
Guidelines for Sizing and Pricing Low Profile Enclosed Flares

ENCLOSED FLARE	GAS NUMBER ONE			GAS NUMBER TWO		
	70 dB(A)	75 dB(A)	85 dB(A)	70 dB(A)	75 dB(A)	85 dB(A)
Max. Gas Flow Rate (million lb/hr)	200	200	200	500	500	500
Overall Height, ft	100	100	90	110	110	100
Outside Diam., ft	38	38	38	57	57	57
Number of Panels	15	15	15	24	24	24
Fence Height, ft	15	15	10	20	20	12
Fence Diam, ft	92	70	56	130	106	86
Number of Burners	216	138	138	468	294	294
Estimated Price (million $U.S.)	2.15	1.56	1.27	3.97	2.96	2.30

ENCLOSED FLARE	GAS NUMBER THREE			GAS NUMBER FOUR		
	70 dB(A)	75 dB(A)	85 dB(A)	70 dB(A)	75 dB(A)	85 dB(A)
Max. Gas Flow Rate (million lb/hr)	700	700	700	1000	1000	1000
Overall Height, ft	120	120	110	150	150	135
Outside Diam., ft	64	64	64	68	68	68
Number of Panels	30	30	30	32	32	32
Fence Height, ft	22	22	15	22	22	15
Fence Diam, ft	140	114	94	148	120	110
Number of Burners	636	402	402	888	564	564
Estimated Price (million $U.S.)	5.00	3.84	3.07	6.34	5.01	4.10

For these examples, the waste gas/offgas has a 40 to 50 molecular weight. Maximum flows of 200-million, 500-million, 700-million and one-billion lb/hr of 20 psig waste stream are designated "Gas Number One" to "Gas Number Four"—with optional maximum noise levels of 70, 75 or 85 dB(A), three feet from the flare. For each example, there is no visible light, no smoke, no nuisance heat, no thermal plume; and there is complete compliance to all federal, regional and local emission requirements.

TABLE 3
Guidelines for Sizing and Pricing Earthen Enclosed Flares

ENCLOSED FLARE	GAS NUMBER ONE		GAS NUMBER TWO	
	70 dB(A)	75 dB(A)	70 dB(A)	75 dB(A)
Max. Gas Flow Rate (million lb/hr)	200	200	500	500
Inside Length and Width, ft	250 × 55	170 × 65	580 × 55	380 × 65
OutsideLength and Width, ft	290 × 95	210 × 105	620 × 95	420 × 105
Depth, ft	18	18	18	18
Inner Fence Length and Width, ft	330 × 135	250 × 145	660 × 135	460 × 145
Outer Fence Length and Width, ft	340 × 145	N/A	670 × 145	N/A
Inner Fence Height, ft	40	30	40	30
Outer Fence Height, ft	10	N/A	10	N/A
Number of Burners	444	282	1110	708
Estimated Price (million $U.S.)	4.29	2.37	7.60	3.86
ENCLOSED FLARE	GAS NUMBER THREE		GAS NUMBER FOUR	
	70 dB(A)	75 dB(A)	70 dB(A)	75 dB(A)
Max. Gas Flow Rate (million lb/hr)	700	700	1000	1000
Inside Length and Width, ft	420 × 85	275 × 110	585 × 85	380 × 110
OutsideLength and Width, ft	460 × 135	325 × 160	635 × 135	430 × 160
Depth, ft	22	22	22	22
Inner Fence Length and Width, ft	500 × 175	365 × 200	675 × 175	470 × 200
Outer Fence Length and Width, ft	510 × 185	N/A	685 × 185	N/A
Inner Fence Height, ft	45	35	45	35
Outer Fence Height, ft	10	N/A	10	N/A
Number of Burners	1554	990	2220	1410
Estimated Price (million $U.S.)	7.69	4.35	9.69	5.37

For these examples, the waste gas/offgas has a 40 to 50 molecular weight. Maximum flows of 200-million, 500-million, 700-million and one-billion lb/hr of 20 psig waste stream are designated "Gas Number One" to "Gas Number Four"—with optional maximum noise levels of 70 or 75 dB(A), three feet from the flare. For each example, there is no nuisance light (no visible flame beyond in-ground flare), no smoke, no nuisance heat; and there is complete compliance to all federal, regional and local emission requirements.

ups and slow cool-downs; and none of the noise that occurs in enclosed ground flares which use firebrick or castable refractory lining.

The very light weight of the ceramic blankets (approximately 5.33 lb/cu ft compared to 80 to 140 lb/cu ft for firebrick and castable refractory) translates, through heat absorption and retention, into very quick heating and cooling responses to changes in flare gas flow rates.

(Very slow cooling responses, typical of heavy castable refractories, tend to decrease flare flow rates, resulting in excessive natural drafts from excessive heat absorbed and retained in refractory walls. This will cause too much quenching or cooling of the combustion process, thus producing nuisance-odored pollutants, such as aldehydes.)

Design with Safety Margin

The relatively soft ceramic lining walls do not reflect noise. Instead, the excellent noise damping characteristic absorbs combustion-chamber pulsations, even when a quality-built enclosed flare is operated in excess of five times its rated design capacity.

Designing a single burner to flare 0 to 700,000 lb/hr, and operate cost-effectively without producing noise, smoke or pollutants would be impossible. A large number of small burners are required to maximize waste destruction efficiency and minimize noise.

The small burners are piped together to form different arrays of staged burners that operate at various flare gas flow rates. Electronic pressure and flow controls trigger pneumatic snap-action (on/off) butterfly valves through explosion-proof solenoid valves to control air to each pneumatic operator. Each group of burners becomes operational at a preset flow or pressure, thus efficiently destroying the waste stream, regardless of its flow rate.

As a backup, built-in/fail-safe mode, each staging valve is bypassed with a rupture disc. If electric or air failure occurs and the butterfly valves don't function, all waste streams will go to the burners through the ruptured discs. While flaring in this fail-safe/ruptured disc mode may not be efficient, it will be completely safe.

The acoustical fence, consists of three layers: A 3/16" thick solid steel outer shell; a middle layer of 3" thick ceramic blanket; and an inner layer of 18 gauge, 304 stainless steel perforated plate. Because the fence requires inlet air to make three ninety-degree turns before contacting the burners, sound waves trying to escape along the inlet-air path are effectively absorbed within the acoustical fence. Free-standing sections of the fence create a tortuous route for inlet air; and because they are not connected to other fence members, these free-standing sections also interrupt and minimize heat conduction from the enclosed flare.

Protect Components from Heat

Pipe headers supplying the burners are buried beneath a gravel layer to protect piping and headers from heat. The gravel further improves heat protection by forming air gaps and voids, which interrupt heat conduction. Gravel air gaps and voids also absorb approximately 80% of the burner noise. Due to the concave bowl shape of the gravel floor, all remaining noise is reflected back into the combustion chamber and not out through the air-inlet openings. (The bowl-shaped gravel bed acts like a sponge to trap and contain any liquid carryover to the flare, thus preventing fire damage to the carbon steel combustion chamber and acoustic fence.)

Although forced draft enclosed flares are smaller for the same flare-gas capacity, they must rely on electric- or diesel-powered fans, which can fail, throwing flare operation into the natural draft mode at a much higher rate than a properly sized and properly manufactured natural draft enclosed flare. Therefore, it is operationally advantageous—and inherently safer—to design for natural draft.

A properly designed natural draft enclosed flare can operate safely and dependably at full capacity under any condition, including power failure.

Operating an enclosed flare with steam is very, very costly. It also presents serious noise and maintenance problems. While steam is commonly used for smokeless operation of elevated open flares, it is not required for state-of-art enclosed flares with properly staged, free-floating vertically fired burners.

These burners ensure smokeless operation through highly turbulent mixing and swirling of even low heating value gases into a vortex, thus eliminating the need for steam, water spray, assist gas and forced air. (A properly designed enclosed ground glare can handle gases with heating values as low as 60 Btu/scf without assist gas.)

Deadly Gases to Elevated Flare

If deadly hydrogen sulfide or carbon dioxide is present, assist gas is required to maintain the waste stream at 100 Btu/scf. However, a hazardous condition can exist if an enclosed ground flare experiences a flame-out when hydrogen sulfide is being fed to the flare. Therefore, hydrogen sulfide normally goes to an elevated emergency flare. (The height of an enclosed flare, such as the relatively short 120-ft high enclosure in our 700,000 lb/hr example, cannot be shortened without putting the flare in danger of being extinguished during high winds.)

For remote locations, where there is no urban congestion, in-ground enclosed flares also provide smokeless, invisible flaring without any costly steam or purge gas. Waste streams are spread out over wide, efficiently staged burner areas to provide optimum mixing with ambient air. Staged rows of flare burner tips, standpipe-mounted above buried pipe manifolds, assure efficient and complete destruction of waste streams. A gravel bed protects the buried carbon steel piping and manifolds from radiant heat and flame damage. It also absorbs noise, while acting as a sponge to trap and dike any liquid spillover, thus preventing fire damage.

The only possible disadvantage of an in-ground, earthen enclosed flare is its thermal plume, which must be allowed to rise up and disperse safely. There must be no elevated structures within the immediate vicinity.

State-of-art in-ground earthen enclosed flares incorporate the same proven components employed in low profile enclosed flares. Reliable operation is assured by fail-safe valves and fully automatic burn-off in the event of a power failure.

The only operating costs for an in-ground, earthen enclosed flare are pilot gas, electricity for the ignitor/control panel (battery or solar backup power is optional), and a small amount of instrument air for the staging system.

Don't Accept Obsolete Designs

Flare burners are the heart of flare systems. Instrumentation and controls, including fully automatic, fail-safe interlocked sensors, pilots and reliable pilot monitors are the nervous systems and the brains.

Flare design, manufacture, installation and service procedures vary widely from manufacturer to manufacturer. Unfortunately, some designs incorporate out-of-date, inherently unsafe technology. For example: Labyrinth-type seals, positioned below elevated flare burners, allow burn-back and burn-thru inside flare tips, stacks and the ineffective seals—which spit out flames and burp out pieces of spalled refractory.

Flare fabrication procedures vary widely. Several flare manufacturers allow improper welds in extremely thin metals; substitute carbon steel components that should be stainless steel or nickel-chromium alloys; and even improperly size elevated flare burners—or utilize exposed burners and headers in enclosed flares.

One very hazardous elevated flare was installed on an offshore platform near Vietnam. Designed with 21 bars of back-pressure, this severely undersized [200mm (8") diameter] and over-pressured [2,168 Kpa (314.5 psig)] flare cannot be operated safely at its "rated capacity." For proper operation, the flare diameter should be 360mm (14"). Severe over-pressure can ruptures the flare header lines, liquid knockout drum and process separator drums, causing destruction of the entire offshore platform.

For detailed information about obsolete and unsafe designs incorporated for open elevated flares and for enclosed flares; ignorance and neglect of flare systems; ineffective and unsafe molecular/labyrinth-type seals; and unreliable pilots and ignition systems, refer to reports in major trade publications, reference 1. The state-of-art enclosed flares described in this paper offer two significant advantages: (1) They provide smokeless incineration of the heaviest, hardest-to-burn waste gases/offgases and watery, low energy wastes without any costly steam, troublesome air blowers, or exposed burners and headers; and (2) the fully enclosed configurations can be upgraded to convert an existing enclosed flare into a true thermal oxidizer flare by adding air-damper controls to the free-floating, vertically fired natural draft, multi-tip, multi-jet burners.

A true thermal oxidizer flare provides a waste gas/offgas and waste liquid destruction removal efficiency (DRE) better than 99.9 percent—for unsurpassed control of sulfur oxides (SO_x), nitrogen oxides (NO_x) and other volatile, carcinogenic emissions.

Heat-recovery systems for an enclosed thermal oxidizer flare may utilize a simple plate or coil heat exchanger to preheat cold waste streams for more efficient incineration and a heat-recovery boiler to create steam from captured heat. Whenever recuperative energy can be put to use, both systems should be considered.

Free-Floating Multi-Jet Burners

Unsurpassed safety and world-proven dependability is maximized in low profile enclosed flares by combining innovative, practical designs with no short-cuts in manufacturing. Fully automatic multi-stage operation of free floating multi-tip, multi-jet burners, for example, assures dependable control of waste gases, offgases and other objectionable waste streams from oil/gas energy fields, refineries, chemical/petrochemical plants, and other manufacturing/processing or energy-storage/loading terminal operations.

Because the patented flare burners installed in state-of-art enclosed flares assure uniform mixing of ambient air with waste streams—without any costly steam—they eliminate very substantial operating costs (often in the neighborhood of more than U.S. $1- or 1.5-million per year) as well as the expensive maintenance/reliability problems associated with exposed burners used in enclosed flares, made by some companies. *Exposed burners have poor design/performance ratings* (Table 4).

With more than 500,000 worldwide burner installations, including units manufactured in the 1920s and 1930s that are still in service, NAO is fully committed to supporting all existing worldwide installations and to maintaining its position as the leading innovator for ultra-reliable combustion, pollution-, safety and energy-control systems.

Unsurpassed safety for state-of-art enclosed flares is further assured by automatic interlocks, liquid seals, ultraviolet (UV) scanners, fail-safe startups and

TABLE 4
Burners Are Key to Performance and Reliability of Enclosed Low Profile (Ground) Flares

	Side Mounted Burners	Jet/Steam Coanda Burners	Matrix Fin Burners	Multi-Tip/Multi-Jet Burners
Visible Flame	No 10	Some 0	No 10	No 10
Thermal Radiation	No 10	Slight 3	No 10	No 10
Thermal Plume	No 10	Yes 0	Yes 0	No 10
Gas Dispersion	Good 10	Poor 0	Poor 0	Very Good 15
Noise	Low 6	Low 6	Very Low 10	Very Low 10
Burner Features				
Complete Combustion	No* 0	No* 0	Yes 10	Yes 10
Uniform Mixing of Waste Gas/Air	No 0	No 0	Yes 10	Yes 10
High Turndown/Flow Range	No 0	Very Poor –5	Yes 10	Yes 10
Steam Required for Smokeless Operation	Yes –20	Yes –20	No 10	No 10
Steam Noise/Pulsation	Yes –10	Yes –10	No 10	No 10
Steam Control Problems	Yes –10	Yes –10	No 10	No 10
Multi-Stage Design	Yes 10	Partial 6	Partial 6	Complete 15
Exposed Burners and Headers	Yes** –10	Yes** –10	Yes** –10	No 10
Long Burner Life	Yes 10	Yes 10	No –10	Yes 10
Expansion Problems	Yes 0	Yes 0	Very Severe –10	None*** 15
Maximum Use of Combustion Chamber Space	No 0	No 0	Yes 10	Yes 10
Ease of Operation	No 0	No 0	Yes 10	Yes 10
Reliability & Low Maintenance	Poor 0	Poor 0	Poor 0	Very Good 15
Comparative Score	**16**	**–30**	**86**	**200**

*Incomplete combustion and quenching due to steam injection and limited number of burners
**Thermal expansion can cause bending and leakage of flare gas.
***Free floating vertical burners with heat/flame shields—eliminate thermal humping (expansion) problems

shutdowns, warning lights, staged burner heads with builtin flame/detonation arrestors, and remote spark pilots with UV scanners. Optional safety devices include gas detectors, video monitors, alarm horns, and automatic fire-suppression systems.

Before a unit can be started, the automatic controls verify all built-in safeguards. Each burner pilot, with its individual UV scanner, must prove ignition before the staged burner heads can be turned on. Automatic fail-safe shutdowns will be initiated if there is a flame failure, or if the system cannot be brought back into adjustment.

This combination of fully interlocked, automatic safety features for ultra-reliable burner operation has been incorporated into enclosed flares for populated areas and shipboard installations. Modular panel, clam-shell and other structural configurations are available to simplify onsite construction.

Although the design and the development of low-profile enclosed flares were pioneered in the United States, it is interesting to note: Most installations of environmentally sensitive, populated area flares are in Southeast Asia, Australia, Japan, China, and Korea.

Bad Designs, Abandoned by Irresponsible Manufacturers

Unfortunately, there have been more than a few unsuccessful installations of poorly designed and/or poorly manufactured thermal oxidizer flares—units with exposed burners, and thermal expansion, steam pulsation, steam noise, steam control and other serious problems. Those bad installations have never operated properly. Several very expensive ground flares, with obsolete and unsafe designs, have been abandoned by irresponsible manufacturers.

In most cases, the principal problem was side-mounted, exposed burners. These burners require complex staging systems and large quantities of very expensive steam with noise and pulsation problems. Since the exposed burners and headers are also subject to thermal expansion, they have a reputation for very high maintenance and very poor performance.

One flare manufacturer has attempted to solve its exposed burner problem by distributing waste gas through a matrix of small holes adjacent to shaped fins. Although this widely publicized exposed matrix fin-burner configuration improves gas/air mixing and combustion efficiency, it also creates very severe thermal expansion, reliability and maintenance problems—and it reduces burner life.

How to Measure Reliability

The performance of enclosed flares depends upon specific design and manufacturing parameters. Key elements are the burners and interlocked controls. There are four types of burner configurations, as indicated by Table 4.

For smokeless operation, two of these designs require complex staging systems and large quantities of costly steam with noise and pulsation problems.

Three incorporate exposed burners and headers. In addition to short burner life, thermal expansion problems, header bending and dangerous leakage of waste gases (with potentially catastrophic explosions), these designs also create very serious maintenance and reliability problems.

Free-floating NAO burners eliminate thermal expansion problems, while also assuring excellent dispersal of waste streams with uniform gas/air mixing for completely enclosed incineration—without steam, purge gas or forced air blowers, and with no visible flame, no smoke, no noise and no odor.

Because burners are the key to the performance of low profile enclosed flares, numbers have been assigned to all variable in table 4. Here are examples:

- *Steam is very expensive.* It also creates pulsation and control problems. Hence, enclosed flares that require steam have been assigned the lowest possible ratings (–20).
- *Exposed burners create maintenance, safety and downtime problems.* Due to very severe expansion problems, matrix fin burners are quickly destroyed. These burners have been assigned a –10 rating.
- *Downtime is expensive.* It is a frequent problem with enclosed flares that use side-mounted burners, jet-steam Coanda burners and matrix fin burners. The reliability and maintenance ratings for enclosed flares with "problem burners" is zero.

Performance Is the True Test

The true measure of reliability is performance in-the-field.

NAO flare systems, elevated and enclosed, have earned a worldwide reputation for reliability, safety and trouble-free operation, day after day, year after year, regardless of variations in waste stream pressures and flows. These systems are built right, to ISO-9001 quality standards, utilizing innovative concepts and practical designs which are based on experience, empirical knowledge and a corporate commitment to treat all customers as valuable long-term friends. The commitment to treat all customers as friends and as partners in joint efforts to solve unique safety and pollution-control problems is taken seriously.

NAO maintains the world's largest fleet of portable pollution solutions to help customers with plant shutdowns, scheduled and unscheduled, and for emergency response to leaks, spills and soil/sludge remediation projects. These portable pollution solutions include self-erecting, trailer-mounted open flares and enclosed flare/thermal oxidizer systems of all sizes.

Using a portable thermal oxidizer, the company developed special tunnel-burner flare tips to replace the very expensive and inefficient incinerators, traditionally used to burn low energy gases from wood wastes (explosive saw dust), carbon black (amorphus carbon) byproducts, and other dangerous wastes with calorific values as low as 2000–2400 KJ/Nm^3 (50–60 Btu/scf).

Because large volumes of assist gases are required by a multi-million-dollar incinerator, its operating cost typically averages U.S. $125,000 per month. Despite a high capital investment and very high operating costs, waste-gas destruction efficiencies do not satisfy worldwide environmental regulations. Safety also leaves a lot to be desired.

NAO's new tunnel-burner tips, developed for elevated (open flares) and low-profile enclosed flares, eliminate the need for expensive assist gas; provide a destruction removal efficiency (DRE) better than 99.9 percent; and significantly reduce harmful, dangerous emissions that contribute to acid rain and global greenhouse warming.

Table 5 summarizes operational advantages of the special tunnel-burner flare tips. It also provides a summary of the principal low energy wastes which are generated as byproducts of carbon-black production for the printing, tire and rubber industries.

Table 6 provides information about a true thermal oxidizer flare, designed to oxidize eight waste gases and a sour-liquid stream. One of the waste gases serves as the pilot gas for this NAO ultra-safe, ultra-efficient and ultra-quiet installation in a densely populated region.

General information about in-ground earthen enclosed flares appears in Table 7. Note the absence of any visible flame outside the chemical/petrochemical plant.

TABLE 5
How Tunnel-Burner Flare Tips Solve Low Btu Waste Control Problems

Consider the watery, low BTU wastes produced as principal byproducts of carbon-black (*amorphus carbon*) production:

Water (H_2O)	42%	Carbon Monoxide (CO)	8%
Nitrogen (N_2)	35%	Carbon Dioxide (CO_2)	3%
Hydrogen (H_2)	11%	Sulfur Compounds	<0.3%

Small amounts of methane are also present in the waste stream. Since the heating value of the wet waste is as low as 2000 to 2400 KJ/Nm^3 (50 to 60 Btu/scf), large volumes of expensive assist gas have been required to burn carbon-black waste streams in conventional incinerators. (Those incinerators represent significant capital expenditures. Operating costs average U.S. $125,000/month.)

The low Btu wastes are efficiently destroyed by tunnel-burner flare tips.

After an initial 5 to 10 minute warm-up on pilot gas, the tunnel-burner flare tips destroy the water-saturated low Btu wastes — without any assist gas.

The destruction removal efficiency is better than 99.9 percent.

Even after a two-minute shutdown, no pilot flame and no assist gas are required to sustain the safe, ultra-efficient destruction removal efficiency.

TABLE 6
Eight Waste Gases Plus Sour-Liquid Stream Are Oxidized in Special Ground Flare

A patented chamber-within-a-chamber design allows a 20-meter (65-ft) high enclosed flare, designed by NAO Inc., to satisfy stringent safety, pollution- and noise-control requirements in a densely populated area of Japan.

This steam-free thermal oxidizer flare efficiently destroys eight different waste gases plus a sour waste-water stream—without any costly steam and with no visible flame.

The eight waste gases range from hydrogen-rich compounds (89% H_2) to extremely heavy, hard-to-burn offgases.

The hydrogen-rich vapors and carcinogenic offgases include:
- tra amounts of ill-smelling hydrogen sulfides
- heavy paraffins, olefins, diolefins, B.T.X. (benzene, toluene, xylene)
- cyclopentadiene, a very heavy, difficult-to-burn gas used in the manufacture of plastics and insececticides.

Molecular weights vary from 4.4 to 68.3.

Pressures range from 27.6 kPa (4 psig) for a heavy C_5 absorber-vent gas to 193 kPa (28 psig) for the hydrogen-rich streams.

Six burner tip nozzles, located around the periphery of the inner combustion chamber vaporize the sour liquid waste for efficient destruction of up to 208 liters per minute (55 gpm) of dirty, contaminated water.

One of the waste gases, an H_2S stripper gas, serves as the pilot gas for this ultra-safe, ultra-efficient installation.

To satisfy extremely tight noise-level requirements, a ceramic blanket-lined combustion chamber absorbs reverberations and combustion noises. Further noise reduction is achieved with an acoustical fence that incorporates a soft absorbent lining to serve as an effective anechoic chamber that absorbs noises over a wide frequency range.

TABLE 7
In-Ground Earthen Enclosed Flares

> For remote locations, where there is no urban congestion, in-ground enclosed flares, also called NAO iso-fluidic oxidizers (*NIFO*), provide safe, ultra-reliable installations for smokeless, invisible flaring without any expensive steam or purge gas. *Recently, for example, an American newspaper reporter asked an engineer at Oxy-Chem's Hutchinson, KS, plant if that facility had been shut down because the reporter noticed the absence of any visible flame or smoke. The engineer replied: "No. Our new in-ground enclosed flare is invisible and smokeless."*
>
> When adequate space is available for an installation in a non-populated area, the only disadvantage of an in-ground enclosed flare is its thermal plume.
>
> The plume must be allowed to rise up and disperse safely with no elevated structures within the immediate area.
>
> For earthen enclosed flares, a gravel bed protects buried carbon steel piping and manifolds from radiant heat and flame damage. The gravel bed also absorbs noise, while acting as a "sponge" to trap and dike any liquid carryover, thus preventing fire damage.
>
> Each in-ground enclosed flare is designed to spread waste gases/offgases out over wide, efficiently staged burner areas for optimum mixing with ambient air. Staged rows of T- or L-shaped flare tips (standpipe-mounted above the buried pipe manifolds) assure efficient and complete destruction of the waste streams. Other advantages include:
> - smokeless operation without steam, water spray or air blowers
> - very low noise levels
> - no thermal radiation
> - very high turndown ratios
> - no purge gas required
>
> Patented *Jet-Mix*^{TM} burners with unique thruster tips may be interspersed between the staged T- or L-shaped flare tips to swirl gas/air mixtures and produce high turbulence—thus assuring smokeless operation, regardless of waste-stream pressure drops.
>
> *The only operating costs for an in-ground NIFO system are pilot gas, electricity for the ignitor/control panel, and a small amount of instrument air for the staging system.*
>
> Fail-safe, iso-fluidic in-ground flares incorporate the same proven components used in the company's low-profile thermal oxidizer flares. Safe, reliable operation is assured by fail-safe valves and fully automatic burn-off in the event of power failure.

Summary

State-of-art enclosed flares, designed, manufactured and installed to ISO-9001 quality standards (onshore and offshore, including energy exploration/production platforms and floating production, storage and off-loading FPSO vessels) conserve energy, reduce maintenance, protect plant/field/platform personnel, surrounding communities and the global environment, while eliminating all noisy, trouble-prone components and costly assist media utilized to reduce smoke, odor and noise from ineffective open or enclosed flares.

PHAzer: An Intelligent System for Automated Process Hazards Analysis

Rajagopalan Srinivasan and Venkat Venkatasubramanian[1]
Laboratory for Intelligent Process Systems, School of Chemical Engineering, Purdue University, West Lafayette IN 47907-1283

ABSTRACT

Process Hazards Analysis (PHA) is an important component of process safety management. It is a labor- and knowledge-intensive process which would gain from automation. Previous work in this area have concentrated on hazard identification, which is just one step during PHA. In this paper, we propose an integrated framework and a knowledge-based system, called *PHAzer*, which takes a more comprehensive approach to the entire PHA process. *PHAzer* can be used to identify hazards using the guideword HAZOP technique, perform detailed hazard assessment to help in consequence evaluation, and synthesize and analyze fault trees. The salient features of *PHAzer* are presented and illustrated using an olefin dimerization plant case study.

Introduction

Process Hazards Analysis (PHA) is the systematic identification, evaluation and mitigation of potential process hazards which could endanger the health and safety of humans and cause serious economic losses. This is an important activity in process safety management (PSM). The importance of PHA was recently underscored by the Occupational Safety and Health Administration's (OSHA) PSM standard Title 29 CFR 1910.119 and Environmental Protections Agency's (EPA) Risk Management Program (RMP) Rule 40 CFR 68 in the United States. A wide range of methods such as Checklist, What-If Analysis, Failure Modes and Effects Analysis (FMEA), Fault Tree Analysis (FTA) and Hazard and Operability (HAZOP) Analysis are available for performing PHA [1]. Of these, HAZOP analysis is widely used and recognized as a preferred PHA approach by the chemical and hydrocarbon process industry.

HAZOP analysis is used for the identification of the hazards and operability problems in a plant that could compromise the plant's safety and productivity. The basic principle of HAZOP analysis is that hazards arise in a plant due to

---

1 Author to whom all correspondence should be addressed

deviations from normal behavior. In HAZOP analysis, a multidisciplinary team of experts systematically examine the process P&IDs to identify every conceivable deviation from the "design intent," and determine all the possible abnormal causes and the adverse hazardous consequences of those deviations. If the protections against the hazards in the plant are not adequate, then recommendations in the form of design or procedure modifications to mitigate the risk are proposed by the team. This analysis is performed for every process unit, process line or process section of the plant depending on the level of detail required. In order to cover all the possible malfunctions in the plant, the process deviations to be considered during HAZOP analysis are generated systematically by applying a set of "guidewords," to the process variables. These guidewords correspond to qualitative deviations of process variables to which they are applied. The guidewords and process variables should be combined in such a way that they lead to meaningful process variable deviations. HAZOP analysis is a tool for identifying *what* undesired event can occur. In order to determine *how* the hazard can occur and the frequency of occurrence, fault tree analysis (FTA) is often used.

A fault-tree is a logic tree that propagates primary events or faults to the top level event or hazard. A fault tree graphically represents the deductive reasoning logic to determine how an undesired event can occur due to various combinations of faults. A fault tree is constructed by asking questions like "what could cause a hazard?" (called a top level event). In answering this question, one generates other events connected by logic nodes. The tree is expanded in this manner until one encounters events which need not be developed further (called primary events). The faults or primary events considered in FTA are equipment failures, control system failures, and operator errors. The tree usually has layers of nodes. At each node different logic operations like AND, OR are performed for propagation. Once the fault tree is constructed, the next step in the analysis is the evaluation of the tree. Qualitative evaluation of the fault tree is concerned with the development of minimal cut sets, defined as "the unique combinations of the primary failures that are necessary and sufficient to cause the top level event leading to system failure". Thus, the minimal cut set identifies the critical component failures. In quantitative evaluation of the fault tree, knowledge about the probability of occurrence of primary events is used to calculate the probability of failure of the top level event. This evaluation of the fault tree enables the safety analyst to rank the ways in which an accident may occur, and focus on implementing preventive measures against faults with high probability of occurrence which are present in the cut sets, to reduce the probability of the top level hazard.

PHA is a laborious, time consuming and expensive activity which requires specialized knowledge and expertise and is typically performed by a team of experts. For PHAs to be thorough and complete, the team cannot afford to overlook even the "routine" causes and consequences which will commonly occur in many plants. The importance of performing a comprehensive PHA is illustrated by Kletz [2, 3, 4] with examples of industrial accidents that could have been prevented if only a thorough PHA had been performed earlier on that plant. A typical PHA can take 1-8 weeks to complete, costing over $13,000 per week. By an OSHA estimate, approximately 25,000 plant sites in the United States require a PHA under the PSM standard. EPA has estimated that approximately 118,000 facilities will be affected by the RMP rule. Given the enormous amounts of time, effort and money involved in performing PHA reviews, there exists considerable incentive to develop intelligent systems for automating PHAs of chemical plants. An intelligent system can reduce the time, effort and expense involved in a PHA review, make the review more thorough and detailed, minimize human errors, and enable the team to concentrate on the more complex aspects of the analysis which are unique and difficult to automate. Also, an intelligent PHA system can be integrated with CAD systems and used during early stages of design, to identify and decrease the potential for hazardous configurations in later design phases where making changes could be economically prohibitive. In addition, it would facilitate automatic documentation of the results of the analysis for regulatory compliance and these PHA results can be made available online to assist plant operators during abnormal situations.

PHAzer: An Intelligent Process Hazards Analyzer

Despite the importance of automating PHA, there has only been limited work in this area. Recently, a digraph model-based expert system, called *HAZOPExpert*, was proposed by Vaidhyanthan and Venkatasubramanian [5, 6] for automating HAZOP analysis. The central ideas in their framework are the separation of the knowledge required to perform HAZOP analysis into process-specific and process-general knowledge. The process-specific knowledge consists of process material properties and the process P&ID. The process-general knowledge consists of the HAZOP-digraph (HDG) models of the process units which are signed directed graph-based qualitative causal models developed specifically for hazard identification. The emphasis in their work has been on application to continuous process plants in steady state operation. Srinivasan and Venkatasubramanian [7, 8] proposed *Batch HAZOPExpert*, which uses a Petri net–digraph based framework for automating HAZOP analysis of batch processes. Here, the operating procedure is represented by high level Petri nets, and the causal relationships between the process variables pertinent to an operation is captured by subtask digraph models. Using this framework process variable deviations as well as process maloperation—wrong sequence of tasks and subtasks, wrong duration of subtasks, addition of wrong materials and occurrence of wrong reactions can be analyzed. *HAZOPExpert* and *Batch HAZOPExpert* have been tested on indutrial-scale petrochemical and pharmaceutical case studies [5, 6, 9] and have been reported to compare favorably with the human teams' results.

HAZOPExpert and *Batch HAZOPExpert* represent the current state-of-the-art in automating hazard identification. During PHA, after hazards have been identified, it is necessary to assess them in order to evaluate their severity and likelihood of occurrence. There are several consequence modeling tools for predicting consequences of accidental release—toxic, flammable and explosive damage—which are often used for hazard assessment. These tools need as input the rate and velocity of release, liquid fraction, temperature and pressure of the release. To this end, quantitative hazard assessment is required. Since *HAZOPExpert* and *Batch HAZOPExpert* use qualitative digraph-based models of process units and operations, they cannot be directly used for performing quantitative analysis. Here, we propose *PHAzer*, a system which can aid in the entire PHA process.

PHAzer is an intelligent multiple models-based system which can help perform a variety of hazards analysis activities. The current capabilities of *PHAzer* include:

1. Automated hazard identification
2. Quantitative hazard assessment
3. Fault tree synthesis
4. Fault tree analysis and reliabilty studies

These various uses of *PHAzer* are described in detail below. The architecture of *PHAzer* is shown in Fig. 1. The knowledge required for PHA has been separated into process specific and process general knowledge as proposed by Venkatasubramanian and Vaidhyanathan [10]. The process generic knowledge comprises of the following models of process units and operations: digraph-based causal model, dynamic mathematical model, and fault tree and reliability models.

Digraph-based causal model of the process equipment is required for HAZOP analysis of continuous operations and subtask digraph model is required for analysis of process transition. In order to perform detailed quantitative hazard assessment for hazards identified by qualitative reasoning, a dynamic mathematical model of process equipment or subtask would be required. The digraph models can be used for fault tree synthesis as well. However, in addition, fault tree models (also called operators) will be required to handle control loops. Once fault trees are synthesized, they can be quantitatively analyzed to evaluate the event probabilities. In order to quantify the fault tree nodes, event probabilities will be required for all the basic events modeled in the abnormal cause nodes of the HDG and the various failure modes of the control loops. Reliability models can also be associated with the various process equipment to perform reliability prediction and analysis. The process specific knowledge comprises of the process P&ID, the operating procedures and the process chemistry.

The *PHAzer* inference engine contains methods for performing hazard identification, hazard evaluation and fault tree synthesis and analysis. Hazard identification is performed as described by Vaidhyanthan and Venkatasubramanian [5] and Srinivasan and Venkatasubramanian [7]. Hence, hazard identifica-

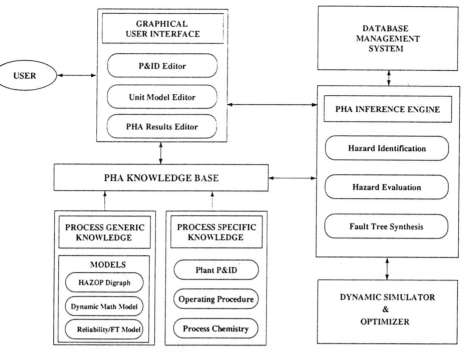

FIGURE 1. Architecture of PHAzer

tion for process variable deviations is by using *HAZOPExpert*'s algorithms and for plant maloperation using the algorithms implemented in *Batch HAZOPExpert*. The inference methods for quantitative hazard assessment and fault tree synthesis are described below. Using these PHA knowledge and inference methods, *PHAzer* can perform PHA for any process plant and generate abnormal causes, adverse consequences, quantitative hazardous scenario evaluation, synthesize fault trees and determine their minimal cut sets, and perform reliability calculations.

Management of Change

A large plant is subject throughout its life-cycle to numerous changes or modifications, both in design and operations. Chemical processes are integrated systems. A change in one part of the process can have unintended effects in other parts of the system. It is therefore, important that all changes be reviewed prior to their implementation to identify any potential hazards that may be created by modifications. Proper documentation and review of these changes are invaluable in assuring that the safety and health considerations are being incorporated into the operating procedures and process. Like the initial PHA, all these involve large quantities of labor, effort and money. In order to help in the management of change, *PHAzer* has the capability to perform PHA on only the changed portion of the plant. The previous PHA results for the unchanged portions can be retained if they are not affected by the change. The ability to propagate deviations all the way upstream and downstream, can help identify any hidden hazards which have been introduced by a change. These changes and their PHA results can also be automatically documented as described earlier.

Quantitative Hazard Assessment

Hazard evaluation using a quantitative process model can determine whether a given hazard is physically realizable, given the dynamic model of the process and ranges on process inputs and disturbances. Such an approach to the safety verification problem was recently proposed by Dimitriadis et al. [11]. In their framework, a state-transition representation of the system was used. Here, the process is deemed unsafe if it reaches an undesirable state under the influence of external inputs and equipment failures. The safety verification problem is formulated as a mixed integer optimization problem, where the objective is to minimize the amount of time during which the process is safe. One disadvantage of this framework is that the size of the resulting optimization problem can be very large for industrial scale processes. In the general case of plants where nonlinearities are present in the process model, the NLP or MINLP suffer from local optima problems. Thus, the safety of the process for such cases cannot always be guaranteed based on the results of this approach.

A hybrid approach for hazard identification and evaluation was recently proposed by Srinivasan et al. [12, 13]. Here, hazard analysis proceeds in two phases. In the first phase, a qualitative analysis of the entire process is performed. From the results of this qualitative analysis, scenarios which require further detailed analysis can be identified. In the second phase, a quantitative safety verification problem is formulated for each of these shortlisted cases. The quantitative analysis of a scenario indicates whether the hazard is realizable, and if so its severity. The first phase of analysis can be performed for all process variable deviations which can occur in the plant because qualitative analysis is computationally efficient. This analysis generates a list of causes and consequences for each deviation. The power of this phase of causal reasoning comes from the fact that it is complete. Most of the hazards identified at this stage are valid and do not need further analysis. However, some of the hazards flagged by qualitative analysis could be unlikely. This conclusion can only be arrived at by a more detailed analysis. Some hazards are therefore shortlisted to be analyzed further using quantitative safety verification. Since only a small number of the total set of hazards identified are ambiguous, detailed analyses need not be extensively used. In the second phase of analysis, quantitative safety verification is performed. For each hazard which has been classified as requiring further quantitative analysis, a mathematical programming problem is generated to identify the worst possible case according to the dynamic process model and its inputs. The dynamic model used at this stage need focus only on the section of the process under consideration and identified by qualitative analysis and not on the entire plant.

A conventional HAZOP analysis as practised considers only single fault scenarios, that is, when a process variable deviation is analyzed, only single faults which can lead to that deviation are considered to determine if that deviation is possible. This approach has the disadvantage that some scenarios which can be extremely hazardous could occur due to two (or more) faults occurring simultaneously. A thorough PHA should consider the occurrence of multiple faults also but without being computationally prohibitive. The integrated approach proposed here provides a framework for considering multiple fault scenarios. The quantitative approach considers *all* possible combinations of inputs which can lead the process to an unsafe state. Thus, multiple faults leading to hazards can be automatically detected.

After an hazard has been identified and assessed, it is necessary to evaluate it in terms of the risk—both probability and consequence—it presents. The quantitative hazard assessment procedure described here along with well established consequence modeling tools can be used to estimate the consequence of the hazard.

Fault Tree and Reliability Analysis

There have been numerous attempts in literature which have addressed the issues in automating fault tree synthesis for process plants. One approach which has received widespread attention is the digraph-based approach of Lapp and Powers [14]. This approach has been implemented in *PHAzer*. The HAZOP-digraph models of process units, used for hazard identification, are also used as the causal model for fault tree synthesis. Special operators have been implemeted to model control loops. Fault trees can be synthesized using the HDG and fault tree models and analyzed to determine the probability of hazards.

One fault trees are synthesized, they can be analyzed qualitatively and quantitatively as discussed above. In order to evaluate fault trees quantitatively, the probabiliites of primary events are required. These are represented in *PHAzer* by associating a probability with each fault in the abnormal cause node in the HDG. Similarly, fault probabilites are also associated with sensor and controller failure modes. Algorithms for identifying cut sets and evaluating top event probabilities are well established [15] and have been implemented in *PHAzer*.

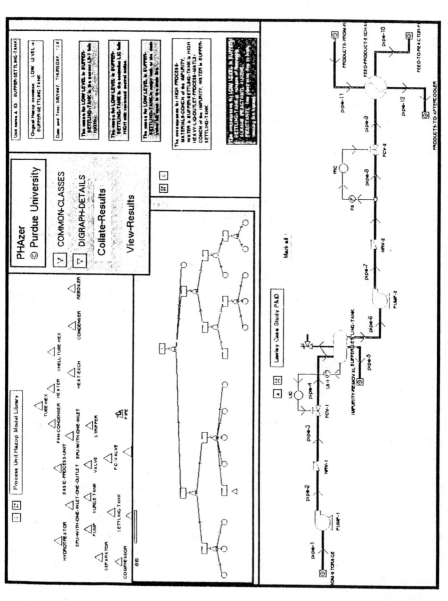

FIGURE 2. *PHAzer* User Interface

PHA Case Study

PHAzer has been tested on an olefin dimerization plant first described by Lawley [16]. The user interface of *PHAzer* for this case study is shown in Fig. 2. HAZOP analysis has been performed for the process. In the interest of space, detailed results are not presented here, however, it was noted that all hazards identified by the human team was identified by the *PHAzer* system. Fault tree analysis to estimate top event probabilities of various failures was also performed. A part of the fault tree for LOW level in settling-tank is shown in Fig. 2.

Conclusions

Process safety management and process hazards analysis are important activities. They are difficult, labor-intensive, and time-consuming, and can benefit by automation. In this paper, we presented *PHAzer*, a framework for automating PHA. One key feature is the separation of PHA knowledge into process-specific and generic components. These two components interact with each other to cover the process-specific aspects of PHA analysis while maintaining the generality of the system as much as possible. The process-generic knowledge of *PHAzer* comprises of multiple models, namely, digraph-based qualitative model, dynamic mathematical model and reliability and fault tree models of process units. Another important aspect of the framework is that the same system can be used to perform HAZOP analysis, quantitative hazard assessment, fault tree synthesis and analysis and reliability studies. The system can also aid in management of change and in plant operation.

References

[1] CCPS. *Guidelines for hazard evaluation procedures.* New York, 1985.
[2] T. A. Kletz. *What went wrong? : Case histories of process plant disasters.* Gulf pub. Co., Houston, 1985.
[3] T. A. Kletz. *HAZOP & HAZAN Notes on the Identification and Assessment of Hazards.* The Institution of Chemical Engineers, Rugby England, 1986.
[4] T. A. Kletz. *Plant design for safety : a user-friendly approach.* Hemisphere Pub. Corp, 1991.
[5] R. Vaidhyanathan and V. Venkatasubramanian. Digraph-based models for automated HAZOP analysis. *Reliability Engineering and System Safety*, 50:33–49, 1995.
[6] R. Vaidhyanathan and V. Venkatasubramanian. A semi-quantitative reasoning methodology for filtering and ranking HAZOP results in hazopexpert. *Reliability Engineering and System Safety*, 53:185–203, 1996.
[7] R. Srinivasan and V. Venkatasubramanian. Petri net-digraph models for automating HAZOP analysis of batch process plants. *Computers and Chemical Engineering*, 20(Suppl): S719–S725, 1996.

[8] R. Srinivasan and V Venkatasubramanian. Automating HAZOP analysis of batch chemical plants—Part I. Knowledge representation framework. *Submitted to Computers and Chemical Engineering*, 1997.

[9] R. Srinivasan and V Venkatasubramanian. Automating HAZOP analysis of batch chemical plants—Part II. Algorithms and application. *Submitted to Computers and Chemical Engineering*, 1997.

[10] V. Venkatasubramanian and R. Vaidhyanathan. A knowledge-based framework for automating HAZOP analysis. *AIChe Journal*, 40:496–505, 1994.

[11] V. D. Dimitriadis, N. Shah, and C. C. Pantelides. Modelling and safety verification of discrete/continuous processing systems. *AIChe Journal*, 43(4):1041–1059, April 1997.

[12] R. Srinivasan, V. D Dimitriadis, N. Shah, and V. Venkatasubramanian. Integrating knowledge-based and mathematical programming approaches for process safety verification. To Appear in the Proceedings of ESCAPE-97, May 26-29, 1997 Trondheim, Norway, 1997.

[13] R. Srinivasan, V. Dimitriadis, N. Shah, and V. Venkatasubramanian. Automating process safety verification using a hybrid knowledge-based mathematical programming framework. *Submitted to AIChe Journal*, 1997.

[14] S. A. Lapp and G. J. Powers. Computer-aided synthesis of fault-trees. *IEEE Trans. Reliab.*, R-26:2–13, 1977.

[15] J. B. Fussell, G. J. Powers, and R. G. Bennetts. Fault trees—a state of the art discussion. *IEEE Trans. Reliab.,*, R-23:51–55, 1974.

[16] H. G. Lawley. Operability studies and hazard analysis. *Chemical Engineering Progress*, 70:105–116, 1974.

Strategic Financial Risk Assessment for Railcar Business Acquisitions

Philip M. Myers and Richard S. Morgan
Four Elements Inc., 355 E. Campus View Blvd., Columbus, Ohio, 43235

ABSTRACT

Liabilities and financial risks can be evaluated to facilitate *strategic business* decision making, risk management, risk control, and risk finance. The risk is presented in terms of a metric clearly understood by business leaders and board level executives—the total *financial* impact, and includes risks to people, the environment, property, market share and the business integrity. Each of these types of risk are benchmarked against real world accidents to validate the risk model. Financial risks are evaluated to provide management with a clear understanding of *risks* to the business. These risks can then be compared against the *rewards*, and used in determining risk finance options as a part of the strategic business management process. A railcar business acquisition case study is provided to illustrate the benefits of a Strategic Financial Risk Assessment approach.

Introduction

Large quantities of chemicals and petroleum products are transported and used throughout the process and manufacturing industries. Safety, the environment, and impacts on business are key concerns to business managers and company executives. A serious accident can have significant, detrimental effects on the profitability and financial stability of any company.

While a number of companies are integrating risk management into their decision and business processes, it is often segmented. Some form of review may be carried out to address potential environmental issues, and another study undertaken by a different team to consider impacts on people. Yet another group may review accident rates, property damage, or lost product concerns and business interruption. Often, each independent group is not be aware of the other's activities. The business managers and company executives are often left with significant information gaps and the chore of trying to equate environmental, human, and business risks. This makes understanding of risks to the business and strategic risk management nearly impossible.

Four Elements, Inc., a member of the Environmental Resources Management (ERM) Group, has developed a methodology for addressing human, environmental, and business risks on a consistent financial basis. Financial Risk Assessments have now been performed for a number of years using this methodology for a variety of industries and applications. These studies have provided company executives and business managers with a complete, understandable picture of the risks posed by the business activity as the risks are presented using the language of business—the total *financial* impact.

Financial Risk Assessment Approach

This financial risk assessment technique was developed to aid corporations in the strategic management of assets, risks, and technologies. The technique is an invaluable step forward from classical Quantitative Risk Analysis (QRA). The approach utilizes the principles of financial risk analysis to measure impacts of hazardous material accidents on a uniform and consistent financial basis. This basis facilitates the comparison of various types of EHS and business risks, and incorporates the total costs of accidents to obtain an estimate of the financial liability from the undesired events. A financial basis naturally lends itself to the risk/reward evaluation of business ventures, alternative approaches, and potential risk mitigation options. This approach is based on the principles of sound risk management, and includes three basic elements: developing risk profiles,

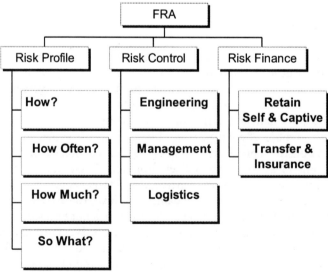

FIGURE 1. The Financial Risk Assessment Approach

examining risk control alternatives, and evaluating risk finance options. These building blocks are illustrated in Figure 1.

The impacts of accidental releases or spills are measured not only in terms of the acute fatalities and injuries, but also associated environmental and longer term impacts, property damage, product loss, and business impact. Finally, these are combined to arrive at an estimate of the overall *financial* risk (or risk profile) for the business or activity. Risk control options are then reviewed and may be re-evaluated. For those risks that remain, finance and risk transfer strategies are devised and evaluated.

Benefits of Approach

The Financial Risk Assessment approach has been applied to a number of diverse industrial operations for several years. These operations include chemical plants, power generation facilities, truck transportation, rail transportation, marine shipping, plastics processing facilities, refineries, petrochemical plants, pharmaceutical companies, terminals, gas processing facilities, pipelines, offshore platforms, and others.

Throughout the different applications, the end results are presented to business managers. The success of this approach lies within the ability to present to these business managers the whole risk picture on a uniform *financial* basis. In traditional QRAs, the results tend to be mathematical abstracts that are difficult to explain to non-"risk assessors". An even greater challenge is for a business to set guidelines based on these results.

Some of the benefits of using the Financial Risk Assessment technique are that it:

- presents the total risk picture using the language of business—the *financial* impacts
- directly combines the risks to people, the environment, property, production, and the business on a common metric without use of convoluted theories or artificial value systems
- produces results that can be easily communicated to board level executives
- provides essential information for executive or management decision making
- is invaluable in developing alternative strategic business options to achieve short and long term goals.
- presents the total liability
- reveals the maximum potential loss for financial planning
- presents the expected annual losses which can then be incorporated in business plans

- can be used to evaluate insurance needs, restructure insurance programs, assess the value for premiums paid, and optimize the insurance portfolio
- can be used to determine if certain risks make a business segment or activity prohibitive
- assists in evaluation of risk transfer
- can be invaluable in assessing alternative technologies
- uses a sound objective basis to evaluate risks

Railcar Financial Risk Assessment Methodology

For each type of business examined, the risk profile is developed by answering the questions shown in *Figure 1—How? How Often? How Much?* and *So What?* The detailed Financial Risk Assessment framework, as it applies to rail transportation analysis specifically, is shown in Figure 2. An extensive amount of data is integrated into the assessment of financial risk. This provides a measure of risk that is highly tailored to the specific operation. A description of the risk profile development follows.

How?

The question of *How?* is the question of "How can an accident happen?" This step includes the collection of accident histories for the specific business and chemicals involved. In this stage, all potential hazards are identified.

How Often?

The next component in the financial risk assessment process is the identification of the sequence of events or failure modes along with their likelihood. Review of rail accident data is undertaken to evaluate the chain of events necessary to result in a chemical release. For each proposed operation and commodity, the frequency of the following events are estimated using the fleet demographics:

- the number of "accidents"—defined as a railcar being derailed or otherwise damaged; and,
- the number of "releases"—defined as the loss of lading from a pressurized car including puncture, shearing of connections, etc.

To this end, the statistical accident rate data from the Federal Railroad Administration (FRA) for the last 20 years were evaluated. This data shows a generally decreasing accident trend since 1981. These data provided the derailment and collision rate for all railcars on mainline track, sidings, and in yards. The probability of chemical spills from pressurized cars as well as the relative proportion of minor, small, medium, and large spills are also determined. The

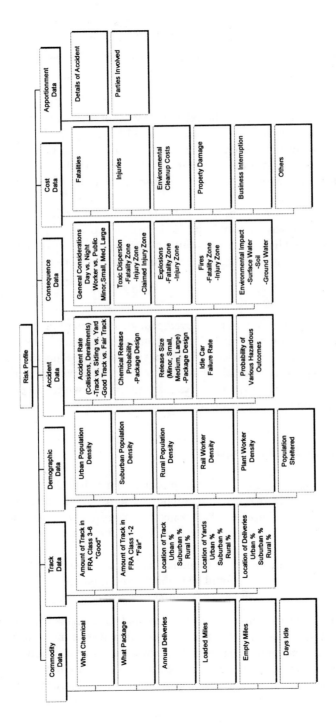

FIGURE 2. Financial Risk Assessment Framework for Rail Transport Analysis

potential for significant 'non-accident' releases involving failure of tankcar equipment is evaluated using data from AAR and other sources. The frequency of releases in various geographic/demographic areas is also calculated.

How Much?

The next question focuses on the extent or impact of the accidental releases. Traditionally, the answers to this question would be in terms of the extent of the resultant consequences from accidents and the number of persons affected, or the amount spilled into the environment. In a Financial Risk Assessment, an additional step is required: to examine the impact in terms of the potential financial liability or risk.

Following a rail accident with a major hazardous material release, a number of organizations may be named in the litigation process. Therefore, degree of responsibility and liability apportionment among potentially responsible parties is a critical factor in assessing the magnitude of financial risk.

So What?

Once the financial risk profile is developed, an assessment of the risks is made and presented to company executives and business managers. The results can be compared to company policies or financial risk guidelines, insurance coverages, and business plans to determine how the business can maximize profits and control and/or finance the risks. If the resultant risk is well within acceptable limits, no further action may be required.

Case Study Background

A large company was considering acquisition of a railcar fleet, involved in the transport of various pressurized commodities across all of North America. The company Board of Directors had concerns regarding the financial risks and would not authorize purchase of the fleet. The acquisition business team needed to:
- determine the increased financial liability to the company
- show how the risks would be managed and financed,
- demonstrate a high level of profitability to overcome potential risks.

The objectives of this work were to evaluate and assess the financial risks, assist with risk management and risk finance plan development, and to assist the business team in gaining board-level approval for the acquisition.

The project results have been altered for this presentation as the focus is the tremendous benefit of this approach rather than the specific results.

Transportation Accidents Involving Hazardous Chemicals

The data on accident statistics and severity of accidents were obtained from the Federal Railroad Administration (FRA), Association of American Railroads (AAR), and a variety of other sources. A review of accident statistics over the past 20 years, covering the pre- and post-hazardous material transportation regulations (HM144 and HM181) was conducted to identify overall trends. The results indicate the following:

- Between 1978 and 1986, the railroads invested over $40 billion in improving the equipment and an additional $30 billion in improving the track and rolling stock.
- During this period, the accident likelihood, as measured in terms of number of railroad accidents per million train miles was reduced by 55%. The number of accidents where a hazardous chemical was released was reduced by 63%.
- Hazardous chemical release related casualties declined by about 90%.
- The FRA, Association of American Railroads (AAR) and Railway Progress Institute (RPI) initiated pressurized tankcar improvements which have enhanced the overall safety performance of all pressurized tankcars. In particular, the 111, 112, 114 and 105 tankcar performance data indicates a significant decline in the likelihood of serious damage resulting in a major release.

The financial risk assessment included review of FRA and AAR data, and determination of accident frequencies, probabilities for tankcar damage, and magnitudes of release. Consequence models addressing the fire, explosion, toxic, and environmental exposure effects were used to determine the extent of hazard zones. The route specific demographic data were used to assess the overall exposure of people, environment, property, and business to such hazards. All of these methods were combined to yield an overall picture of the financial risk associated with the railcar fleet.

Financial Liability Associated with Transportation Accidents

In order to determine the potential financial liability of accidents, extensive data was collected and reviewed. A statistical analysis was performed to determine the costs associated with fatalities, injuries, environmental spills and damage, property damage, and other business impacts. Generally, distributions are used for the cost data, covering both the average and extreme (worst case) values.

The apportionment of potential liabilities was also studied. The liability of any given company is a function of a number of factors. For railroad accidents, the liable parties may include:

- railroad company,
- chemical manufacturer
- chemical shipper
- chemical receiver
- chemical owner
- tankcar owner
- tankcar manufacturer
- tankcar lessor
- tankcar lessee
- tankcar maintenance company
- others

One or more of the above categories may be applicable to any given company, increasing the potential liability with increased involvement.

Following review of the role of the case study company, and identification of additional potentially liable parties, two risk apportionment cases were developed. The first case was defined as "Most Likely" and is based on the data gathered. This study case represents the liability that the company can reasonably expect a majority of the time should an accident and associated impacts occur. The second study case is termed "Worst Credible" case and is the greatest liabilility the company may incur taking into account its involvement in the transport process. While not presented in this paper, an absolute worst case, with the company bearing 100% of the liability, was also evaluated.

Business Opportunity—Pressurized Tankcar Fleet Acquisition

To assess the potential financial liabilities, the accident likelihood and potential consequences for six selected chemicals, representing more than 80% of the "highly hazardous" chemical movements were modeled. The six selected representative chemicals were chlorine, ammonia, LPG, butadiene, ethylene oxide and vinyl chloride monomer. The proposed railcar fleet consists of about 7000 pressurized tankcars carrying a wide range of commodities all across North America. The Financial Risk Assessment technique was used to evaluate the financial risks posed by the fleet in terms of the likelihood and magnitude of potential losses, the average annual predicted loss, and the maximum loss.

Following is a brief summary of the financial risk assessment study results:

- The chance of the fleet causing a single fatality is about 1 in 10 in a given year. The serious injury likelihood is about 1 in 5 in a given year. A major accident involving multiple fatalities and injuries is predicted at less than 1 in 5,500 in a given year.
- The predicted worst case accident involves a large release of chlorine in an urban setting, in the middle of the night. This scenario, estimated at a

likelihood of about 1 in 175,000 years, is considered highly improbable when compared to the worst case accident to date in the US.
- In the event of a major accident, nearly all of the time other parties carry a large portion of the risk due to the nature of the company's involvement.
- In the "Most Likely" case, the company's share of the maximum liability is expected to be less than $150MM. For the "Maximum Credible" case, where the company may be identified as contributing to the accident, it is estimated that the share of the liability will increase to about $500MM.
- The expected annual average risk of the fleet is about $300,000 for the most likely scenario and $1.0 MM for the maximum credible scenario.

These results are presented in Figure 3 and Figure 4. Figure 3 shows a comparison of the various predicted financial liabilities with the corresponding recurrence periods. This figure contains a wealth of information about the financial risks posed by the fleet. This information can be assessed and presented in a number of ways to meet the company's needs. Specifically, the most important point this figure illustrates is that the "Maximum Credible" and "Most Likely" worst Case accident scenarios will be covered by the company's current insurance which has a deductible of $20 MM and a cap of $800 MM. Therefore, the company will not be put out of business even in the event of a large accidental release. In *Figure 4*, a comparison is made between the potential liabilities for the two risk apportionment cases and the insurance coverage of all potentially responsible parties. It is clear that the insurance coverage of parties likely to be named in a legal action is substantial, and would be sufficient to cover a very large event. Therefore, the insurance of other insured parties provides an added measure of protection against the full liability burden of an accidental release.

Once the risk profile was developed for the fleet, work was conducted to provide assistance in developing plans for risk management and risk finance. These plans were presented, along with the business plan, and ultimately board level approval of the acquisition was achieved.

Summary

Financial Risk Assessment is a very powerful tool, as it directly combines the risks to people, the environment, property, production, and the business on a common metric—the financial impact. As a result, financial risk assessment results can be easily communicated to board level executives whereas "classical" measures of risk are often not understood, complete, or considered entirely relevant. Therefore, a financial risk assessment provides essential information for executive or management decision making regarding strategic business plans to achieve short and long term goals. This technique presents the total risk picture using the language of business—the financial impacts—making it an ideal tool for evaluating mergers and acquisitions.

Figure 3   Comparison of Insurance Coverage with Predicted Liability

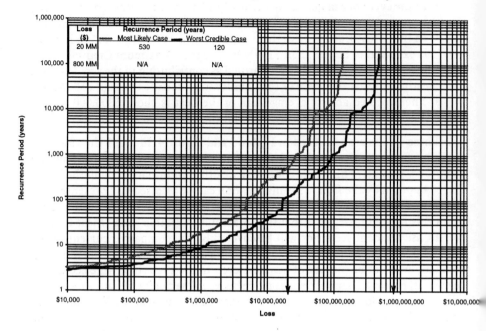

Figure 4   Worst Case Liability Distribution and Insurance Coverage

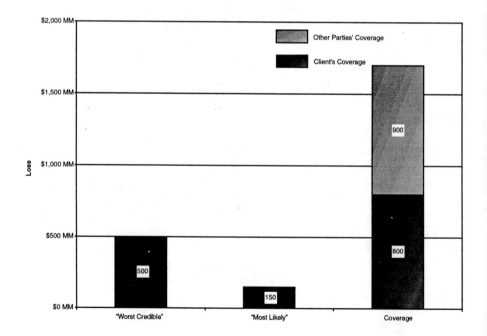

References

[1] Mudan, Dr. Krishna S., Shah, Jatin N., and Myers, Philip M., "Financial Risk Assessment A Uniform Approach to Manage Liabilities," AIChE Summer National Meeting, Boston, MA, August, 1995.
[2] Myers, Philip M. and Mudan, Krishna S., "A Financial Risk Assessment Case Study—Tank Truck Transportation of Chemicals and Petroleum Products," AIChE Spring National Meeting, New Orleans, LA, 1996.

Risk-Based Decision Making for Fire and Explosion Loss Control Strategies: A Practical Approach

Thomas F. Barry
HSB Professional Loss Control, Kingston, Tennessee 37763

Introduction

Every chemical plant faces risk in its daily operations. This risk often involves potential loss events—fire, explosion, power loss—which could halt processing operations. Completely eliminating risk is simply impossible, both physically and financially. Even minimizing the risk of experiencing a loss event will consume a certain amount of budgetary resources. Clearly, plants need a method to balance minimizing the risk of a loss event which could stop their production lines with maintaining a profit margin which can sustain their own viability.

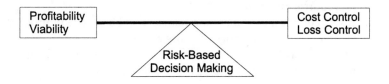

The risk-based decision making approach described in this paper outlines such a method. This method provides a basis for quantifying both the likelihood of a potential loss event and the consequences of that event, such as life safety exposure, property damage or business interruption. It thus provides a sound basis for prioritizing loss prevention budgetary allocations. By focusing a plant's limited resources to specific areas of significant risk as identified using this method, the plant's decision makers can optimize their investments in those risk reduction alternatives which will prove most cost-effective.

This paper applies the risk-based decision making methodology to one area of risk faced by chemical processing plants: loss of electric power through a fire or explosion loss event. To mitigate this area of risk, many chemical plants have their own power generation facilities on-site. Even those who use outside sources of power can benefit from this methodology by recommending it to their power providers and by applying its strategies to other processes within their facilities.

The paper uses a real-life risk-based evaluation of an electrical power generation process and associated fire/explosion protection equipment conducted by the author. The paper presents the benefits of the risk-based approach and its application to critical processes and the risk and reliability evaluation methodology, including risk analysis, event and fault tree modeling, and risk reduction cost/benefit analysis.

Quantitative Measurement: Better Management and Control

Risk-based quantification efforts provide measurement and monitoring to assist in the determination of loss control systems performance and business planning objectives. Figure 1 depicts the general benefits of applying fire and explosion (F&E) risk-based decision making to power and chemical production processes.

Benefits of the Risk-Based Approach

General benefits of risk-based decision making as applied to fire and explosion loss control include

- Ability to support the cost-effective solutions of complex fire and explosion loss control alternatives.
- Facilitated communication among managers. The decision process can make employees more supportive of organizational decisions.
- Improved management control over loss control expenditures and improved performance and profitability of the company.

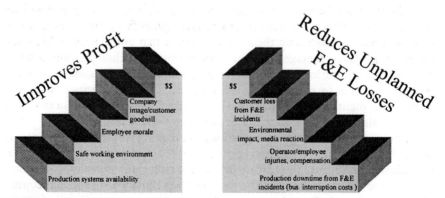

FIGURE 1. Risk-Based Assessment Benefits

- Cost savings from routine application of risk-based decision making, which usually results in considerable cost reduction or reduction/elimination of the cost of wrong decisions.

CASE STUDY—THE XYZ POWER PLANT PROJECT

Risk-based decision making techniques recently were applied at a large fossil fuel power plant to quantitatively assess business interruption exposures and the costs/benefits of risk reduction alternatives, including automatic fire protection system upgrades and emergency response program modifications. The following provides an overview of this project. To maintain client confidentiality, this project is referred to as the XYZ Plant.

Project Background

The XYZ Plant has downsized its operating and maintenance staff over the last few years. Plant management has limited human resources for supporting a fully trained structural fire brigade (i.e., it takes 55-75 trained people for a structural brigade) and limited monetary resources for making major automatic protection system upgrades. Production equipment in many areas of the plant is nearing the end of its useful life, and many plant update project proposals are competing for corporate funding. At the same time, corporate marketing is targeting customers for uninterruptible power contracts, as industry deregulation unfolds. The XYZ Plant has property damage insurance but not business interruption coverage.

Risk Evaluation Methodology

The primary focus of the XYZ Power Plant Project was the evaluation and prioritization of fire risk reduction alternatives as related to production downtime minimization. The methodology, which is graphically illustrated in Figure 2, included a four-step risk-based decision making process:

1. Appraisal
2. Analysis
3. Assessment
4. Actions

The decision making team included engineering, operations, and management representatives from the XYZ Plant. The author of this paper acted as the risk assessment team leader.

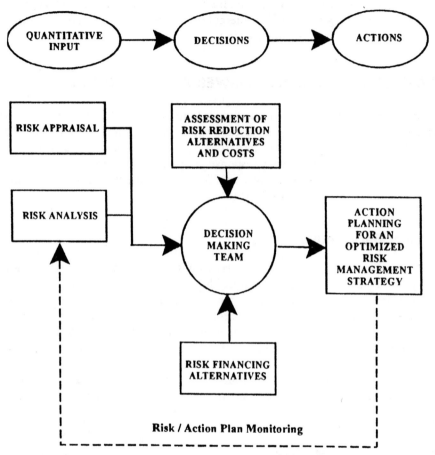

FIGURE 2. Risk-Based Decision Making Framework for XYZ Power Plant

1. Appraisal

During the appraisal segment, the project team members accomplished the following:

- Setting objectives with XYZ Plant Engineering, Operations, and Management
- Providing an equivalent monetary basis for property damage (PD) and business interruption (BI) loss potentials
- Establishing Risk (i.e., likelihood and consequence) Tolerance Limits

Objectives included

- Minimizing business interruption exposure from fire incidents

- Minimizing plant fire brigade utilization
- Maximizing risk reduction/cost ratios for fire risk reduction alternatives

Equivalent monetary basis involved

- Estimating property damage at current repair and replacement dollar values
- Estimating business interruption in terms of dollar loss per day of downtime based on an equation provided by the XYZ Plant

Fire Risk Tolerance Limits were set as follows:
- For the operations screened as critical to business continuation, an annualized residual monetary risk [$RISK = CONSEQUENCE\ (\$PD + \$BI) \times LIKELIHOOD\ (events/yr)$] of $25,000 per year per defined plant production unit was established as a comparative risk tolerance "benchmark."
- For fire brigade intervention, a relative conditional limit of 15% was used as a "benchmark" for needing structural fire fighting backup (the 15% is a relative limit, following the conditions of a fire occurring and automatic protection failing).

2. Risk Analysis

Risk analysis involved the estimation of the existing or baseline fire risk. This estimation included the frequency or likelihood of a fire incident occurring and the potential consequences.

Risk packages were developed to quantify several areas/operations which were screened during this project. Screening included a detailed plant site survey, interviews with key plant personnel, review of industry and plant-specific loss data, and agreement among the decision making team. These risk packages included event trees, fault tree analysis, evaluation of risk reduction alternatives, and cost/benefit analysis of selected risk reduction strategies. Risk packages were developed for the following plant units:

Risk Package 1—Coal Conveyors
Risk Package 2—Boiler Front Fuel Oil Fire
Risk Package 3—Turbine Generator Bearing Fire
Risk Package 4—Cable Spreading Room Fire
Risk Package 5—Main Station Transformers

Startup, operation, and controls for coal pulverizers, boilers, and pollution control systems were evaluated previously by the plant and were not evaluated as part of this project.

Over 20 event and fault trees were structured for this project. Point probabilities were used as input, not probability distributions. Sensitivity analysis of input was conducted in only a few cases. Overall, the results were felt to be within an order of magnitude.

EVENT TREE ANALYSIS

An event tree analysis (ETA) was developed for each risk package. ETA provides an inductive, forward-thinking process which conveys fire and explosion initiating events, intermediate events which contribute to the propagation or mitigation of the initiating event—such as the performance of loss control systems—and potential consequence levels.

ETA also provides a model to quantify risk levels by relating the likelihood of initiating events and conditional probabilities of "time-related" intermediate loss control events to consequence levels. Because fire and explosion exposure is very time-dependent, the event-time relationship capabilities of ETA are very important.

To structure event tree modeling logic, the risk evaluation team evaluated potential initiating events (e.g., equipment failures, human errors, external events which could lead to fires or explosions) and intermediate events such as fire detection and protection system responses and emergency procedures.

Figure 3 provides an example of an event tree developed in this project for the evaluation of coal conveyers. To simplify this presentation of the risk-based methodology, all examples are drawn from the same risk package, the one developed for the coal conveyers. The event tree depicted in Figure 3 displays the development of accident sequences, beginning with the initiating event and proceeding to the loss control systems responses. The results are clearly defined fire and explosion incident outcomes that can result from the initiating event. The loss control system responses are depicted chronologically.

Microsoft's® Excel computer spreadsheet program was used to develop event trees for all the risk packages. The example event tree shown in Figure 3 depicts the headings Initiating Fire Likelihood [A] and those for Intermediate Loss Control Events [B] → [E]. Event Tree logic is from left to right. This event tree indicates systems success in the upward (YES) branch segments. Probability of Success = 1 - Probability of Failure (i.e., from fault tree analysis). Branch probabilities are calculated by multiplying the Initiating Event Likelihood [A] and the Intermediate Loss Control Event probabilities [B] → [E].

The time line in Figure 3 establishes a frame of reference for estimating loss control system success probabilities and the magnitude of the consequences at various time intervals. For example, a 10-minute time limit could be established as the point where structural steel damage could start to occur from a fire exposure. Therefore, to be successful in minimizing damage, XYZ Plant needs to detect, shut down, and suppress before the 10-minute interval.

EVENT TREE SCENARIOS

Numerous possible scenarios may exist for the various fire and explosion initiating events and incident outcomes. It would be time- and cost-prohibitive to con-

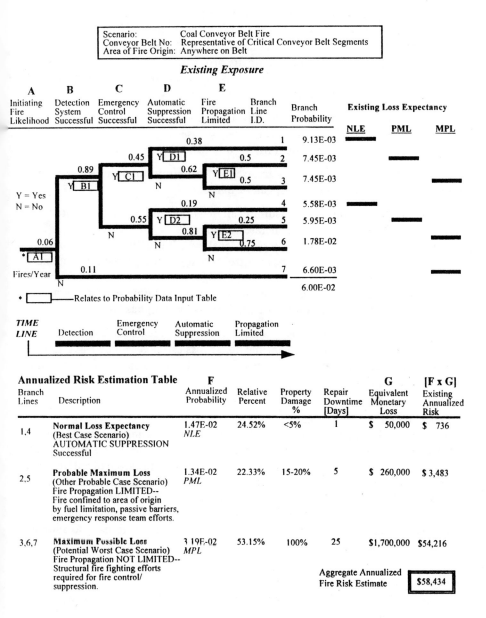

FIGURE 3. Example of Event Tree Model

struct event trees for all possible situations. Therefore, in many risk assessment projects, a focused approach must be conducted to identify and evaluate defined scenario situations of interest.

Of special interest in the fire and explosion risk assessment process are those event tree scenario branch lines which represent the best case scenario situation, the worst possible case scenario situation, and other likely case scenarios which may be of interest. Traditionally, in the fire protection industry, loss expectancy analysis has related these scenario situations to the following definitions, which initially were established by the fire insurance industry:

- Normal Loss Expectancy (NLE)—best case
- Maximum Possible Loss (MPL)—worst case
- Probable Maximum Loss (PML)—other likely scenarios

Although different variations are used in the definition of these loss expectancy levels, in general terms they can be defined as

- NLE   Loss expectancy assuming all detection and protection features are in service and operate as designed.
- MPL   Worst case loss expectancy level, which usually assumes that all detection, emergency control and automatic protection systems have failed
- PML   Loss expectancy assuming the primary automatic protection system (e.g., automatic water spray system) has failed.

An equivalent monetary loss (EML) basis was established to combine property damage and production downtime dollar loss estimates. The EML multiplied by the annualized probability [F x G] provides the estimate of the existing annualized fire risk.

PROBABILITY DATA INPUT TABLE

A Probability Data Input Table, such as the example in Table 1, was developed for each event tree, indicating the probability data source, method, and selection basis. In this table, the event tree provides an estimate of the annualized fire risk. Consequence levels were determined by using historical incident data—both industrywide and plant-specific—and engineering judgment based on current plant values. In many cases computerized fire models, such as FPETool from the National Institute of Standards and Technology (NIST), were used to support consequence evaluation efforts.

FAULT TREE ANALYSIS

As indicated in Table 1, under the column labeled *Selection Basis*, many of the frequencies and probabilities were developed using fault tree analysis (FTA). Figure 4 provides an example of an FTA developed for an initiating conveyor belt fire. A Probability Selection Basis Table (see Table 2) was provided for each

TABLE I
Example Probability Data Input Table

XYZ Plant XYZ Station 1996		Surveyed by: Thomas F. Barry, P.E. XYZ Plant Representatives	
Event Tree Scenario:	Scenario: Conveyor Belt No.: Area of Fire Origin:	Coal Conveyor Belt Fire Representative of Critical Conveyor Belt Segments	

EXISTING EXPOSURE
EVENT TREE PROBABILITY INPUT TABLE

Event	Description	Branch I.d.	Frequency (F)/ Probability (P)		Selection Basis
Initiating Fire	Conveyor belt fire	A1	0.06	F	Fault Tree Analysis A—Initiating Fire Likelihood
Detection Successful	Fire detection within 3–5 min.	B1	0.89	P	Fault Tree Analysis B—Fire Detection
Emergency Control Successful	Shutdown of conveyor belt within 3–5 min.	C1	0.45	P	Fault Tree Analysis C—Emergency Shutdown
Automatic Suppression Successful	Automatic sprinkler system: Operation within 3–5 min. Suppression within 10–20 min.	D1 D2	0.38 0.19	P P	Fault Tree Analysis D—Automatic Suppression Estimated at 50% D1 probability based on plant survey and engineering judgment
Fire Propagation Limited	Success is defined as action by the plant emergency team to • Limit flame spread along belt to 10–20 ft • Control fire with two 1.5-in. hose lines within 10–20 min.	 E1 E2	 0.5 0.25	 P P	Engineering judgment based on plant survey, review of conveyor belt and conveyor enclosure design, and existing manual fire fighting capabilities Estimated at 50% E1 probability based on plant survey and engineering judgment

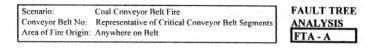

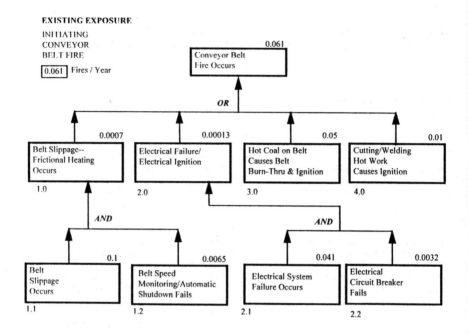

FIGURE 4. Example Fault Tree Analysis

fault tree, indicating the probability data source and selection basis. Some other items to note concerning these tables are as follows:

- Fault trees were developed in the Microsoft® Excel computer spreadsheet program; therefore, standard FTA symbols such as those for logic gates (e.g., AND, OR) and undeveloped and basic event symbols are not shown in the FTA presentation.
- The FTAs were developed on a "first-order" approach, meaning contributing failure events were only decomposed to a point where valid data input was available.
- In only a few cases was sensitivity analysis conducted. The majority of major contributing failure data was estimated from plant records and interviews with plant personnel.

TABLE 2
Probability Selection Basis Table For FTA-A

Event I.D.	Probability	Failure Probability Description	References/ Remarks
1.1 Belt Slippage Occurs	0.1	Plant indicated belt slippage primarily occurs during startup under wet conditions. Very rarely occurs under loaded operating conditions. Based on plant information and conservative engineering judgment, assigned a 0.10 probability of annualized belt slippage exposure.	Plant information/ Engineering judgment
1.2 Belt Speed Monitoring/Auto Shutdown Fails	0.0065	Belt speed monitoring switches automatically shut down the conveyor belt when a 15% variance in speed is detected. Belt misalignment switches are installed. Maintenance and calibration on this equipment appears to be good. Assigned a failure probability of 0.0065 based on upper bound failure rate data for electric speed switches.	Reference 1, Tax. No. 2.1.4.7
2.1 Electrical System Failure Occurs	0.041	Primary electrical system failure of interest is failure of drive pulley motors, causing electrical overheating and ignition of coal dust on the conveyor belt. Assigned a failure probability based on upper bound AC-Motor failure data: 0.41 failures x 0.10 (assumed overheating probability) = 0.041 failures resulting in overheating per year.	Reference 1, Tax. No. 1.1.1
2.2 Electrical Circuit Breaker Fails	0.0032	Assigned an upper bound failure probability, per demand, for AC-Circuit Breakers: 0.0032 failure to operate per emergency demand.	Reference 1, Tax. No. 1.2.3.1
3.0 Hot Coal On Belt Causes Burn-Thru and Ignition	0.05	Proprietary plant specific fire data	Plant Information/ Past fire incidents
4.0 Cutting/ Welding Hot Work Causes Ignition	0.01	Plant indicated occurrence of past incidents where cutting and welding sparks have ignited coal dust. Majority of these incidents are quickly extinguished. Based on this information and conservative engineering judgment, assigned an annualized probability of 0.01 for this event contributing to a critical conveyor belt fire. The plant has a formal cutting-welding permit procedure which was reviewed during the plant survey.	Plant Information/ Engineering judgment

Reference 1: Center for Chemical Process Safety, *Guidelines For Process Equipment Reliability Data, with Data Tables*, AIChE, New York, 1989.

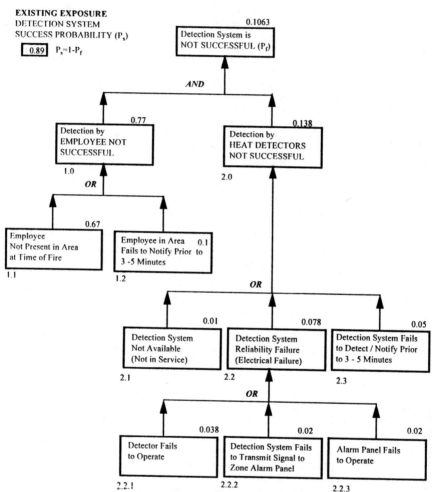

FIGURE 5. Example Fault Tree Analysis for Detection System

Fault trees were developed to quantify the conditional probabilities of loss control performance success. Several fault trees were developed to evaluate the following systems:

- Detection systems (primarily heat and air pilot line detection devices)
- Emergency control systems (mainly emergency shutdown devices; e.g., to shut down conveyor belts, oil pumps, electrical power)
- Automatic suppression systems (primarily deluge water spray system and pre-action sprinkler system)
- Flame spread limitation and manual suppression systems

Figure 5 presents an example of an FTA for evaluation of a conveyor detection system. The top failure event is the conditional probability of system failure (P_f). The needed input into the event tree is a performance success probability (P_s), which is obtained by: $P_s = 1 - P_f$. The time line related performance success response is 3–5 minutes by either manual or automatic means, as depicted in events 1.2 and 2.3 in this figure. In many cases, computerized fire models were used to estimate the time response for automatic detection and suppression, based on the initiating fire scenario. Table 3 provides an example of the probability data source and selection basis table used for the FTA depicted in Figure 5.

PROBABILITY DATA SOURCES

Probabilistic failure data were extracted from plant-specific data sources. This is the best source of data, and this type of data was obtained from operational and maintenance logs and records and from numerous interviews conducted with experienced plant personnel. Engineering judgments were based on plant surveys, available data, interviews, and experience and were made by risk evaluation team consensus. Published failure rate data sources included those in the following list:

Center for Chemical Process Safety, *Guidelines for Process Equipment Reliability Data*, With Data Tables, AIChE, New York, 1989.
Edison Electric Institute (EEI) Fire Incident Data, Proprietary Data.
EPRI NP-4144, *Turbine Generator Fire Protection by Sprinkler System*, Section 4-18, Fire Protection System Reliability, July 1985.
Federal Emergency Management Agency, Handbook of Chemical Hazard Analysis Procedures (ARCHIE Manual), Washington, D.C., 1989.
Fire-Induced Vulnerability Evaluation (FIVE), EPRI Report TR-100370, Final Report, Apri 1992.
Hathaway, Leonard R., *A 30-Year Study of Large Losses in the Gas and Electric Utility Industry*, Fifth Edition, Marsh and McClennan, March 1995.

TABLE 3
Probability Selection Basis Table for FTA-B

Event I.D.	Probability	Probability Selection Basis For Fta-B	References/ Remarks
1.1 Employee Not Present	0.67	Based on plant information, on an average people are present in the majority of conveyor system areas 33% of the time.	Plant estimate
1.2 Employee In Area, Fails to Notify Prior to 3–5 Min	0.1	Employees are trained in notification procedures. Under emergency conditions, assigned a 0.10 probability that detection and notification by an employee would not occur until approx. 5–10 minutes following a conveyor belt fire.	Plant information/ Engineering judgment
2.1 Detection System Not Available	0.01	Plant approximates systems out of service for 40–48 hours per year. 48 hours/8,760 hours per year = approx. 0.01	Plant information
2.2.1 Detector Fails to Operate	0.038	Based on heat detector reliability data. Because of detector location in harsh outside environment, used upper bound heat detector failure probability from Reference 1.	Reference 1, Tax. No. 4.2.2
2.2.2 Det. Sys. Fails to Transmit Signal to Zone Alarm Panel	0.02	Based on generic DC Relay Failure Rates in Reference 2.	Reference 2, Table 3.2, Relays
2.2.3 Alarm Panel Fails to Operate	0.02	Alarm fails to operate and notify plant personnel. 0.23 failures per 1,000,000 hours x 8.760 hours per year = 0.002 annual failure probability. Based on plant survey, some panels are in constant alarm condition; increased failure probability to 0.02.	Reference 3, Table 4-1. Modified via plant survey, engineering judgment
2.3 Det. Sys. Fails to Detect/ Notify Prior to 3–5 min	0.05	Based on field survey, detector type, location, spacing, assigned a low probability of failure to this event.	Plant information/ Engineering judgment

Reference 1: Center for Chemical Process Safety, *Guidelines For Process Equipment Reliability Data, With Data Tables*, AIChE, New York, 1989.

Reference 2: IEEE Standard 500-1984.

Reference 3: EPRI NP-4144, "Turbine Generator Fire Protection by Sprinkler System," Section 4-18, *Fire Protection System Reliability*, July 1985.

Hathaway, Leonard R., *A 30-Year Study of Large Losses in the Gas and Electric Utility Industry*, Fifth Edition, Marsh and McClennan, March 1995.

IEEE Standard 500-1984.

Matthews, Peter B., *A Decision Analysis Method for Evaluating Fire Loss Potential From Oil-Filled Outdoor Transformers*, Thesis, WPI, Worcester, Massachusetts, May 1988.

NFPA Fire Technology, Data and Expert Opinions in the Assessment of the Unavailability of Suppression Systems, May 1988.

Steciak, J., and Zalosh, R. G., "A Reliability Methodology Applied to Halon 1301 Extinguishing Systems in Computer Rooms," *Fire Hazard and Fire Risk Assessment*, ASTM STP 1150, Marcelo M. Hirschler, Ed., American Society for Testing and Materials, Philadelphia, 1992, pp. 161–182.

3. Risk Assessment

Risk assessment involved the evaluation of alternatives to reduce the existing risk levels estimated in the previous Risk Analysis step to the Fire Risk Tolerance Levels established in the Appraisal (i.e., $25,000 annualized residual risk per operation, and 15% or less structural fire fighting backup needed). Figure 6 provides graphical presentation of the conveyor Risk Analysis results. The top graph in Figure 6 indicates an established 43% risk reduction needed to meet the established annualized risk tolerance limit of $25,000.

Based on the results of the Risk Analysis, if the existing estimated risk exceeds the Risk Tolerance Limits, then risk reduction analysis should be conducted. *Risk reduction* can be defined as follows:

> *the identification and selection of cost-effective options for reducing or mitigating intolerable risks, including technological measures such as fire protection systems and/or management safety controls such as loss prevention programs, operator training, and emergency procedures.*

The information provided through the Risk Analysis must support the needs of the risk management decision makers. Table 4 summarizes some primary decision support issues of the XYZ Plant.

In each risk package, numerous fire risk reduction alternatives and combinations of alternatives were examined by computerized simulation in the fault tree of What-If changes in failure probabilities. This simulation quantified the changes in risk which could be achieved by the various alternatives. The best risk reduction strategies in terms of risk reduction benefits were identified.

Risk Reduction Analysis Worksheets

Risk Reduction Analysis Worksheets were developed for each risk package. These worksheets summarize the results of conducting What-If simulations for

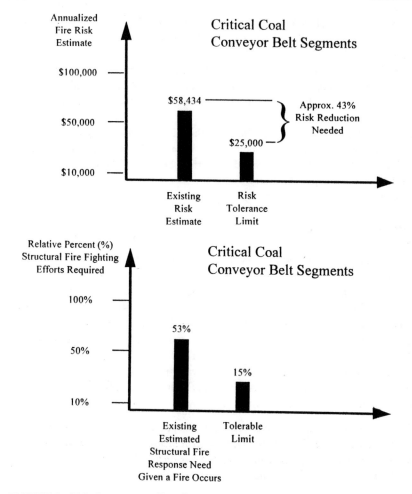

FIGURE 6. Risk Assessment Results

selected risk reduction improvements. Table 5 presents an example worksheet for conveyor fire risk reduction options which included

1. No action
2. Automatic conveyor belt interlocks
3. Sectional control valve monitoring
4. Electric booster pump improvements
5. Primary fire pump system improvements
6. Emergency response program upgrading
7. All risk reduction features (alternatives 2–6)

TABLE 4
Risk Reduction Decision Support Issues

DECISION CONCERNS	DECISION SUPPORT
How serious are the fire and explosion exposures (on-site, off-site)?	Graphical fire and explosion risk tolerance profiles
What regulatory implications (OSHA, EPA, building codes, etc.) are involved?	Code compliance reviews
How beneficial are the proposed risk reduction recommendations?	Estimated change in risk versus risk reduction options
How expensive are the risk reduction recommendations	Estimated initial and annual costs associated with each risk reduction option
How long will it take to implement the risk reduction recommendations? Can the work be done while the facility is operating, or will it require an extended shutdown?	Short summary of potential installation ramifications associated with risk reduction options
Do the recommended risk reduction changes represent an exception to corporate policy? Will company safety standards have to be modified? Will other similar facilities be affected?	Identification of any potential conflicts with present policies or procedures

The table in Figure 7 presents the resulting modified risk in terms of changes to event probabilities, changes in the relative percent of structural fire fighting efforts, and changes in the estimated annualized fire risk. The graph at the bottom of Figure 7 illustrates the estimated annualized fire risk versus the risk reduction alternatives. Risk reduction alternative 1 is to do nothing (retain the existing fire risk). Risk reduction alternatives 2–6 reflect the What-If improvements. Risk reduction alternative 7 is the modified risk for incorporating all the risk reduction features (i.e., alternatives 2–6). In this example, risk reduction alternative 7 would have to be implemented to reduce the existing risk below the $25,000 Risk Tolerance Limit.

Risk Reduction Cost Evaluation Worksheets also were developed in each risk package. These summarized the estimated costs associated with risk reduction improvements in terms of initial installation cost, acceptance testing cost, annual maintenance cost and estimated useful life. Cost estimates for risk reduction alternatives were provided by the XYZ Plant's engineering staff.

	1	2	3	4	5	6	7
RISK REDUCTION ALTERNATIVES	Do Nothing Maintain Existing Fire Risk	Automatic Conveyor Belt Interlocks	Sectional Control Valve Monitoring	Electric Booster Pump Improvements	Primary Fire Pumps System Improvements	Emergency Response Program Upgrading	All Risk Reduction Features (2 to 6) Added
Rel. % - Structural Fire Fighting Efforts	53.15	40.05	49.7	49.31	44.32	37.32	8.1
Estimated Annualized Fire Risk	$58,434	$46,047	$52,729	$48,272	$48,259	$52,988	$14,559
% Change in Structural Fire Fighting Response Needs	0	13.1	3.45	3.84	8.83	15.83	45.03
Change in Annualized Fire Risk	0	$12,387	$5,705	$10,162	$10,175	$5,446	$43,875

ANNUALIZED FIRE RISK REDUCTION

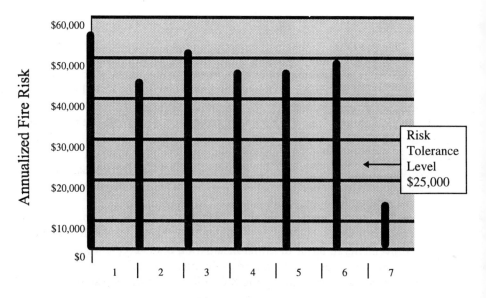

Risk Reduction Alternatives

FIGURE 7. Example Risk Reduction Analysis Worksheet

Actions

The information generated in the Risk Assessment was used to develop a ranking and prioritization of risk reduction opportunities. Ranking of risk reduction opportunities was done in three ways:

1. Risk Reduction Benefit (ΔR)
   ΔR is the estimated annualized change in risk for the risk reduction alternatives.
2. Risk Reduction Benefit (ΔR)/Cost (C) Ratios
   C is the estimated annualized cost, which includes the initial installation and acceptance test costs divided by the estimated useful equipment life, plus the estimated annual testing and maintenance costs. This estimate does not include the time value of money over the useful life of the asset.
3. Reduction in structural fire fighting response potential, presented as a relative percent change.

Table 5 provides an example of ranking risk reduction opportunities for coal conveyors in terms of annualized Risk Reduction Benefit (ΔR) listed in Column [D]. The items listed as common cause improvements indicate that these improvements provide a risk reduction improvement in areas of the plant beyond just the coal conveyor segments. The rankings in Column [D] represent an overall plantwide ranking.

Table 6 provides an example of Risk Reduction Benefit/Cost ratios for some conveyor items. The larger this ratio, the greater risk reduction benefit is achieved for the cost, and hence the more desirable the risk reduction opportunity.

Following reviews of the ranking information and cost estimates, a team from the XYZ Plant prioritized risk reduction opportunities over a 3- and 5-year implementation schedule based on a projected budget for fire and explosion loss control improvements.

Results

As part of XYZ Power Plant's decision making process, risk-based quantification was provided for fire and explosion loss potentials related to business interruption factors and fire brigade needs. This risk-based consulting services support provided the XYZ Plant with several tangible and useful results.

The project provided quantitative insight, screening, and ranking of fire and explosion exposures which could create unplanned shutdown of critical production systems at the plant. This provided a better understanding of fire and explosion impacts on the overall plant's production availability.

TABLE 5
Example of Risk Reduction Benefit Ranking Worksheet

[A] Risk Reduction Opportunities	[B] Estimated Annual Risk Reduction Benefit ΔR	[C] Estimated % Change in Structural Fire Brigade Backup Needs	[D] Overall Plantwide Ranking of Risk Reduction Benefit* ΔR	[E] Remarks
1.1.1 Upgrade detection to linear thermistor cable in critical conveyor segments 1.1.2 Provide automatic belt shutdown interlocks with manual override	$12,387	13.1	6	
1.2.1 Improve availability/reliability of fire protection control valves	$5,705	3.5	7	Common cause improvement
1.3.1 Improve electric booster pump (in boiler house) reliability	$10,162	3.8	9	
1.4.1 Improve diesel fire pump reliability	$10,175	8.8	2	Common cause improvement

NOTES:
*Risk reduction benefits for common plantwide improvements were added together for risk ranking:
Total Annual Risk Reduction Benefit
Improve availability/reliability of Items 1.2.1, 2.3.1, 3.2.1, 5.2.1 $11,371
fire protection control valves
Improve diesel fire pump reliabilityItems 1.4.1, 2.4.1, 3.3.1, 5.3.1 $24,612

The project quantified *risk*, which involved evaluating the likelihood of event occurrence and the consequences. By this approach, major opportunities for *risk* reduction were identified and evaluated. For example, instead of just focusing on fixed fire suppression systems, such as sprinkler systems for hazard reduction, major risk reduction opportunities were found in making design changes (e.g., design changes to contain combustible oil by piping and flexible hose modifications, tank high level alarms and pump interlocks) and program changes (e.g., monitoring critical fire protection valves, improved emergency response training) Also, major risk reduction was found in providing reliable

TABLE 6
Example of Risk Reduction Benefit/Cost Ratios

[A] Risk Reduction Improvements	[B] Estimated Annual Risk Reduction Benefit	[C] Estimated Annualized Risk Reduction Cost	[D] Risk Reduction Benefit [B]/ Cost [C] Ratio	Remarks
1.1.1 Upgrade detection to linear thermistor cable in critical conveyor segments 1.1.2 Provide automatic belt shutdown interlocks with manual override	$12,387	$10,866	1.14	
1.2.1 Improve availability/ reliability of fire protection control valves	$5,705	$2,350	2.43	Valves in coal yard for conveyor systems
1.3.1 Improve electric booster pump (in boiler house) reliability	$10,162	$3,175	3.20	

detection-emergency control system interlocks—such as quick shutdown of conveyors, oil pumps, and electrical equipment or operation of a smoke removal system in the control block-cable spreading rooms.

Prioritization of fire-related business interruption risk reduction strategies based on cost/benefit analysis allowed the plant to target major risk reduction opportunities and to optimize a 3-year budgeting plan to implement the highest priority system improvements. Since the primary focus of this project was to provide a *relative* ranking of risk reduction improvements, the consistency in the methodology applied in this project provided confidence in the final rankings and in the prioritization of action plans.

Conclusion

The risk-based methodology described in this paper offers a practical decision making approach to fire and explosion loss control strategies, especially given the current trend in industry to downsize staff and streamline budgets. The case study presented here provided the XYZ Plant with a quantifiable reduction in risk and with other cost-saving measures at a relatively low cost.

Over a 3-year period, it is estimated that the potential annualized risk of unplanned production downtime expenses (i.e., potential business interruption costs) from fire impact will be reduced approximately 40%–50%. In addition, the XYZ Plant received a detailed plan and cost estimate for implementing a highly trained plant emergency team response as a backup to automatic protection systems in lieu of a fully trained structural fire brigade. The cost savings to the plant is estimated to be in the range of $60,000 to $80,000 per year.

The project took approximately 3 months to complete and was performed using a risk assessment team approach. The best source of failure data was plant-specific data. Extracting and compiling this data involved conducting a detailed plant survey; reviewing operational, maintenance, and safety logs; and conducting interviews with experienced plant employees. The project used a Microsoft® Excel computer spreadsheet program to develop and quantify event and fault tree analyses.

The cost of conducting the fire and explosion risk-based evaluation was approximately equal to the estimated expense of 3–5 days of unplanned total production shutdown. This estimate is based on the plant's equation related to an uninterruptible power supply contract which assumed that contingency power backup may not be immediately available.

The opinion of the plant was that this risk-based project provided solid risk management decision support information so that better informed and justified decisions on fire and explosion risk reduction could be made.

Pressure Relief System Documentation: Equipment Based Relational Database is Key to OSHA 1910.119 Compliance

P.C. (Pat) Berwanger, R.A. (Rob) Kreder, and A. A. (Aman) Ahmad
Berwanger, Incorporated Houston, Texas

ABSTRACT

Under OSHA 1910.119, all Process Safety Management (PSM) facilities are required to keep their pressure relief system design information current. This article demonstrates why a pressure relief system design verification effort must be based on an equipment list, rather than a relief device list, in order to ensure that every piece of equipment is adequately protected. The formerly common practice of simply checking the design bases of all existing relief devices is deficient since this technique does not systematically ensure that every piece of equipment is protected.

The "Berwanger Method" is a step by step process for designing or analyzing a pressure relief system to meet OSHA 1910.119 Process Safety Information (PSI) and Process Hazard Analysis (PHA) mandates. The method uses a relational database which tracks the relationships between protected equipment, potential overpressure scenarios, and protective devices.

The challenge facing an operating company does not end once the design basis has been "verified"—the design basis information must also be maintained and be readily accessible to avoid costly reinvention of the wheel down the road. The "Berwanger Method" also addresses these maintenance issues.

Pressure Relief Systems and OSHA 1910.119: Three Goals

As process safety managers, we generally have three goals for a pressure relief system as it relates to OSHA 1910.119:

Goal #1: *Ensure the plant is adequately protected against all potential overpressure scenarios.*
To ensure protection, we must address all the potential overpressure scenarios associated with every piece of equipment.

Goal #2: *Maintain the safe design as changes are made to the plant.*
To maintain the safe design, we must have a realistic and reliable information management system.

Goal #3: *Implement the verification and maintenance effort at low cost.* To minimize costs, we must carefully define tasks and then execute each task with the most efficient people and tools.

Why the Old Pressure Relief System Data Structure Does Not Work

To understand why the old pressure relief system data structure, which is based on a relief device list and a filing cabinet full of relief device data sheets (see Figure 1) does not work, take a look at the typical distillation column example shown in Figure 2.

Five potential overpressure scenarios have been identified:

- inlet control valve failure
- blocked outlet
- loss of overhead coolant
- reboiler tube rupture
- external fire

For these five potential overpressure scenarios, we have a single PSV.

Thus, this example helps illustrate the natural relationships (i.e. data structures) that exist between pieces of equipment, overpressure scenarios, and relief

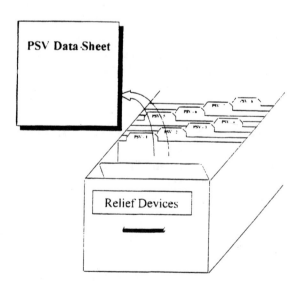

FIGURE 1. Traditional Relief System Documentation

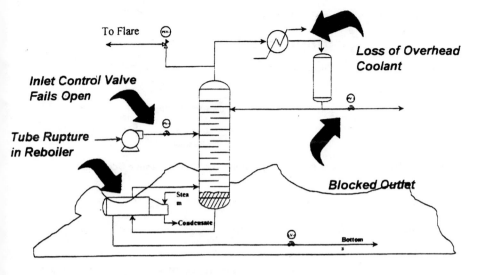

FIGURE 2. Example Distillation System

devices. Figure 3 shows that, in general, for each piece of equipment, there can be one or more overpressure scenarios and relief devices.

Now, let us remove the relief device from the distillation column example - as illustrated in Figure 4. And then, ask ourselves a few questions:

- If we are verifying overpressure protection in our plant by working our way down a PSV list, at what point in this "process" would we discover that a piece of equipment without a relief device (such as the distillation column) was unprotected?

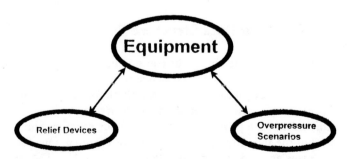

FIGURE 3. Relationships between Equipment and Overpressure Scenarios and Relief Scenarios

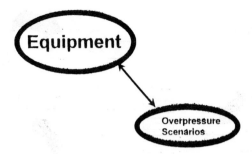

FIGURE 4. Relationships between Equipment and Overpressure Scenarios

Answer: There is not any point in this "process" for determining that a piece of equipment is without protection.

- Where would we store information about potential overpressure scenarios (that could effect the distillation column) if there was no relief device?

  Answer: We certainly could not store the information with the relief device data sheet. However, we could store the information with the equipment file.

- If we changed a relief device, does this change the potential overpressure scenarios?

  Answer: No, the potential overpressure scenarios would still be there.

- In light of the above, does it make sense to use a relief valve list, rather than an equipment list, as the basic checklist for relief system verification? Does it make sense to store potential overpressure scenarios with the relief device data sheets rather than the equipment?

  Answer: No, it doesn't—but, this is what a lot of folks have been doing.

An Equipment Based Data Structure Solves the Problem

The key to resolving our dilemma is the equipment information form in Figure 5, which lists protective devices as well as overpressure scenarios. If we prepare such a form for each piece of equipment in the plant, we have a means of tracking all the overpressure scenarios as well as the protective devices.

With this in mind, a physical metaphor for our relief system information would consist of the two filing cabinets shown in Figure 6. One for equipment information forms and the other for relief devices.

Pressure Relief System Documentation 755

Berwanger, Inc.
Equipment Information Form

Revision Record
Tag No: V-1
No: 01 Date: 04/19/95 Initials: AAA
Propylene Purification Vessel

Reference Information
Manufacturer:	Vessels, Inc.	**PFD No:**	100-1A
Nat'l Board No:	4235	**P&ID No:**	200-1A
		Iso No:	300-1A

Equipment Description
Type:	Vertical Vessel (Cylinder)	**Height/Length:**	20.0	ft
Material:	Carbon Steel	**Width/ID:**	5.0	ft
Mode:	Continuous	**Equip Volume:**	400.0	ft3
Contents:	Oxo-Purge	**Fluid Volume:**	400.0	ft3

Conditions
Normal Operating Pres:	350 psig	**Normal Operating Temp:**	150	degF
Max. Operating Pres:	360 psig	**Max Operating Temp:**	150	degF
Max. Working Pres:	400 psig	**Max Working Temp:**	150	degF

Other
Is equipment insulated?	Yes	**Insul. Mat'l:**	1" Ca Silicate
Is equipment cooled?	No	**Capacity:**	
Is fluid agitated?	No		

Equipment Notes

The vessel could be overpressured in the event of thermal expansion due to ambient temperature rise while the fluid is blocked in. The relief requirement is nominal compared to the external fire relief case.

In the event of an external fire, the initial relief will be flashing liquid. Upon formation of a vapor space, the relief will be vapor. The relief device should be sized for the case requiring the largest orifice area.

Protective Devices	Causes of Overpressure		
PSV-1	External Fire: V. Vessel w/ Liquid	57469	lb/hr
PSV-2	Failure of Automatic Controls: Inlet	25093	lb/hr
	Thermal Expansion	416.9	lb/hr

FIGURE 5. Equipment Information Form

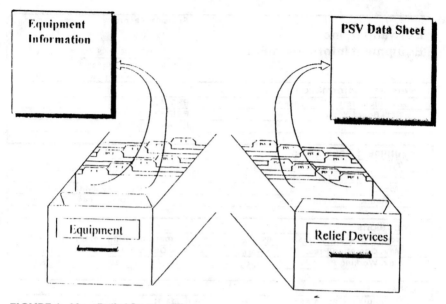

FIGURE 6. New Relief System Files

Of course, it would be even more efficient to store the relief system design information electronically in a relational database with the basic structure shown in Figure 7. In electronic format, we do away with the paper, and the information can be accessed from a PC and/or over a computer network.

A Step by Step Procedure for Analyzing a Relief System

OK, now we know how we can store and access the relief system design information. The next question, how do we create the information in the first place? The

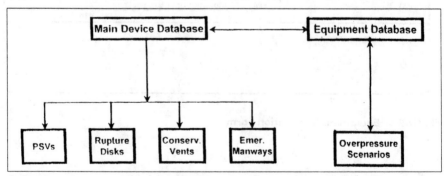

FIGURE 7. Electronic Relational Database Structure (patent pending)

four step process described below has proven most efficient for both new designs and verification of existing facilities. Berwanger has filed for patent protection of certain aspects of this process.

Step 1: *For each piece of equipment, an "expert" identifies all potential overpressure scenarios.*

This is unquestionably the most critical step in the process. Only after an overpressure scenario has been identified by the "expert" can it be addressed. Table 1 of API 520, "Bases of Relief Capacities Under Selected Conditions" provides a widely recognized checklist for this purpose (see Figure 8).

For the most part, the remaining three steps can be reduced to computerized algorithms. Except for quality control, the "expert's" work is largely done.

Step 2: *For each overpressure scenario identified in Step 1, quantify the required relief rate.*

The required relief rates for the majority of overpressure scenarios, such as tube rupture, external fire, control valve failure, etc. . . . can be "canned." The use of computerized algorithms, so long as their applicability are subject to the "expert's" approval, not only greatly improves efficiency, but also improves quality and consistency.

Step 3: *For each required relief rate in Step 2, verify that there is a relief device of adequate capacity.*

Again, relief valve sizing can largely be accomplished with a computer. For years, API 520 has provided the basis of most relief valve sizing equations. More recently, the Design Institute of Emergency Relief Systems (DIERS) has provided more accurate, and complicated, correla's) are concerned, OSHA 1910.119 does not limit itself to the "engineering"

API 520 Table I

1. Closed outlets on vessel
2. Cooling water failure to condenser
3. Top tower reflux failure
4. Sidestream reflux failure
5. Lean oil failure to absorber
6. Accumulation of noncondensables
7. Entrance of highly volatile material
8. Overfillong storage or surge vessel
9. Abnormal heat input
10. Failure of automatic controls
11. Split exchanger tube
12. Internal explosions
13. Chemical reaction
14. Hydraulic expansion
15. Exterior fire
16. Power failure

FIGURE 8

analysis which has been described above. Other elements which must be addressed are:

- hazards of the process (paragraph e.3.i)
- previous incidents (e.3.ii)
- engineering and administrative controls (e.3.iii)
- consequences of failures (e.3.iv)
- facility siting (e.3.v)
- human factors (e.3.vi)
- qualitative evaluation of safety and health effects (e.3.vii)
- team/employee input (e.4)

The author has addressed these other elements by:

- developing a written Process Hazards Analysis procedure, approved by company management, which clearly sets forth the scope, requirements, relevant documents, materials and equipment, and procedure for performing the PHA.
- performing operating employee interviews, and otherwise inviting employees to express their opinions, to ensure that their concerns are addressed
- review of previous incident reports
- review of previous PHA reports
- conducting a plant management review session, during which, for each concern raised, a qualitative risk is assessed, a timely plan for resolution developed, and responsibility for resolution assigned.
- issuing a detailed report which is then used to effectively communicate the substance and findings of the PHA.

Summary

Potential overpressure scenarios exist whether or not relief devices are there to protect the effected equipment. To ensure that every piece of equipment is protected it is essential that the pressure relief system design verification be based primarily on the equipment list. The formerly common method of simply checking existing relief devices is inadequate since this technique does not systematically ensure that every piece of equipment is protected.

It is most efficient to store relief system design information in a relational database where overpressure scenarios and relief devices are associated with the effected equipment. This system can be implemented electronically and made available across a computer network. Most of the necessary calculations can be "canned" which improves efficiency and quality control.

Relief systems can be designed and documented using the four step procedure which forms the heart of the "Berwanger Method."

Other OSHA mandated Process Hazard elements such as facility siting, consequence evaluation, and human factors have been integrated into the "Berwanger Method," so that in its entirety, the method can be used to meet OSHA 1910.119 PHA requirements.

Integrated Safety Analysis Project

Robert W. Johnson
Battelle, 505 King Avenue, Columbus, Ohio 43201

Mark Elliott
Babcock & Wilcox Navy Nuclear Fuel Division, P.O. Box 785, Lynchburg, Virginia 24505

ABSTRACT

Babcock & Wilcox operates its Navy Nuclear Fuel Division (NNFD) facility near Lynchburg, Virginia. The primary focus of the NNFD facility is the fabrication of uranium fuel elements for small nuclear reactors. NNFD also has process-oriented operations such as a recovery facility which dissolves uranium-containing scrap metal with concentrated hydrofluoric and nitric acids, followed by countercurrent organic extraction of the uranium.

A combination of What-If/Checklist Review, Hazard and Operability Study, and Failure Modes and Effects Analysis methods have been used for process hazard analysis (PHA) of NNFD operations, depending on the nature of each operation. An integrated approach has been used that allows the same order-of-magnitude estimates of cause likelihood, safeguard effectiveness, and impact severity to be used regardless of the PHA method employed. This also allows a single Risk Matrix to be used for action item decisions.

A key feature of the integrated safety analysis is the evaluation of all scenario impacts on a consistent order-of-magnitude scale. This allows the combination of diverse impact types (hazardous material release, fire, nuclear criticality, radiation exposure) when considering the scenario risk.

In addition to presenting the above methodology, this paper describes NNFD's needs that led to the development of the integrated safety analysis approach; presents example PHA worksheets; and discusses use of the order-of-magnitude risk results beyond the PHA action item decisions, such as for showing total risk reduction.

Integrated Safety Analysis Project History

Concerns regarding the concept of an Integrated Safety Analysis in a nuclear fuel fabrication facility began to surface in the U.S. Nuclear Regulatory Commission (NRC) soon after the January 4, 1986 accident at the Sequoyah Fuels Corporation's uranium hexafluoride (UF_6) conversion facility. In that accident, a cylinder of UF_6 ruptured, exposing the cylinder's contents to water vapor in the air and leading to the release of clouds of hydrofluoric acid vapor and uranyl fluoride particulates. Although the UF_6 and uranyl fluoride were radioactive, a more serious condition was the release of hydrofluoric acid which resulted in the death of a worker. This led to the conclusion that the NRC should expand its licensing and inspection procedures to ensure a comprehensive review of all aspects of fuel cycle licensees' activities important to safety (FR, 1989). Thus the terms "Integrated Safety" and "Integrated Safety Analysis" began to show up in NRC conversations with fuel fabrication licensees.

At the Babcock & Wilcox Naval Nuclear Fuel Division (NNFD), located near Lynchburg, Virginia, these conversations, and the issuance of four NRC Branch Technical Positions on March 21, 1989, began to capture management's attention. Meanwhile, the NRC's growing concern about integrated safety was heightened by incidents at Nuclear Fuel Services, Erwin, Tennessee in 1990 and the General Electric facility in Wilmington, North Carolina on May 29, 1991. This prompted NNFD to form several independent task forces to investigate and address, as necessary, similar safety issues. Realizing the need for coordination of these activities and the development of comprehensive and structured safety documentation, the Integrated Safety Analysis Project (ISAP) task force was established on December 15, 1991. Headed by one part-time person acting as project manager and several part-time team members, this newly formed group incorporated the scope and work of the other task forces.

During the next several months, the ISAP team developed a safety analysis methodology based primarily on the Hazard/Barrier/Target concept described in an Accident/Incident Investigation Workshop conducted by EG&G InterTech Inc. (Woodstock, Georgia) on October 8–12, 1990. The requirements for drawings, Process Descriptions, Block Flow Diagrams, and Risk Assessment as well as the details for performing individual safety analyses were also developed and described in written procedures. Documents such as Military Standard MIL-STD-882B, "System Safety Program Requirements"; Los Alamos National Laboratory document LA-11661-MS, "Guidelines for the Preparation of Safety Analysis Reports"; the NRC Branch Technical Positions discussed above; and NUREG-75/014, "Reactor Safety Study" were used for guidance. Also during this time, NNFD's Chemical Safety Engineer and member of the ISAP team began using a methodology known as hazard and operability (HAZOP) studies for the chemical safety part of the analysis.

On September 15, 1992, representatives of NNFD's ISAP team participated in the NRC's Fuel Cycle Workshop in Bethesda, Maryland, one day of which was devoted to Integrated Safety Analysis. Speakers representing several NRC licensees, including NNFD, and U.S. Department of Energy (DOE) contractors discussed their views on integrated safety. As was pointed out by one member of the NRC's Office of Nuclear Material Safety and Safeguards staff, NNFD's methodology, while on the right track, appeared to lack "integration." As a result of this meeting, several DOE documents were obtained, a representative from NNFD visited the Westinghouse–Hanford facility, and the pursuit of integration continued.

As part of NNFD's ongoing methodology refinement, the facility was divided into processes and prioritized according to the highest potential for a nuclear criticality accident. The Uranium Scrap Recovery facility, a chemical processing operation, was identified as the highest potential and became the first area ISAP analyzed. In November 1992, it was decided that HAZOP meetings involving all the safety disciplines be used for preliminary hazard identification. It was a natural fit for evaluating the chemical operations in Uranium Scrap Recovery. HAZOP studies forced the ISAP team to identify hazards as a group, but safety analyses and risk assessments were being performed individually. True integration still somewhat eluded the process. Work continued in this mode for the next year as evaluations were conducted on a part-time basis. Interesting findings were identified, and recommendations were made for safety improvement.

In January of 1994, it was recognized by NNFD management that the only way to complete the project was to have a full-time staff working on it. A manager was assigned in April 1994, and a support group, including engineers from chemical, criticality, fire, and radiological safety, was added. Additional methodology enhancements were implemented over the next year, and much good work was accomplished in spite of two temporary suspensions of the ISAP. During the week of February 6, 1995, NNFD ISAP personnel and certain other safety and operations engineers attended a training class at the NNFD, conducted by Battelle's Process Safety and Risk Management group. After the training and subsequent meetings with the instructor, NNFD's integrated safety methodology was merged with Battelle's integrated risk analysis approach. This resulted in a more tightly structured approach to the ISAP work, and brought true integration into the process.

ISAP Methodology: Scenario Development

The fundamental order-of-magnitude risk analysis approach used in the ISAP studies is given in a parallel presentation at this conference (Johnson et al., 1997). All risk parameters are estimated on an order-of-magnitude basis, and documented by only using the exponent of the risk parameter. This allows simple addition of the exponents to calculate scenario risk, since addition of exponents is the same as multiplication of the numbers themselves. For example, multiplying 100 times 1,000 is the same as multiplying 10^2 times 10^3, with the resulting product of 100,000 being equal to $10^{(2+3)} = 10^5$.

The underlying accident scenario structure is the same, regardless of the PHA methodology used to analyze it. This scenario structure can be represented

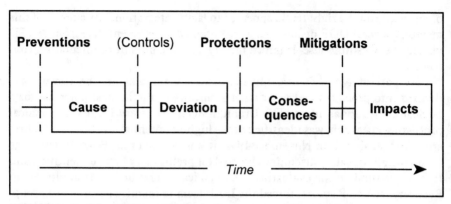

FIGURE 1. Accidemtal Event Scenario Sequence and Terminology

in simplified form as in Figure 1. This figure shows the terms in the accident progression from Cause (initiating event of the scenario) to Deviation (deviation from intended design or operation) to Consequences (irreversible accidental event) to Impacts (injuries, damages and losses resulting from the Consequences). The safeguards used in ISAP are Preventions (which keep Causes from occurring), Protections (which keep Consequences from occurring, given that a Deviation has taken place), and Mitigations (which reduce the Impacts, or severity of consequences).

The basic difference between PHA methodologies is where they start in the analysis of this accidental event sequence:

- HAZOP studies start with the Deviation, then look both backward (to identify Causes for the Deviation) and forward (to identify Consequences)
- Failure Modes and Effects Analyses (FMEAs) start with the Cause, which is identified by cataloging the failure modes of each component in the system
- What-If and What-If/Checklist reviews likewise start with the Cause, which is identified by the "What-If" question, and then develop the scenario as the "What-If" question is answered
- Fault Tree Analyses (FTAs), which ISAP has not employed to date, start with a specific Consequence and trace the accidental event sequence backward to its basic causes.

Hence, by keeping this common accidental event scenario structure in mind, and using the same terminology for safeguards (Prevention, Protection, and Mitigation), scenario risks can be analyzed by an integrated approach within the framework of any scenario-based PHA method. Examples of ISAP analyses using three different PHA methodologies but with a common scenario structure are shown in Appendix A (HAZOP studies), Appendix B (FMEA), and Appendix C (What-If).

ISAP Methodology: Risk Analysis

Order-of-magnitude risk analysis of ISAP accidental event scenarios is very similar, regardless of the PHA method employed. This is again due to the common accident scenario structure that is employed (Figure 1).

Cause Frequency

The first step in the risk analysis is the estimation of the Cause likelihood. Table 1 illustrates the Cause frequency scoring used in the order-of-magnitude risk analysis. The Score representing the frequency magnitude is entered onto the PHA worksheet in the corner of the Cause, Failure Mode, or What-If cell.

Protection Effectiveness

The next step in the risk analysis is to estimate the order-of-magnitude effectiveness of existing Protection safeguards. Table 2 gives typical Protection magnitudes and descriptions. A Protection Score of 2, for example, would correspond to a Protection that reduces the scenario frequency by two orders of magnitude (10^2 or 100-fold). In other words, the Protection is expected to be inadequate to keep the Consequences from occurring only once out of every 100 times the Protection system is challenged by a Deviation. Hence, if the Deviation occurs at a frequency of once a year, the Consequence would occur at a likelihood of only one chance in 100 per year of operation due to the Protection effectiveness.

For correctable systems such as hardware alarms and interlocks and configuration controls, protection effectiveness is a strong function of the duration of failure, or how long it takes to detect and correct a protection system failure. For the ISAP reviews, protection effectiveness magnitudes for such systems were

TABLE 1
Cause Frequency Magnitudes

Score	Frequency	Qualitative Description
+1	10/yr	Likely to occur repeatedly during the system life cycle
0	1/yr	Likely to occur several times during the system life cycle
−1	1/10 yrs	Likely to occur sometime during the system life cycle
−2	1/100 yrs	Not likely to occur during the system life cycle
−3	1/1000 yrs	Very unlikely to occur during the system life cycle
−4	1/10,000 yrs	Not expected to be possible

TABLE 2
Protection Effectiveness Magnitudes

Score	Frequency Reduction	Example
0	1×	No protection
1	10×	Protection by a single trained operator with adequate response time
2	100×	Protection by a single hardware system, functionally tested on a regular basis
3	1,000×	Protection by a single passive safety device, functionally tested on a regular basis; or single, tested hardware system with trained operator backup
4	10,000×	Protection by two independent, redundant hardware systems, each functionally tested on a regular basis

determined using a matrix of failure frequency and failure duration, such as shown in Table 3.

An example using Table 3 would be a protection system that failed about once a year (Failure Frequency Magnitude=0), and stayed in the failed state for a few days each time before being detected and corrected (Duration of Failure=Days). This would be assigned a protection effectiveness of 2 from Table 3, indicating that this system is in the failed state about 1% of the time (3 or 4 days out of each 365 days).

TABLE 3
Protection Effectiveness Magnitudes Based on Detectability and Correctability of Control Failure

	Failure Frequency Magnitude				
	−3	−2	−1	0	1
Duration of Failure	PROTECTION EFFECTIVENESS MAGNITUDE				
Indeterminate	0	0	0	0	0
Weeks	4	3	2	1	0
Days	5	4	3	2	1
Hours	6	5	4	3	2
Minutes	7	6	5	4	3

Mitigation

For ISAP, consequence severities (Impacts) are evaluated in four categories: chemical exposures (Chm), radiation exposures (Rad), fire exposures (Fire), and nuclear criticality effects (Crit). Mitigation measures, which act to reduce the severity of consequences, are specific to each impact type. For this reason, mitigation measures were documented on the ISAP worksheets using a numerical identifying system, as shown in Table 4. Criticality mitigators are individually specified.

Impacts

The impacts, or severity of consequences, in each of the four impact categories are assessed by the ISAP review team using an integrated Severity of Consequences Table that gives descriptions of consequence severities by impact magnitude. For example, an impact magnitude of 5 ("Very High" qualitative severity) is ten times more severe than an impact magnitude of 4 ("High"). The Severity of Consequences Table used for the ISAP reviews is shown in Appendix D. This Table is a key feature of the ISAP methodology, since it allows the comparing and combining of diverse impact types on an integrated magnitude scale.

For each impact type, a range is often specified to show the minimum and maximum expected severities. Following this, a single impact magnitude is

TABLE 4
Accident Mitigators

#	Chemical	Radiation	Fire
1.	Containment	Containment	Area-wide automatic fire suppression system
2.	Ventilation	Criticality monitoring and alarm system	Area-wide fire detection system
3.	Personal protective clothing	Personal protective equipment	Local coverage automatic fire suppression system
4.	Emergency response program	Ventilation	Local coverage fire detection system
5.	Personnel training	Ventilation failure alarm	Fire extinguisher(s)
6.	Spill procedure	Emergency response procedures	Qualified fire barriers
7.	Manual emergency alarm	Spill procedure	Dikes, liquid retention barriers
8.	Warning property of chemical (e.g., smell)	Radiation Protection Program	Other (specify)
9.	Other (specify)	Other (specify)	

selected by the review team as representing a most likely, or average, magnitude across all impact types. This single value is put into the "Severity" column of the worksheet and used in the determination of the scenario risk.

Risks

The "Severity" magnitude, as described above, is combined with the scenario frequency magnitude to determine the *risk magnitude* for each scenario. The scenario frequency magnitude is found by subtracting the Protection effectiveness from the Cause frequency.

For determining whether corrective action is required, the risk matrix in AppendixD is used to find the "Risk Zone" for each scenario. Scenarios in Risk Zone 1 require immediate corrective action. Scenarios in Risk Zone 2 require corrective action within a limited period of time. Risk Zone 3 represents an area of tolerable risk where no corrective action is required.

This type of risk matrix is in common use within the chemical and petrochemical industries, although the consequence severities are generally in qualitative categories, or ranges that are not on an order-of-magnitude basis. The order-of-magnitude risk analysis has the potential for value far beyond scenario-by-scenario decisions of the need for action. For example, scenario risks can be combined to determine a total risk for each process under study, and total process risks can then be ranked and compared. The quantitative risk-reduction benefit of safety improvements can also be determined using the risk magnitude values. True cost-benefit analyses can be performed by using the dollar-equivalent impacts shown on the Severity of Consequences table and inferring an annualized liability before and after the safety improvement.

It should also be noted that the use of risk matrices by themselves can lead to nonconservative risk decisions. For example, if a process has a large number of potential accidental event scenarios, and many of these scenarios have the same predicted frequency and severity (which is very common), then the total risk of all the scenarios combined may be well into the next risk zone. Using the order-of-magnitude risk analysis approach, it can easily be seen that ten scenarios with the same risk magnitude will combine to give the next higher risk magnitude. Likewise, the combining and ranking of the risks in this way lends itself readily to computerized database functions, as described in the paper by Johnson et al. (1997) in this conference.

References

Johnson, R.W., McSweeney, T.I., and Yokum, J.S. (1997). "Resource Optimization by Risk Mapping." International Conference and Workshop on Risk Analysis in Process Safety, Atlanta, Georgia, October 21-24.

FR (1989). *Federal Register*, Vol. 54, No. 53, March 21/Notices.

APPENDIX A
Example ISAP Worksheets:
Hazard and Operability (HAZOP) Studies   770

APPENDIX B
Example ISAP Worksheet:
Failure Modes and Effects Analysis (FMEA)   772

APPENDIX C
Example ISAP Worksheets:
What-If/Checklist Analyses   773

APPENDIX D
Severity of Consequences
and Risk Assessment Tables   777

PHA-3

OPERATION/PROCESS: Scrap Recovery/Main Extraction   **SESSION:**

NODE/ITEM: FCU/primary feed columns

DESIGN INTENTION: Provide favorable geometry equipment for storage, circulation and transfer of feed solution to the primary extraction column. Normal operating parameters for the feed solution are: 0.02 - 20 grams U-235/liter, density of 1.19-1.29 grams/cc, free acid of 2-5 molar, and pH of 0.9-1.3. (Chemical adjustment of the feed solution is accomplished by addition of nitric acid, aluminum nitrate, process water and (rarely) HF through the Add Column.)

Parameter/ Guide Word/ Deviation	CAUSE or FAILURE MODE (Human Factors) (Prevention)	EXISTING PROTECTION (Human Factors)	ACCIDENTAL EVENT CONSEQUENCES	SEVERITY ASSESSMENT					ACTION		
					Chm	Rad	Fire	Crit	Scenario #	Severity	Risk
storage/ no/ no storage (spill)	1. column break due to impact fire (-2)	chem/rad - none criticality: (1) operator intervention; (2) no sumps or stairwells where fissile material could collect in unfavorable geometry; (3) unfav. size containers not allowed in the area w/o rings (detection: hours) 0/4 -2	spill of acidic feed material from 4 or 5 columns (765 liters approx.) criticality if collected into unsafe geometry	Mitig Range	4,6,8, 9* 2-4	3,6,7 / 2,6 0-2/5	n/a	6	Bounds all other spill scenarios for criticality. 9* = eyewash. FC1-3a	3/6	3/3
storage/ no/ no storage (leak)	1. column break/crack due to wear and tear (P - material integrity, flange installation procedure) fire (-2)	chem/rad - none criticality: (1) operator intervention; (2) no sumps or stairwells where fissile material could collect in unfavorable geometry; (3) unfav. size containers not allowed in the area w/o rings (detection: hours) 0/4 -1	leak of acidic feed material criticality if collected into unsafe geometry	Mitig Range	4,6,8, 9* 1-3	3,6,7 / 2,6 0-2/5	n/a	6	Bounds all other spill scenarios for criticality. 9* = eyewash. FC1-3aa	2/6	3/3
storage/ no/ no storage (leak, spill)	flange/valve break P - maintenance, procedures, training, testing) fire (-2)	chem/rad - none criticality: (1) operator intervention; (2) no sumps or stairwells where fissile material could collect in unfavorable geometry; (3) unfavorable size containers not allowed in the area w/o rings (detection: hours) 0/4 -2	leak or spill of acidic feed material of 4 or 5 columns criticality if collected into unsafe geometry	Mitig Range	4,6,8, 9* 2-4	3,6,7 / 2,6 0-2/5	n/a	6	9* = eyewash. FC1-3b	3/6	3/3

page 15-2

PHA-3

OPERATION/PROCESS: Scrap Recovery/Main Extraction

NODE/ITEM: PPE3/primary evaporators phase separators and reflux columns

SESSION:

DESIGN INTENTION: Provide favorable geometry columns for separating liquid concentrated evaporator product from the vapor phase.

Parameter/ Guide Word/ Deviation	CAUSE or FAILURE MODE (Human Factors) (Prevention)	EXISTING PROTECTION (Human Factors)	ACCIDENTAL EVENT CONSEQUENCES	SEVERITY ASSESSMENT					ACTION		
					Chm	Rad	Fire	Crit	Scenario #	Severity	Risk
separation/ no/ no separation	blockage in phase separator portion [-1]	vent in constant head tank protects against overpressurization [1]	pressurized phase separator and breakage of the Pyrex reflux column	Mitig	4,6,8	3,7,8					
				Range	1-3	0-2	n/a	n/a	PPE3-1a	2	3
separation/ no/ no separation	blockage in reflux column portion [-2]	raschig rings in the condensate tanks [6]	liquid in phase separator and condenser, high U-235 in condensate criticality in condensate tank	Mitig		2,6					
				Range	n/a	5	n/a	6	PPE3-1b	6	3
transfer/ no/ no transfer	operator error - pump off, transfer valve closed (P - procedures, training) [-1]	sight glass and operator intervention (1) operator response to level in columns [1]	spill from primary product tank	Mitig	4,5,6, 9*	3,7			9* = operator intervention * Spills at PPE3 = 1. Spills at SPE3 and TPE3 = 2		
				Range	1-3	0-2 *	n/a	n/a	PPE3-2a	2	3
transfer/ no/ no transfer	operator error - drain valve left open (P - procedures, training) [-1]	none [0]	spill at the drain valve	Mitig	4,5,6	3,7					
				Range	1-3	0-2	n/a	n/a	PPE3-2b	2	3
transfer/ no/ no transfer	pump failure, plugged line [-2]	sight glass and operator intervention [1]	spill from primary product tank	Mitig	4,5,6	3,7					
	fire (-2)			Range	1-3	0-2	n/a	n/a	PPE3-2c	2	3

PHA-6 Uranium Recovery Ventilation Systems
FMEA Worksheets

FMEA6.DOC   05/09/97

Operation/Process: Uranium Recovery/Ventilation Systems
Node No.: DEU1- Dry Exhaust Unit
Description: Provide efficient removal of contaminants from dry process air streams (Recovery, CRF, and Conversion) by pulling contaminated air through prefilters and HEPAs before exhausting it out the roof unit. Prefilters have a 60-65% efficiency and the HEPAs have an efficiency of 99.97%. Air flow out of the unit was last measured at 25,000 CFM.

SCENARIO NO.	FAILURE MODE (Prevention)/(Human Factors) (C)ause or (B)arrier (Frequency)	EXISTING PROTECTION (Barrier) (Failure Detection Method) (Protection Score)	ACCIDENTAL EVENT CONSEQUENCES (or Barrier Against) (Severity Range) Crit	Chem	Fire	Rad	MITIGATION Crit	Chem	Fire	Rad	RECOMMENDATIONS/ NOTES Final Severity	Risk Zone
DEU1-10	dry exhaust fan failure (e.g. belt/pulley breakage(-1), power failure (0), fan blades disintegrate (-2), bearing seizure (-1)) (P - preventive maintenance for belt replacement and bearing lubrication) fire: [-2] 0	loss of ventilation alarm (supervised circuit) and evacuation of the area local differential pressure gauges coupled with operator action to shut down processes 2	NA	1-3	NA	0-2		4,5,8		8	2	3
DEU1-11	violating H/X limits due to water leaking into a dry glovebox via the dry ventilation system (P - separation of wet and dry system exhausts, ductwork typically attaches to side or top of gloveboxes - not the bottom) -2	size of containers or mass limit within glovebox unattended containers are capped within gloveboxes 2	criticality in the glovebox 6	NA	NA	5				2,6		
DEU1-12	U-235 build-up in dry ventilation ductwork in excess of 100 kg (P - local HEPA and pre-filters) -2	design of ductwork critical mass in excess of 700 grams lack of full reflection spacing of ductwork with respect to other fuel accumulations 3	criticality in duct 6	NA	NA	5				2,6	Add the LEU dissolver glovebox, LEU product enclosure, maintenance shop hoods to annual dry ventilation inspection survey. 6	3
DEU1-13	dry ventilation alarm does not annunciate when required (when dry ventilation fails) (P - preventive maintenance on dry vent. fans) fire: [-2] -1	enclosures [1] magnehelic gauge with operator action [1] dry ventilation supervised alarm circuit (trouble signal when circuit is not maintained) [1] 3	potentially high airborne leading to chemical and radiological personnel exposure NA	1-3	NA	0-2		4,5,8		8	Proceduralize the periodic testing of the Dry Exhaust System alarm. 2	3

Page 9-42

PHA-4
What-If Scenarios

WHATFWHF.DOC 05/09/97

Operation/Process: Scrap Recovery/Waste Handling Process
Node No.: 1
Description: In-Line Monitor No. 1

SCENARIO NO.	WHAT IF... (Prevention)/(Human Factors)	EXISTING PROTECTION (Human Factors)	POTENTIAL CONSEQUENCES (Severity Range)				Mitigation (Refer to codes on last page)				RECOMMENDATIONS/ NOTES	
			Crit	Chem	Fire	Rad	Crit	Chem	Fire	Rad	Final Severity	Risk Zone
1-3	What if In-Line Monitor #1 (and In-line #2 and #3) is not reference and background checked after recovering from a loss of power?	in-line monitors at retention tanks on UPS/emergency power and single automatic valve closure [3] normal operation of the raffinate extraction system results in low U-235 concentrations [1] favorable geometry at the WCS and LLRWD	criticality in the retention tanks										
			6	NA	NA	5				2,6	6	3
	Frequency: 0	Protection Score: 4									Safety Enhancement: Proceduralize requirement of reference checks and background checks when recovering from a loss of power.	
1-6	What if the operator fails to close one or more valves IM-2, IM-3, IM-4, or IM-8 during normal operation of In-line Monitor #1? (P - procedures, training) (H: procedures, valve location/tags)	none	small spill of raffinate waste from IM #1										
			NA	0-1	NA	0-1		5,6,8		3,7	1	3
	Frequency: 0	Protection Score: 0										
1-7	What if the operator fails to open one or more valves R-13, IM-4, or IM-5 and IM-6, IM-16, or IM-9 during normal operation of In-line Monitor #1?	operation response to alarm at overflow columns.	spill of raffinate waste at overflow columns										
			NA	1-2	NA	0-1		5,6,8		3,7	2	3
	Frequency: -1	Protection Score: 1									Safety Enhancement: Post the overflow columns with instructions noting what systems are automatically shut down due to an alarm and what systems that are not automatically shut down that could be overflowing into the columns. List in order of priority which systems to check first to stop the overflow.	

Page 10-2

PHA-4
What-If Scenarios

Operation/Process: Scrap Recovery/Waste Handling Process
Node No.: 7
Description: Dry Waste Handling
WHATFWHP.DOC 05/09/97

SCENARIO NO.	WHAT IF... (Prevention)/(Human Factors)	EXISTING PROTECTION (Human Factors)	POTENTIAL CONSEQUENCES (Severity Range)				Mitigation (Refer to codes on last page)				RECOMMENDATIONS/ NOTES	
			Crit	Chem	Fire	Rad	Crit	Chem	Fire	Rad	Final Severity	Risk Zone
7-1	What if there is a fire in the Maintenance Shop as a result of cutting operations used to disassemble and dispose of contaminated equipment, considering the 55-gallon combustible and 55-gallon noncombustible DLLW drum allowed in the area? (P: IH&S general procedures)	[1] fire: none [0] [1] rad: nature of process: contamination spread over volume of content, no SNM accumulations [1] [2] chem: exterior wall separating Maintenance Room from hydrofluoric acid pipe [2]	[1] fire causing radiological exposure [2] potential breach of the hydrofluoric acid line outside the wall [2]	[1]	1-3	0-2			2,5	3,4,6,8	2/4	3/3
	Frequency: -1	Protection Score: 0/1/2	NA	3-4								
7-2	What if the operator fails to identify that the maintenance hoods have inadequate or no airflow during their use for waste reduction/disposal (laundry sorting, etc.)? (P: procedures, training)	none	radiological exposure								2	3
	Frequency: -1	Protection Score: 0	NA	NA	NA	0-2				3,8		
7-3	What if the operator fails to identify that the Do-All saw has adequate lubricant flow over the contaminated cutting surface? (P: procedures, training)	nature of process: saw blade fails rapidly without lubricant	radiological exposure								2	3
	Frequency: -2	Protection Score: 1	NA	NA	NA	0-2				3,8		
7-4	What if the operator fails to identify that the General Purpose and UNH Sampling Hood has inadequate or no airflow during neutralization of liquid filters in the hood? (P: procedures, training)	filter not removed from container during liming	radiological exposure chemical exposure						5,8		3,8	1	Note: Neutralization can also be done in the Low Level Dissolver Hood. 3
	Frequency: -1	Protection Score: 1	NA	0-1	NA	0-1						

PHA-8
What-If Scenarios

05/09/97

Operation/Process: Scrap Recovery/Uranium Oxide Preparation Process												
Node No.: UOPF2												
Description: Dry and Calcining Furnace Glovebox												
SCENARIO NO.	WHAT IF... (Prevention)/(Human Factors)	EXISTING PROTECTION (Human Factors)	POTENTIAL CONSEQUENCES (Severity Range)				Mitigation (Refer to codes on last page)					RECOMMENDATIONS/ NOTES	
			Crit	Chem	Fire	Rad	Crit	Chem	Fire	Rad	Final Severity	Risk Zone
UOPP2-1	What if there is a non-process related exposure fire near the Furnace Glovebox that threatens nuclear material inside the glovebox? (P - ignition source control, combustible material control)	none	effects of fire		personnel exposure/contamination				2,5	4,6,8		* low combustible loading. Most probable fire unlikely to breach glovebox containment.
	Frequency: -2	Protection Score: 0	NA	NA	0-2*	0-2*					1	3
UOPP2-2	What if greater than 5 kg ^{235}U is placed in a furnace location? (Note: The box has a total limit of 10 kg ^{235}U.) (P - procedures, training, posting) (-H- "furnace location" on NCS sign is not defined nor marked in glovebox)	geometry of boats and less than or equal to 2.5 liter containers spacing in glovebox (pertaining to less than or equal to 2.5 liter containers) lack of full reflection total volume of containers < 10 liters	criticality in the glovebox							2,6		Recommendation: Mark the A and B calcining furnaces with tape on floor of glovebox to delineate where 5 kg limit applies. Note: Double batch in all containers is subcritical. Minimum critical volume is 10 kg which is > 10 liters.
	Frequency: 1	Protection Score: 5	6	NA	NA	5					6	3
UOPP2-4	What if there is a partial or complete loss of ventilation? (local causes: plugged HEPA filter on the exhaust side)	glovebox containment	area contamination personnel containment and exposure							7,8		
	Frequency: -1	Protection Score: 2	NA	NA	NA	0-2					1	3
UOPP2-5	What if a glovebox glove develops a pinhole leak? (P - glove integrity)	PPE negative pressure negative pressure	personnel contamination personnel exposure							7,8		
	Frequency: 0	Protection Score: 2/1	NA	NA	NA	0-1					2	3

Page 10-5

PHA-8
What-If Scenarios

Operation/Process: Scrap Recovery/Uranium Oxide Preparation Process
Node No.: UOPP3
Description: Drying and Calcining Furnace
Date: 05/09/97

SCENARIO NO.	WHAT IF... (Prevention)/(Human Factors)	EXISTING PROTECTION (Human Factors)	POTENTIAL CONSEQUENCES (Severity Range)				Mitigation (Refer to codes on last page)				RECOMMENDATIONS/ NOTES	
			Crit	Chem	Fire	Rad	Crit	Chem	Fire	Rad	Final Severity	Risk Zone
UOPP3-17	What if water is used to cool the door? (P - change control) Frequency: -3	integrity of the cooling system Protection Score: 2	criticality in glovebox if fuel is re-arranged into bottom of furnace by cooling water leaking into furnace	NA	NA	NA					2,6	6	3
			6	NA	NA	5							
UOPP3-20	What if furnace A and/or B is over-pressurized by the argon compressed gas supply? (P - 3" discharge line, exhaust ventilation) Frequency: -3	furnace (shell and retort, door) integrity [2] glovebox containment/ventilation [1]	personnel contamination and exposure if furnace fails area contamination		5,8	2,5	8					2/4/5	3/3/3
		furnace (shell and retort, door) integrity [2] oxygen monitor and operator action [2] area ventilation [2]	argon in room leading to personnel asphyxiation										
		furnace (shell and retort, door) integrity [2] Protection Score: 3/6/2	fire from breach of furnace door seal	NA	5-6	3-4	0-2						
UOPP3-21	What if excessive temperature (>370° ± 28° C for 5 minutes) in the exhaust from furnace A and/or B (non-contact cooling air) damages the final HEPA and/or pre-filter? (P - normal operations with programmed temperature control in furnace and air flow, chiller on exhaust) Frequency: -2	glovebox and HEPA filters on all other operations vented through the damaged HEPA filter except on High Temp. Furnace (contact air) Protection Score: 1	degraded HEPA filter leading to elevated radiological release	NA	NA	NA	1-3				8	2	3
UOPP3-22	What if excessive temperature (>370° ± 28° C for 5 minutes) in the exhaust from furnace A and/or B (contact cooling argon) damages the local HEPA and/or pre-filter)? Frequency: -2	final HEPA filter [2] large dilution volume in final discharge [1] Protection Score: 3	degraded HEPA filter leading to elevated radiological release	NA	NA	NA	0-2					1	3

SEVERITY OF CONSEQUENCES TABLE

On-Site							Off-Site	
SEV	Qualitative Descriptor Consequence of Event Order of Magnitude Loss in Dollars	Example Effects from Chemical Hazards Exposure	Example Effects from Fire Hazards Exposure	Example Effects from Criticality Hazards Exposure	Example Effects from Radiological Hazards Exposure	SEV	Qualitative Descriptor
6	Extremely High Multiple fatalities $10,000,000+	Fatalities as a result of contact with hazardous material; Environmental contamination requiring remediation of > $10,000,000	Third degree burns over large fraction of body surface of multiple individuals, smoke inhalation	Lethal radiation dose, plant shutdown	Lethal radiation dose	7	Catastrophic
5	Very High Fatality or multiple permanent health effects $1,000,000+	Significant chemical burns/respiratory damage as a result of contact with a hazardous material; release which requires offsite evacuation; Environmental contamination requiring remediation of $1M - $10M	Third degree burns over large fraction of body surface, smoke inhalation		Lethal radiation dose, acute radiation syndrome, evacuation of off-site population, administration of stable iodine, an offsite exposure exceeding 1 rem effective dose equivalent or an intake of 2 milligrams of soluble uranium	6	Extremely High
4	High Permanent loss of function/limb or multiple lost-time injury $100,000+	Reduced respiratory capacity/chemical burn requiring hospital stay; Offsite release of hazardous material which requires filing of federal or state reports; Environmental contamination requiring remediation of $100K - $1,000,000	Extensive third degree burns		Exposure of multiple workers in excess of 10 CFR 20 limits, remediation of on-site environmental contamination, off-site release causing doses in excess of 10 CFR 20 limits but would not exceed 1 rem effective dose equivalent or an intake of 2 milligrams of soluble uranium	5	Very High

	On-Site				Off-Site
3	Intermediate Restricted/lost time work injury or multiple medical treatment cases $10,000+	Respiratory distress/chemical burn requiring multiple medical treatments; Environmental contamination requiring remediation of >$10K - $100K	Third degree burn and/or extensive second degree burns		4	High Exposure of worker in excess of 10 CFR 20 limits, temporary impairment of kidney function from inhalation of soluble uranium, personnel contamination > 500k dpm / 100 cm^2
2	Low Medical treatment or multiple first-aid cases $1,000+	Respiratory irritation or minor chemical burn; Environmental contamination requiring remediation of $1K-10K	Second degree burn or extensive first degree burns		3	Intermediate Radioactive material spill, personnel contamination < 500k dpm / 100 cm^2
1	Very Low Minor injury; first aid case <$1000	Nuisance odor; minor respiratory irritation; Dermatitis; Environmental contamination requiring remediation of >$1000	First degree burn		2	Low Radioactive material spill, personnel contamination <50k dpm / 100 cm^2 exposure of workers below 10 CFR 20 limits, offsite release causing doses below 10 CFR 20 limits.
0	Negligible Probably no health effects; nuisance odor, visible plume	No noticeable effects			0-1	Negligible to Very Low Personnel contamination < 5k dpm / 100 cm^2, exposure of workers below license action levels, off-site release below license action levels

Risk Assessment Table

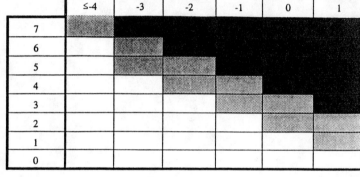

= Risk Zone 1 (immediate corrective action required)

= Risk Zone 2 (corrective action required within time limited waiver)

= Risk Zone 3 (no corrective action required)

Air Modeling Issues Associated with the Risk Management Program

Geoffrey D. Kaiser
Science Applications International Corporation, Reston, Virginia

ABSTRACT

On June 20, 1996, the Environmental Protection Agency (EPA) published its Risk Management Program (RMP) Rule, 40 CFR Part 68. This rule requires Hazard Assessments that include worst-case scenario analyses in atmospheric stability category F with a windspeed of 1.5 m/s. There are numerous uncertainties associated with atmospheric dispersion calculations in these conditions. These uncertainties include (but are not limited to): (a) deposition mechanisms for vapors and droplets; (b) time-varying toxic endpoints; (c) puff versus continuous release models; (d) plume meandering; and (e) limited persistence of the stated weather conditions.

Overall, these uncertainties make it difficult to make credible predictions of the distances to which a release travels before the airborne concentration falls below the toxic endpoint. In addition, it makes it difficult to develop generic guidance that is both simple to use and is believable. The purpose of this paper is to discuss how these issues were addressed in developing the EPA's "Model Risk Management Program and Plan for Ammonia Refrigeration'" which was published as draft guidance in June 1996. The approach adopted included: (a) a full and frank discussion of the uncertainties and the effect they have on predictions; (b) continual consultation with industry organizations and EPA in order to understand concerns; and (c) the development of caveats and guidance that will help owners and operators to put the worst-case scenario into perspective when communicating with the public.

Dry Deposition Mechanisms for Vapors and Droplets

Vapors and droplets are removed from the base of a plume by dry deposition. One simple model of dry deposition [1] states that

$$C_D(x,y) = v_d\, C(x,y,z) \tag{1}$$

where C_D is the rate at which material is deposited onto the unit area of the ground at position (x,y), measured from the point of release; v_d is the dry deposition velocity (m/s); and C is the airborne concentration at point (x,y,z). The variables x, y and z are the respective distances downwind of the source, across the wind and above the ground, measured in meters. There are three pertinent observations:

a. There are numerous factors that affect v_d [2, 3]. For reactive vapors, these include interactions with vegetation and moisture on the ground. In general, a vapor that causes toxic health effects will be reactive. Dry deposition velocities as high as several cm/s are not out of the question. However, the range of uncertainties is very large, with v_d being uncertain within a range of more than an order of magnitude.
b. Many of the chemicals on the EPA's RMP list are denser-than-air (loosely speaking, heavy). Heavy vapor clouds slump to form broad, low clouds, which should, in principle, favor dry deposition since the vapor cloud presents a very large surface area to the ground. However, the author is not aware of any validated models of the dry deposition of vapors from the base of a heavy vapor cloud.
c. If a gas is stored as a liquid under pressure and then released as a liquid jet, there can be the formation of fine liquid droplets that remain airborne for considerable distances, until they either fall out or evaporate. This phenomenon has been observed in a great number of experiments, including those involving materials on the EPA's RMP list, such as chlorine, ammonia and propane. Theoretical models or computer programs that address this issue are quite complex and not suitable for use in simple guidance.

In developing the model ammonia plan, sensitivity studies were performed, an example of which is shown on Figure 1. Case I represents a "base case" model, in which there was no dry deposition. The details of the particular model are not very important. In summary, it contains an initial module in which ammonia is treated as a mixture of ammonia vapor, droplets and cold air that is heavy. As the plume travels downwind, it trends asymptotically towards a Gaussian model that is like the rural-site model in EPA's "Technical Guidance for Hazards Analysis" [4]. The releases are assumed to be worst-case ones that are released at a uniform rate over ten minutes, as required by the rule. Figure 1 presents the predicted distance downwind to the toxic endpoint, which, for ammonia, is 200 ppm, as a function of total mass released.

Case III on Figure 1 presents a calculation in which the ammonia is assumed to have a dry deposition velocity of 0.01 m/s (1 cm/s) at distances downwind at which it behaves like a passive gas. For the reasons stated above, dry deposition was not included near the surface, where the mixture of ammonia and air is heavy. That is, Case III may understate the effect to be expected if dry deposition is taken into account, and yet it makes a very large difference to the predicted outcome of the releases.

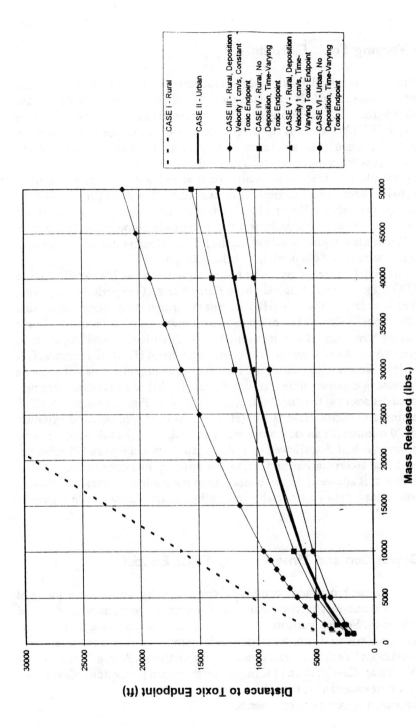

FIGURE I. Worst-Case Ammonia Release over 10 Minutes: Sensitivity Studies

Time-Varying Toxic Endpoints

As required by the rule, the toxic endpoint for ammonia is 200 ppm. This is independent of exposure time. It is, in fact, the ERPG-2 (Emergency Response Planning Guideline [5], as published by the American Industrial Hygiene Association [AIHA]), and it is valid for an exposure time of one hour. However, the worst-case scenario has a duration of only ten minutes and, at least near the source, will pass by in much less than an hour.

In principle, the airborne concentration that will cause a given health effect is likely to be a decreasing function of exposure time. One of the simplest expressions of this principle is Haber's law [4], which states that, for a given health effect, the tolerable airborne concentration is inversely proportional to exposure time. If Haber's law applies to ammonia, the ERPG-2 for an exposure duration of ten minutes would be 1200 ppm, rather than 200 ppm.

Unfortunately, there is generally not enough data available to tell whether the ERPG-2 for a given chemical obeys Haber's law. (One perhaps surprising exception is hydrogen fluoride [HF]. Recent experiments on rodents show that the ERPG-2 obeys Haber's law down to an exposure time of two minutes. These experiments were carried out in support of a Quantitative Risk Assessment [QRA], in which Mobil examined the continuing use of HF at its Torrance, CA, refinery [6].) Nevertheless, assuming that the toxic endpoint does not increase with decreasing exposure time introduces considerable conservatisms into predictions of distances to the toxic endpoints. Case IV on Figure 1 shows the effect of assuming a modified Haber's law (Haber's law is obeyed down to exposure times of 20 minutes; at an exposure time of five minutes, the tolerable airborne concentration is half that allowed by Haber's law; between five and 20 minutes, there is a linear interpolation; and, below five minutes, no further increase in the toxic endpoint is allowed.) As can be seen, such plausible time-varying behavior of the toxic endpoint can considerably reduce the predicted distances on Figure 1.

Dry Deposition and Time-Varying Toxic Endpoints

Case V on Figure 1 shows the combined effects of a dry deposition velocity of 0.01 m/s and a time-varying toxic endpoint. Even greater reductions in predicted distances result. For comparison, Case II shows the results of a worst-case analysis on an urban site in which the model tends asymptotically to the urban Gaussian model from "Technical Guide for Hazards Analysis". As can be seen, Cases II and V overlap. Case VI shows the impact on the urban predictions of assuming a time-varying endpoint in the urban case.

There are two pertinent comments:

a. Taking account of dry deposition velocity and/or time-varying toxic endpoint greatly reduces the difference between rural-site and urban-site predictions. This is an important result. It shows that predictions of extremely large worst-case scenario distances to toxic endpoints at rural sites are likely grossly exaggerated.
b. Including dry deposition and/or time-varying toxic endpoints has a much smaller effect on urban-site prediction than it does on rural-site predictions. This is because the urban model incorporates much more rapid dilution than does the rural model, thereby considerably lowering the concentration at the ground and reducing the rate of dry deposition. In addition, the much more rapid dilution means that the distance over which the predicted concentration will fall from (say) 1200 ppm to 200 ppm is much less in the urban case than in the rural case.

Puff versus Continuous Releases

In practice, a vapor cloud that is released over a period of only ten minutes will begin to spread in the alongwind direction, as well as in the acrosswind and vertical directions. Far enough downwind, it will look the same as if it had been released as a puff. Case IP on Figure 2 shows the predicted distances to the toxic endpoint at a rural site, using a simple puff model. For comparison, Case IIP presents the results of a puff model at an urban site. For a worst-case release taking place over a period of ten minutes, at the right-hand end of Figure 2, Case IP is a better prediction than is Case I on Figure 1. The other curves on Figure 2 show how dry deposition and a time-varying toxic endpoint affect the predictions of the puff model.

Other Uncertainties

There are some phenomena that have not yet been addressed at all. For example, atmospheric stability category F conditions, with a windspeed as low as 1.5 m/s, are unlikely to persist as the plume travels long distances. Therefore, worst-case predictions of distances of several tens of miles are likely unrealistic for this reason alone. In addition, this low windspeed weather condition is characterized by large fluctuations in the wind direction. It is quite possible that, even over a period as short as ten minutes, different portions of the plume will travel in different directions, so that much of the released material will not be closely concentrated about the nominal plume centerline.

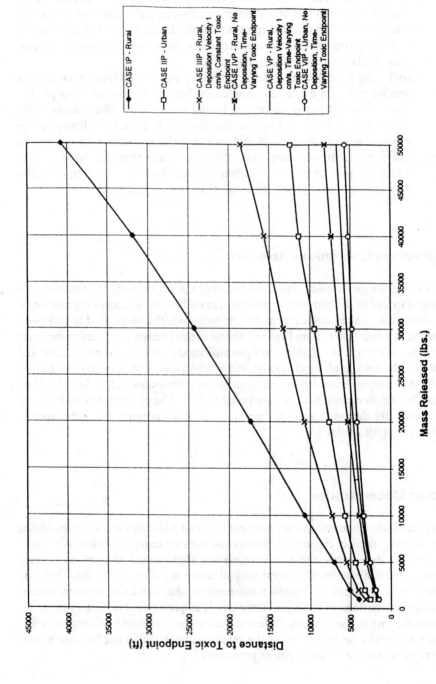

FIGURE 2. Worst-Case Instantaneous Ammonia Release: Sensitivity Studies

Observations

Uncertainties

Even though Figures 1 and 2 present only a very limited fraction of the sensitivity studies that would be needed to characterize the full range of uncertainties that exist in worst-case atmospheric dispersion modeling, they illustrate the difficulty that exists in attempting to provide easy-to-use guidance for owners and operators of facilities such as ammonia refrigeration systems, chlorine delivery systems in water treatment facilities, etc. "Easy-to-use" guidance would be, for example, a single curve on Figure 1. In addition, using other atmospheric dispersion models (e.g., HGSYSTEM, DEGADIS, ALOHA, SLAB, etc.) would introduce further uncertainties due to differences on the models themselves, as well as uncertainties arising from attempts to assess the range of potential values of input parameters, such as the dry deposition velocity, etc.

In principle, a comprehensive uncertainty analysis should be performed, using a technique such as Latin Hypercube sampling. This would enable the analyst to determine confidence measures such as the 95th percentile. To the author's knowledge, such an analysis has not been attempted for the types of atmospheric dispersion models that are appropriate for the RMP, including such phenomena as heavy vapors, dry deposition and time-varying toxic endpoints. Considerable effort has been devoted to examination of uncertainties in consequence modeling for nuclear applications [7], where dispersion and health effects models are better established than in the chemical field. These nuclear uncertainty analyses required many man-years of effort, which has not yet been made available for model guidance in support of the RMP. Therefore, the preparer of generic guidance generally has to make decisions based on limited sensitivity studies.

Conservatisms

It is apparent from even a cursory review of the sensitivity studies carried out above that it would be very easy to grossly overpredict the consequences of a worst-case release, particularly at a rural site. Even though none of the sensitivity studies performed can be said to be based on an unambiguous choice of model and/or input parameters (and, indeed, some of the above discussion is only qualitative), the overall conclusion is incontrovertible. It is very easy to incorporate far too much conservatism into worst-case models. Therefore, the would-be developer of guidance must be careful to exclude at least the grossest of conservatisms.

In the "Model Risk Management Program and Plan for Ammonia Refrigeration," this problem was addressed by taking a range of plausible sensitivity studies such as those on Figures 1 and 2 and enveloping them, leading to easy-to-use guidance, as displayed on Figure 3. That is, the ammonia guidance was produced

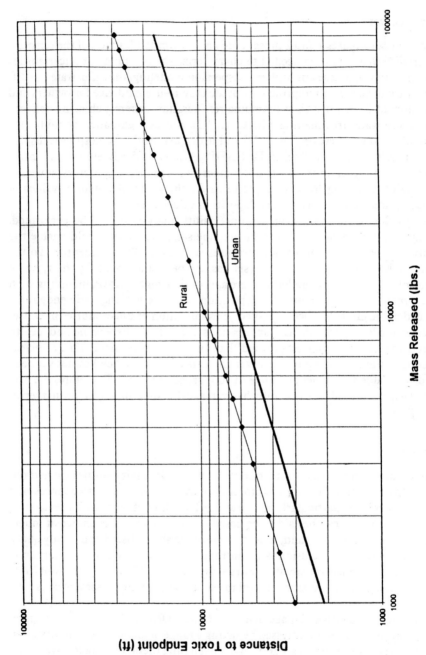

FIGURE 3. Worst-Case Instantaneous Ammonia Release: Generic Guidance on Distance to Toxic Endpoint

by using *judgment* informed by a limited range of sensitivity studies. Therefore, Figure 3 is not unique. Other analysts can and have derived different predictions, including both larger ones and smaller ones.

Questioning the validity of Figure 3 is, of course, entirely legitimate. However, if any proposed alternative to Figure 3 is based entirely upon a single run of a computer model, with no attempt to identify a plausible range of uncertainties, then the credibility of the alternative is in question. In particular, performing calculations with a number of different computer models and a number of different input parameters until one finds the shortest possible (or longest possible) predicted distances to the toxic endpoints should not be regarded as credible unless the predictions are backed up by experimental data.

Presentation of Worst-Case Modeling Results

The RMP requires that the results of worst-case modeling should be made available to local organizations such as the Local Emergency Management Committee (LEPC) and to the public. In light of the above discussion of uncertainties, it is not possible to say that "a worst-case release at Z facility will lead to the toxic endpoint being exceeded out to an accurately predicted distance x downwind." One can only state that "using such-and-such a model, with specifically stated assumptions leads to a predicted distance $x_,$, or that "guidance published by the EPA (or some specified trade association such as the American Petroleum Institute or the American Water Works Association Research Foundation [AWWARF]) gives a predicted distance x." It is clear that, depending on the source of the predicted values, x can lie within a large range of values.

Some possible comments and caveats that may help in conveying the information to stakeholders are as follows:

a. Atmospheric dispersion models that are suitable for worst-case analyses have generally not been tested against experiments in the worst-case conditions of atmospheric dispersion category F and windspeed 1.5 m/s.

b. There are large uncertainties on the predicted distances to the toxic endpoints.

c. It is easy to apply the models in such a way that there are large conservatisms in the predictions. It is not so easy to apply the models in such a way that all conservatisms have been removed. Therefore, in cases where very large distances are predicted, it is more than likely that the predictions are very conservative.

d. Generic guidance and generic models may not represent what would actually happen at a real site. From the model ammonia plan, examples are (i) that releases would often take place in an engine room, and, while it is not possible to prove generically that the release would be passively contained and mitigated by the room, in practice, there may be many specific cases in which such containment and mitigation would, in fact, occur, and (ii) generic models of flashing liquid ammonia from a high-pressure receiver have to assume generically that all of the ammonia becomes airborne as a mixture of vapor and fine liquid droplets, whereas, in practice, many of the liquid droplets may be removed by impaction on obstacles. No doubt, similar caveats can be identified for other types of facilities and other chemicals.

e. The worst-case scenarios are extremely unlikely. For example, the likelihood that there will be a large rupture in an ammonia vessel that has been designed and installed according to code and has been operated and maintained in safe ways that are well understood by the ammonia industry could be as low as one chance in a million per year. The probability that this rupture will occur at the same time as the worst-case weather condition is also very small, so that, overall, the worst-case scenario is not a meaningful measure of the risks posed by the facility. More important are the steps taken to prevent severe accidents at the facility and/or mitigate their consequences. This then leads to a discussion of such preventive and mitigative measures. Given that the law and regulations require the analysis of the worst-case scenario and the communication of the results, the most constructive use of the worst-case scenario must be to communicate the measures that are being taken to control the hazardous material in question and to prevent accidents.

Consultation with Industry and EPA

Finally, it almost goes without saying that guidance that is intended for use by owners and operators of hundreds or thousands of similar facilities will not be used if they do not find it acceptable. It is, therefore, necessary to keep open a constant dialog with appropriate industry organizations. Similar comments apply to EPA. It is only with this constant dialog that there is a chance of providing useful and accepted guidance.

References

1. Chamberlain, A. C., and R. C. Chadwick, 1953. "Deposition of Airborne Radio-Iodine Vapor," *Nucleonics*, Vol. 8, pp. 22–25.
2. Sehmel, G. A., 1980. "Particle and Gas Dry Deposition—A Review," *Atmospheric Environment*, Vol. 14, pp. 983–1011.
3. Hannah, S. R., and R. P. Hosker, 1980. "Atmospheric Removal Processes for Toxic Chemicals," ATDL Contribution File No. 80/25, Air Resources, Atmospheric Turbulence and Diffusion Laboratory, National Oceanic and Atmospheric Administration, Oak Ridge, TN.

4. Environmental Agency, Federal Emergency Management Agency and U.S. Department of Transportation, 1987. "Technical Guidance for Hazards Analysis—Emergency Planning for Extremely Hazardous Substances," Washington, D.C.
5. American Industrial Hygiene Association, Various Dates. "Emergency Response Planning Guidelines," Fairfax, VA.
6. Maher, S. T., and G. D. Kaiser, 1995. "Torrance Refinery Safety Advisor Project—Evaluation of Modified HF Alkylation Catalyst," Report 59111.01, EQE International, Irvine, CA.
7. Gorham, E. D., et al., 1993. "Evaluation of Severe Accident Risks: Methodology for the Containment, Source Term, Consequence, and Risk Integration Analyses," NUREG/CR-4551 (SAND86-1309), Vol. 1, Rev. 1, Washington, DC.

LUNCHEON ADDRESSES

Meeting the Needs of Our Stakeholders

Jack Weaver
Director, CCPS

The vision and mission of CCPS go back to its founding in 1985. They have served us well for more than a dozen years. However, to fulfill our vision and satisfy our stakeholders, we need to be attentive to the changing needs of industry, academia and our other stakeholders.

Our vision is to provide leadership in chemical process safety to prevent major incidents associated with processing, storing, using and distributing hazardous materials. Our mission is to develop both engineering and management practices to fulfill that vision, with four major objectives:

(1) to advance the state-of-the-art of process safety technology and management practices,
(2) to serve as a premier resource for information on process safety,
(3) to foster process safety in chemical and related engineering education, and
(4) to promote process safety as a key industry value.

Today I'd like to share with you some of the more exciting things we've been doing in the past year, particularly with regard to the first and third elements of our mission statement.

Advancing Technology

CCPS projects are determined largely by our 90 sponsors utilizing a 9-month selection process to identify the best and most urgently needed projects. As a result a variety of promising projects are currently underway or about to be initiated. These fall into two basic categories - technical and management projects. The following are some of the most promising technical projects:

- *Flammable Mass of a Vapor Cloud:* the aim of this project is to develop a method for calculating the actual flammable portion of a vapor cloud in order to assess more accurately the consequences of vapor cloud releases. This project received strong industry support and government support at our meeting in January and is already well underway.

- *Process Safety Management in Batch Operations*: this project will focus on the unique aspects of batch operations, as experienced by many CCPS sponsors, especially those in specialty chemical and pharmaceutical operations. This project was also launched recently, and a Concept book is planned.
- *Process Equipment Reliability Database:* this project was started two years ago and has grown in scope as 25 international companies have enrolled as participants. This database will gather data on reliability of various categories of equipment (e.g., compressors, heat exchangers, pressure relieving devices, piping, pumps, instrumentation, vessels, and furnaces) with a primary goal of improving maintenance practices among the participating companies. Other goals include determining overall failure rates and equipment mean time between failures.
- *CD-ROM Publication of Loss Prevention Conferences:* this project is a response to a ground swell of interest from individuals in various CCPS sponsor organizations, as well as the AIChE Health and Safety Division. A CD-ROM publication will include proceedings from all of the AIChE Loss Prevention symposia from 1980 until 1996, as well as proceedings from many of the CCPS annual international conferences. The publication is being produced and made available at prices below our actual cost, thanks to the support of several generous organizations and companies.
- *RELEASE:* This is a computer based model to predict aerosol formation and rain-out in vapor clouds resulting from the accidental release of hazardous liquids under pressure. The model was developed from data gathered over the past decade in large-scale outdoor experiments conducted by CCPS in Oklahoma and Nevada on a range of materials including cyclohexane, chlorine and mono-methyl amine.

Advancing Management Practices

In addition to these technical projects, a number of projects are underway to provide tools or methods for better management of process safety:

- *Inherently Safer Operations:* A series of projects and workshops has been carried out over the past two years, starting with a popular and provocative workshop held at the Technical Steering Committee meeting in Chicago in May of 1995. Last year the Concept book on Inherently Safer Operations was published (the "gold book"), and this year another workshop was held in May on providing economic justification for inherently safer design and operations. As a follow-up to this workshop, other projects are being planned.
- *Lessons Learned Database:* this project has progressed well in the past year, and major chemical and petroleum companies are now enrolling to par-

ticipate. Our sponsors have also expressed interest in a common database of lessons learned from process safety incidents. In response, we've initiated a project—the CCPS Lessons Learned Database—that will pool available information on process safety incidents so participating companies can learn from each other's experiences. The database will cover major process safety incidents, including fires, explosions, fatalities, injuries, and significant releases of hazardous materials.

- *PSM Measurement:* Measuring performance in process safety is a difficult undertaking. The third phase of an ambitious project, aimed at developing a system for measuring PSM performance, is nearing completion, and a fourth and final phase is being planned. CCPS has developed a methodology to measure the performance of any company or operating unit with regard to the 12 elements of process safety management. Our first PSM measurement product will address 2 of the 12 elements: *management of change and training.* The full product addressing all 12 elements will be available the following year.
- *Process Safety for Small Enterprises:* CCPS has continued to develop awareness and training programs targeting smaller companies and public enterprises, consistent with our mission. Reaching beyond Florida, CCPS has now introduced introductory training for small industry and the public sector in Iowa and Michigan, with plans to extend this to South Carolina and Indiana. Engineers from local industry are conducting five half-day seminars on the basics of PSM and RMP, and the reaction has been very enthusiastic.
- *The Sloan Workshop:* CCPS participated with MIT's Sloan School in an NSF sponsored workshop on Managing Process Safety Resources in the Chemical Process Industries. The focus of this management workshop was on organizational behavior and its impact on safety performance and PSM management. Roughly equal numbers of academics and CCPS industrial experts participated and identified opportunities for further learning, including improved communications techniques, learning more from "near-miss" incidents, and learning how to make regulations work to advantage for PSM.

We are pleased that you decided to participate in this important and exciting conference. We hope that you all will contribute to the meeting, and that you will take away some excellent new ideas on risk analysis and risk management. Thanks for your participation.

Luncheon Presentation

Hans Pasman
TNO, Netherlands

The present conference is another step in the evolution of Loss Prevention. When one looks back and considers where we started in the sixties, tremendous achievements have been made. The performance indicators in the form of Lost time injury (LTI) frequencies and LTI Severity rates keep-on coming down. Already an order of magnitude has been gained, and some companies are aiming for a state of zero accidents.

A number of conditions has made this possible. Where initially these were mainly technical measures, the last ten years the attention devoted to human factors starts to pay off. When risk analysis was introduced in studying process safety, human factor seemed to spoil the new technique. Besides the political effects the unpredictiveness of the human operator was making the application of risk analysis very controversial and so reduced the grip on safety assessments. However, I must say, that in the end the study of human behavior in all its aspects has been very rewarding. I enjoyed an earlier CCPS conference with human factor in the central theme. The studying of the human individual and his interface with the machine, the functioning in a team, the team spirit, the organization and its rules, the management system, the systematic ploughing back of experience of accidents and near misses into the organization, the awareness of small failures and the importance of group morale and culture has paid dividend. It also starts now to have its impact on the design process. Another factor that has an impact there is the full exploitation of the concept of inherent safety. By the way this contains also a strong human factor element in the sense that mental barriers have to be taken down between e.g. departments in the company to really make break-throughs in thinking how things can be made more inherently safe.

When participating in the preparation of this conference, it struck me that CCPS relatively got many papers from industry, and also with case studies. I think this is a favorable situation. It also struck me that the concepts of risk management is evolving. The trend seems to go in the direction of safety as good business. I remember from a paper selection committee meeting perhaps in the early seventies that a paper with the title "Safety as a money spinner" was deemed a little bit as suspect and perhaps not fully decent. In the process of competition that made the West strong, fortunately this has become now an important factor and risk analysis for cost-benefit consideration becomes a common means.

At TNO, over the years, I spent at one hand much time in defending peace (research for defence) and at the other hand in developing methods for improving industrial safety. Both are trades that do not make one rich, but it s all very satisfying if you see results. It takes however a long time before you see any. TNO has many different branches of expertise and technology and is an ideal place for doing safety work, were it not that it is very hard to get researchers together long enough to work out an acceptable proposal to industry or government agencies with the perspective to solve the problem. Cooperation and team work are things that take constant attention, in particular when budgets of various groups get tighter. Also, over the years the effort to get innovative ideas financed is getting more difficult, and safety research with a sophisticated content suffers. However, last year the TNO board gave sufficient leeway for considering a restructuring. In the eighties much work has been done on hazard identification and effect and damage modeling. Right now reliability and availability, also in the context of maintenance and cost saving, are important issues. Despite the fact that in the old topics there is plenty of space for further improvements: Think, for example, about test methods, many are primitive and inadequate—new research items get on board. Information technology, decision support, hi tech control in intimate interfacing with the human factor open new possibilities. So I am very enthusiastic about what the future can bring in designing and managing chemical processes.

Finally I would like to mention the European Federation of Chemical Engineering and its Working Party on Loss Prevention. The equivalent to CCPS the European Process Safety Centre got on its way. It had good contributions in developing thoughts with respect to Safety Management Systems and in the preparations for EU's SEVESO-2 Directive. The parent Working Party is at present organizing the 9th International Symposium on Loss Prevention and Safety promotion in the Process Industries at Barcelona, Spain, from 4 to 8 May 1998. It has as a motto "Safety as a Factor in Business and Operation." The main themes will be:

- Safety as a factor in the management of business
- Safety in the management of operations
- The achievement of risk reduction through risk assessment and related techniques
- Safe design of processes, installations and equipment
- The use of programmable electronic systems in safety related applications
- Research into the prevention and modeling of accidental releases to the air, water and soil
- The impact of legislation and industry initiatives

In arranging the programme we shall try to organize some sessions such that business and plant line managers can get what interest them most by attending only parts of the conference. Due to the unfortunate illness of our liaison person the announcement of the symposium in the United States has been weak and late. By the way, Barcelona is the city of the finest architects, and many magnificent buildings can be admired in the city and its immediate vicinity.

Quality Assurance, Uncertainty, and Expert Judgment in Risk Analysis

Jim McQuaid
Director of Science & Technology, Health & Safety Executive, London, UK

ABSTRACT
In this lunchtime address, I propose to discuss some outstanding issues central to the actual and perceived quality of risk decisions.

Introduction

The quality of risk assessments and of decisions to which they contribute is a subject of much current interest. There is a need for greater transparency especially to those who are expected to live with or tolerate the decisions.

By quality in this context, I mean the rigor and scrutability of the processes of risk analysis, from framing of the problem through to the making of decisions on risk control. This interpretation of quality has both actual and perceived dimensions. It is not sufficient to do things right to ensure rigor but necessary to do the right things to ensure scrutability.

The outcomes of risk analysis are often presented as having the status of hard, objective information. This is especially the case if they come off a computer accompanied by pretty pictures. But those with a stake in the decisions though not directly involved in the decisionmaking process are increasingly sceptical. Perhaps the most devastating criticism of a sceptic is that of Richard Feynman on the risk analysis of the Challenger missions.

The public or their representatives are not interested in or do not understand the figures with which they are presented. They want to be assured about the quality of the processes so that they can have more trust in the outcomes. What do we need to do to try to engender that trust?

That is to be the subject of my talk. I will range across a variety of issues that affect the trust that can be placed in the outcomes of risk analysis. I will pose questions and make suggestions. But my main message will be that we must give greater attention to the question of trust rather than submerging ourselves in ever more complicated analysis. We have already left the public behind and we have had the inevitable reaction. You have had two influential publications in the US seeking to reverse the trend. And we have had similar reactions in Europe.

Quality Assurance

Quality Assurance of processes seeks to ensure that reliance can be placed on the outcomes as an alternative to quality control by checking and if necessary reworking the outcomes. In the area of risk, quality assurance will embrace operating systems and management systems as well as the formalities of predictive modeling of the risks. I will concentrate on predictive modeling in all its forms.

A predictive model provides a forecast of what could happen in defined circumstances. It will be based on a representation of the problem and will incorporate empirical evidence in the form of data and working assumptions.

How do we assure ourselves on the quality of a predictive model? The above outline description refers to many sources of potential ambiguity and divergent opinion.

- The forecast may represent an extrapolation beyond the limits of feasible validation so that different models may agree within the range of validation but produce conflicting predictions which cannot be separated into the good and not so good.
- The definition of circumstances may expose justifiable differences of view on the way an accident situation can arise. For example (from another field), I understand that there are over 60 different explanations of the circumstances in which an ice age can develop, each consistent with the available evidence.
- The representation of the problem, however defined, in mathematical terms or in a physical simulation will be subject to assumptions about what and how particular characteristics can be approximated or neglected in the representation.
- The data incorporated in the model will be subject to errors in measurement and/or to questions of relevance.
- The working assumptions in the solution procedures will differ, depending on the structure of the model and it may be difficult for an independent user to weigh their relative validity.

But, at the least, we should be able to expect of a predictive model that:

- it will be transparent as to how it deals with the above issues;
- it will give consistent results - that different users presented with the same problem will produce broadly the same result, whether right or wrong;
- it will account for all the available evidence, particularly that evidence not itself incorporated in the model, and
- it will state clearly what level of confidence can be placed in the results.

Furthermore, the quality of a model will be suspect if new evidence emerges which is in conflict with its predictions. This does not mean that the model is wrong since it is never going to be right in any absolute sense. Rather it means that the confidence that can be placed in the model predictions is reduced though the predictions may well be still fit-for-purpose depending on the requirement of the application and also, of course, on the relative performance of competing models.

Uncertainty

All of the issues I have mentioned arise because of the uncertainty that inevitably exists in the problems with which we have to deal. So there is great room for contention amongst the experts. This contention does nothing for trust by non-experts. So we owe it to ourselves to reduce the potential for contention.

One way is to insist on "sound science." It is a common nostrum that policies and procedures in risk assessment should be based on sound science. But these simple words mean different things to different people. I will say something about what I understand it to mean in our context. We need to be wary of "group think" and "discipline capture" which can undermine the perceived soundness of the scientific basis of risk assessments.

Uncertainty is itself a catch-all word. It covers a multitude of different origins of the state of mind where we are not sure of the result of an event. We need to be clear about the different kinds of uncertainty. What can be done to reduce uncertainty depends on its nature.

Expert Judgment

As will be clear from what I have said so far, many of the issues that arise in risk assessments require the exercise of expert judgment. This exercise of judgment is informed by data and information derived from research and from statistical and scientific/technological evidence of various kinds on hazards and risks. It is often the case that different conclusions can be drawn by experts from the same base of data and information. The final judgment also draws substantially on knowledge and understanding derived from experience and awareness of good practice. These various inputs have to be aggregated into a composite decision, whether as a single course of action or as a formally elaborated range of options. There is considerable public interest in the ways in which this aggregation is performed and in the factors influencing individual and group exercise of judgment. The need for information on this subject arises because judgments in areas of scientific and engineering uncertainty are not perceived by the public to be value free. Methodologies exist to formalize the ways in which data are processed in the formulation of judgments so that they can be said to be based on "sound science." The procedures by which the more intractable influences on judgments are brought to bear are not so well developed, if at all.

The same issues arises in many other fields where differences in expert judgments can arise. We have carried out some research on practices in other fields to see if we can learn from them. Ultimately, we hope to develop some guiding principles on procedures. The research comprised a short survey to collect information from decision makers. The decision makers were selected from various walks of life and their cooperation secured on the basis of their shared interest in developing a better understanding of this socially important phenomenon. The survey consisted of carefully structured face-to-face or telephone interviews with 19 decision makers. I will outline the results from this preliminary research.

Concluding Remarks

My main conclusion is that we in the process safety community need to give greater attention to the considerations I have mentioned. We have tended to spend our time preaching to the converted. This is very worthy and has served to advance the safety of the industry. But we need to turn our attention to the unconverted. What is it that worries them? How should we tailor our approaches to ensure their conversion? How do we engage them constructively? We need to maximize health and safety in a cost-effective way. We need to do this if society is to gain the benefits of technological advance and not be hampered by unwarranted fears and suspicion.

Publications Available from the
CENTER FOR CHEMICAL PROCESS SAFETY
of the
AMERICAN INSTITUTE OF CHEMICAL ENGINEERS

CCPS Guidelines Series

Guidelines for Postrelease Mitigation in the Chemical Process Industry
Guidelines for Integrating Process Safety Management, Environment, Safety, Health, and Quality
Guidelines for Use of Vapor Cloud Dispersion Models, Second Edition
Guidelines for Evaluating Process Plant Buildings for External Explosions and Fires
Guidelines for Writing Effective Operations and Maintenance Procedures
Guidelines for Chemical Transportation Risk Analysis
Guidelines for Safe Storage and Handling of Reactive Materials
Guidelines for Technical Planning for On--Site Emergencies
Guidelines for Process Safety Documentation
Guidelines for Safe Process Operations and Maintenance
Guidelines for Process Safety Fundamentals in General Plant Operations
Guidelines for Chemical Reactivity Evaluation and Application to Process Design
Tools for Making Acute Risk Decisions with Chemical Process Safety Applications
Guidelines for Preventing Human Error in Process Safety
Guidelines for Evaluating the Characteristics of Vapor Cloud Explosions, Flash Fires, and BLEVEs
Guidelines for Implementing Process Safety Management Systems
Guidelines for Safe Automation of Chemical Processes
Guidelines for Engineering Design for Process Safety
Guidelines for Auditing Process Safety Management Systems
Guidelines for Investigating Chemical Process Incidents
Guidelines for Hazard Evaluation Procedures, Second Edition with Worked Examples
Plant Guidelines for Technical Management of Chemical Process Safety, Revised Edition
Guidelines for Technical Management of Chemical Process Safety
Guidelines for Chemical Process Quantitative Risk Analysis
Guidelines for Process Equipment Reliability Data with Data Tables

Guidelines for Safe Storage and Handling of High Toxic Hazard Materials
Guidelines for Vapor Release Mitigation

CCPS Concepts Series

Inherently Safer Chemical Processes. A Life--Cycle Approach
Contractor and Client Relations to Assure Process Safety
Understanding Atmospheric Dispersion of Accidental Releases
Expert Systems in Process Safety
Concentration Fluctuations and Averaging Time in Vapor Clouds

Proceedings and Other Publications

Proceedings of the International Conference and Workshop on Risk Analysis in Process Safety, 1997
Proceedings of the International Conference and Workshop on Process Safety Management and Inherently Safer Processes, 1996
Proceedings of the International Conference and Workshop on Modeling and Mitigating the Consequences of Accidental Releases of Hazardous Materials, 1995.
Proceedings of the International Symposium and Workshop on Safe Chemical Process Automation, 1994
Proceedings of the International Process Safety Management Conference and Workshop, 1993
Proceedings of the International Conference on Hazard Identification and Risk
Analysis, Human Factors, and Human Reliability in Process Safety, 1992
Proceedings of the International Conference and Workshop on Modeling and Mitigating the Consequences of Accidental Releases of Hazardous Materials, 1991.
Safety, Health and Loss Prevention in Chemical Processes: Problems for Undergraduate Engineering Curricula